Springer Series on

ATOMIC, OPTICAL, AND PLASMA PHYSICS 68

Springer Series on

ATOMIC, OPTICAL, AND PLASMA PHYSICS

The Springer Series on Atomic, Optical, and Plasma Physics covers in a comprehensive manner theory and experiment in the entire field of atoms and molecules and their interaction with electromagnetic radiation. Books in the series provide a rich source of new ideas and techniques with wide applications in fields such as chemistry, materials science, astrophysics, surface science, plasma technology, advanced optics, aeronomy, and engineering. Laser physics is a particular connecting theme that has provided much of the continuing impetus for new developments in the field. The purpose of the series is to cover the gap between standard undergraduate textbooks and the research literature with emphasis on the fundamental ideas, methods, techniques, and results in the field.

Please view available titles in *Springer Series on Atomic, Optical, and Plasma Physics* on series homepage http://www.springer.com/series/411

Viacheslav Shevelko
Hiro Tawara
Editors

Atomic Processes in Basic and Applied Physics

With 226 Figures

Springer

Editors
Viacheslav Shevelko
P.N.Lebedev Physical Institute, Optical Department
Leninskii prospect 53, 119991 Moscow, Russia
E-mail: shev@sci.lebedev.ru

Hiro Tawara
National Institute for Fusion Science, Plasma Physics
Oroshi 332-6, Toki, 509-5292 Gifu, Japan
E-mail: tawara.hiroyuki@nifs.ac.jp

Springer Series on Atomic, Optical, and Plasma Physics ISSN 1615-5653
ISBN 978-3-642-25568-7 ISBN 978-3-642-25569-4 (eBook)
DOI 10.1007/978-3-642-25569-4
Springer Heidelberg New York Dordrecht London

Library of Congress Control Number: 2012939524

Printed on acid-free paper

Springer is part of Springer Science+Business Media (www.springer.com)

Preface

This book is aimed at providing an overview of modern atomic, plasma, and accelerator physics and their applications such as inertial thermonuclear fusion, tumor therapy, industrial plasmas, and others. It is a comprehensive edition which considers the interactions of atoms, ions, and molecules with charged particles, photons, and laser fields and reflects the present understanding of atomic processes such as electron capture, ionization, recombination, and other processes occurring in most sources of laboratory and astrophysical plasmas.

Although atomic physics and related basic atomic processes have a very long history in the developing course of general modern and precision physics, they are still providing, through newly developed techniques, important new and the most accurate information in basic physics itself as well as related fields such as astrophysics which includes hitherto unknown and unanalyzed phenomena, giving a new understanding and new material production. Some of the most striking applications of the modern atomic physics are associated with cancer/tumor therapy, which ensures curing some illnesses, incurable so far with even modern medicines/medical techniques, with fusion reactions for the future power-generating industrial tokamak devices and with industrial plasmas used for effective production of microchips and integrated circuits.

In some respect, this book is a continuation of a previous series of the books such as *Physics of Highly Charged Ions* (R.K. Janev, L.P. Presnyakov, and V.P. Shevelko, Springer, Berlin, 1985), *Atomic Physics with Heavy Ions* (H. Beyer, V.P. Shevelko, eds., Springer, Berlin, Heidelberg, 1999), *Introduction to the Physics of Highly Charged Ions* (H. Beyer and V.P. Shevelko, IOP, Bristol, 2003), and *The Physics of Multiply and Highly Charged Ions* (F.J. Currell, ed., Kluwer Academic Pub., Dordrecht, Boston, London, 2003). However, the present book deals not only with highly charged ions but also with low-charged positive as well as negative ions and neutral atoms.

The book consists of 4 parts including about 18 chapters presented by active specialists from Germany, Russia, USA, Japan, France, Brazil, and Korea.

In Part I, entitled *Atomic Processes in Laboratory and Astrophysical Plasmas*, the importance of atomic processes in plasmas is considered. The *ball lightning* phenomenon has a very long history of observation which has been seen by many people since the human was born on Earth. However, it is very complicated and involves not only various atomic processes but also a series of unexplained and unknown phenomena a deep understanding of which is still far from complete although new modeling experiments and analysis are being performed.

Through recent precision measurements of new electron recombination processes are investigated in laboratory plasma sources, many mysteries of the dark matter in and near *black holes* and in the *supernova* are being unraveled presently and thus providing new astrophysical information and understanding.

Naturally, understanding of *hot coronal plasmas* in the Sun and astro-plasmas needs the exact and new knowledge of various collision and radiative processes involving highly ionized atoms based upon new spectroscopic observations.

Various atomic processes are investigated in the *laboratory plasmas* required for injection of a neutral beam into the core of magnetically confined plasmas as well as for investigations of dusty plasmas including strong collective plasma interactions.

In Part II, *Atomic Heavy-Particle Collisions*, atomic charge-changing processes in collisions of heavy ions with neutral atoms are considered (electron capture, loss, ionization, and excitation) over a wide collision energy range including relativistic energies. The data on the corresponding cross sections of these processes are required in many fields of atomic accelerator, and plasma physics such as heavy-ion fusion (HIF), heavy-particle tumor therapy, and heavy-ion probe beam (HIPB) diagnostics in plasma devices as well as for design of accelerator machines. In particular, the international facility for antiproton and ion research (FAIR) project started in 2011 at GSI, Darmstadt, requires benchmarks for the cross sections of such collision processes because these reactions play a major role in the ion-beam loss processes during acceleration/storage.

Part II also includes recent new studies of ion-pair formation and resonant-quenching processes in slow collisions between the highly excited (*Rydberg*) atoms and the ground-state atoms with small electron affinities. The theory presented is based on the general approach for transition matrix elements between the ionic and Rydberg-covalent states of a diatomic quasimolecular system using the momentum representation for highly excited electron wave functions and the technique of the nonreduced tensor operators. The results are illustrated by numerical calculations of the ion-pair production processes in slow collisions of Rydberg Ne(ns) and Ne(nd) atoms with the ground-state alkaline-earth atoms.

Part III, *Atomic X-Ray Physics for Laboratory and Astrophysical Plasmas*, deals with atomic processes involving X-ray radiation.

Diagnostic methods for hot laboratory (tokamak) and astrophysical (solar corona) plasmas are considered based on X-ray and extreme ultraviolet (XUV) emission spectra of highly charged ions in plasmas and the modern methods of atomic data on spectral and collisional ion characteristics which allows one to determine various physical parameters of the emitting plasmas (density, temperature, ion-charge states and fractions, etc.).

Accurate data of *dielectronic recombination* (DR) reactions are presented on the basis of experimental studies recently carried out at the heavy-ion storage rings,

ESR, in Darmstadt and TSR in Heidelberg, and latest progress in applications of the DR processes as a tool for precision spectroscopy required in astrophysics, plasma physics, fundamental interactions, atomic, and nuclear physics is also presented. The observed data of the accurate DR cross sections and energies for a few-electron ions presented can be explained using calculations performed with a high precision up to the level of the QED theory.

Recent experiments carried out by the new X-ray *free-electron laser* at Free-electron LASer Hamburg (FLASH) as one of the first soft X-ray FEL sources open possibilities to completely new fields on photo–matter interactions such as sequential and nonsequential multiphoton ionization of the gas phase targets and linear and nonlinear photoionization processes which can be used for online photon diagnostics at new radiation sources.

Finally, in Part IV, entitled *Atomic Data Applications and Databases*, some important applications are considered such as heavy-ion radiotherapy using high-energy carbon-ion beams and the use of industrial plasma for production of the electronic integrated circuits (IC). The clinical results of high energy carbon *tumor therapy*, performed at heavy ion medical accelerator in Chiba (HIMAC), Japan, are presented, and some developments of a new scintillation counter system are discussed for simultaneous measurements of the radiation dose and quality of heavy-ion beams.

New industrial materials are being investigated and produced through *industrial plasma sources* using precise knowledge of atomic and molecular processes and collision processes, particular for producing microchips and IC.

A review on various theories for plasma diagnostics based on the broadening of spectral lines in *magnetized plasmas* using the Stark and Doppler broadenings is also presented.

A production of an *ultracold ion beam* with very low longitudinal and transverse temperatures is studied using stochastic, electron, and laser cooling to realize an antiproton beam in order to create a weak boson beam contributing very much to new elementary particle physics.

Detailed information on existing atomic and molecular data banks, that is, about radiative and collisional properties of atoms, ions, and molecules interacting with atomic/ionic particles (electrons, atoms, and ions) and photons, can be found in the last chapter of the book.

We are grateful to all the contributors to the book who presented a recent progress in atomic process physics and its applications achieved in the last 10 years, both experimentally and theoretically.

Moscow, Toki

Viacheslav Shevelko
Hiro Tawara

Contents

Part IV Atomic Data Applications and Databases

Contributors

S.A. Bogachev P.N. Lebedev Physical Institute, Leninskii prospect 53, Moscow, Russia

C. Brandau Institut für Atom- und Molekülphysik, Justus-Liebig-Universität, Leihgesterner Weg 217, 35392 Giessen and GSI Helmholtzzentrum für Schwerionenforschung, Planckstr.1, 64291 Darmstadt, Germany

V.L. Bychkov Moscow State University, Building 2, Leninskiye Gory, Moscow, Russia

R.E.H. Clark Department of Physics and Astronomy, The University of Texas at San Antonio, One UTSA Circle, San Antonio, TX, USA

R.D. DuBois Department of Physics, University of Missouri-Rolla, Rolla Missouri, USA

F. Goryaev P.N. Lebedev Physical Institute, Leninskii prospekt 53, Moscow, Russia

M. Grieser Max-Planck-Institut für Kernphysik, Heidelberg, Germany

D. Humbert Laboratoire de Recherche en Informatique, UMR 8623, Université Paris 11, Orsay, France

S.-H. Hwang National Fusion Research Institute, 113 Gwahangno Yuseong-Gu, Daejeon, Korea

M. Imai Department of Nuclear Engineering, Kyoto University, Sakyo, Kyoto, Japan

W.-S. Jhang National Fusion Research Institute, 113 Gwahangno Yuseong-Gu, Daejeon, Korea

Y.-D. Jung Department of Applied Physics, Hanyang University, 1271 Sa 1-dong, Sangrok-gu, Ansan-si, Gyunggi-Do, Korea

D.-H. Ki Department of Applied Physics, Hanyang University, 1271 Sa 1-dong, Sangrok-gu, Ansan-si, Gyunggi-Do, Korea

T. Kanai Heavy Ion Medical Center, Gunma University, 3–39–22, Showa, Maebashi, Gunma, Japan

Y.-K. Kang National Fusion Research Institute, 113 Gwahangno, Yuseong-Gu, Daejeon, Korea

Ch. Kozhuharov GSI Helmholtzzentrum für Schwerionenforschung, Planckstr.1, 64291 Darmstadt, Germany

S.V. Kuzin P.N. Lebedev Physical Institute, Leninskii prospect 53, Moscow, Russia

D.-C. Kwon National Fusion Research Institute, 113 Gwahangno, Yuseong-Gu, Daejeon, Korea

V.S. Lebedev P.N. Lebedev Physical Institute, Leninskiy prospekt 53, Moscow, Russia

M.S. Litsarev Uppsala Universitet Institutionen for fusik och astronomy Materialteori, Lägerhyddsvägen 1, Uppsala, Sweden

O. Marchuk Institute of Energy and Climate Research, Plasma Physics Forschungszentrum Jülich GmbH, Partner in Trilateral Euregio Cluster, Jülich, Germany

M. Martins Universität Hamburg, Institut für Experimentalphysik, Luruper Chaussee 149, Hamburg, Germany

M. Meyer European XFEL GmbH, Notkestrasse 85, Hamburg, Germany

A. Müller Institute for Atomic and Molecular Physics, Justus-Liebig-University Giessen, Leihgesterner Weg 217, Giessen, Germany

B. Najjari Max-Planck-Institut für Kernphysik, Saupfercheckweg 1, Heidelberg, Germany

A.A. Narits P.N. Lebedev Physical Institute, Leninskiy prospekt 53, Moscow, Russia

A. Noda Institute for Chemical Research, Kyoto University, Gokano-sho Uni-city, Kyoto, Japan

E. Oks Physics Department, 206 Allison Lab, Auburn University, Auburn, AL, USA

S.N. Oparin P.N. Lebedev Physical Institute, Leninskii prospect 53, Moscow, Russia

J.-H. Park National Fusion Research Institute, 113 Gwahangno, Yuseong-Gu, Daejeon, Korea

Yu. Ralchenko National Institute of Standards and Technology, Gaithersburg, MD, USA

A.A. Reva P.N. Lebedev Physical Institute, Leninskii prospect 53, Moscow, Russia

M. Richter Physikalisch-Technische Bundesanstalt, Abbestr. 2–12, Berlin, Germany

A.C.F. Santos Instituto de Fi'sica, Universidade Federal do Rio de Janeiro, CEP 21941–972, Rio de Janeiro, RJ, Brazil

S. Schippers Institute for Atomic and Molecular Physics, Justus-Liebig-University Giessen, Leihgesterner Weg 217, Giessen, Germany

S.V. Shestov P.N. Lebedev Physical Institute, Leninskii prospect 53, Moscow, Russia

V.P. Shevelko P.N. Lebedev Physical Institute, Leninskii prospect 53, Moscow, Russia

T. Shirai National Institute of Radiological Sciences, 4–9–1 Anagawa, Inage-ku, Chiba-shi, Japan

V.A. Slemzin P.N. Lebedev Physical Institute, Leninskii Prospect 53, Moscow, Russia

M.-Y. Song National Fusion Research Institute, 113 Gwahangno, Yuseong-Gu, Daejeon, Korea

A. Sorokin Deutsches Elektron-Synchrotron, Notkestrasse 85, Hamburg, Germany

Th. Stöhlker GSI Helmholtzzentrum für Schwerionenforschung, Planckstrasse 1, Darmstadt and Physikalisches Institut, Universität Heidelberg, Philosophenweg 12, Heidelberg and Helmholtz-Institut Jena, Helmholtzweg 4, Jena, Germany

H. Tawara National Institute for Fusion Science, 322–6 Oroshi-cho, Toki, Japan

K. Tiedtke Deutsches Elektron-Synchrotron, Notkestrasse 85, Hamburg, Germany

I.Yu. Tolstikhina P.N. Lebedev Physical Institute, Leninskii Prospect 53, Moscow, Russia

A.B. Voitkiv Max-Planck-Institut für Kernphysik, Saupfercheckweg 1, Heidelberg, Germany

G. Weber GSI Helmholtzzentrum für Schwerionenforschung, Planckstrasse 1, Darmstadt and Helmholtz-Institut Jena, Helmholtzweg 4, Jena, Germany

J.-S. Yoon National Fusion Research Institute, 113 Gwahangno, Yuseong-Gu, Daejeon, Korea

Chapter 1
Unsolved Mystery of Ball Lightning

V.L. Bychkov

Abstract Ball lightning is an unusual phenomenon always drawing attention of people. There are still questions about its origination, features, interaction with environment, and phenomena related to it. On a way of studying this phenomenon, there are a lot of difficulties, the basic of them is insufficiency of authentic, scientific data. The chapter sets as the purpose to interest the reader in the problem, to describe conditions of ball lightning occurrence, theories, and its hypotheses explanation, to include readers in a circle of experimental searches in creation of a ball lightning and its analogues, and to describe fascination of a problem and difficulty of its solution.

1.1 Introduction

The term ball lightning (BL) is usually applied to an autonomous, stably shining ball-like object, which is observed in the atmosphere, and is connected usually with the thunderstorm phenomena, and the natural linear lightning. Under the autonomous feature, we understand its capability to move in space keeping its form, size, and color during a time compared with its lifetime. It is one of the most surprising natural phenomena, which on an extent of 1,000 years amazed the imagination of people. One of the first mentions about this phenomenon can be found in the Flavius Josephus's book "War of the Jews," written soon after 70 AD. In it, the garnet balls on the clothes of priests symbolized BL. First meeting of the scientists, investigating an atmospheric electricity, with BL was in 1753 [1–3]. Prof. G.W. Richmann was killed by stroke during a thunderstorm. He was measuring a potential of a metallic bar, placed on a roof of his house with a help of an electrometer—a metallic rod, which was placed inside the house. The rod was isolated from the ground by a crystal glass, and a silk thread was attached to the rod by one of its ends. He judged by an angle of the thread deflection about the value of the rod potential. In Fig. 1.1, one can see a picture of this event.

This event was soon investigated by M. V. Lomonosov. He revealed that the fireball was formed outside the house, and penetrated into the room through either

V. Shevelko and H. Tawara (eds.), *Atomic Processes in Basic and Applied Physics*,
Springer Series on Atomic, Optical, and Plasma Physics 68,
DOI 10.1007/978-3-642-25569-4_1,

Fig. 1.1 Death of professor G.W. Richmann after BL impact, the picture is taken from [1]

the door, or the window. The report of Lomonosov can be considered as the first qualified description of traces left by BL. Besides this event, Lomonosov knew also other cases of BL observations; about this, he wrote in his article [4] "combustion of fats, gathered together in air." In fact, this was the first model of BL with a chemical source of energy. Unfortunately, Richmann's death had for a long time retarded studies of the atmospheric electricity (including investigations of BL).

We can consider an activity of Francois Arago to be the next expression of interest in BL. He had collected and published 30 evidences of BL observations, explaining them as "lightning energy condensation" [5]. Later, BL observation cases and its models were discussed at Paris Academy of Sciences sessions.

An important milestone in BL investigation had become the Walter Brand's book "Ball Lightning," published in 1923 [6]. Brand in his book had represented 215 BL observations from 1665 to 1919 as described by eyewitnesses and indicated main 14 features of BL.

In the twentieth century after W. Brand worked and continue to work on BL problem many well-known scientists. They collect data on BL, analyze its properties, and carry out experiments on generation of its analogues. First of them was the Nobel Prize winner P.L. Kapitsa. He thought of its nature, and made the first plasma experiments on its origination. A hypothesis according to which BL is fed by the radio-frequency radiation energy of thunderstorms, P.L. Kapitsa [7] has proposed in 1955. Kapitsa with his coworkers realized a high-frequency constricted discharge in the atmosphere of helium, argon, carbon dioxide, and air at pressure from ten Torr to several atmospheres [8], but it proved to be a plasma, not BL.

J.D. Barry, G.C. Dijkhuis, M.T. Dmitriev, G. Egely, Ya.I. Frenkel, A.I. Grigor'ev, A. Keul, J.R. McNally, I.V. Podmoshenskyi, W.D. Rayle, S. Singer, B.M. Smirnov, I.P. Stakhanov, and many others made a considerable contribution in a solution of this problem. Nowadays, there exists an international Committee on ball lightning (ICBL), which was created in 1990 by S. Singer, B.M. Smirnov and Y.-H. Ohtsuki, and which holds International Symposia each 2 years.

During the twentieth century, collecting and analysis of observations by different investigators and research teams [9–15] was realized; many theories and experiments were made.

However, the problem of BL is still unresolved. There are many theories of BL, but none of them can explain all BL features. Many experiments and experimental approaches to this phenomenon have been realized. Some of them reproduce separate features of BL but cannot copy the phenomenon in the whole.

This chapter presents a short review of the state of the art in BL investigations.

1.2 General Ball Lightning Features

Intensive collecting of data and its analysis made in the twentieth century allowed to create a ball lightning image with averaged characteristics. These characteristics of BL are represented in Table 1.1.

Table 1.1 Average parameters of BL [16, 17]

Parameter	Value
Probability of spherical form	$89 \pm 1\%$
Diameter	24–32 cm
Lifetime	9–16 s
Motion velocity	3–5 m/s
Energy	12.6–31.6 kJ
Energy density	1.6–15.8 J/cm^3
Color	White ($24 \pm 2\%$), yellow ($24 \pm 2\%$), red ($18 \pm 2\%$), orange ($14 \pm 2\%$), blue and violet ($12 \pm 2\%$), others
Light flux	800–2,200 lm
Light output ratio	0.14–2.82 lm W^{-1}
Correlation with electric phenomena	$70 \pm 10\%$ BL appear during thunderstorms
Season	80% of BL appear in summer months (June–August)
Decay	$50 \pm 20\%$ explosion, extinction, decay
Probability of appearance	$10^{-9} - 10^{-8}$ km^{-2} min^{-1}

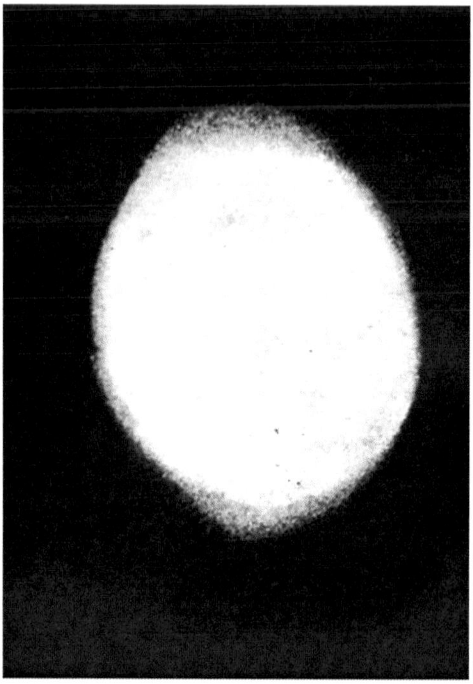

Fig. 1.2 Still photograph of falling-down BL taken during a thunderstorm in 1935 by Schneidermann [14]

To them, most known authors of the end of the twentieth century [11, 14, 16, 17] added the following. BL is observed indoors and outdoors, in airplanes and near them. It moves in horizontal and vertical directions (mostly downward). Its existence is often accompanied with hissing, whistling, and cracking. In most cases, BL thermal impact to objects is not revealed.

In Fig. 1.2, one can see a photo of a falling-down BL [14]. In [15], it is shown that an oval image of BL is caused by the motion of BL during shooting. In Fig. 1.3, we represent a still photograph of BL luminous trace from [14].

However, the collected data showed that BL can have an internal structure (as if it is made of honeycombs, or caviar) [11–13], sparks were observed inside and outside of it [13], and it can be surrounded by a halo much greater in size than BL itself [14]. It has also gray and black colors. Sometimes, it can appear in a form of a tape or a thread and transform it to the ball [13, 18]. In Fig. 1.4, one can see a photo of BL with luminous filaments (maybe emitted luminescent particles) [14].

BL exposes electric properties, making harmful influence on people and animals (similar to affection of the electric current). It often destroys electrical circuits and devices (computers, TV, and radio sets) [11–13]. In Fig. 1.5, we represent a photo of a falling-down BL [14], the validation that this was a real BL discussed in [15].

There are well-documented cases of pepole's death [13, 19] in the result of the electric or the thermal impacts caused by BL. Usually, these cases are biased, questioned [14], or disregarded [11, 14, 16, 17]. According to the authors of these papers, BL creates a path for a stroke of a linear lightning, which is the real reason

Fig. 1.3 Still photograph of a luminous trace of a moving-outdoors BL, taken by Bird [14]

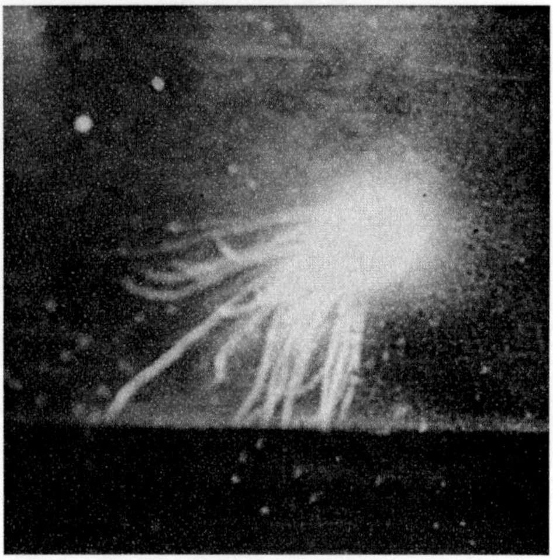

Fig. 1.4 A fragment photo of BL with luminous filaments, the photo is taken from [14]

for the harmful events. However, there are reliable evidences that these cases were not connected with the linear lightning [13, 19].

BL can penetrate through window glasses in three ways: through existing holes, making a hole, and without any holes, leaving small or no traces [20–22]. A photo of the hole in the glass made by BL is represented in Fig. 1.6.

New analysis of BL properties made in the twenty-first century has shown new BL features, and emphasized the already known, but usually disregarded. In [20, 22] was shown that in about of 25% cases of BL observations from the close distance, the BL impact lead to arising of fires, injuries of people, and heating or melting of objects.

Fig. 1.5 Still photo of a falling BL taken by Shagin during a thunderstorm over the black sea near Sochi City [14]

Fig. 1.6 A *hole* in the glass made by BL [21]. To the *right*, there is a *hole* in the glass; to the *left*, there is the *disc* that had fallen out the *hole*. *Disc* (*hole*) axes sizes are 8.2 and 7.6 cm, respectively

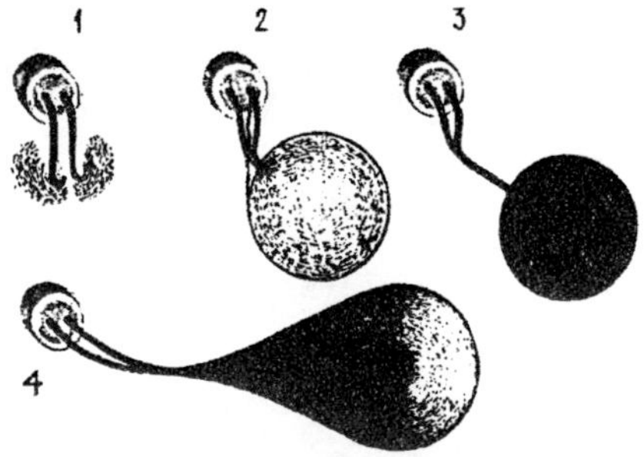

Fig. 1.7 An appearance of BL from a socket [19]

The idea about a hot flexible surface of BL leads also to the picture of BL in Fig. 1.7 where the BL blowing off from the socket is represented [19]. This BL capability is often described by the observers.

The photo in Fig. 1.8 shows a place of BL impact to the birch tree bark, which reveals the evident thermal effect.

Basing on the available data analysis, and especially on the analysis of works [12, 13, 15, 22–24], it was concluded that BL can store large amount of energy with the energy density up to 3×10^6 J/cm^3 and more in conditions of the linear lightning absence in immediate vicinity to BL.

Another outstanding BL feature [19,23,25] is its capability to drag heavy objects, such as metallic roofs (with weight about 10^4 kg) and metallic constructions (with weight about 10^2 kg), and to break large trees [26].

One has also to take into account that in a number of cases [12, 24, 26], tens or hundreds of kilograms of slad, carbonaceous steel with slag, and rocks at the places of large fireball explosions or falls were found.

These last two features first of all speak about a possibility of BL large momentum and mass. These cases were known long ago and were absolutely ignored in the twentieth century.

BL properties, described above, and a variety of its parameters define a complexity of BL phenomenon starting from the question "where does it from?" and "what is its composition?."

However, parameters of this phenomenon averaged over a large number of events are not impressing, see Table 1.1. To say more, the given parameters are usually useless when one wants to have a reliable picture of the BL phenomenon, and prepare a new experiment on BL modeling. Collection and analysis of more than 3,000 cases of BL observations, and establishing of more than 80 BL features, put forward a multitude of questions. Some important and interesting of them are formulated in [15]. Majority of these features can not be included into the Table 1.1 because of its average character, as we have already indicated.

Nonimpressive average parameters of BL in the Table 1.1 have played an unkind joke with the BL problem in the twentieth century: the most interesting BL feature—its possibility to accumulate high values of energy stopped to be considered seriously [11, 14, 16, 17], and new investigators have lost their interest in BL researches. There still appear voices claiming that BL is an optical phenomenon or the effect of human perception.

Fig. 1.8 Still photograph of BL burn trace on a birch tree, taken by Yu. Kurilenkov in 2007 (with a permission of the author)

1.3 Energy Storage in the Ball Lightning

1.3.1 Hypotheses About BL Energy Sources

From the facts of BL observation presented above, one can have a general idea of this phenomenon. A desire of people to explain BL's nature and realize a similar object is understandable. Since BL observations have rich history, then there is a large number of hypotheses about its nature, many of them are discussed in [1,2,11–15]. On their basis, theoretical models of BL as a physical phenomenon are under a construction. In their basis lies the information on processes, which take place in air excited by a thunderstorm and the surrounding nature.

Concerning the hypotheses, it is necessary to note that their number is great. With time, some of them are forgotten—and then they appear again in works of new authors with new shades. Therefore, for an explanation of BL nature, following [16, 17], there is no sense to search for new physical principles inherent its nature, but is reasonable to use already existing ones. It is necessary to note that BL is a complex phenomenon, which combines at first sight inconsistent properties, so many existing models usually describe only the separate sides of the phenomenon.

One can stand on a view point of the corresponding model, and critically estimate other sides of the phenomenon, using the modern scientific information on the processes, the phenomena in the thunderstorm-excited air, and the nature surrounding the observed object. If it leads to principle contradictions between the used theoretical model and the observable facts, then it is necessary to conclude the inconsistency of the model.

Let us carry out the critical analysis of the existing hypotheses connected with BL energy accumulation. We concentrate our attention to this property most attractive for the autonomous objects. In the agreement with the conclusions of [11], BL has no constant external feeding with energy. Each of these hypotheses should explain, first of all, the energy of BL, and where it comes from. First, we divide existing hypotheses according to the supposed energy source. Then, following [16, 17] we can attribute possible hypotheses to one of the following types:

(1) Exotic
(2) Plasma energy
(3) Energy of excited molecular particles in the gas
(4) Electrical energy
(5) Chemical energy

Let us explain each of these types. We begin with the first one, which we conditionally name "exotic." To this type, we have included such assumptions, in which the BL energy is connected with dark and antimatter, X-ray radiation,, etc. Such type of the hypotheses is discussed in [1, 11, 15], where their inconsistency is shown.

1.3.2 BL Plasma Models

The plasma hypothesis is quite natural since BL is connected with electric phenomena in the thunderstorm atmosphere, with the usual lightning, by which a plasma is formed. An internal energy of such BL is reserved in different charged particles. This energy is released in the processes of their recombination. Depending on a type of the charged particles in the plasma—electrons, ions, complex and cluster ions, or aerosol-charged particles—there can be different versions of BL plasma models.

We consider that the energy in the BL plasma is not constantly delivered from the outside and the internal energy source exists. Thus, the BL temperature is insignificant (about of 1,000–2,000 K).

Let us consider that according to the observation data, the noticeable density of the internal BL energy is $W_E = 10^2 - 10^6$ J/cm^3, and the average value of its lifetime is about of 12 s (see Table 1.1).

The energy reserved in such a BL plasma is connected with the ionization of atoms and molecules. In rough approach, the plasma energy density is

$$E = N_t I, \tag{1.1}$$

where I denotes the ionization potential of atoms or molecules and N_t is a number density of the charged particles. At the concentration of the charged particles in the plasma of $\sim 10^{19}$ cm^{-3}, created by the lightning, and the ionization potential of atoms, or molecules of ~ 15 eV, from the expression (1.1) for the energy density reserved in the unit of the BL volume, we obtain the following value for the energy density of BL

$$W_E \sim 24\,\mathrm{J/cm^3}.$$

As is known, the lifetime of the plasma is defined by the time of the charged particles elimination. The analysis of [16, 17] has shown, that the recombination time of the plasma consisting of the electrons and the positive ions, and consequently of BL, consisting of these particles (e.g., ions of nitrogen N_2^+ and oxygen O_2^+), is about of 5×10^{-13} s; the recombination time of the plasma, consisting of the positive and the negative ions, appears to be of the order of 5×10^{-14} s; and the recombination time of the plasma containing complex ions appears to be of the order of 10^{-13} s. The recombination time estimations [16] of the oppositely charged aerosol particles in air, when they approach each other due to the Coulomb attraction, and this motion is decelerated by the friction forces in the gas, show that the recombination passes quickly enough, with the typical time of ~ 0.2 s.

From these estimates follows the inconsistency of the BL plasma models. Really, a process of the charged particles energy transformation in heat at the positive and the negative charges recombination in the plasma occurs too quickly, so appreciable energy cannot be kept in the plasma long enough.

1.3.3 The Long-Lived Excited Atoms and Molecules in Air

Another way of the energy storage in the linear-lightning-excited gas can be connected with a creation of a large number of the excited atoms or molecules. Two types of the excited particles—the metastable electronically excited atoms or molecules and the vibrationally excited molecules- are possible in air.

The energy store in them is about the given in the formulae (1.1), or smaller. The smallness of this energy density is true for any excited states in any gas. Even the increase of the energy by several times, does not lead to the observed numbers for real BL. At that, the processes with the participation of the excited atoms and molecules at the atmospheric air pressure proceed rather quickly [27].

Therefore, those models of BL, in which the excited particles are used as the energy source, appear to be also inconsistent by two parameters—the stored energy and the lifetime.

1.3.4 An Electric Way of Energy Storage

To electric hypotheses, we attribute such in which the BL internal energy is connected with the electric fields created by the charged particles. In this case, we initially have a system of the charged particles (electrons, ions, or aerosol-charged particles) collected in the set element of space. The energy used for placing there the charged particles, having overcome the Coulomb interaction forces between them, is the internal energy of the system interesting for us.

From the BL observation data follows, that it possesses rather high electric charge. An electric field created by the object's charge, can cause a luminescent discharge in air. So, we have to estimate an energy of such a system.

Let us consider that the full charge of BL is equal to q, and is located in a sphere of a radius R_0. Then, in the case the charge is uniformly distributed over the volume of the sphere, the electric energy of the sphere is equal to [16, 17]:

$$E = \frac{3}{5}\frac{q^2}{4\pi\epsilon\epsilon_0 R_0}, \qquad (1.2)$$

where ϵ is the dielectric constant of BL material and ϵ_0 is the dielectric constant of vacuum. If the charge is uniformly distributed over the sphere surface, its electric energy is equal to [15–17]:

$$E = \frac{q^2}{4\pi\epsilon\epsilon_0 R_0}, \qquad (1.3)$$

where ϵ is the dielectric constant of air. One can see that they are of the same order of magnitude. At that, an electric field strength value F is maximal at a surface of the sphere and is

$$F = \frac{q}{4\pi\epsilon\epsilon_0 R_0^2}. \qquad (1.4)$$

As is known, at F $= 27 - 30\,$kV/cm, the air breakdown takes place at the atmospheric pressure in dry or humid air. For such an electric field strength, the energy density of the surface-charged sphere is $W_E = 2 \times 10^{-4}\,\text{J/cm}^3$. Even rise of the electric field on the ball surface by several orders of magnitude [15] does not save the situation. In this case, we also come to the conclusion, that by the electric interactions, it is impossible to explain the observable BL energy values.

Recently, Nikitin in [15] has published the concept of BL as a dynamic capacitor. As well as other authors of modern models, he had a predecessor—de Tessan—who proposed in 1859 a BL model in a form of a spherical electric capacitor. Even at replacement of air by an insulator with the electric field strength $E_{\mathrm{brp}} = 10^9\,\mathrm{V/m}$ the energy stored in the capacity is insufficient for an explanation of the high energy density of BL. However, a strong dependence of the energy density on the electric field ($W_{\mathrm{E}} \sim E^2$) makes it very attractive to use, namely, this electric capacitor as the keeper of energy. The complexity consists in that in real electric capacitors the field pulls out electrons at the superstrong electric fields from the metal electrodes, and in the case of BL there are no solid electrodes. Therefore, it is necessary to search for a configuration, which existence would be supported by the moving separated electric charges of a different sign. The author has come to the model named "the dynamic electric capacitor." In it, the electrons are located in the center, moving on a ring orbit, and round them rotate the positive ions under the influence of the electric field, created by the charge of the electrons. The author met a problem of a cover of such BL, which protects moving electrons and ions from the collisions with the molecules of air. He supposes, that it is created of the raindrops water during the lightning discharge in the thunderstorm atmosphere. However, a possibility of such a cover creation simultaneously with the starting of the charged particles torsion is doubtful.

1.3.5 *A Chemical Way of Energy Storage*

The most ancient hypothesis of BL is connected with a chemical way of the energy storage [4, 5].

In the case of the chemical way of the BL energy storage, its energy is released at the chemical reactions. An elementary act of the chemical process is connected with a tunnel transition of atomic particles, and a reconstruction of an atomic system at a moment of the particles rapprochement. At the thermal energies, the probability of such a transition can be very small, so there are many examples with the large storage time of the chemical energy. The chemical way of the energy storage has one more advantage before others, providing the high energy density, considerably exceeding those of the electric processes and the plasma.

For a comparison, we represent values of the air energy density at the atmospheric pressure: the energy density of the completely dissociated and half ionized air is $W_{\mathrm{E}} = 0.7\,\mathrm{J/cm^3}$ [16, 17], and the energy density of air with a coal dust concentration greater than 0.08 g per 1 g of air is $W_{\mathrm{E}} = 3.6\,\mathrm{J/cm^3}$ [16, 17]. One can see high chemical energy density store with respect to those of the plasma. Even more energy is released at combustion of the solid materials. For example, if the silicon oxidation takes place in BL of 10 cm in radius filled with 1.0 g of the silicon, then the energy density released at the combustion would be about

$W_E = 2.2 \times 10^6$ J/cm^3, as it can be obtained from the data of [28], and it is considerably closer to the observed BL with high energy density.

In the case of the BL huge energy parameters, one can suppose a heavy weight of BL, and it in turn can explain the observed mechanical impacts of BL.

As the result of our analysis, in agreement with [16, 17], we come to the conclusion that the unique way of the energy conservation in BL is the chemical way.

At definite application, one can start at some model features of BL, so from the represented BL analysis we have to consider initially the ball-like, the high-energy, rather heavy, and the long-lived object.

One of the most consistently developed model was created in works of Smirnov [16, 17, 29]. According to his model, BL has the light aerosol framework with the specific weight of the atmospheric air. This framework is charged, that maintains its stability and rigidity. In the framework pores, there is a small amount of the active substance, which represents a mixture of a fuel and oxidizer. A weight of the active substance is by several times smaller than the weight of the framework. The active substance inside the framework has a fractal structure, and can be represented in a form of a large number of thin threads. The energy density of such BL is about $W_E = 30 - 50$ J/cm^3, this shows that such a model cannot be applied to the observed high-energy BL.

Articles [30, 31] have appeared to be important for understanding of a soil role in BL origination. In [30], it was proposed that BL appears as a result of the linear lightning stroke in the Earth. It releases high energy and creates the fulgurite area—a cavity in the Earth. In it, chemical reactions with participation of soil consisting of sand, organic, and other components take place. A sand includes silica (SiO_2) in its composition. A recovery of silica to metal silicon takes place, namely, in the reaction:

$$SiO_2 + 2C \rightarrow Si + 2CO. \tag{1.5}$$

Here, C is carbon and Si is silicon atoms.

At high temperature, this leads to the formation of metallic chains, and those, appearing in air, create a ball-like formation of the type predicted by Smirnov [29]. However, this model gives BL energy of the order of the model [29], so it cannot explain the features of the high-energy and heavy BL.

The model recently represented in [15, 32] continues the chemical approach to BL with the high-energy content, and it develops an approach of [30,31]. BL appears at an impact of either a linear lightning, or a high-power electric discharge to some melting and evaporating material. At this, hitting to some melting object a cavern (or a fulgurite area in case of the Earth) is created. There chemical processes occur at high temperature with the participation of metal, dust (mixture of sand and organic particles), rust, and water vapors in the case an impacted subject is over the Earth. Inside the soil components—silica [16, 17], atoms, and molecules of the dissociated organic—macromolecules including C and H_2 inside the Earth are participating.

At high temperature, the recovery processes of oxides take place in the cavern; they lead to creation of a metallic powder and accompanying gases. For example, inside the Earth they are [30, 31, 33, 34]:

$$SiO_2 + 2C \rightarrow Si + 2CO;\ \ SiO_2 + 2H_2 \rightarrow Si + 2H_2O. \tag{1.6}$$

Here, H_2 denotes hydrogen molecules and H_2O are water molecules. Over the earth rust, Fe_2O_3, and other oxides can participate in the recovery processes. The cavern with the oxide surface filled with a metallic powder and gases is created: in the Earth—due to melting of SiO_2 on the surface of the cavern; over the Earth—due to oxidizing of the melted metallic surface at interaction with oxygen of air.

At the same time, the transportation of noncompensated charge to the particles of this cavern or a bubble takes place from the linear lightning or discharges. It is evident in the case of the linear lightning, which carries a noncompensated charge [35]. So, in the result of these processes, the charged sphere with the metallic powder and gases, such as CO_2 and H_2O inside, and the oxide cover layer outside is created.

Being charged, this bubble, created over the ground, separates from the place of its origination, if it has the same charge as the Earth. The linear lightning with average parameters transfers to the Earth about 10^{10} J [35], creating high pressure up to several hundreds of atmospheres in the cavity. After this, an ejection of a modified material part takes place. So, the bubble created in the earth is ejected into air. Bubbles can be light, heavy, hot, and highly charged objects, with respect to their composition. The oxide layer strongly decelerates an internal metal particle's oxidation due to prevention of oxygen penetration inside. The film has a rather high tensile strength [33, 34], so the cover can withstand development of hydrodynamic instabilities on the BL surface. The pressure of charges on the surface is initially compensated by the air pressure. Charged heavy object—the ball lightning moves in air. It does not fall down due to the Coulomb repulsion from the charged surface of the Earth. Due to the large charge of the object, a plasma layer originates on its surface [15]. The surface layer warms up the ball surface.

Slow combustion processes take place inside the reservoir of BL. They ensure not only an illumination of the ball, but its explosion, when the pressure of created gases inside the ball becomes comparable with the atmospheric one. Oxygen propagation into the reservoir sharply increases at break of the surface. Oxidation and release of energy takes place in different chemical reactions such as [30, 33]

$$Si + O_2 \rightarrow SiO_2,\ \ SiO_2 + Si \rightarrow 2SiO. \tag{1.7}$$

The combustion takes place; the object explodes, and tears apart. For estimation of such a BL energy density, consider BL with the mass of 10 g ensured by the Si powder. It is in a sphere of 10 cm radius. The energy W_{ch} released at its oxidation in air is $W_{ch} = m_{Si} \cdot \Delta Q$, where m_{Si} is the mass of the silicon and $\Delta Q = 8.8 \times 10^6$ J/kg is the oxidation enthalpy of Si in air at typical combustion temperature $T = 2{,}000$ K, as it follows from [28]. So the chemical energy, which can be released at its combustion, is about $W_{ch} \sim 2.2 \times 10^7$ J/cm^3, that is even larger than the estimates of the BL energy density given above. BL lifetime can be limited by a number of processes, one of them the metallic powder material combustion. Absence in the literature of necessary data for Si combustion rate constants makes us use the data for the Al powder combustion [34]. The radial combustion velocity of Al powder (in H_2O and CO_2) is $V_{com} = 5-10$ mm/s. So, the BL lifetime estimate is ~75–150 s, which well fits to the observed lifetime of the large BL. We see that such BL can have high level of the chemical energy and live rather long.

This model considers BL as the object which can be hollow and can contain charged metallic particles inside it, that is can have the metallic core inside the oxide cover. The object's internal substance can represent a powder. The powder can be charged its motion inside the cover can lead to mechanical effects on BL motion and form. This object is alike a soft flexible cushion that can penetrate through holes. Its main energy is connected with a metal oxidation, but it also possesses a sufficiently large amount of the electrical energy.

Presence of the electric charges in this object represents its essential feature, since it determines its capability to fly, and to realize harmful impact to different electrical devices and people. So BL represents a unique joint of chemical, electrical, and mechanical features. BL studies require additional experiments, collection, and analysis of observation data, and field investigations to the places of its impact to Earth, objects, and people.

From the point of view of possible experiments, the presented model shows the productivity of the experiments with Si, rocks, metals, and metallic particles, as well as with natural organic materials in plasmas; some of them are discussed below.

1.4 Experimental Modeling

An attempt to understand the BL "construction" and create some working hypothesis on its nature was accompanied with attempts to realize a laboratory model of BL, or artificial ball lightning (ABL). In the past, there were many experiments undertaken on BL modeling in a laboratory. Some of them are described in [1, 2, 15, 36]. Experiments were realized by different types of gas discharges or plasma jets. It was defined by a convenience of the energy put by these means and the appropriate energy of existing discharges and plasma jets.

Following the accepted approach, we consider experiments that model the chemical nature of BL, when the realized long-lived object had a spherical form, which did not change into other forms (vortex or so).

In [1, 2] there is a description of H. Nauer's experiments on creation of luminescent balls at gas discharges in air with admixture of hydrogen and hydrocarbons, such as methane, propane, and benzene in concentrations somewhat smaller than those necessary for the inflammation of mixtures. The greatest effect was observed at the application of benzene. Bright luminescent balls were observed even at presence of tiny traces of it in the discharge chamber. At an explanation of the experiments, the author made a supposition that the glow appears on the tangle of thin threads, produced at benzene combustion. At that, the luminescence was considered as the result of an organic material combustion.

One can find descriptions of Barry's experiments with the pulsed discharge in air with propane admixture (with inter electrode voltage of 10 kV and energy in the pulse ~250 J) in [2]. Air pressure was the atmospheric one, a volumetric propane concentration was 1.4–1.8%; it is smaller than the concentration necessary for an inflammation of the mixture. In the result a yellow–green ball appeared which lifetime was 1–2 s.

In the reference [37], that ideologically continued Barry's experiment [2], the pulsed discharge in air, was created between the copper electrodes, being at the distance of 3 mm. At that, the discharge voltage was changed in the range 8–10 kV, and energy in the pulse was 350 J. At Ethane content of 2.1% in mixture with air red balls with a diameter ~4 cm and the lifetime of 0.3 s were observed sometimes. ABL of white color and 3 cm in diameter was observed in the discharge afterglow during 0.8 s at addition of 100 cm^3 of cotton fibers to the gas mixture. Appearance of the luminescent objects were observed several times in the mixture consisting of ethane 2.7% and 100 cm^3 cotton fibers processed into small particles. The maximum diameter of such an object was 5 cm, and its lifetime was 2 s.

Streamer discharges were used in [38] for a generation of ABL. For this purpose, a Tesla generator at the frequency of 67 kHz was used. A setup of Tesla [39] was reproduced in the smaller size [36]. The mean power delivered to the high-voltage electrode was 3.2 kW. This electrode was covered with a piece of wax or charred wood. It was experimentally revealed that a large number of hydrocarbon and metallic particles, evaporated from the electrode, appeared during the discharge. Their maximum number was observed in a small area near the electrode region with strong electric fields (of 10–20 kV/cm). ABL appeared near the high-voltage electrode as if "from nowhere" because they were absent at previous frames of video recoding. Their colors were different: red, yellow, blue, and white; their lifetime reached 2 s and size 1–5 cm. In the end of the existence, ABL often exploded with a loud bang, this effect we have to specially mark.

Tesla [39] had realized fireballs by himself, their sizes reached several centimeters. The high-voltage electrode in his case was covered with the isolation made of a natural rubber, and a layer of graphite powder was between the electrode and the isolation. Therefore, particles of metal, soot, carbon, and polymer evaporated from the electrode could take part in complicated processes of a structure formation and combustion.

To another popular investigations, one can attribute a formation of balls out of a metallic vapor, realized at closing of contacts of powerful electric batteries [40, 41]. Sizes of ABL were of several cm, and their lifetime reached tens part of a second.

Luminescent balls of 2–4 mm in diameter, and the lifetime of 2–5 s were obtained in [42] at short-circuit of two electrodes in water. The balls resembled drops of melted metal burning in air. After cooling, they represented metallic spheres of 2 mm in diameter with traces of oxides on the surface in a form of thin hair.

Experiments described in [43] were devoted to creation of regions with plasma conditions analogous realized at the linear lightning stroke to the soil, or at the explosion in a closed space. The discharge and the plasma were created in the closed volume of the organic tube. Then the created luminescent formation, consisting of a mixture of the vapor, particles of the electrode material melt, and the ionized air, went outside through a specially prepared hole, or a hole broken through in the tube due to the pressure rise during the discharge. The lifetime of such formations reached 5–7 s. They exploded after hitting an obstacle standing in front of it, leaving traces as from a multilayered object with the melted core. Sometimes, it was possible to find their remains, that resembled oxidized metallic shells. This was one of few experiments when there were detected an explosion and remains of ABL.

This direction had a development in the work [44]. It was discovered there that different ABL structures were formed at the application of organic plasma-forming materials and at injection of plasmas into air saturated with organic vapors. A plasma jet from the plasma generator passed over a cuvette, keeping melted paraffin or wax at undertaking of these experiments. At that, the plasma generator energy of about 200 J was delivered to the gas mixture during ~6–8 ms. A video recording had showed that the spheroid or mushroom formations were formed over the cuvette immediately after the "shoot" and rose up. Their size was 10–20 cm, and the temperature was $T \sim$ 1,500–2,000 K (as it was estimated by their rising velocity), they existed up to 0.5 s. Luminescent objects in experiments [44] were created by the plasma stream. They appeared over a cuvette with a melted wax and a paraffin even 10 min after stopping of the cuvette heating. Such a long time of the vapors high-concentration existence proved that in real conditions gaseous hydrocarbons can be accumulated in some local places and ignited later. Appearance of such objects allows to explain a nature of the luminescent objects that lead to summer forest fires and are observed in coniferous forests in clear weather.

In the another experimental series in [44], with use of the capillary of 1–2 mm made in a material of a complex organic mixture consisting of paraffin, colophony and milled wood with average sizes of the "seed" $1 \times 0.3 \times 0.3$ mm were obtained luminescent objects with typical visible sizes 1–2 cm and the lifetime of 1–2 s. Several times, it was possible to find remains of these ABL. They were investigated with the electronic microscope, which showed their polymer fractal-porous structure. These ABL practically completely copied descriptions of the BL motion. They unexpectedly appeared from the channel of the plasma generator, usually 30 ms after the plasma flow. Visually, they unexpectedly disappeared, though the video record showed their gradual starvation during 0.3–0.4 s. In the reference [45], it was experimentally and theoretically was shown that analogous in sizes and the lifetime objects were observed at burning of small particles of wood with sizes of ~1–2 mm. Such particles could be formed in [44] at erosion of the channel wall, and later they could be ignited.

Abrahamson [31] has undertaken experiments on imitation of the lightning stroke to the Earth in order to verify his hypothesis [30], as described above. He passed a discharge current from the capacity 204 μF, charged to the voltage of 20 kV, through a layer of the humidified soil of 3 mm in the thickness. The soil filled the bottom graphite electrode; the graphite rod of 15 mm in diameter was used as the upper electrode located at a distance of 22–36 mm from the surface of an Earth layer on the bottom electrode. Energy released during the discharge was 110 kJ, and the charge of 1.3–3.4 C passed through the soil. At the highest power experiments, the negative potential was applied to the upper electrode. Products of chemical reactions were sucked away by a pump and deposited on the filter of quartz fibers and on a lattice of nickel. Fiber threads of 100 nm in diameter and to 7 microns in length were found on the filter by means of an electron microscope. Chains from 25 to 120 nm in length consisting of balls of 25 nm in diameter were found on a surface of a nickel lattice. However, authors of [31] did not manage to realize an appearance of large autonomous spheres similar to BL.

After papers [30, 31] several works [46–49] devoted to the combustion and explosion of porous silicon with the formation of ABL were published.

The work [46] was devoted to the explosion and combustion of nanostructured silicon in the presence and absence of the hydrogen on its surface. Special attention was paid to the investigation of spherical luminous structures formed in these processes. Porous silicon layers created as a result of the electrochemical anodizing of the single-crystal silicon were used as a source of the silicon nanoparticles. The explosion and combustion of porous silicon samples were initiated thermally, mechanically, electrically, and optically. For example, thermal initiation was made by a contact heating of the sample to 900°C. Visible differences in explosive reactions with respect to the way of their initiation were not revealed. The explosion and combustion of porous silicon were accompanied by the origination of luminous balls with a diameter of 0.1–0.8 m moving with a velocity up to 0.5 m/s. The lifetime of such objects could reach 1 s.

In [46], it was proposed a hypothesis of the appearance and development of BL. A linear lightning striking the Earth leads to creation of silicon particles including those whose sizes are equal to several nanometers. These nanoparticles appear in air in the form of a fractal cluster, and are not visible until their combustion or explosion is initiated by some impact. The hydrated silicon and humid air ensure the formation of the hydrated plasma in the processes of the combustion and the explosion, which increases the lifetime of such a plasma to several seconds.

However, this hypothesis based on the described experiments and the model [29] cannot explain the high energy of the real BL and its heavy weight.

In Fig. 1.9, one can see an example of realized luminescent structures obtained in [46].

In the work [47], were also tested the ideas of [30, 31] by vaporizing at normal atmospheric pressure of small pieces of highly pure Si wafers by an electric arc. The arc was generated by the interruption of the electric circuit. It realized high temperatures sufficient to melt and vaporize the Si pieces. Authors generated luminous balls that have long lifetime and several properties typical for natural BL. Their initial diameters were in the range from 1 to 4 cm, and the lifetime was up to 8 s.

We have to emphasize that the considered experimental works correlate with recent theoretical approaches [15, 32], though they did not represent high-energy

Fig. 1.9 Artificial ball lightning produced by thermal ignition of 100 μm thick 1 cm in diameter doped porous silicon: 0.5 s after formation [46]

objects and did not reveal a mechanism of BL flying or levitating in air, and there are doubts concerning the origination of the activated silicon in nature.

As we can see, our approach to the experimental revealing of the BL chemical nature has allowed us to select works devoted to the analysis of the combustion processes of different materials that can be under the lightning, discharge, and the plasma impacts. Their combustion is accompanied with complex gas-dynamic and radiation processes, which in many features resemble appearance of real BL. However, there was practically absent explosions of the created ABL typical for real BL.

1.5 Difficulties of BL Investigations

There is a complexity of BL events collecting since it is connected with several reasons. BL appearance is usually unexpected even if an observer is prepared to it theoretically: the observer becomes excited, so he can make mistakes in the observation details and believe in them. He can analyze it by himself, put it to some scheme, use some known hypothesis, and to misrepresent facts. It is true for events manifesting high BL energy, heavy weight, large momentum, and capability to take away golden adornments, change of BL form. So, the continuation of BL observation is required in order to eliminate ambiguity of observations, hence to simplification of experimental and theoretical modeling.

Difficulties of BL science development is strongly connected with the experimental realization of the observation events. It is well understood that the BL problem will be considered as solved only after the experimental realization of this object, having all indicated above features. Nowadays, there are problems with copying of BL impact to window glasses. Especially, it is true with their penetration through glasses without clearly detected traces. Available information about BL impact on glasses is insufficient for understanding of physical and chemical processes that take place.

There are difficulties with the experimental modeling of magnetic impact of BL on metallic subjects. There are only few descriptions left [3, 19] and no analysis of metals influenced by BL. So, we do not definitely know what to model. It is also difficult to imagine how to model BL impact on golden rings and bracelets leading to their disappearance. We understand that we can use high-frequency electric fields, but what else have we to foresee in experiments in order to realize these subjects disappearance? We have to understand a mechanism of BL transformation into a ball from a tape and back.

Generally speaking, there is a shortage of detailed observations showing how these effects are realized by nature. So, we have to continue collection of detailed descriptions of BL unusual events and analyze approaches to their experimental modeling.

New investigations are to be undertaken in modeling of BL appearance and creation of luminescent objects on a base of burning metals inside oxide covers. We have to understand and realize mechanisms of BL explosions. It is difficult to model BL levitation, though we can charge a rubber ball by the Van der Graaf accelerator. This can be another type of experiments. We have to continue experiments with combusting powders and jets, especially creating them in charged and excited states and so on.

1.6 Conclusions

In the given chapter, we have shortly looked at the BL problem as a whole. For better understanding of this phenomenon, we have represented its photos and descriptions of its properties.

We have presented an approach to hypotheses of BL origin which showed that BL most probably has a chemical nature.

Some known experiments on reproduction of BL analogues are described. They show various approaches to the experimental realization of this object and their results. These experiments show a complicated character of origination and elimination of the ABL of the plasma chemical nature.

Analysis of experiments demonstrates existence possibility of the ABL on a basis of heterogeneous structures (a mixture of a gas with solid or melted components) or an object with solid framework. They are formed at the destruction both of organic polymeric and inorganic materials, such as Si, SiO_2, SiC, Fe_2O_3, etc. In the plasma conditions, these objects can be highly excited and represent nonequilibrium electrically charged structures and melts that combust in air. Some models of this phenomenon, which have drawn the attention of the investigators, are named.

Unlike other directions of physics where theory development is based on the analysis of a large set of experiments with unequivocal and regularly reproduced results, such a possibility for BL theory construction at the moment is absent. Theories are born often in a random way, and are under the construction without accounting of all achievements of the predecessors. In these circumstances, the weakness consists of the BL physics.

One can conclude that BL is not only the physical phenomenon, but the geophysical, and the geochemical as well. BL appears over and from the Earth mainly after the lightning impact. It can be formed at its impact to metallic and organic subjects. It can be formed during the earthquakes and the volcano activity as well.

Therefore, laborious work on the further ball lightning data collection, on experimental and theoretical modeling of its properties, is necessary. As a whole the science of ball lightning has passed the long way which has led to accumulation of this phenomenon observations, results of many experiments, and theories. It has shown that this complex physical phenomenon is connected mainly with chemical and electric processes accompanying them in the atmosphere, over and in the Earth.

References

1. S. Singer, *The Nature of Ball Lightning* (Plenum, New York, 1971)
2. J.D. Barry, *Ball Lightning and Bead Lightning* (Plenum, New York, 1980)
3. R.A. Leonov, *Ball Lightning Enigma* (Nauka, Moscow, 1965) (In Russian)
4. M.V. Lomonosov, in *Selected Works on Chemistry and Physics* (AS USSR Publishers, Moscow 1961), pp. 220–256 (In Russian)
5. F. Arago, Annuaire au roi par le bureau des longitudes, Notices Scient. 221 (1838)
6. W. Brand, *Der Kugelblitz* (Verlag von H. Grand, Hamburg, 1923)
7. P.L. Kapitsa, Phys. Blatter. **14**, 11 (1958)
8. P.L. Kapitsa, Sov. Phys. JETP. **30**, 973 (1970)
9. J.R. McNally Jr, Oak-Ridge Nat. Lab. Report N. ORNL3938 (1966)
10. W.D. Rayle, NASA Techn. NOTE-D- 3188 (1966)
11. I.P. Stakhanov, *The Physical Nature of Ball Lightning* (Atomizdat, Moscow, 1979) (CEGB translation (p. 227) CE 8224)
12. G. Egely, Hungarian ball lightning observation. Budapest: Centr. Research. Inst. Physics. Hung. Acad. Sci. KFKI-1987–10/D, (1987)
13. A.I. Grigorjev, *Ball Lightning* (YarGU publishers, Yaroslavl, 2006) (In Russian)
14. M. Stenhoff, *Ball Lightning. An Unsolved Problem in Atmospheric Physics* (Kluwer, New York, 1999)
15. V. Bychkov, A. Nikitin, G. Dijkhuis, in *The Atmosphere and Ionospher: Dynamics, Processes and Monitoring*, ed. by V. Bychkov, G. Golubkov, A. Nikitin (Springer, Heidelberg, 2010), pp. 201–373
16. B.M. Smirnov, Phys. Rep. **152**, 178 (1987)
17. B.M. Smirnov, *Problem of Ball Lightning* (Nauka, Moscow, 1988) (In Russian)
18. W.R. Corliss, *Nandbook of Unusual Natural Phenomena* (Gramercy Books, New York, 1995)
19. G.U. Likhosherstnykh, Tekhnika Molodezhi. N.3. 38 (1983) (In Russian)

20. V.L. Bychkov, in *Proceedings of 9th International Symposium on Ball Lightning (ISBL06)* (Eindhoven, The Netherlands, 2006), pp. 18–25
21. G.P. Shelkunov, A.I. Nikitin, V.L. Bychkov, T.F. Nikitina, A.M. Velichko, A.L. Vasiliev, in *Proceedings of 10th International Symposium on Ball Lightning (ISBL08)* and *3rd International Symposium on Unconventional Plasmas (ISUP 08)* (Kaliningrad, Russia, 2008), pp. 127–134
22. J. Abrahamson, A.V. Bychkov, V.L. Bychkov, Phil. Trans. Roy. Soc. **360**(1790), 11 (2002)
23. G. Egely, in *Science of Ball Lightning (Fire Ball)*, ed. by Y.-H. Ohtsuki (World Science, Singapore, 1989), pp. 19–30
24. M.T. Dmitriev, B.I. Bakhtin, B.I. Martynov, Sov. Phys. Tech. Phys. **26**, 1518 (1981)
25. N.G. Nikolayev, Scientific vacuum. NPK "RITAL", Tomsk (1998) (In Russian)
26. A. Ol'khovatov, *Tunguska Phenomenon of 1908* (Binom, Moscow, 2008) (In Russian)
27. B.M. Smirnov, *Excited Atoms* (Energoatomizdat, Moscow, 1982) (In Russian)
28. V.V. Gorskyi, P.Ya. Nosatenko, *Mathematical Modeling of Thermal and Mass Exchange Processes at Aero-Thermal-Chemical Destruction of Composite Heat Protecting Materials on Silica Base* (Nauchnyi mir Publishers, Moscow, 2008) (In Russian)
29. B.M. Smirnov, Phys. Rep. **224**, 151 (1993)
30. J. Abrahamson, J. Dinniss, Nature **403**, 519 (2000)
31. J. Abrahamson, Phil. Trans. Roy. Soc. **360**, 61 (2002)
32. V. Bychkov, IEEE Trans. Plasma Sci. **38**(12), 3289 (2010)
33. N.N. Greenwood, A. Ernshaw, *Chemistry of Elements* (BINOM, Moscow, 2008) (In Russian) (available in English)
34. P.F. Pokhil, A.F. Belyaev, Yu.V. Frolov et al., *Combustion of Powder – Type Metals in Active Media* (Nauka Publishers, Moscow, 1972) (English translation FTD-MT-24–551–73(1972))
35. Y.P. Raizer, *Gas Discharge Physics* (Springer, Berlin, 1991)
36. R.F. Avramenko, V.L. Bychkov, A.I. Klimov, O.A. Sinkevich (eds.), *Ball Lightning in a Laboratory* (Khimiya, Moscow, 1994) (In Russian)
37. H. Ofuruton, Y.-H. Ohtsuki, Il Nuovo Cimento **13C**(4), 761 (1990)
38. K.L. Corum, J.F. Corum, ICBL Article Series N. 1992/1. For Submission to 2nd International Symposium on Ball Lightning, Budapest, Hungary, 26–29 June 1990
39. N. Tesla, *Colorado Springs Notes 1899–1900* (Nolit, Beograd, 1978)
40. P.A Silberg, Appl. Phys. **49**, 1111 (1978)
41. G. C. Dijkhuis, in *Proceedings IX International Wroclaw Symposium on Electromagnetic Compatibility*, Wroslaw, 1988, P. 166
42. R.K. Golka Jr., J. Geoph. Res. **99**(D5), 10679–10681 (1994)
43. S.E. Emelin, V.S. Semenov, V.L. Bychkov, N.K. Belisheva, A.P. Kovshyk, Tech. Phys. **42**(3), 269 (1997)
44. V.L. Bychkov, A.V. Bychkov, I.B. Timofeev, Tekh. Phys. **49**(1), 128 (2004)
45. E.V. Konev, *Physical Grounds of Plant Materials Combustion* (Nauka, Novosibirsk, 1977) (In Russian)
46. S.K. Lazarouk, A.V. Dolbik, V.A. Labunov, V.E. Borisenko, JETP Lett. **84**(11), 581 (2006)
47. G.S. Paiva, A.C. Pavo, E.A. de Vasconcelos, O. Mendes Jr., E.F. da Silva Jr, Phys. Rev. Lett. **98**, 048501-1 (2007)
48. V. Dikhtyar, E. Jerby, Phys. Rev. Lett. **96**, 045002-1 (2006)
49. K.D. Stephan, N. Massey, J. Atmos. Solar-terr. Phys. **70**, 1589 (2008)

Chapter 2
Unravelling the Mysteries of Matter Surrounding Supermassive Black Holes

A. Müller and S. Schippers

Abstract Uncertainties of atomic data can compromise the interpretation of astronomical observations. Here, we discuss the case of low-temperature dielectronic recombination of iron ions in connection with X-ray spectra from active galactic nuclei, and we explain how new dielectronic recombination rate coefficients from storage-ring experiments have helped to remove previous inconsistencies in the astrophysical modeling of such spectra.

2.1 Introduction

Black holes have captured the imagination not only of physicists since their existence was suggested by the theory of general relativity [1]. Today, it is widely accepted that a black hole is produced, for example, by the gravitational collapse of a star which is a few times more massive than our Sun. A much more massive black hole of more than two million solar masses occupies the center of the Milky Way. The existence of this supermassive black hole (SMBH) has been inferred from the observed motions of stars orbiting the SMBH at close distances [2].

SMBHs reside at the centers of many if not all galaxies. In some cases, they induce extremely violent phenomena in their vicinity which lead to the emission of energetic electromagnetic radiation with a luminosity exceeding the contribution from all stars in the host galaxy by up to orders of magnitude. The central region of such an "active galaxy" is called "active galactic nucleus" (AGN).

In recent years, AGN have been intensively studied by astronomical observations, and a unified model for AGN has emerged [3,4] (Fig. 2.1). Accordingly, the SMBH accretes gaseous matter and dust from a surrounding torus. In the accretion disc matter is subject to strong friction forces and heated to elevated temperatures such that it becomes ionized. Much of the gravitational energy gain of matter spiraling into the black hole is converted into bremsstrahlung. In addition, relativistic gas jets are emitted in both directions of the SMBH rotation axis.

V. Shevelko and H. Tawara (eds.), *Atomic Processes in Basic and Applied Physics*, Springer Series on Atomic, Optical, and Plasma Physics 68,
DOI 10.1007/978-3-642-25569-4_2, © Springer-Verlag Berlin Heidelberg 2012

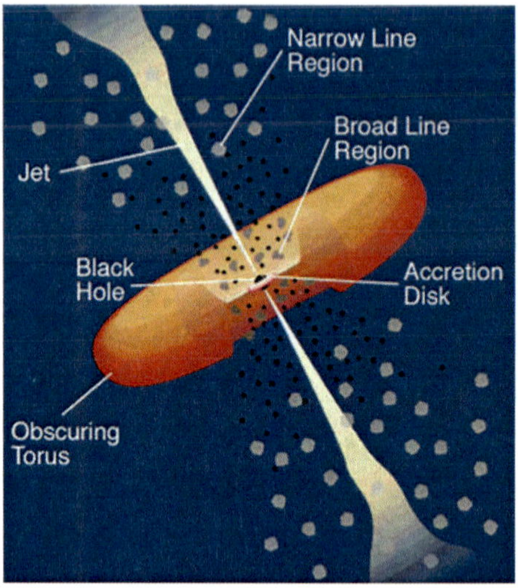

Fig. 2.1 Illustration of the unified AGN model (see text). Credit: Urry and Padovani [4, 5]; Copyright: Astronomical Society of the Pacific, reprinted with permission from the authors

The broad bremsstrahlung continuum of X-rays backlights the gas which is in the line of sight between the accretion disc and the observer. Consequently, absorption lines are frequently recorded in the X-ray spectra from AGN. Since the absorption line-width is determined by Doppler broadening, the temperature of the absorbing gas can easily be deduced from the observations. Cool narrow-line regions and hot broad-line regions have thus been identified in AGN (Fig. 2.1). Information about the chemical composition of the absorbing gas can be obtained if one manages to identify the measured absorption lines by chemical element and ionic charge state. For example, in the AGN X-ray spectrum in Fig. 2.2 absorption lines have been identified which are caused by multiply charged carbon, nitrogen, oxygen, neon, and iron ions with charge states ranging from 4 to 19. As is exemplified by Fig. 2.2, iron ions play a particularly prominent role in X-ray astronomy [6]. Iron is the most abundant heavy element [7] and still contributes to line emission or absorption in astrophysical plasmas when lighter elements are already fully stripped.

More detailed insight into the properties of the absorbing medium is provided by astrophysical model calculations (see, e.g., [9]) which aim at reproducing the measured spectrum by optimizing model input parameters, such as chemical composition, density, and temperature of the absorbing medium. Next to the astrophysical assumptions about the studied object, these model calculations require atomic data as additional input. The atomic data needs for astrophysical modeling are vast [10, 11]. Energies and strengths of spectral lines of atoms, molecules, and their ions must be known as well as cross sections or rate coefficients for various atomic collision processes as, for example, photoionization or electron ion recombination. An entire discipline, that is, "laboratory astrophysics" [12] is devoted to providing accurate atomic data from laboratory experiments and theoretical calculations.

It is clear that the accuracy of the astrophysical model results depends critically on the accuracy of the underlying atomic data. In this chapter, we explain how

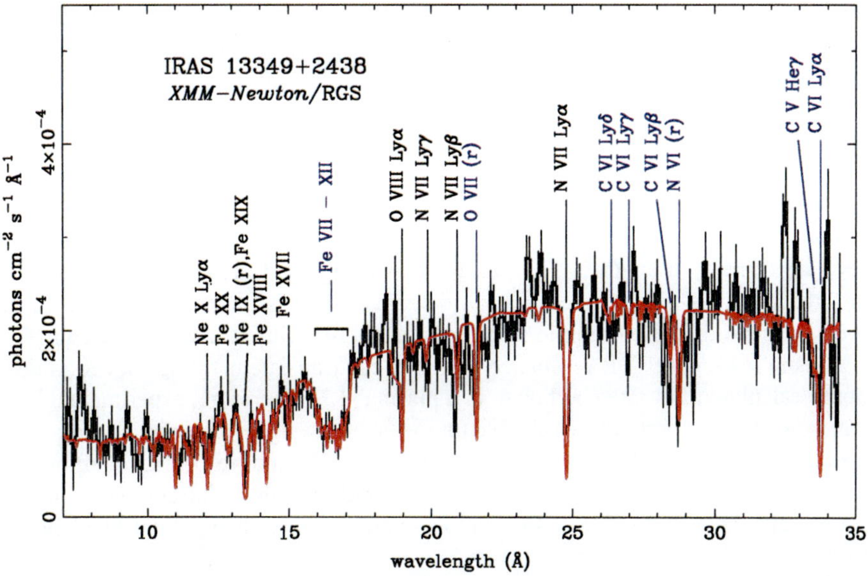

Fig. 2.2 X-ray spectrum from an AGN recorded by the X-ray telescope XMM-Newton (*full black line*). Absorption lines which have been identified are labeled by the respective chemical element. The *roman numbers* denote the ion charge state with "I" representing charge state 0 (neutral), "II" representing charge state 1 (singly charged), etc. The *red full line* is the result of an astrophysical model calculation (figure taken from [8]). Credit: Sako et al. , A&A 365, L168, 2001, reproduced with permission © ESO.

the limited knowledge about the atomic process of dielectronic recombination (DR) affected the interpretation of AGN spectra in the past and how better DR rate coefficients became available from laboratory astrophysics experiments using heavy-ion storage rings. First, we will discuss the limitations of the previously available recombination rate coefficients. Subsequently, we will briefly describe the storage-ring method for measuring accurate DR rate coefficients. Finally, we will present some recent experimental and theoretical DR results in particular for iron ions and their impact on astrophysical modeling of matter in the vicinity of SMBHs.

2.2 Dielectronic Recombination in Cosmic Atomic Plasmas

Dielectronic recombination (DR) is an important electron–ion collision process governing the charge balance in atomic plasmas [13, 14]. In DR, an initially free electron excites another electron, which is initially bound on the primary ion, and thereby looses enough energy such that it becomes bound, too. The DR process is completed if in a second step the intermediate doubly excited state decays radiatively to a state below the ionization threshold of the recombined ion. The initially bound core electron may be excited from a state with principal quantum number N to a state with principal quantum number N'. There are an infinite number of excitation channels. In practice, however, only the smallest excitation steps with $N' = N$ ($\Delta N = 0$ DR), $N' = N+1$ ($\Delta N = 1$ DR), and sometimes also $N' = N+2$ ($\Delta N = 2$ DR) contribute significantly to the total DR rate coefficient.

Energy conservation dictates that the DR resonance energies E_{res} are given by $E_{\mathrm{res}} = E_{\mathrm{d}} - E_{\mathrm{i}}$ with E_{d} and E_{i} being the total electron energies of the doubly excited resonance state and the initial state, respectively. Both E_{d} and E_{i} can amount to several 100 keV. In contrast, E_{res} can be less than 1 eV. Calculating such a small difference of two large numbers with sufficient accuracy is a considerable challenge even for state-of-the-art atomic structure codes [14]. Unfortunately, small uncertainties in low-energy DR resonance positions can translate into huge uncertainties of the calculated DR rate coefficient in a plasma [15]. Therefore, experimental DR measurements are particulary valuable for ions with strong resonances at energies of up to a few eV which decisively determine the low-temperature DR rate coefficients important for near neutrals in an electron-ionized plasma or for complex ions in photoionized plasmas. Almost all iron ions more complex than helium-like belong to this latter class.

Cosmic atomic plasmas can be divided into broad classes: collisionally ionized plasmas (CP) and photoionized plasmas (PP) [13]. In CP, ionization is predominantly caused by electron impact, and high-energy DR is the main recombination channel for most ions. Ions form at temperatures which correspond roughly to one half of their ionization potential [16]. Examples for cosmic CP are the solar corona, supernova remnants, and the interstellar medium. In PP, ionization is caused by an external radiation field and is balanced by low-energy DR. Since the ionization mechanism does not directly depend on the electron temperature, ions in PP form at lower temperatures as compared to CP, that is, at temperatures corresponding to about one twentieth of their ionization potential [17]. Examples for cosmic PP are H II regions, planetary nebulae, X-ray binaries, and AGN (Fig. 2.1).

Historically, most theoretical recombination data were calculated for CP. At the rather large temperatures where ions exist in CP, DR rate coefficients are largely insensitive to low-energy DR resonances. Consequently, the theoretical uncertainties are much smaller at higher than at the lower plasma temperatures which are typical for PP. Until recently, theoretical recombination rate coefficients were mainly calculated for the CP temperature ranges. If these DR data are used for the astrophysical modeling of PP, inconsistencies arise.

Such inconsistencies have been noted by Netzer [18] and Kraemer et al. [20] in the astrophysical modeling of X-ray spectra from AGN. On the basis of their model calculations, these authors have suspected that the DR rate coefficients for iron ions with an open M-shell (Fe^{1+}–Fe^{15+}) from the widely used compilation of Arnaud and Raymond [19] are much too low in the PP temperature range. Figure 2.3 shows the observed X-ray spectrum from an AGN in the wavelength range of the unresolved transition array (UTA) associated with $2p \to 3d$ excitations in iron M-shell ions (see also Fig. 2.2). Using the DR rate coefficients for these ions from [19], the astrophysical model calculation fails to reproduce the shape of

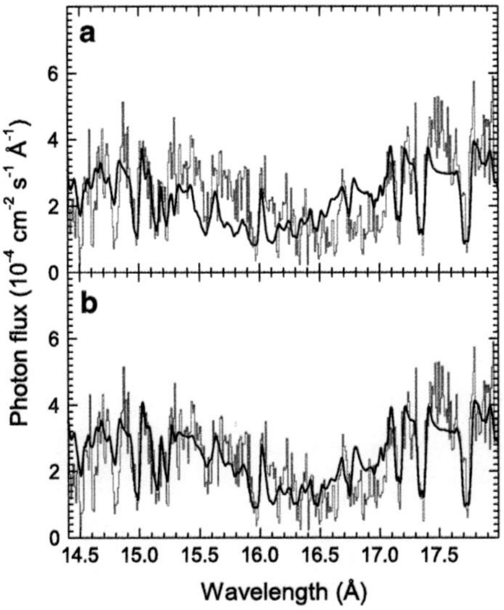

Fig. 2.3 Measured X-ray spectrum (*thin gray curves*) from an AGN (NGC 3783) in the wavelength range of the UTA associated with $2p \rightarrow 3d$ excitations in iron M-shell ions (see also Fig. 2.2). The observed spectrum is compared with results of astrophysical model calculations [18] (*thick black curves*) using (**a**) DR rate coefficients for iron ions from [19] that were calculated for CP and using (**b**) deliberately adjusted low-temperature DR rate coefficients

the UTA (Fig. 2.3a). This discrepancy was traced back to the uncertainties of the recommended DR rate coefficients at low plasma temperatures. When increasing the low-temperature DR rate coefficients deliberately, the astrophysical model calculation yielded a better agreement with the astronomical observation [18] (Fig. 2.3b). Clearly, this finding provided a strong motivation for the laboratory astrophysics work on DR of iron M-shell ions which will be discussed in the following section.

2.3 Storage-Ring DR Experiments

Heavy-ion storage rings equipped with electron coolers serve as an excellent experimental environment for electron–ion collision studies [21–24]. In electron–ion merged-beams experiments at heavy-ion storage rings a fast-moving ion beam is collinearly merged with a magnetically guided electron beam with an overlap length L of the order of 1–2 m. Recombined ions are separated from the primary beam in the first bending magnet downbeam of the interaction region and directed onto a single-particle detector (Fig. 2.4). Since the reaction products are moving fast and are confined in a narrow cone, they can easily be detected with an efficiency η of nearly 100%.

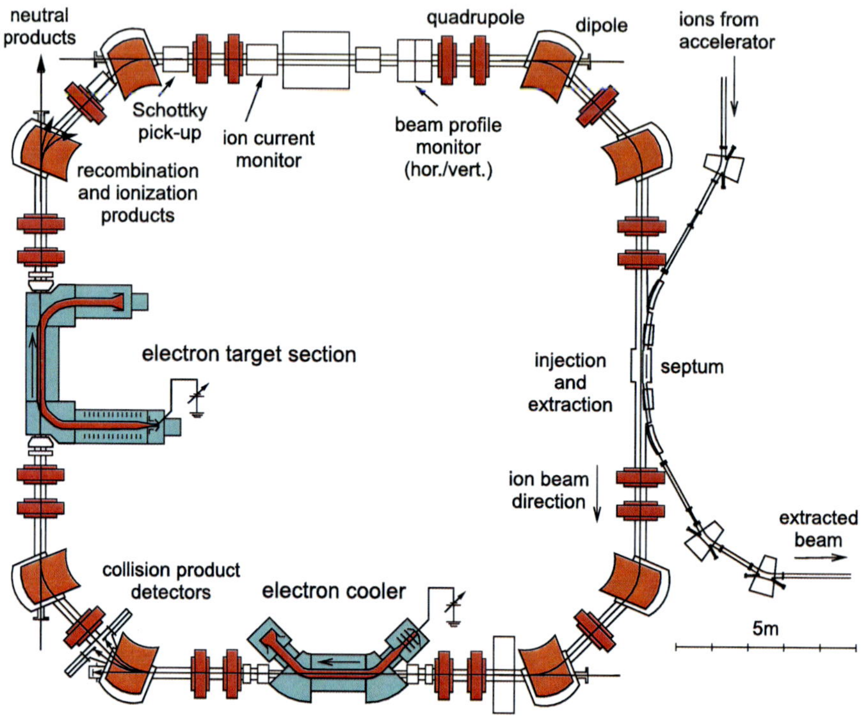

Fig. 2.4 Layout of the Heidelberg test storage ring (TSR) with the electron cooling device and the additional electron target. Product ions that have changed their charge state in electron-ion collisions can be detected behind the first bending magnet following the electron target (or the electron cooler) in ion-beam direction

Storage rings measure the electron–ion recombination cross section times the relative velocity convolved with the energy spread of the experiment, called a merged beams recombination rate coefficient (MBRRC). This differs from a plasma recombination rate coefficient (PRRC) for a Maxwellian temperature distribution. From the measured count rate R, the stored ion current I_i, and the electron density n_e of the electron beam, the MBRRC is readily derived as

$$\alpha_\mathrm{MB}(E_\mathrm{cm}) = R\frac{eqv_\mathrm{i}}{(1-\beta_\mathrm{i}\beta_\mathrm{e})I_\mathrm{i}n_\mathrm{e}L\eta}. \tag{2.1}$$

Here, eq is the charge of the primary ion, $v_\mathrm{i} = c\beta_\mathrm{i}$ and $v_\mathrm{e} = c\beta_\mathrm{e}$ are the ion and electron velocity, respectively, with c denoting the speed of light in vacuum. The center-of-mass energy E_cm can be easily calculated from the laboratory ion and electron energies. In a storage-ring experiment, β_i is kept fixed and E_cm is varied by changing β_e via the cathode voltage at the electron gun. Various aspects of the experimental and data reduction procedures at the Heidelberg heavy-ion storage ring TSR have been discussed in depth in several publications which are referenced

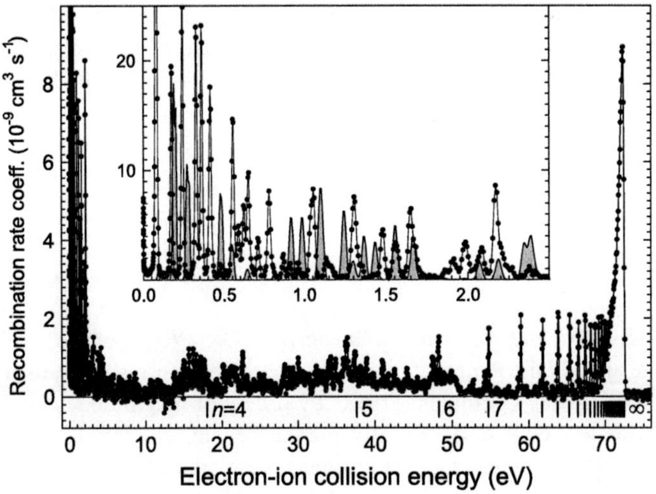

Fig. 2.5 Merged-beams recombination rate coefficient of Fe^{8+} measured at the heavy-ion storage ring TSR in the energy range of DR resonances mainly associated with $3p^6 \rightarrow 3p^5\,3d$ $\Delta N = 0$ core excitations [25]. The vertical bars denote resonance positions of the dominating $3s^2\,3p^5\,3d\ (^1P_1)nl$ Rydberg series as expected on the basis of the hydrogenic Rydberg formula (2.2). The *inset* shows the low-energy region of the measured spectrum in more detail and in comparison with state-of-the-art theoretical calculations (*shaded curve*) of low-energy DR resonances [25]

in recent review papers [26, 27]. The systematic experimental uncertainty of the measured MBRRC is typically 20–25% at a 90% confidence level.

As an example, Fig. 2.5 shows results for the recombination of Fe^{8+} ions [25]. The spectrum consists of DR resonances at specific energies sitting on top of the monotonically decreasing continuous rate coefficient due to radiative recombination (RR) which is barely visible on the rate coefficient scale of the figure. The resonances are mainly associated with $3p^6 \rightarrow 3p^5\,3d\ \Delta N = 0$ core excitations. The resonance positions of the dominating $3s^2\,3p^5\,3d\ (^1P_1)nl$ Rydberg series can be estimated by applying the Bohr formula for hydrogenic ions of charge q ($q = 8$ for Fe^{8+}), that is,

$$E_n \approx E_\infty - 13.606\,\mathrm{eV} \times \frac{q^2}{n^2}. \tag{2.2}$$

For $n \rightarrow \infty$, the Rydberg series of DR resonances converges to the series limits at $E_\infty = 72.47\,\mathrm{eV}$. Toward low n values, the strength of the interaction of the Rydberg electron with the excited ionic core increases. This leads to a considerable energy spread of the DR resonance positions within the low-n Rydberg manifolds. Additionally, weaker Rydberg series also contribute to the DR resonance structure at the lowest energies.

For the purposes of astrophysical modeling, the measured MBRRC has to be converted in to a PRRC. After subtraction of the continuous RR "background" from the measured recombination spectrum the DR PRRC is derived by essentially convoluting the DR MBRRC with a Maxwell–Boltzmann electron energy distribution (see [15, 29] for details). The systematic uncertainty of the PRRC is basically the 20–25% uncertainty of the experimental MBRRC.

2.4 Astrophysical Impact of the Experimental DR Rate Coefficients

Figure 2.6 displays the experimentally derived DR PRRC of Fe^{8+} along with theoretical results. In the temperature range, where Fe^{8+} forms in PP, the previously recommended DR rate coefficient [19] is by orders of magnitude smaller than the experimentally derived one. Apparently, the theoretical calculations have neglected the low-energy DR resonances. Such resonances are present in the DR spectra of all iron ions measured so far, that is, of Fe^{q+} ions with $q = 7$–10 and 13–22 [27, and references therein]. The situation for Fe^{8+} DR PRRC is thus symptomatic for all iron ions. The previous recommended DR rate coefficients are too small by orders of magnitude in the temperature ranges that are relevant for PP. This experimental

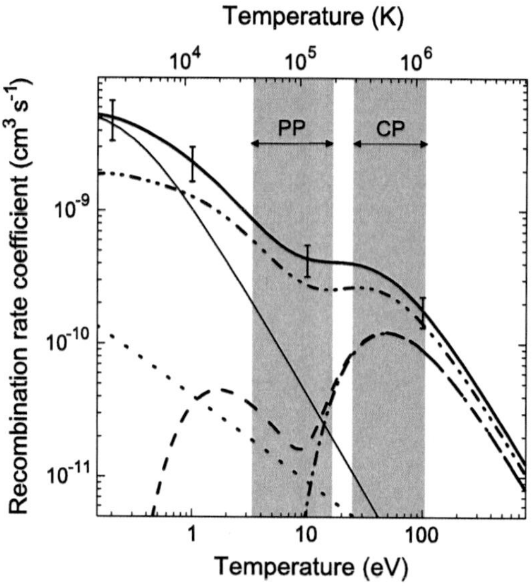

Fig. 2.6 Comparison of the experimentally derived Fe^{8+} DR PRRC with the previous recommendation [19] (*dash-dotted curve*) and with the deliberate modification by Netzer [18] (*dashed curve*). Also shown is the state-of-the-art calculation of Badnell [28] (*dash-dot-dotted curve*). The contribution of RR to the latter is shown as a *dotted curve*. The *thin full curve* shows the contribution of the DR resonances below 0.45 eV. The temperature ranges where Fe^{8+} exists in PP [17] and CP [16] are highlighted

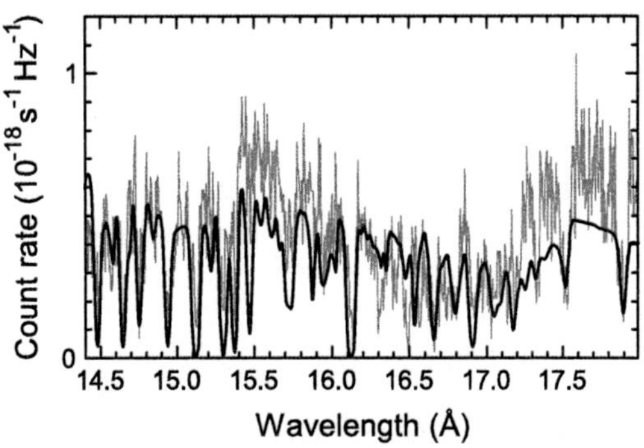

Fig. 2.7 Measured X-ray spectrum (*thin gray curve*) from an AGN (NGC 3783) compared with results of astrophysical model calculations [17] (*thick black curve*), using new theoretical DR rate coefficients for iron ions from [28]. Note that a different quantity is plotted here as compared to Fig. 2.3

finding confirms the above-discussed suspicion of Netzer [18] and Kraemer et al. [20]. It should be noted that the deliberately modified Fe^{8+} DR PRRC by Netzer is still more than an order of magnitude lower than the experimentally derived one.

Figure 2.6 also shows the result of a state-of-the-art theoretical calculation of the Fe^{8+} DR rate coefficient [28] which includes low-energy DR resonances as shown in Fig. 2.5. This comparison illustrates the above-discussed theoretical difficulty to calculate low-energy DR resonance positions with sufficient accuracy. Consequently, the theoretical low-temperature DR PRRC deviates from the experimentally derived one. In the PP temperature range, it is smaller by up to 32%, that is, by more than the experimental uncertainty. Although similar remaining discrepancies between new theoretical calculations and experimental results exist also for other charge states of iron (see, e.g., [25, 30]), the overall situation has greatly improved by the partly joint experimental and theoretical efforts.

Recently, Kallman [17] has revisited the astrophysical modeling of the NGC 3783 X-ray spectrum using the new theoretical Fe DR rate coefficients [28, 30]. In Fig. 2.7, his result is compared with the X-ray spectrum from the AGN NGC 3783. The new rates yield a much better representation of the observed UTA than the previously recommended rates (Fig. 2.3a), because the ionization balance has been changed considerably. The much-increased recombination rate coefficients lead to a lower average charge state of the plasma, i.e., the UTA absorption feature is caused by iron ions in lower charge states than previously thought. Conversely, in photoionized plasmas iron ions form at higher temperatures than previously assumed.

2.5 Conclusions

The astrophysical interpretation of astronomical observations relies heavily on atomic data. Uncertainties of these data directly translate into uncertainties of astrophysical models. As an example for this situation, we have discussed discrepancies between recent observations and model calculations of X-ray spectra from the photoionized gas in the vicinity of SMBHs. The suspicion that the previously recommended rate coefficients for dielectronic recombination of iron ions were to small has initiated experimental and theoretical atomic physics work on this problem. The resulting new low-temperature dielectronic recombination rate coefficients are indeed much larger than the previously recommended ones and yield astrophysical model results in much better agreement with the astronomical observation. It turns out that the photoionized gas in the vicinity of SMBHs is less ionized than previously thought.

Providing accurate atomic data for astrophysical and plasma physical application, for example, for fusion research, is an ongoing effort. Regarding the enormous costs for modern astronomical equipment such as space-borne X-ray telescopes any investment in laboratory astrophysics, which operates on comparatively low budget, should be considered as well spent.

Acknowledgements We would like to thank our colleagues for fruitfully collaborating on the subject of the present review, in particular, Michael Lestinsky, Daniel Wolf Savin, Eike Schmidt, and Andreas Wolf. We thank Tim Kallman for providing his data in numerical form. Financial support by the Max-Planck-Society and the Deutsche Forschungsgemeinschaft is gratefully acknowledged.

References

1. K. Schwarzschild, Über das Gravitationsfeld eines Massenpunktes nach der Einsteinschen Theorie. Sitzungsberichte der Königlich-Preussischen Akademie der Wissenschaften, Sitzung vom 3. Februar 1916; S. 189–196, ed. by Deutsche Akademie der Wissenschaften zu Berlin (Reimer, Berlin, 1916)
2. A. Eckart, R. Genzel, Nature **383**, 415 (1996)
3. R. Antonucci, Annu. Rev. Astron. Astrophys. **31**, 473 (1993)
4. C.M. Urry, P. Padovani, Publ. Astron. Soc. Pac. **107**, 803 (1995)
5. http://heasarc.nasa.gov/docs/objects/agn/agn_model.html
6. F.B.S. Paerels, S.M. Kahn, Annu. Rev. Astron. Astrophys. **41**, 291 (2003)
7. M. Asplund, N. Grevesse, A.J. Sauval, P. Scott, Annu. Rev. Astron. Astrophys. **47**, 481 (2009)
8. M. Sako, S.M. Kahn, E. Behar, J.S. Kaastra, A.C. Brinkman, T. Boller, E.M. Puchnarewicz, R. Starling, D.A. Liedahl, J. Clavel, M. Santos-Lleo, Astron. Astrophys. **365**, L168 (2001)
9. G.J. Ferland, Annu. Rev. Astron. Astrophys. **41**, 517 (2003)
10. T.R. Kallman, P. Palmeri, Rev. Mod. Phys. **79**, 79 (2007)
11. A. Müller, Adv. At. Mol. Phys. **55**, 293 (2008)
12. P. Beiersdorfer, Annu. Rev. Astron. Astrophys. **41**, 343 (2003)
13. D.W. Savin, J. Phys. Conf. Ser. **88**, 012071 (2007)
14. N.R. Badnell, J. Phys. Conf. Ser. **88**, 012070 (2007)
15. S. Schippers, M. Schnell, C. Brandau, S. Kieslich, A. Müller, A. Wolf, Astron. Astrophys. **421**, 1185 (2004)
16. P. Bryans, N.R. Badnell, T.W. Gorczyca, J.M. Laming, W. Mitthumsiri, D.W. Savin, Astrophys. J. Suppl. Ser. **167**, 343 (2006)
17. T. Kallman, Space Sci. Rev. **157**, 177 (2010)
18. H. Netzer, Astrophys. J. **604**, 551 (2004)

19. M. Arnaud, J. Raymond, Astrophys. J. **398**, 394 (1992)
20. S.B. Kraemer, G.J. Ferland, J.R. Gabel, Astrophys. J. **604**, 556 (2004)
21. A. Müller, A. Wolf, in *Accelerator-Based Atomic Physics Techniques and Applications*, ed. by J.C. Austin, S.M. Shafroth (AIP Press, Woodbury, 1997), p. 147
22. A. Wolf, R. von Hahn, M. Grieser, D.A. Orlov, H. Fadil, C.P. Welsch, V. Andrianarijaona, A. Diehl, C.D. Schröter, J.R. Crespo López-Urrutia, M. Rappaport, X. Urbain, T. Weber, V. Mallinger, C. Haberstroh, H. Quack, D. Schwalm, J. Ullrich, D. Zajfman, in *Beam Cooling and Related Topics, AIP Conference Series*, vol. 821, ed. by S. Nagaitsev, R.J. Pasquinelli (American Institute of Physics, Melville, New York, 2006). *AIP Conference Series*, vol. 821, pp. 473–477
23. R. Schuch, S. Böhm, J. Phys. Conf. Ser. **88**, 012002 (2007)
24. D. Schwalm, Progr. Part. Nucl. Phys. **59**, 156 (2007)
25. E.W. Schmidt, S. Schippers, D. Bernhardt, A. Müller, J. Hoffmann, M. Lestinsky, D.A. Orlov, A. Wolf, D.V. Lukić, D.W. Savin, N.R. Badnell, Astron. Astrophys. **492**, 265 (2008)
26. S. Schippers, J. Phys. Conf. Ser. **163**, 012001 (2009)
27. S. Schippers, M. Lestinsky, A. Müller, D.W. Savin, E.W. Schmidt, A. Wolf, Int. Rev. At. Mol. Phys. **1**, 109 (2010)
28. N.R. Badnell, Astrophys. J. **651**, L73 (2006)
29. S. Schippers, A. Müller, G. Gwinner, J. Linkemann, A.A. Saghiri, A. Wolf, Astrophys. J. **555**, 1027 (2001)
30. N.R. Badnell, J. Phys. B **39**, 4825 (2006)

Chapter 3
Large Hot X-Ray Sources in the Solar Corona

S.V. Kuzin, S.A. Bogachev, A.M. Urnov, V.A. Slemzin, S.V. Shestov, and A.A. Reva

Abstract Energy release in solar corona is used to attend with plasma heating. For the most powerful processes, temperature of plasma increases up to millions of Kelvin. For this reason, the study of hot coronal plasma is the key for understanding mechanisms of heating of the corona and nature of solar flares. In this chapter, we analyze observation of Mg XII line (10 MK) in series of experiments onboard CORONAS satellites. We studied configuration, dynamics, and temperature distribution in hot plasma structures. The possible mechanisms of heating are discussed.

3.1 Introduction

The solar corona is outer part; of solar atmosphere [8, 40, 79]. It is placed just above the chromosphere; the physical properties of plasma changes dramatically in extremely thin transient layer: the plasma density drops from 10^{13}–$10^{15}\,\mathrm{cm}^{-3}$ to 10^{8}–$10^{9}\,\mathrm{cm}^{-3}$, and temperature jumps from 6,000 K to one million K. Both density and temperature show significant spatial nonuniformity and indicate out coming magnetic fields which have both close and open configuration.

These specific conditions (the plasma is named "coronal" one) provide efficient excitation of multicharged ions of heavy elements, and for this reason, the maximum of coronal radiation is in soft X-ray and extreme ultraviolet (EUV) part of spectra, from 1 to 1,000 Å. The excitation processes could be both thermal and nonthermal one.

The corona is the most dynamic part of the Sun: the coronal processes have typical duration from a hundredth of a second up to days. Most of them are connected with accumulation and dissipation of energy in different forms. The basic energy for most coronal processes is energy of magnetic field, but they differ by the mechanisms of energy transformation, duration of the energy release process, their power, etc.

The most powerful short processes named flares are developed with energy release as X-ray and EUV radiation and particle acceleration. The nature of these

V. Shevelko and H. Tawara (eds.), *Atomic Processes in Basic and Applied Physics*, Springer Series on Atomic, Optical, and Plasma Physics 68,
DOI 10.1007/978-3-642-25569-4_3,

processes is one of basic question of Solar physics. One of the feature of flares is plasma heating up to temperatures of tens millions degrees.

In this chapter, we study the properties of compact hot sources in solar corona by means of X-ray imaging spectroscopy.

3.2 Properties of Impulsive and Long-Lived Hot X-Ray Sources in Corona

Physical characteristics of the high-temperature plasma in the corona were studied in the CORONAS series of experiments involving Mg XII X-ray spectroheliometers as part of instrumentation of RES/CORONAS-I, SPIRIT/CORONAS-F, and THESIS/CORONAS-Photon complexes.

The ion excitation temperature of Mg XII ($\lambda = 8.42$ Å) is $T = 5$–15 MK, which provides direct observation of high-temperature plasma in the corona [116], whereas according to SXT/Yohkoh experimental data the hot-temperature plasma was detected only by indirect techniques based on comparison of SXT images obtained with different filters. The Mg XII spectroheliograph is a transient informative data channel between EUV telescopes recording coronal plasma (SPIRIT, TESIS, 171, 175, 195, 285, and 304 Å; SOHO/EIT, TRACE) and instruments registering superhigh-temperature plasma such as, for example, RHESSI. Comprehensive analysis of events using data from all these instruments allows the coronal plasma to be diagnosed in a wide temperature range.

The Mg XII spectroheliograph observation carried out during a long period of time at different solar cycle stages made it possible to register the presence of the high-temperature plasma of the temperature of no less than 5 MK even in the solar minimum and detect an entire class of new phenomena characterized by specific form and dynamics [116]. New types of hot coronal structures were revealed that are characterized by lifetime of from minutes to days and by dimensions of from several arc seconds to several tenths of the solar radius. The following regularly observed formations were marked out:

- "Hot clouds"—diffusive formations variable in shape and having characteristic dimensions, height above the limb of up to $0.4R_o$, and lifetime of up to several hours.
- "Spiders"—large structures associated with active regions and characterized by having a spider shape for long periods of time (up to several days), a bright spherical "body" at 0.1–$0.3R_o$ heights and less bright "legs," the morphology of which is a system of giant arcs that does not coincide with the cold magnetic loop system (Fig. 3.1).
- Complex events including the appearance of clouds with the further formation of spiders and giant arcs and accompanied by flares, coronal mass ejections, and other eruptive phenomena.

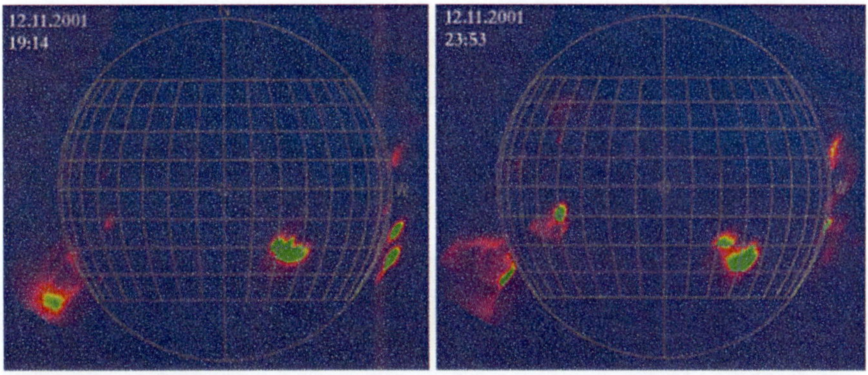

Fig. 3.1 Dynamics of hot ($T \sim 5$ MK) plasma structures—“spiders” observed in Mg XII 8.42 Å line on 12 November 2001

- “Waves”—phenomena in the form of divergent wave fronts or successive evolution of magnetic arches that were observed during high solar activity.

The discovery of a new class of high-temperature coronal objects, that is large-scale regions of high-temperature plasma having the lifetime much longer than the period of their heat-conductive and emission cooling, was the most prominent result of the operation of the Mg XII spectroheliometer. The observations demonstrated that the regions were positioned at a substantial height in the corona, of up to $0.3R_o$ (100,000 km), and had the temperature of ~10 MK. As opposed to pulsed sources detected in flares by Yohkoh, this class of objects is not directly related to flare activity.

The temperature content of soft X-ray emission (SX) in the solar corona plasma was determined using data obtained simultaneously from the RES apparatus operating in the SPIRIT experiment onboard CORONAS-F, the GOES X-ray monitor, and channels of softest X-ray emission on the RHESSI satellite. Absolute calibration was performed for fluxes in the magnesium channel based on the data of emission time profiles in two channels (1–8 and 0.5–4 Å) of the GOES X-ray monitor, as well as spatial correlation of flare sources according to the RHESSI data.

The relationship between Mg XII and GOES fluxes may be generally represented as follows:

$$I_{G_I}(t+\tau) = A * I_{Mg}(t) + B + C(t), \tag{3.1}$$

where $I_{G_I}(t)$, $I_{Mg}(t)$ are current fluxes, respectively, in GOES (1–8 Å) and Mg XII channels, designated below as G_I and Mg; A is proportionality coefficient; B is a constant determined by the least square method; $C(t)$ is a time-dependent portion of the flux that characterizes deviation from a linear relation and substantially differs from zero only during the maximum phase of flare events. The relative value of this flux as related to $I_{G_I}(t)$ varies depending on the type of the event within the limit of $\approx 5\%$ for the flare rise and decay phases, and achieves the maximum value of

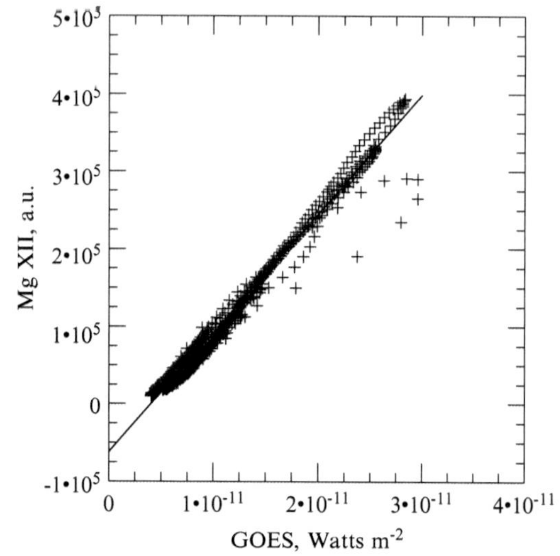

Fig. 3.2 The comparison of full emission fluxes (from the entire Sun) in the 8.42 Å line (Mg XII) and in the 1–8 Å band (GOES-8)

$\approx$15–20% for IEs. The time shift τ nominally introduced into (10.39) is zero for the time resolution used in this chapter.

The comparison of full emission fluxes (from the entire Sun) in the 8.42 Å line (Mg XII) and in the 1–8 Å band (GOES-8) revealed good coincidence, within statistical error limits (on the order of 10%), of their time profiles characterized by substantial change in intensity during long periods of time excluding relatively short intervals during the maximum phase of the flare events (Fig. 3.2).

The observed quasilinear relation between monochromatic emission fluxes in the Mg XII line and in the broad spectral band 1–8 Å (GOES-8) can be explained, provided two conditions are met: a physical one associated with the generation, during the flare growth and attenuation phase, of the transient plasma showing a major emission measure in the 5–15 MK temperature range and an instrumental one caused by weak temperature dependence in this interval of the ratio of transmission factors in each Mg XII/GOES channel.

Figure 3.3 represents temperature dependences of the ratios of lines in such ranges to fluxes registered in the channels, as well as temperature responses from such channels. The values of line emissions are calculated using CHIANTI (Version 4.2) database for coronal abundances and calibration data according to the research work [113]. The figure shows that within the 4–10 MK temperature range, the ratio of rated line intensities to observations in the GOES-8 channel has the smoothest dependence with a maximum of about 8 MK.

In the presence of plasma with a dominating emission measure in this temperature range, the fluxes in both channels will be mutually proportional, the proportionality coefficient depending on the abundance of elements because the relation of line contribution to the total luminosity is maximal for temperatures of 4–12 MK.

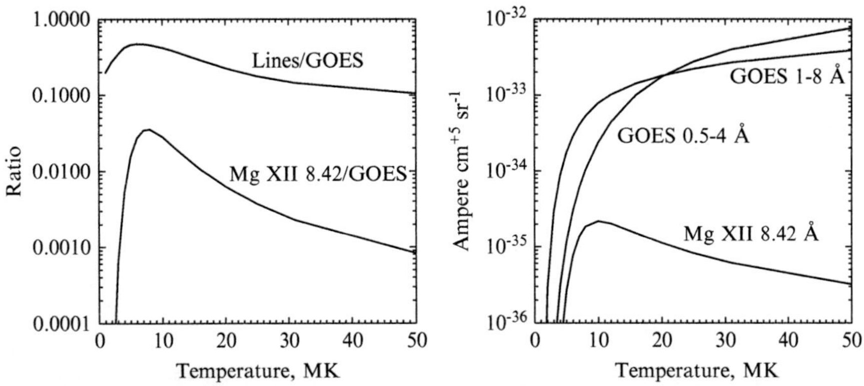

Fig. 3.3 *Left*: flux ratio in different spectral ranges—line emission to total GOES 1–8 Å range emission and Mg XII 8.42 Å to total GOES 1–8 Å range emission. *Right*: temperature response of GOES 0.5–4, 1–8 and Mg XII 8.42 Å channels

The results of a series of continuous observations carried out with the aid of RES X-ray spectroheliometer in the period between February 6, 2002, and February 28, 2002, were used for a more detailed study of correlation in X-ray flux dynamics. Flare and other transient events were observed during this period of time in Mg XII line images, as well as the development of a complex large-scale and long-term phenomenon earlier referred to as "the spider."

The comparison of dynamic characteristics of flux time profiles in the 8.42 Å line and the 1–8 Å range allowed all events to be conditionally grouped into three types: short impulsive events (IE), long-duration "gradient" events (LDE) related to the formation of "a spider," and other complex events, for example, impulsive ones having a long decay (impulsive long-lived events, ILE) or LDEs accompanied by impulsive bursts.

Figure 3.2 shows high correlation of fluxes for the three types of events in the Mg XII and GOES (1–8 Å) channels. Deviation from the linear relationship of these fluxes is observed only during the maximum phase of flare events and varies depending on the type of the event within 5% for the flare rise and decay phases, reaching its maximum of 15–20% for IEs.

A method based on the multitemperature parametric model (MTP) [104] was employed to restore the temperature distribution of the volumetric differential emission measure (DEM) of the hot flare plasma from X-ray fluxes. At the same time, coronal abundances of elements were used.

Figure 3.4 shows flux time profiles for the Mg XII and GOES (0.5–4 and 1–8 Å) channels and their decomposition into components corresponding to three temperature ranges: relatively cold plasma of 2–4 MK, an intermediary component of 4–10 MK, and a hot component of 10–20 MK. Figure 3.5 presents respective EM profiles and compares average-temperature and EM profiles calculated within this multitemperature model against the results provided by a single-temperature model (STM) based on the GOES data.

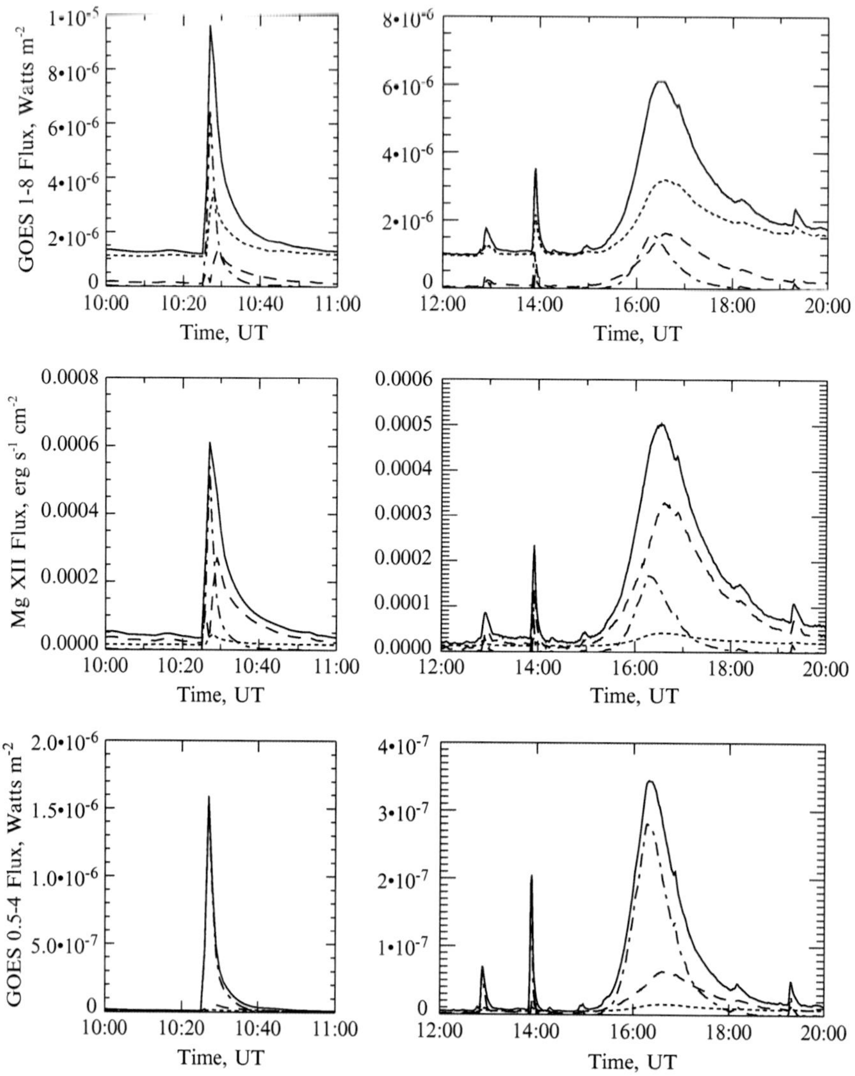

Fig. 3.4 Time profiles in GOES 0.5–4, 1–8, and Mg XII 8.42 Å fluxes and their decomposition to the temperature ranges: relatively cold plasma of 2–4 MK (*dotted*), an intermediary component of 4–10 MK (*dashed*), and a hot component of 10–20 MK (*dash-dotted*)

A significant difference should be marked out in the distribution between the components during the development of flare events of different types, that is, IE and LDE: the contribution from the hot component is dominating in IE, while the main contribution to LDE comes from the transient plasma of intermediary temperatures, the development of the intensity of which is faster and lasts much longer than the growth of the hot component. A substantial difference between the temperatures

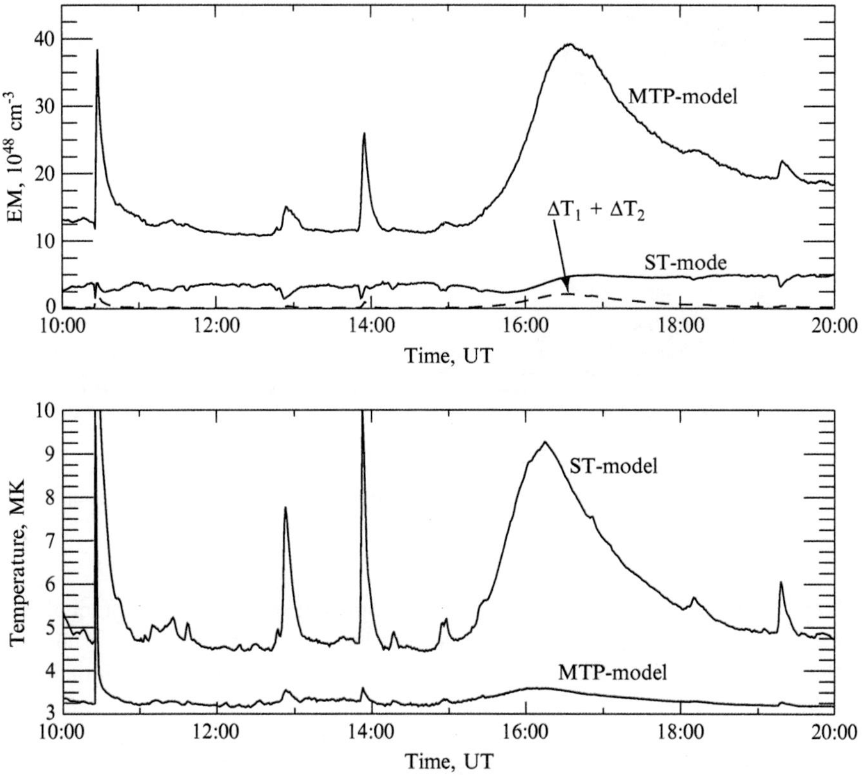

Fig. 3.5 Comparison of the EM and mean temperature profiles in the MTP and STM models

and EM calculated using the MTP and STM should as well be noted: the rise of EM in the MTP model almost coincides with the temperature rise, as opposed to the STM characterized by a considerable time delay in the EM growth as compared to the temperature. As the figure shows, the average temperature for many events is considerably lower, while the EM exceeds the respective values of the STM.

Correlation between the results calculated by MTP model and the RHESSI data demonstrates the following [117]. For impulsive events (IEs) at 10:26 14:00 UT, the EM according to the MTP model and RHESSI data was equal to 2.9 and $0.5 \cdot 10^{48}$ cm^{-3}. Such agreement between MTP model results and RHESSI data indicates experiment data consistency, high-quality modeling, and correct determination of intercalibration constants. Figure 3.6 compares images and temporal fluxes in the Mg XII and RHESSI channels. Another important result is that the MTP model calculations performed with photospheric abundances do not provide satisfactory agreement with the RHESSI data.

The comparison of the calculated time dependencies of fluxes with the model-based calculations demonstrates that the deviation from a strictly linear dependence is based on the contribution from the luminosity of the cold component during the

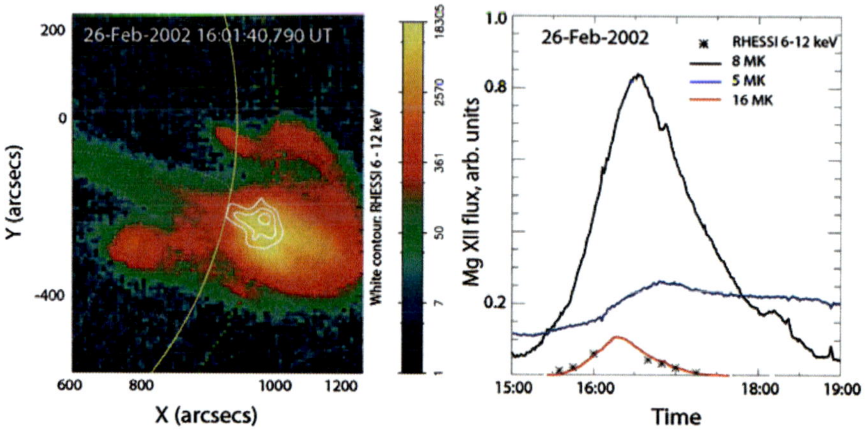

Fig. 3.6 *Left*: simultaneous images in Mg XII 8.42 Å line and RHESSI (contour) channels registered at 16:01 UT 26 February 2002. *Right*: flux decomposition to different temperature components

intensity rise and decay periods and the hot component at the flare maximum phase, these luminosities being related to the proton bremsstrahlung mechanism. Thus, the observed relation of the Mg XII and GOES (1–8 Å) intensities was quantitatively interpreted in terms of the MTP model.

The principal fraction of the source EM during the flare rise and decay phases is contained in the transient plasma having the temperature of 4–10 MK, which ensures the proportionality of the total GOES channel intensity to the Mg XII channel intensity due to a weak dependence of the ratio of the temperature coefficients in this range. The GOES channel intensity at the maximum phase substantially exceeds the magnesium channel intensity due to a significant contribution of the hot component (10–20 MK) to the total intensity during this time period, which is related to the determining contribution of the proton continuous (bremsstrahlung) emission to the GOES channel luminosity function.

The above deviation from the linear law for fluxes in the line and spectral band may result from the following physical causes: a temperature change (e.g., when plasma heats up at the impulse stage of the flare or cools down at the decay stage, when the flux in the line is weaker than that in the 8–12 MK range at low $T < 8$ MK or high $T > 12$ MK), or a change in the luminosity function related to the nonequilibrium luminosity mechanism (e.g., in the presence of epithermal electrons). It should be pointed out that the hard X-ray profile registered by RHESSI should correlate with the injection rate of accelerated electrons. Thus, there is a probability that the hot coronal sources of emission visible in the Mg XII line could have been heated by electrons accelerated during the flare event. This assumption is based on the fact that the Mg XII line emission begins to increase almost at the same time (RHESSI has missed the beginning of the flare) when the electrons start accelerating in the flare. Upon the electron injection, the Mg XII line emission drops down to the previous level during approximately 4 h. The emission delay in the Mg XII line with respect to the hard X-ray burst may be the demonstration of the Newpert effect.

Both total fluxes and Mg XII channel images were used to model temporal dynamics of the spatial electron density and temperature distributions. The plasma of LDEs ("spiders"), which differ from IEs by much larger sizes, was modeled. A spatial region having the intensity with a time profile similar to that of fluxes in the 4–10 MK temperature range and soft RHESSI channels was determined in a time series of Mg XII images. It thus became possible to estimate the size of the region and to reference the magnesium images to the solar disc (see Fig. 3.6). Analysis of the "spider" images recorded in the Mg XII channel demonstrates that the intensity distribution in their images has a bright quasispherical core surrounded by a much less bright region. The boundaries of these regions vary significantly during a flare, reaching a maximum at the maximum phase of the flare event.

In order to model the spatial and temporal structure of the "spider" emission on February 26, 2002, N_e and T_e distributions were plotted as functions of R radius (see Fig. 3.7). The corresponding parameters for the maximum phase (16:30 UT) were determined by comparing the emission from the modeled source with the experimental measurements. It is important to note that the N_e and T_e distributions are determined unambiguously due to strong temperature dependence of $G_{Mg}(T)$ coinciding with the luminosity function. Slight variations both in form and absolute values of the parameters lead to significant disagreement with the observed distributions.

Calculations of total fluxes in the GOES and Mg XII channels for these distributions exhibited good quantitative agreement with the experimental fluxes and emission measures in temperature ranges obtained in the MTP model. Analysis of the image time dependence shows that the intensity during the evolution of the flare event increases due to a change in the emission measure of the cold and intermediate regions enlarged by an increase in their size.

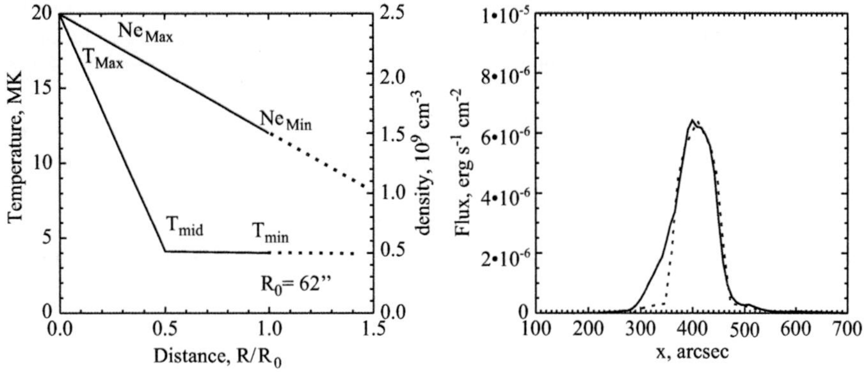

Fig. 3.7 Distribution of N_e and T_e (*left*) and flux for the spider observed on 26 February 2002

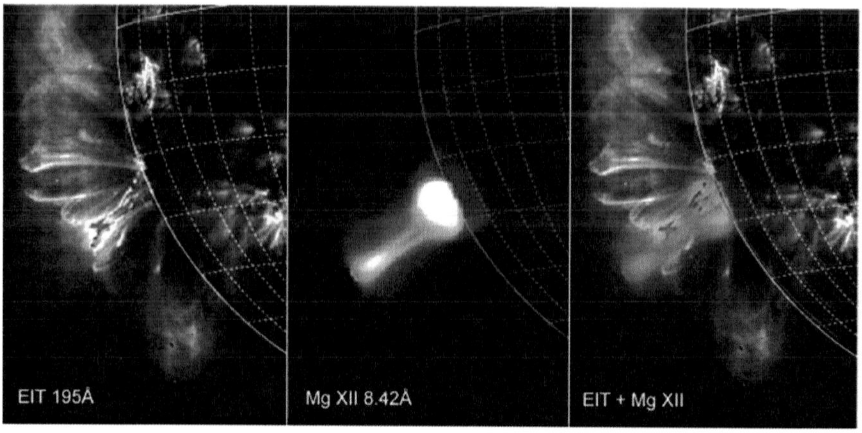

Fig. 3.8 "Spider" seen on 29 December 2001 in different spectral bands: 195 Å SOHO/EIT, 8.42 Å Mg XII and common image

The density distribution in the analyzed flare event differs significantly from that in IEs. The density in the "spider" plasma is actually constant and close to the coronal density ($2 \cdot 10^9\,\mathrm{cm}^{-3}$) as opposed to the impulsive flare at 10:26 UT on February 26, 2002, where the density reaches $2.6 \cdot 10^{11}\,\mathrm{cm}^{-3}$ [82]. At the same time, the peak temperatures in both events turn out to be close, 16 and 21 MK, respectively.

The DEM was calculated for "spiders" and other hot structures in the corona with the use of data from a EUV spectroheliometer [121]. An event recorded on 28–29 February, 2001 was selected for the analysis. Figure 3.8 presents its images in a "hot" (8–12 MK) Mg XII channel and in a "cold" (about 2 MK) telescopic channel at 195 Å EIT/SOHO. The figure shows a radially elongated (along the Sun radius) "spider" structure in the RES X-ray ("hot") image and a post-eruptive ("cold") arcade of magnetic loops grouped along the perpendicular direction. Appropriate orientation of the EUV spectroheliometer in the 280–330 Å channel (the "spider" was oriented along the axis perpendicular to the dispersion direction) made it possible to study the dependence of the DEM temperature distributions along the Sun radius (see Sect. 5.3).

Figure 3.9 shows the DEM distribution at different heights of the "spider" and posteruptive ("cold") arcades. The diagrams demonstrate substantial difference of the DEM of the "spider" from that in the active region: the "spider" generally consists of hotter plasma. At the same time, both distribution (low- and high-temperature) maxima are shifted to the "hot" region. The DEM of the "spider" varies with height toward relative increase of a high-temperature plasma component of about 8 MK temperature [117].

As a significant portion of the plasma in the corona is concentrated in the magnetic flux, the following question arises: what are the values of the magnetic

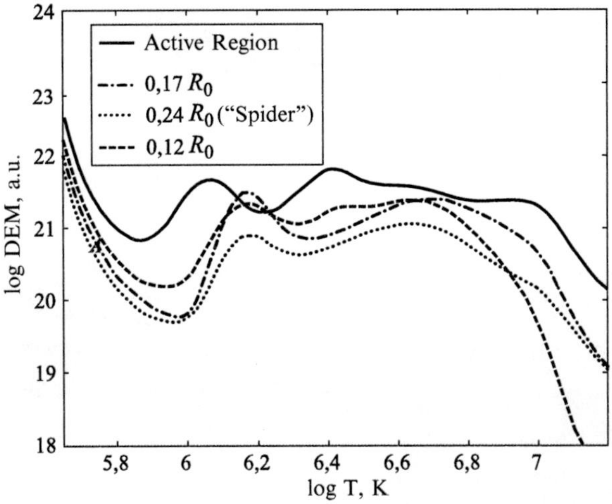

Fig. 3.9 DEM distribution at different heights of the "spider" and posteruptive ("cold") arcades

field strength in the arcades, and how do the magnetic and plasma pressures correlate?

The current conceptions that the value of the corona plasma pressure is much lower than the magnetic pressure $\beta = \frac{2nkT}{H^2/8\pi} \ll 1$ and that, on the other hand, the strength of the magnetic fields in the corona rapidly decreases with height contradict each other and have been doubted in recent research works. Some works [38] make a conclusion that the $\beta < 1$ condition is noncompulsory for giant coronal loop systems at a later stage of flare events. This finds support in researches of analytical and observational character [89,90]. On the other hand, some research papers [9,51] demonstrate that the decrease in the magnetic field strength with altitude may be less than expected. Both possibilities, $\beta \ll 1$ and $\beta \sim 1$, may probably be implemented under different conditions. Parameters of the plasma in coronal arcades may be evaluated using soft X-ray emission data. But the magnetic field strength in the corona may be assessed only by radio astronomy methods.

In order to determine these parameters, a comprehensive study of a posteruptive arcade and a "spider" observed on October 22, 2001, was conducted [19]. The arcade resulted from an eruptive event that occurred at about 00:40 UT on October 22, 2001, and was accompanied by 1SF/M1.0 subflare. The ejection evidently took place in the region 9,658 beyond the western limb. The arcade could be observed in the radio-frequency band, soft X-rays (Yohkoh/SXT), and extreme UV emission [42].

The SPIRIT complex observed the arcade in several ranges, particularly, in 8.42 Å magnesium doublet (Fig. 3.10). The height of the bright part of the "spider" is about 10^5 km. Characteristics of a radio source observed by RATAN-600 at 1.9–10 cm wavelengths correspond to optically thin thermal bremsstrahlung emission. The temperature and emission measure were assessed according to

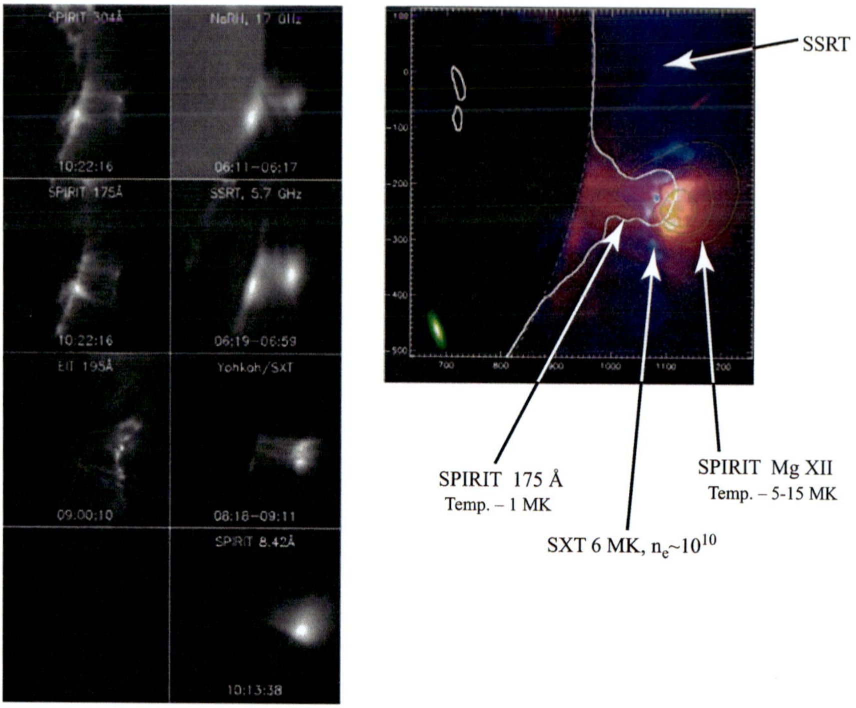

Fig. 3.10 Evolution of spider observed in different spectral bands: EUV, SPIRIT/CORONAS-F and EIT/SOHO; Soft X-ray, SXT/Yohkoh and Mg XII/CORONAS-F; radio, NoRH and SSRT

Yohkoh/SXT data 8 h after the event. The temperature of the major portion of the arcade including its brightest part amounts to 6 MK, and the temperature of its upper edge reached 8 MK. The emission measure in a $2.46'' \times 2.46''$ column for the bright part is $2.5 \cdot 10^{45}$ cm^{-3}, and for the resolvable less-bright lowering loop legs $\approx 5 \cdot 10^{44}$ cm^{-3}, with the minimum value of $2.5 \cdot 10^{44}$ cm^{-3}. The volume of the bright region is typically assumed as $V = A^{3/2}$, where A is its area, or its depth is considered to be equal to the lateral dimension. Thus, the plasma density in the bright part is $5 \cdot 10^9$ cm^{-3}.

The high brightness of the central part of the arcade may as well be an optical effect: it is possible that the brightness of the loops is equal in height, but when the loops are oriented along the view axis the DEM of the loop peaks is collected along their upper portion. With this assumption, the emission measure along the downward portion of the loop may be evaluated, and the depth of the emitting layer in this case is equal to the lateral dimension of the loop. Such evaluation results in $9 \cdot 10^9$ cm^{-3} density. Both evaluations are similar, but the second method seems more feasible. Values of the magnetic field are $B_d > 7$ gauss and $\beta > 1$. Extrapolation of the magnetic field is not possible because the region is located close to the limb.

3.3 Periodic Oscillations of X-Ray Sources

In solar flare emissions, oscillations with periods ranging from centiseconds to several thousand seconds were observed in addition to the regular components [7]. The search for and analysis of such oscillations contain information about the mechanism of the emission generation, in some cases providing diagnostics for the physical condition of the emitting plasma. The latter is possible when stable oscillations in the plasma are excited only at a specific temperature, density, and values of the magnetic field [34, 81].

The periodic component of emission has been studied more thoroughly in two spectral ranges: hard X-rays and radio-frequency band. Observation of the Sun is carried out in this case with a high time resolution (up to centiseconds), allowing periodic processes to be studied within a wide range of periods and frequencies. Oscillations of hard X-rays emitted by Solar flares can be explained in terms of a thick target model: they may result from variations in the acceleration rate of electrons and the rate of their injection from a reconnection region into the target.

As regards radio emission, its variations with periods on the order of more than a minute are probably related to oscillations of magnetic flare loops. This refers to a part of radio emission formed by electrons captured inside the magnetic tubes in the solar corona, the bases of the tubes bearing upon the chromosphere and functioning as mirrors. If the cross section of an oscillating tube changes (radial oscillations), so does the concentration of the captured electrons and thus the intensity of the emission they produce. In case of Alfven oscillations, which actually do not compress the plasma, variations of gyrosynchrotron emission may be generated by variations in the magnetic field strength [96]. Loop oscillations during flare events are an observable phenomenon. They are registered in up-to-date experiments, due to both the Doppler effect and the intensity variations [109–111], and can also be observed directly by solar telescopes of EUV range [10].

The study of soft X-ray oscillations has been limited until recently by two factors: low time resolution of imaging devices incapable of singling out oscillations, with a frequency better than several minutes, and the absence of spacial resolution in X-ray photometers, which did not allow the selected oscillations to be associated with a specific process in the corona. These drawbacks, however, were eliminated in the experiment with a soft X-ray spectroheliograph onboard the CORONA-F satellite [78].

To determine the power spectrum of flare events, two active regions were selected: NOAA 9825 and NOAA 9830, which at that period of time had the highest activity [18]. The regions were studied for 78 h, from 17:44 UT on February 19, 2002 until 23:59 UT on February 22, 2002.

Images of NOAA 9825 and NOAA 9830 regions were cut out from the images of the full Sun disc. The fragments cut out had the size of 50 pixels at each side, which approximately corresponds to 200,000 km. Background was removed from each section, after which a signal from each fragment was integrated. The resulting profiles (time dependence of the integral emission of the region in the Mg XII line) are given in Fig. 3.11. The emission profile for the entire solar disc measured in the 1–8 Å range at the GOES satellite is given as a reference in the bottom panel of Fig. 3.11.

The frequency spectra were studied through application of the Fourier transform to emission time profiles obtained in the Mg XII line. The distinction of experimental findings is that they are always determined in discrete points corresponding to different moments of observation. For their frequency basis expansion, a so-called discrete (or fast) Fourier transform expressed by the formula below was applied:

$$F_k = \frac{1}{n} \sum_{m=0}^{n-1} f_m \exp\left(-i\frac{2\pi km}{n}\right). \tag{3.2}$$

Here, f_m is the discrete emission profile (m may take values from 0 to $n - 1$), and F_k is the frequency dependence of the emission intensity, or the emissive power spectrum. F_k is a discrete function each point of which corresponds to frequency $\omega = k/n\Delta t$. As the time period of observation Δt must be constant throughout the series, the experimental emission profiles were interpolated to a uniform time scale with a 1-min interval. A cubic interpolation method based on four nearest points was applied for this purpose.

The function F_k was constructed with the help of a sliding window of 120 min width. This means that the value of the function F_k at the point of time t is determined from the observation data obtained during the period of time from $t - 60$ to $t + 60$ min. On passing to each next point of time, this range shifted to the right by one interval (1 min) and so on for all images of the series. Time variation of frequency characteristics of emission was thus analyzed.

The method described above produced in total almost ten thousand functions F_k that characterize the power spectra of two active regions observed for 78 h at an interval of 1 min The entire dataset was divided into several simple types so as to establish which frequency distributions were most typical of the hot solar plasma. In total, three such distributions were defined. They are referred to hereinafter as type I, II, and III power spectra.

Figure 3.12a gives an example of type I power spectrum. This distribution has no distinctive features and resembles noise. It can nominally be defined in the following way: the type I power spectrum is frequency distributions when neither of the peaks exceeds the adjacent ones by a value of more than 3σ (where σ is a mean-square variation of the distribution). As the research has shown, such distributions are very unstable and may substantially change when the sliding window is moved through just a single interval. If each of their peaks is interpreted as a result of the periodic process, then a conclusion should be made that these processes are frequency-distributed in a random manner, none of them dominating the others.

Figure 3.12b gives an example of type II spectrum. On transition from low to high frequencies, this spectrum does not decrease monotonously, but passes through one or several maxima. The frequencies at which such maxima are observed shall further be referred to as characteristic emission frequencies. There are three features

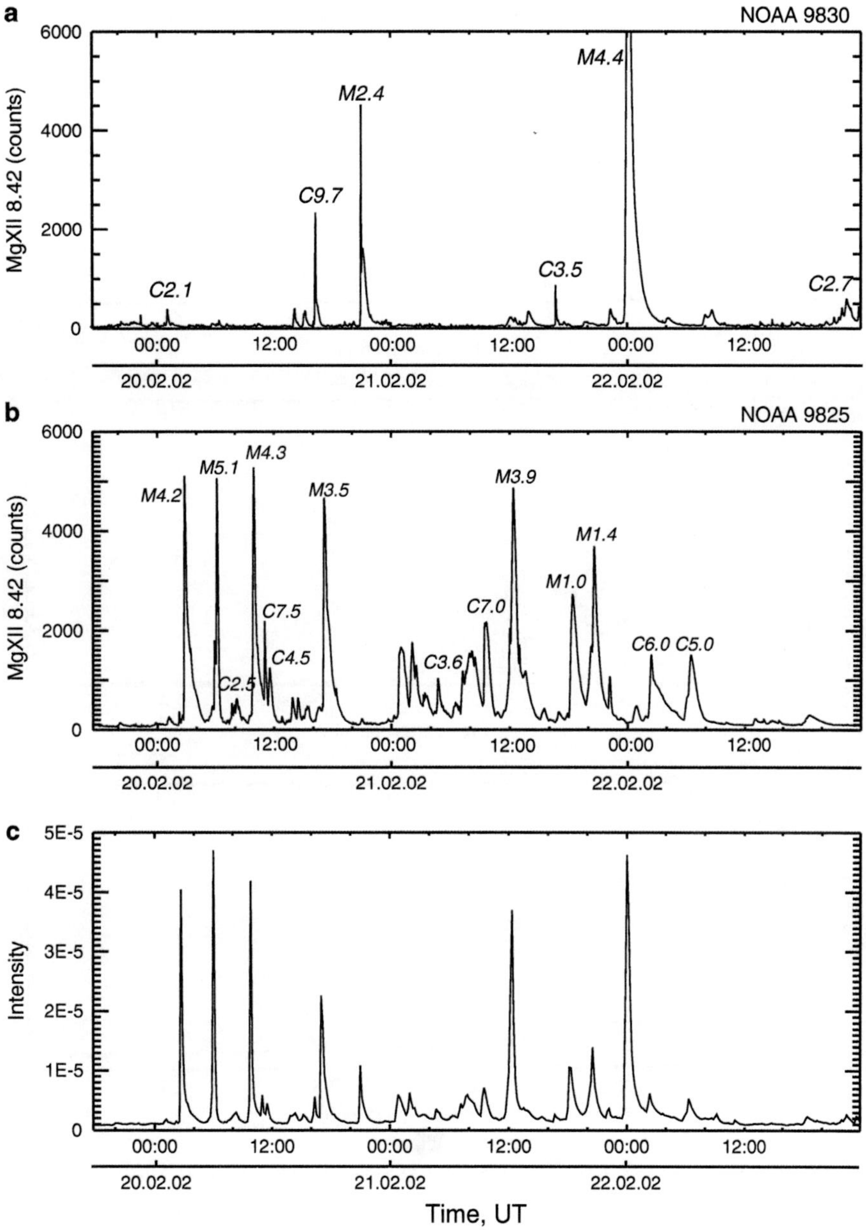

Fig. 3.11 Soft X-ray flux profiles: (**a**) NOAA 9830 active region (Mg XII 8.42 Åline); (**b**) NOAA 9825 active region (Mg XII 8.42 Å line); (**c**) full Sun (GOES 1–8 Å)

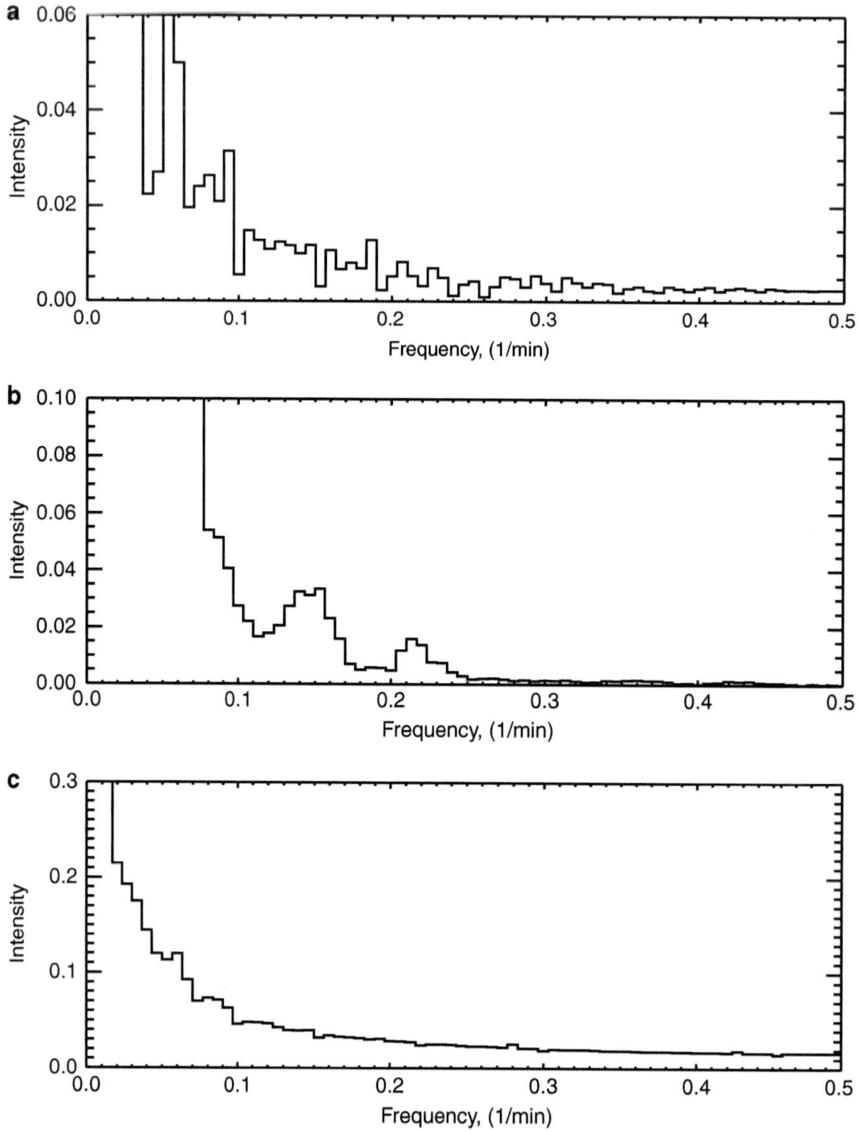

Fig. 3.12 Three type of power spectra in hot plasma oscillations: (**a**) type I, (**b**) type II, (**c**) type III

that distinguish the characteristic frequencies from peaks observed in distributions of type I:

- Their amplitude exceeds the threshold of 3σ, that is, the chance variation level of a signal.
- The peak width is much larger than in type I spectra.

- Frequencies at which the maxima are observed are stable, which means that they do not disappear when the sliding window is shifted, but exist for a long time.

It therefore may be assumed that the type II spectra contain periodic components dominating over random noise. Respective periods may be determined on the basis of maxima in the frequency spectra.

Figure 3.12c shows in its bottom panel an example of type III spectrum. In this distribution, all possible oscillation frequencies are full, without any visible domination of one any of them. The type III spectra can be assumed to correspond to emission generated due to a thermal mechanism that substantially differs from the emission generation procedure in the case of type II spectrum.

The question was studied as to whether the peculiarities in frequency spectra are caused by solar flare and, if so, then whether it is possible to establish a correspondence between different types of spectra and different types of flares. To solve this question, all obtained spectra were transformed into frequency charts. A frequency chart is a diagram where time is plotted against the X-axis, and the position of all peaks in a frequency spectrum corresponding to a given point of time is plotted against the Y-axis.

Thus, if F_k is a power spectrum obtained at the time point t from formula (3.2), then the following values will be plotted against the Y-axis of the frequency chart:

$$Y_k = \begin{cases} 1, & \text{if } F_k > F_{k-1} \text{ and } F_k > F_{k+1}, \\ 0, & \text{otherwise.} \end{cases} \tag{3.3}$$

The values of $Y = 1$ show the frequencies at which the F_k spectrum contains local maxima. Equation (3.3) defines the frequency chart in digital representation. The chart in this form is a matrix filled with ones and zeroes. The number of the matrix columns is equal to the number of processed spectra, and the number of lines—to the number of points (frequencies) in each spectrum. The digital representation is useful when the chart is intended for mathematical analysis.

Graphic representation is more suitable for visual study of frequency characteristics of emission. Time in this case is plotted against the X-axis, while the Y-axis is divided into segments according to the number of points in the F_k spectrum. The number k segment is shaded black if $Y_k = 1$ and remains blank if $Y_k = 0$.

The computation results, that is, graphic frequency charts for NOAA 9830 and NOAA 9825 regions, are presented in Figs. 3.13 and 3.14, respectively. Two frequency chart fragments are given for each active region. Simultaneous emission profiles in the Mg XII line are shown along with the charts.

The graphic frequency charts allow easy determination of the frequency spectrum type dominating in the active region. Sections randomly filled with black and white dots correspond to time periods during which the emission exhibited frequency characteristics of type I. Time periods during which the type II spectra dominate can be detected in the frequency charts by solid horizontal lines $Y = \text{const}$ that indicate the presence of stable characteristic frequencies in the spectrum. The periods of corresponding oscillations are within a 5–20-min range. Empty regions on the frequency charts correspond to type III spectra.

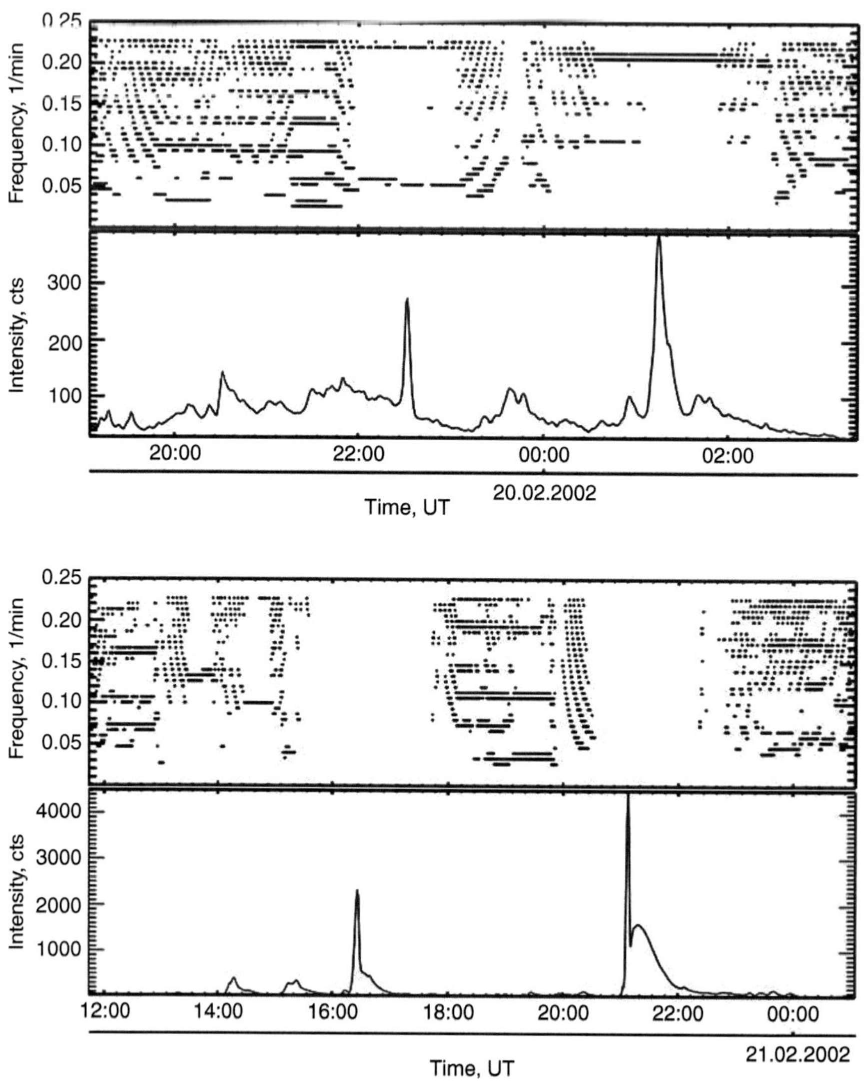

Fig. 3.13 Frequency charts and flux profiles in Mg XII 8.42 Å line for AR 9830

Figures 3.13 and 3.14 show that the type I spectra are predominantly observed in the emission of "quiet" active regions, whereas type II and III spectra are generated during flare events. At the same time, type II spectra are predominantly observed in impulsive flares having a symmetric emission profile whose decay phase lasts almost as long as the rise phase does. Phenomena with type III spectra belong to LDE class described in Sect. 4.2 as having a long-term decay phase, which is typically several times longer than the rise phase.

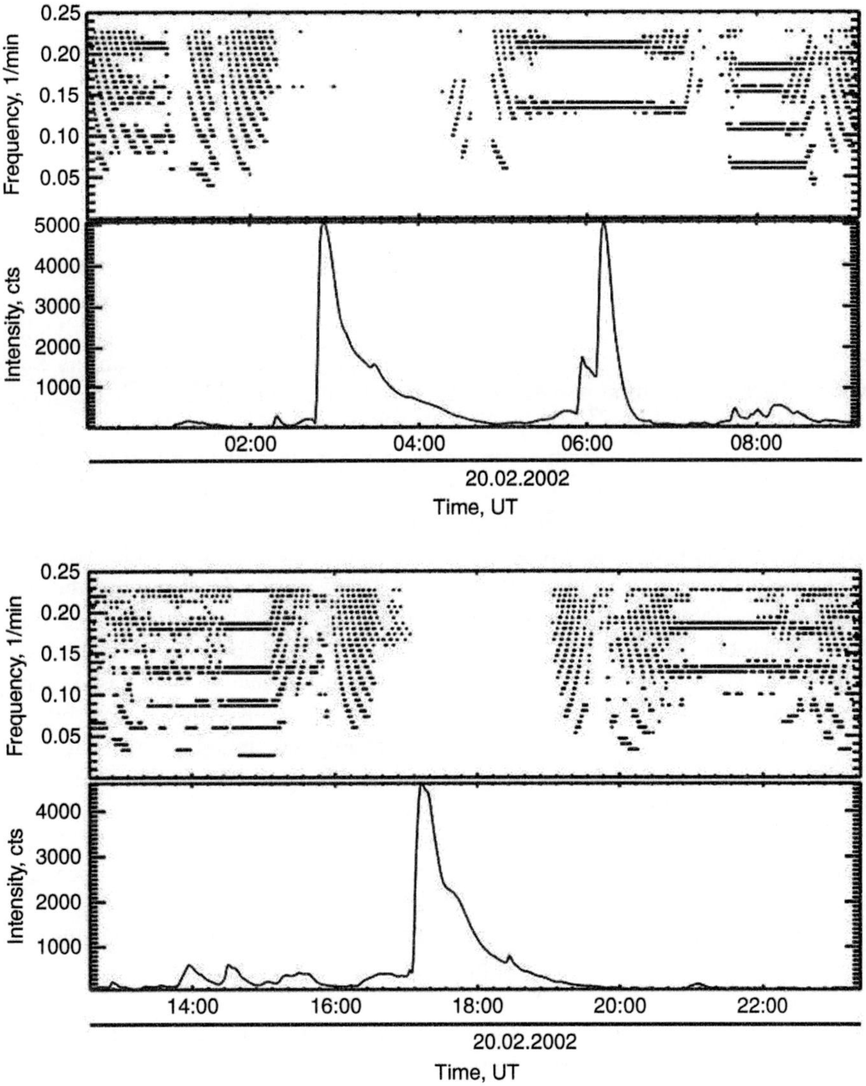

Fig. 3.14 Frequency charts and flux profiles in Mg XII 8.42 Å line for AR 9825

The measured oscillation periods in type II flares, which fall within 5–20-min range, are generally consistent with other measurement results, in particular, the SUMER/SOHO data [108]. The oscillation nature of hot sources, however, differs from the SUMER recordings. The SUMER device detected rapidly damping oscillations of high amplitude that apparently were connected with loop oscillations induced in the flare event. The emission oscillations detected in this research are more stable (visible for dozens of minutes), have much lower amplitudes extracted only by frequency analysis methods, and, above all, they are spatially connected to other objects, that is, high-temperature emission sources localized above the loops.

As fast processes, the flares are evidently characterized by oscillations of relatively high frequencies. At the same time, it is a known fact that on the Sun, there exist rather long-period oscillations caused first of all by processes taking place in the internal areas of the Sun [10]. A portion of the energy should apparently be transferred from such oscillations to the upper coronal layers as well. But because of the low degree of localization of the coronal sources in the coronal plasma, they are rather difficult to locate and study. A research was therefore carried out to search for long-period oscillations in well-localized structures of the coronal plasma [2].

Five active regions (AR) were selected for the research: NOAA 9825, 9830, 9835, 9837, and 9840. Other high-temperature formations observed on the Sun within the analyzed time period were short-term and thus did not suit the study of periodic changes in the intensity. The ARs 9825 and 9830 represented developed sunspot groups having a complex magnetic configuration of and classes, respectively. Many flares were registered in these regions in Í and X-ray ranges according to solar geophysical data (SGD, http://sgd.ngdc.noaa.gov/sgd/jsp/solarindex.jsp). The 9835 region had a β-configuration in the period under study, while AR 9837 was developing from a single sunspot to a β-configuration. The onset of the AR 9840 at the photospheric level was recorded on February 21 at 9:25 a.m. (SGD). Hot coronal plasma above the place of the AR onset was observed in the Mg XII 8.42 Å line starting from February 19. After the appearance of the sunspots, the brightness and area of the AR 9840 in the X-ray images considerably increased.

Totally 3,530 images were processed. Time profiles computed for each AR were gridded at an interval of 1.748 min. Time sequences for the ARs 9830, 9837, and 9840 cover approximately 89 h (3,044 values). The sequences for ARs 9825 and 9835 are somewhat shorter. This is caused by the fact that the regions were located during the period under study close to the western limb, and their heliocentric longitudes amounted on February 23 to 90°.

As the dynamic range of X-ray bursts is of several orders, their periodicity was studied with the help of an emission flux logarithm for each active region.

Figure 3.15 presents time dependence for integral emission fluxes of the ARs under study in a logarithmic plot. As is seen from the figure, AR integral flux variations have the nature of random processes with the presence of regular components. The quasiperiodicity of the flux variations changed with time even in a single active region, for example, the AR 9840 curve on 19–21 February notably differs from variations in the interval of 22–23 February.

To evaluate the spectral power of emission variations of the ARs under study, Fourier analysis was applied to autocorrelation functions of the logarithms of the AR integral emission fluxes. The autocorrelation functions were calculated by the formula:

$$k(m) = \frac{1}{N \cdot s^2} \cdot \sum_{n=1}^{N-m} [u(n) - s] \cdot (u(n+m) - s), \tag{3.4}$$

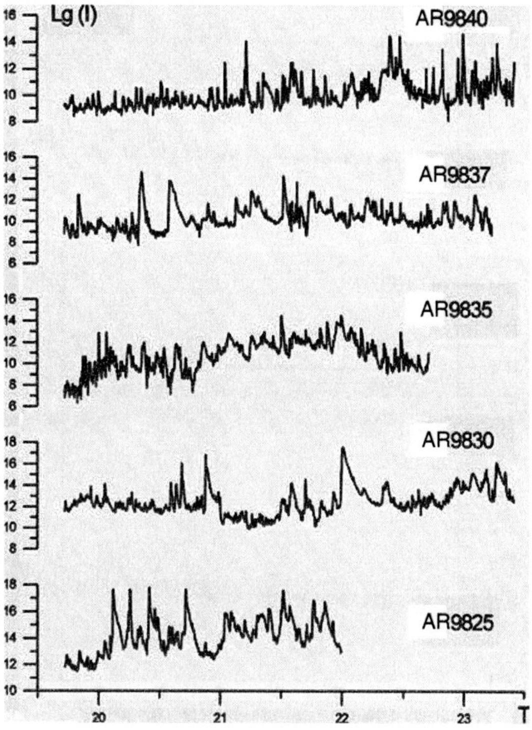

Fig. 3.15 Flux profiles for different active regions during 19–23 February 2002

where $u(n)$ is the value of the integral flux logarithm, s is the average value, and N is the number of points in the array. This method of calculating the spectral density provides for a substantially smaller influence of the sample size.

A typical autocorrelation function at low argument values is similar in form to an exponent, which should be expected due to the Poisson nature of distribution of instants of the pulsed process bursts, random pulse length, and stepwise rise of the bursts. Therefore, the power density in the high-frequency spectral region should reduce inversely to the squared frequency, which is generally observed. However, against the general background of the power reducing with the frequency, peaks are isolated that are characteristic of processes in the given active region. Statistically, significant peaks in the high-frequency region are identified through subjecting the spectra to averaging with a sliding window. The entire data array was divided for this purpose into intervals having the length of 1,027 values (1,795.2 min). The intervals were shifted to a halfwidth, and the resulting power spectra were averaged for all intervals. The selected size of the window allows the statistically significant peaks to be isolated with confidence in the range of periods of 4–40 min.

Figure 3.16 presents the averaged power spectra, obtained for all analyzed active regions, in the range of periods of 4–30 min (the period interval of 30–40 min does not contain statistically significant peaks). Periods are plotted against the horizontal axis in the order of their increasing. As Fig. 3.16 shows, the power

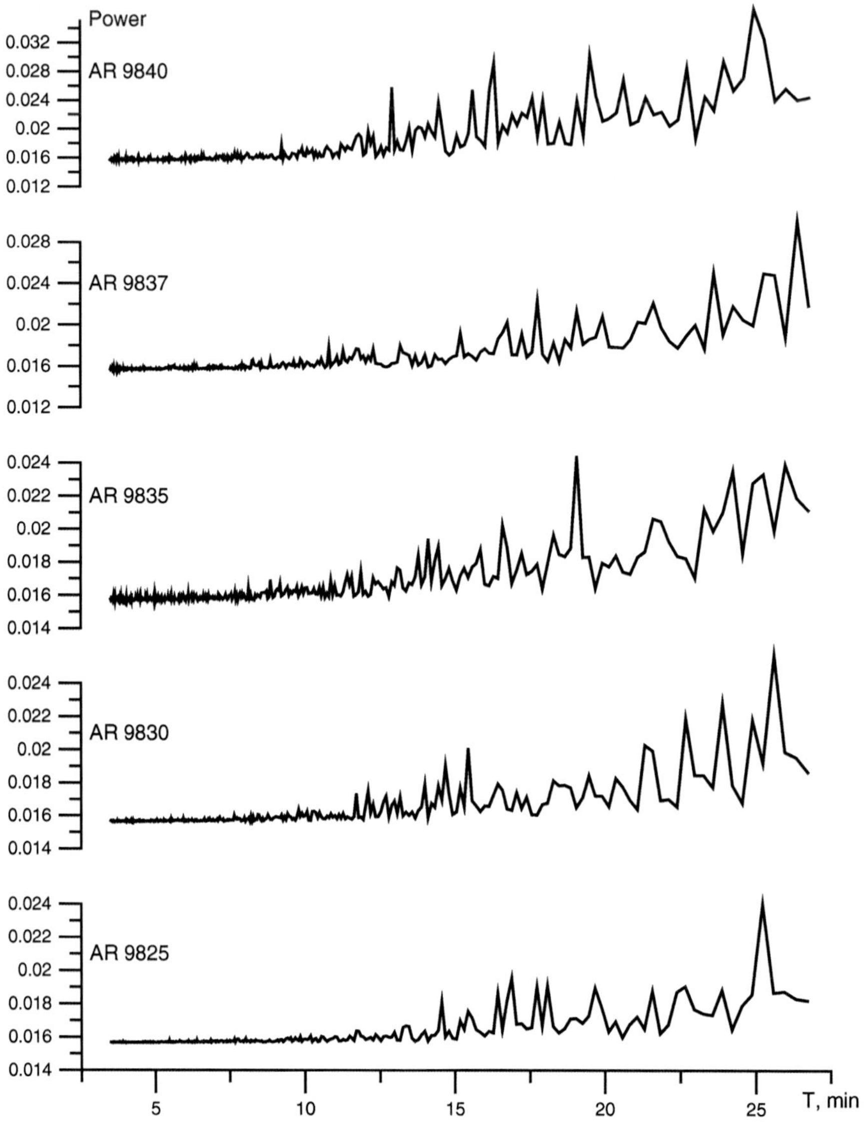

Fig. 3.16 Averaged power spectra, obtained for all analyzed active regions, in the range of periods of 4–30 min

spectra of all active regions feature the presence of peaks having characteristic quasiperiods in the interval of 12–30 min and the absence of significant peaks in the interval of periods of 4–10 min. The powers of peaks corresponding to these periods are approximately equal for the spectra of different ARs, but the values of the periods do not quite coincide. Thus, a significant peak having the quasiperiod of

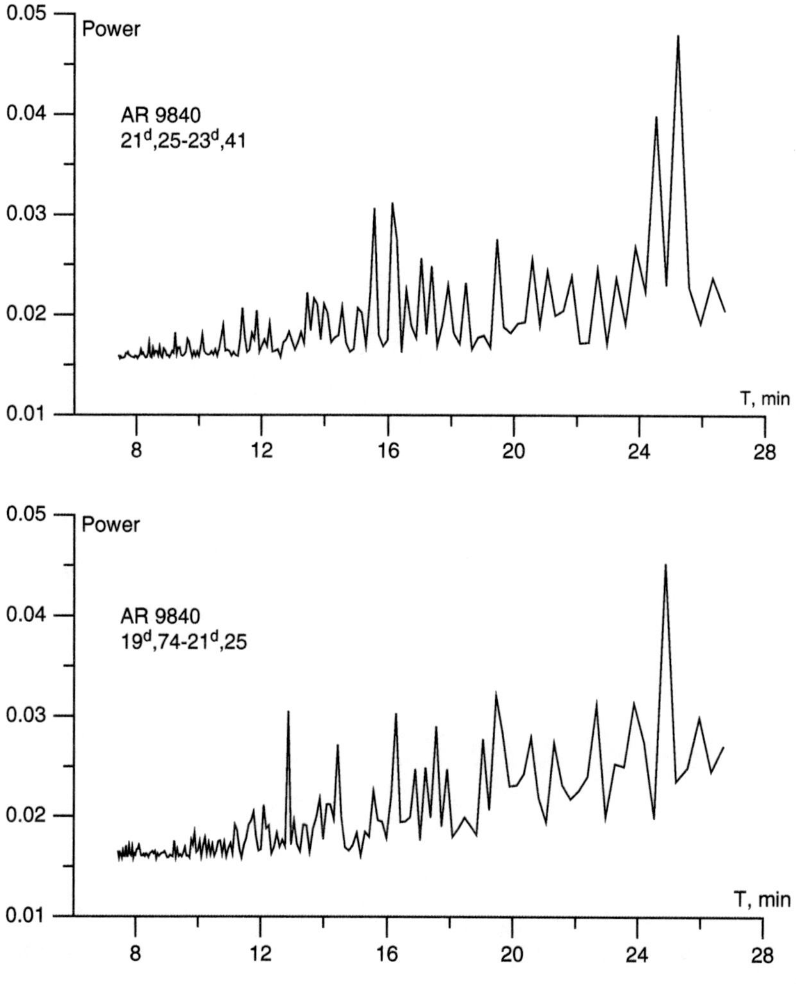

Fig. 3.17 Averaged power spectra for AR 9840 separately for the time when the photosphere was still free of the sunspots ($19^d.74 - 21^d.25$) and after their appearance ($21^d.25 - 23^d.41$)

12.5 min is present only in the spectrum of AR 9840 and only at its initial stage of development before the appearance of sunspots. This result is illustrated in Fig. 3.17, which gives the averaged power spectra for AR 9840 separately for the time when the photosphere was still free of the sunspots ($19^d.74 - 21^d.25$) and after their appearance ($21^d.25 - 23^d.41$).

As was mentioned above, the observed short-period oscillations in AR most probably reflect periodic processes or waves existing in hot coronal loops that cannot be resolved in these observations.

In a low-frequency region (the period interval of 40–200 min), the obtained power spectra at the same time include a number of maxima the amplitude of which

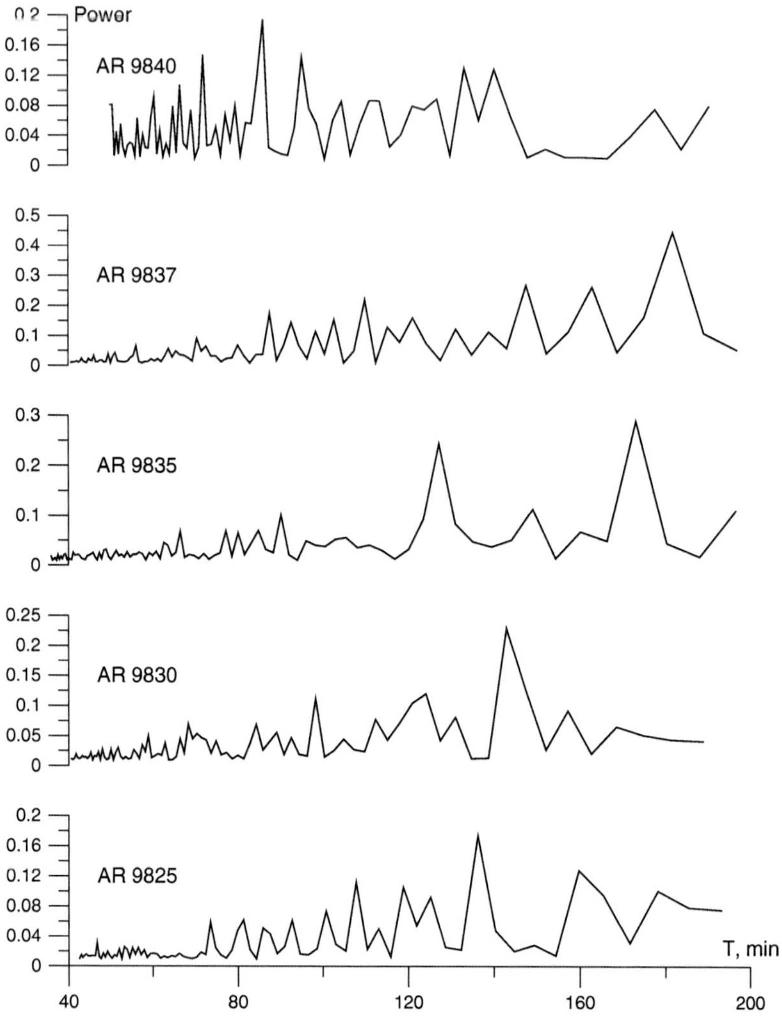

Fig. 3.18 Averaged power spectra in the range low-frequency region of 40–200 min

is 3–5 times greater than the average values (see Fig. 3.18). The power spectra in the low-frequency region shown in Fig. 3.18 were obtained without averaging by the sliding window.

We should point out that in a low-frequency region the peaks in the power spectra differ for different ARs. For example, characteristic quasiperiods are 72 and 84 min for the developing compact region 9840, and about 138 and 160 min for the flare-active region 9825. The difference in the position of low-frequency region maxima connected with the frequency of the rise of large bursts in different ARs is apparently determined by the structure and strength of the AR magnetic fields.

3.4 Quasisynchronous Bursts in High-Temperature Plasma

It is currently a reliable fact that the topology of the magnetic field in the solar corona solar corona is quite diverse. Both microloops characteristic of weak solar activity manifestation and structures of medium solar-scale size characteristic of active regions are observed. An assumption was made rather long ago about the presence in the corona of magnetic loops connecting regions spaced apart by several hundred thousand kilometers, but for the first time they were observed probably by the Mg XII spectroheliometer onboard the CORONAS-F satellite (Fig. 3.19). So, a question arises as to the extent to which these connections are effective and whether they can propagate disturbances.

In the search for such phenomena, bursts observed in the Mg XII line in the period between March 3, 2002, and March 4, 2002, 06:20 UT, were analyzed [4]. The series of observations consists of 932 images of 512 × 512 size, the first 452 images obtained at a 0.6-min interval and the rest of them at 1.75 min. This series is distinguished by the fact that many brightness bursts in different areas of the solar disc occurred almost simultaneously or at short delays.

In addition to flares, all visible brightness bursts recorded in images of the high-temperature plasma were analyzed. The bursts are identified as short-term (from several minutes to dozens of minutes) increase in the brightness of AR fragments if such increase is no less than 20% of their stationary brightness level.

The processing produced a time dependence of the surface brightness for fragments of ten ARs observed at that time on the disc. Figure 3.20 shows the position of the analyzed ARs on the solar disc that were observed in the EUV range: in 175 Å (Fe IX-XII) and 284 Å (FeXV).

The processing resulted in a time dependence of the surface brightness for eleven fragments of ten ARs. The most extended coronal region was observed above a sunspot group 9845 NOAA/USAF. It consisted of two distinctively separated elements: 9845(N) and 9845(S) having the centers of their surface brightness spaced

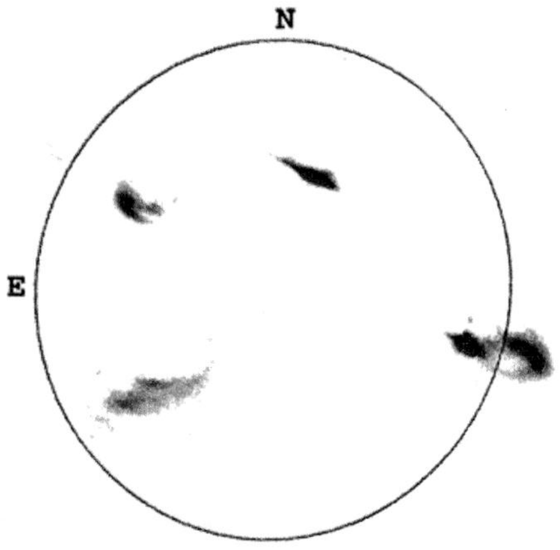

Fig. 3.19 Solar corona in the Mg XII 8.42 Å line observed by Mg XII spectroheliograph aboard CORONAS-F. The image obtained on 21 February 2001 at 19:35 UT

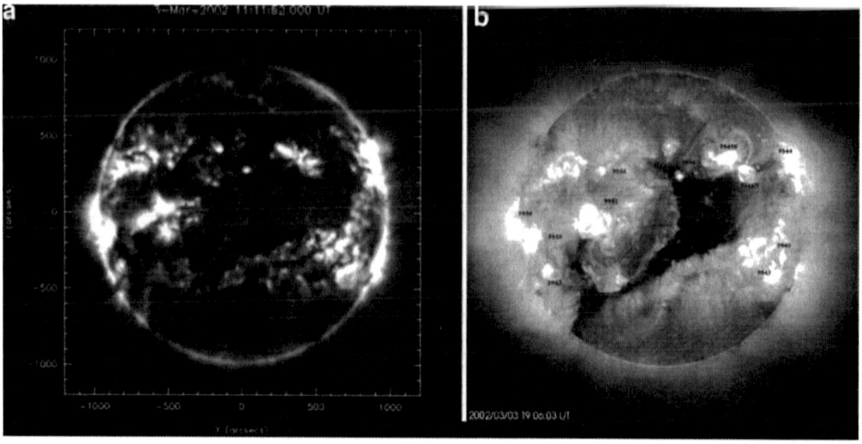

Fig. 3.20 Solar corona observed in 175 Å by SPIRIT/CORONAS-F and 284 Å by EIT/SOHO. On the EIT/SOHO image NOAA active region numbers are given

by approximately 10°. Photometry of fragments of these elements was therefore performed. Figure 3.21 presents the photometric results; the AR surface brightness is expressed in relative units. The brightness of the strongest burst in AR 9856 was 6,100 at its maximum. In relative units, the surface brightness of the weakest bursts reoccurring at the locations of fragments totally faded to the background level was about 100. The diagrams show that the maxima of many brightness bursts were almost simultaneous in different ARs.

According to the SGD catalogue, two flares were registered in the soft X-ray range during the period of measurements. The first flare of Ñ2.3 class lasted from 18:16 to 18:51 UT (with its maximum at 18:33 UT) on March 3. Observations in the Mg XII line demonstrate that this flare consisted of two bursts taking place almost simultaneously in two active regions and having the surface brightness of the same order. The burst maximum of AR 9856 located near the eastern solar limb was at 18:27:43 UT. In AR 9845, the burst maximum occurred at 18:45:11 UT, and the brightness rise in both active regions practically coincided in time with the flare onset (SGD data). These ARs were spaced apart by approximately 106°.

The second flare of C1.4 class was observed on March 4, 2002, from 06:04 to 06:16 UT (the maximum at 06:10 UT). According to the Mg XII line observations, two bursts occurred at the same time in the corona. The first, weak burst was recorded in the developing AR 9855 (the maximum at 06:04 UT), and the major maximum—above AR 9845 at 06:11:32 UT.

Images of the coronal sources in the Mg XII line thus show that either of the flares consisted of bursts in two active regions spaced apart.

The delay time between the burst maxima in different ARs was about 15 and 7 min for AR pairs 9856 → 9845 and 9855 → 9845, respectively. This may be an argument in favor of interpreting the phenomena as sympathetic ones. To finally solve this issue, physical connection between active regions spaced apart should be proved and a mechanism providing secondary bursts should be revealed.

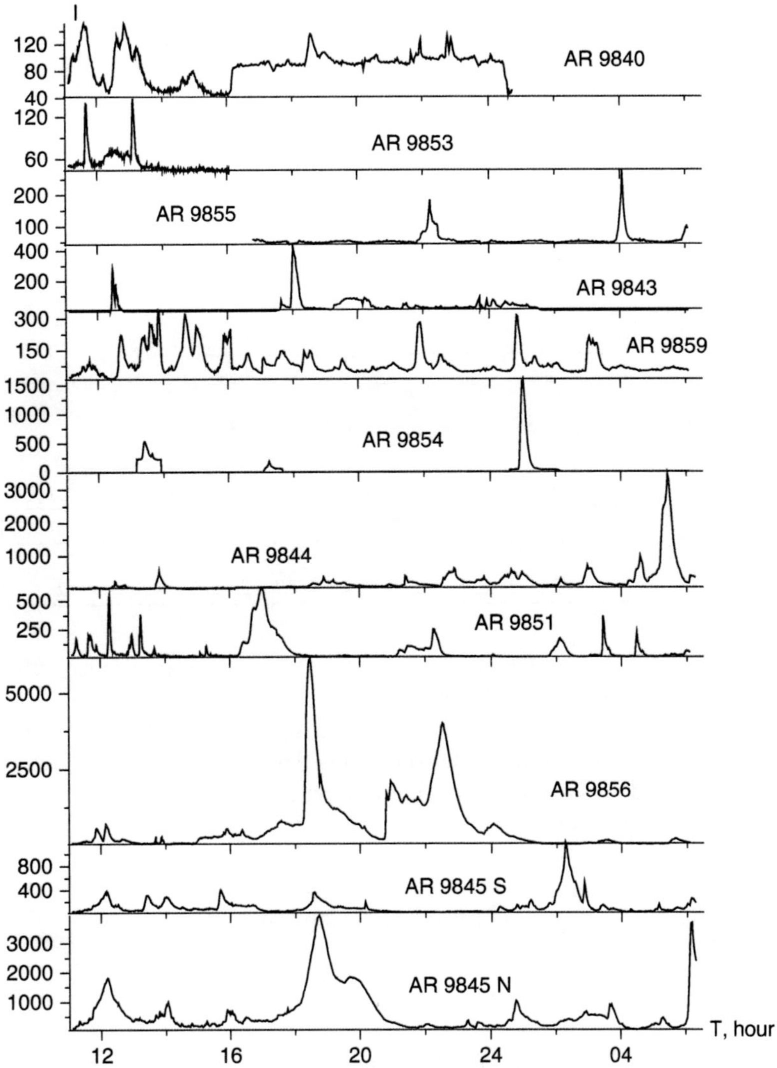

Fig. 3.21 Surface brightness of active regions during the 3–4 March 2002

At the first research stage, the question of the existence of sympathetic bursts was considered statistically, without taking into account the probability of physical connection between the active regions. The time of the maximum brightness bursts in the selected 11 fragments of ARs were used for this purpose (Fig. 3.22). In the figure, the time of the burst maxima in minutes is plotted against the X-axis (the moment of the first burst as the reference), and the numbers of the active regions according to the descending order of the number of bursts recorded in the regions are plotted against the Y-axis. A large amount of bursts may be observed in certain intervals, and at the same time there are long time periods without any bursts.

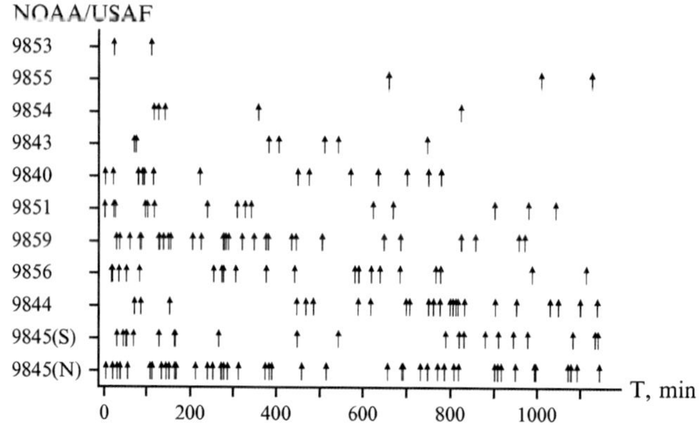

Fig. 3.22 Brightness maximum chart of the active regions during 3–4 March 2002

Table 3.1 Number of bursts registered in each active region during the 3–4 March 2002

9845(N)	9845(S)	9844	9856	9859	9851	9840	9843	9854	9855	9853	Total
1(N)	1(S)	2	3	4	5	6	7	8	9	10	11
43	23	26	20	27	17	14	7	5	3	2	187

Temporal distribution of the number of bursts in 11-min intervals also indicates a considerable heterogeneity of the process.

We should find out whether such heterogeneity can be observed at random Poisson distribution of the times of burst maxima. The probability theory states that in the case of random distribution of times of events the probability that no event occurs in the time period t is:

$$P_0 = \mathrm{e}^{-\lambda \cdot t}. \tag{3.5}$$

The probability of k events occurring in the time period t is:

$$P_k = \frac{(\lambda \cdot t)^k}{k!}\mathrm{e}^{-\lambda \cdot t}. \tag{3.6}$$

In this case, λ is the number of bursts in unit time.

A question arises during this study as to the extent to which the statistical properties of the burst maximum time distribution correspond to the Poisson process model. Table 3.1 gives the number of bursts registered in each active region during the observation period.

Statistical properties of six AR fragments, from 1(S) to 6, were analyzed since the numbers of bursts in them during the observation period are approximately of the same order. The burst rate λ for this group of active regions is about 0.11 bursts per minute (127 bursts in 1,150 min). The temporal distribution of the times of burst maxima includes seven time periods longer than 30 min during which no bursts were observed (empty intervals). Durations of the empty intervals in minutes may be listed in descending order: 54, 41, 38, 37, 34, 33, 27....

Table 3.2 Estimated and observed probabilities of empty intervals and of several bursts rising in an 11-min interval

AR 9845(S)—AR 9840 (six parts)						All parts of AR					
P_0 (54 min)		P_0 (41 min)		$P_5(3)$		P_0 (36 min)		P_0 (30 min)		$P_6(6)$	
Theor.	Exp.	Theor.	Exp.	Theor.	Exp.	Theor.	Exp.	Theor.	Exp.	Theor.	Exp.
1/390	1/21	1/85	1/15	1/465	1/34	1/350	1/32	1/132	1/38	1/786	1/17

Table 3.2 presents estimated and observed probabilities of empty intervals and of several bursts rising in an 11-min interval. P_0 (41 min) is the probability of a 41-min empty interval; $P_5(3)$ is the probability that in the distribution of the times of burst maxima, five bursts are observed thrice in an 11-min interval; similarly, $P_6(6)$—probability of six bursts—occurring six times in the same interval.

Thus, in six analyzed AR fragments, the observed probabilities of empty intervals substantially exceed the estimated probabilities, that is, they are too long as compared to a Poisson process. In addition, the observed probability of five bursts occurring thrice in an 11-min interval is by more than an order higher than the probability of such events for a random process.

Table 3.2 also shows that for the maxima of bursts in all 11 AR fragments, the statistical picture changes insignificantly, though the AR 9845(N) fragment increases to some extent the average rate of bursts. The list of empty intervals in this case consists of the following sequence: 36, 31, 28, 27… min, and in the distribution of times of the burst maxima in all measured ARs, six bursts were observed six times in an 11-min interval, and five bursts were observed six times.

It should be noted that the obtained results demonstrate predomination of bursts observed on the solar disc in different ARs quasisimultaneously, within a chosen narrow time period. Such bursts may include both sympathetic, induced, events and events triggered simultaneously by some global process. The detailed analysis of the time dependence of surface brightness in fragments of different ARs has shown that a number of regions produce the same picture.

Figure 3.23 presents cross-correlation functions of surface brightness of two fragments AR 9845 and AR 9845(S) from AR 9856 observed on March 3, 2002 (continuous 13-h observation). The cross-correlation function was calculated using surface brightness values normalized to an average value.

As the figure shows, the correlation coefficient for brightness variations in ARs 9845(N) and 9856 is $k_{xy} = 0.68$ at zero and reaches the maximum ($k_{xy} = 0.84$) at a shift by approximately 15 min. Bursts on the average occur in this case first in AR 9856 and then in AR 9845(S). The correlation is somewhat lower for the pair of fragments AR 9845 (N) and (S): $k_{xy} = 0.55$ at zero and $k_{xy} = 0.58$ at a shift of approximately 5 min. Bursts in this pair occur earlier, mainly in the fragment AR 9845(S).

As was mentioned above, the delay time between bursts in different ARs may support the hypothesis of the existence of actually induced, sympathetic events.

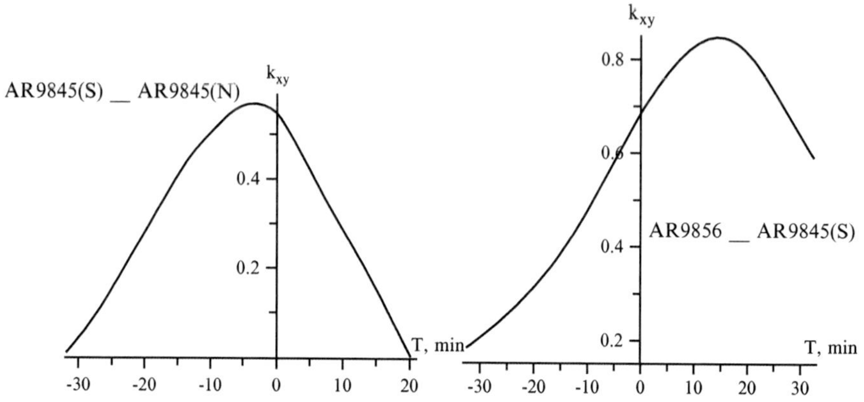

Fig. 3.23 Cross-correlation functions of surface brightness of two fragments AR 9845 and AR 9845(S) from AR 9856 observed on March 3, 2002

For a more complete analysis, from among the measured fragments pairs were separated having an experimentally observed physical connection in the form of coronal magnetic loops. Additional data is evidently needed for this purpose because only coronal plasma regions of $T \sim 5$–20 MK are recorded on solar images in the Mg XII line. Solar images obtained on March 3, 2002, in EUV range by SPIRIT/CORONA-F in $\lambda = 175$ Å (Fig. 3.20a) and by EIT/SOHO in $\lambda = 284$ Å (Fig. 3.20b) were used additionally. As it was already noted in the previous chapter, the hot compact sources in the upper corona approximately coincide with most bright fragments of AR in the lower corona in the $\lambda = 175$ Å and 284 Å lines ($T \sim 1.5$ MK and $T \sim 2$ MK, respectively).

It is also easily seen in the image that in $\lambda = 284$ Å the magnetic loops connect fragments of AR 9845 of different polarities. Magnetic loops connecting regions of different polarities are not so vivid in other AR pairs, but the analysis of consecutive images recorded during March 1–5 demonstrates that AR pairs 9845 and 9844, 9851 and 9856 and 9856 and 9859 may be connected with each other by loops of low brightness. No magnetic loops are noticed between ARs 9845 and 9856 spaces apart by 106°, but according to the above said, a significant correlation was observed in the time dependence of their surface brightness, especially in events of March 3. Therefore, the possibility that a giant loop connecting ARs 9856 and 9845 exists should not be ruled out. Cross-correlation curves plotted in Fig. 3.23 show that the delay time of bursts tends to increase with a larger distance between ARs. The average delay time was evaluated using histograms of the distribution of time periods between the burst maxima for the chosen pairs of AR fragments.

Histograms in Fig. 3.24 demonstrate the distribution of time periods between the burst maxima for the following pairs of AR fragments: 9845(S) $\rightarrow$ 9845(N), (59); 9845(N) $\rightarrow$ 9844, (44); 9856 $\rightarrow$ 9859, (26); 9859 $\rightarrow$ 9851, (27); 9856 $\rightarrow$ 9851, (21); and 9845(N) $\rightarrow$ 9856, (36). Parentheses following the names of fragments in the chosen pairs include the number of intervals between the bursts by which the

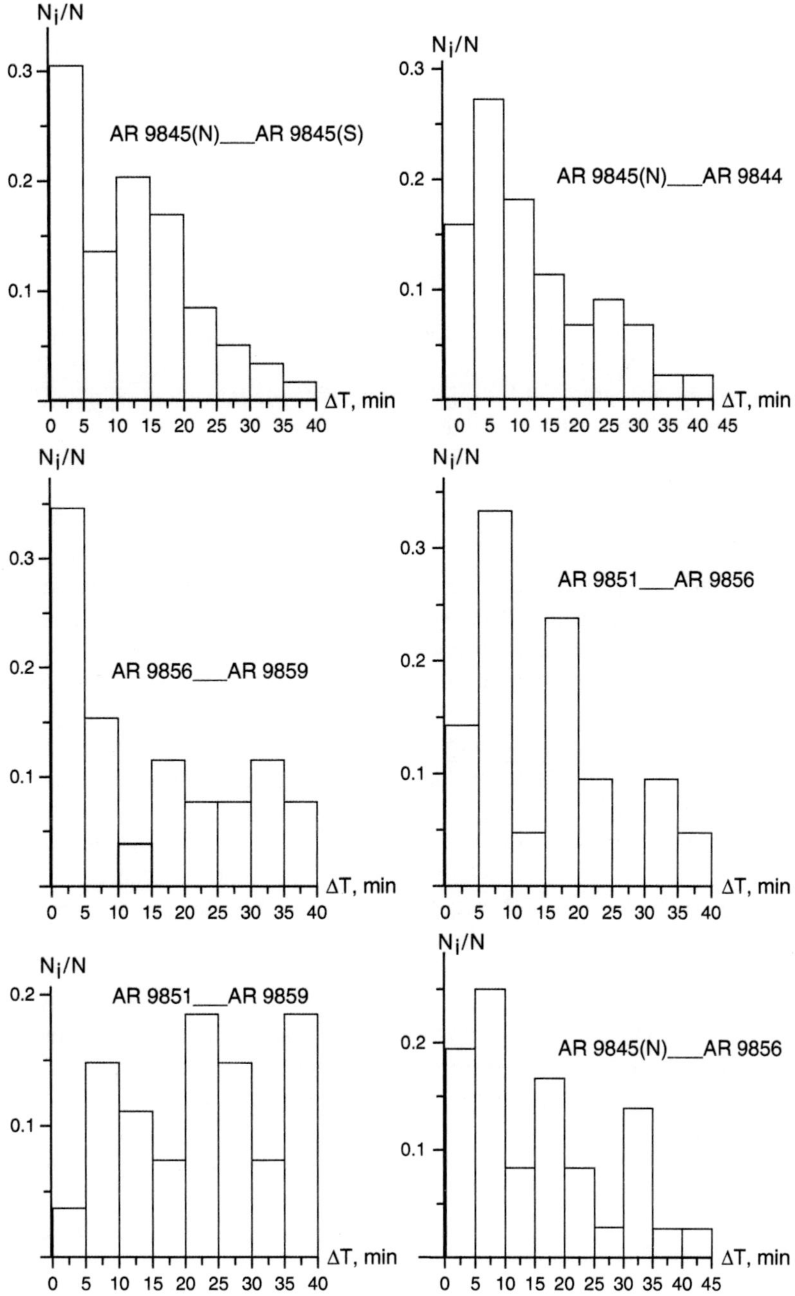

Fig. 3.24 Distribution of time periods between the burst maxima for AR pairs

histograms were constructed. As these numbers are insufficient for strict statistical estimates, the histograms were constructed without taking into account the sign of the interval, that is the propagation direction of the disturbance.

A number of conclusions can be drawn from the histogram analysis. A reliable fact of the presence of magnetic connection between AR 9845 fragments having different polarities serves this purpose (individual images show the rise of the loops connecting these fragments). According to the distribution histogram of these two fragments, there are 18 intervals between bursts lasting from 0 to 5 min. This amounts to slightly more that 30% of the total number of intervals. Therefore, magnetic connection between other pairs of fragments can be considered as reasonably probable if in their histograms about 30% (or more) values lie within any of the 5-min intervals.

Of six, three histograms in Fig. 3.24 demonstrate the maximum number of time periods between bursts within an interval of 0 to 10 min, the value of the maxima amounting to 30–34% of the total number. It is thus probable that the fragment pairs 9856 $\leftrightarrow$ 9859 and 9856 $\leftrightarrow$ 9851 are interconnected, as well as 9845(S) $\leftrightarrow$ 9845(N).

The result of the connection is not so evident in the 9845(N) $\leftrightarrow$ 9,844 fragment pair because the maximum number of time periods between bursts in a 5-min interval is only 27.3% of the total number. However, the connection between these fragments is highly probable. If the sign of disturbance propagation is taken into account in the analysis of the connection between these fragments, 10 time periods between bursts indeed fall within the interval of -10 to 0 min, the average value amounting to -5.6 min. At the same time, eight intervals having the average value of $+5.7$ min fall within the interval of 0 to $+10$ min. Such a coincidence of average values is hardly accidental and most likely indicates a physical interaction between these active regions.

The distribution histogram for 9,859–9,851 pair does not show any prevalence of short periods (5–15 min) between burst maxima. The number of values in each of the intervals is less than 20% of their total number, that is, the histogram reflects a random Poisson process.

The histogram for the pair of fragments AR 9845(N) and AR 9856 shows a maximum number of values in a 5–10-min interval, which is only 25% of their total number. This value is lower here than in other pairs in spite of high correlation between the time dependences of the surface brightness recorded for these fragments on March 3. This contradiction is explainable because the high correlation coefficient ($k_{xy} = 0.84$) is caused by a fact that specifically in these fragments very intense bursts was observed simultaneously. The resulting histogram may indicate the absence of connection between AR 9845(N) and AR 9856 fragments. The pair of fragments AR 9845(N) and AR 9840, the magnetic connection between which is also clearly absent, has a similar histogram.

A conclusion can thus be made that real collective processes were observed on March 3–4, 2002, in the high-temperature coronal fragments of active regions widely spaced apart. A common cause of such phenomena in the corona during the observation period had to be identified. The behavior of the photospheric magnetic field during the period under study was analyzed for this purpose. The sunspot activity was rather high in early March 2002, when over ten sunspot groups were observed on the solar disc simultaneously. The developing group AR 9845 (magnetic class $\beta\gamma$) was the most intense among them, reaching one of the maxima of its development by the end of March 3. The group then consisted of 32 sunspots and extended for about 20°. The sunspot group AR 9844 was located near the western limb and had passed the maximum of its development by the time of observation. Other sunspot groups were relatively small, bipolar or unipolar. An interrupting group AR 9840 and a small bipolar group AR 9854 formed at midday on March 2 inside an extended equatorial coronal hole can be singled out. It is an interesting fact that the group AR 9840, the coronal regions above which were observed during February 20–24, 2002 [2], also reappeared on the solar disc on March 2. The sunspot group AR 9855 emerged on the solar disc at midday on March 3. Figure 3.25 shows

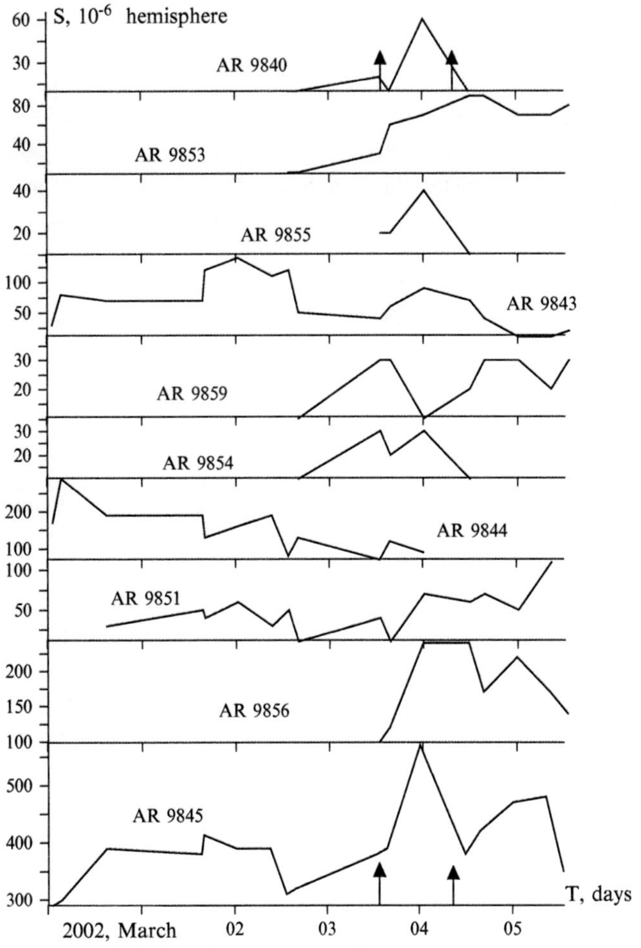

Fig. 3.25 Dependence of the corrected area of sunspot groups from February 27 till March 5, 2002

the time dependence of the corrected area of sunspot groups for a period of time from February 27, 2002 to March 5, 2002 according to the SGD catalogue.

An AR magnetic flux is known to change similarly to the area of a sunspot group, the flux maximum coinciding with the area maximum to within a day [107]. Vertical arrows on the X-axis in Fig. 4.4.7 mark out the period of observing X-ray bursts. The area of the majority of the sunspot groups was increasing during this period, that is, new magnetic fluxes emerged there. The only exception was AR 9851 in which the area (magnetic flux) was increasing from the afternoon of March 2 until 14:00 UT on March 3.

Therefore, it might be assumed that the basic reason for quasisynchronous bursts is the simultaneous change in the condition of magnetic fields in different active regions, particularly the increase of their magnetic fluxes, and that the reasons for the parallel increase of the magnetic fluxes in active regions widely spaced apart (to up to 100°) should be searched for in the evolution of large-scale and global fields on the Sun because specifically these fields determine the energetics and spatial localization of fields of any levels.

The research results speak in favor of the evolution of a large equatorial coronal hole during March 3–4, 2002 (Fig. 3.20) being the global reason for an almost simultaneous increase of the magnetic flux in the active regions. It was the emergence of new magnetic fluxes that induce the quasisynchronous bursts in the coronal areas of the active regions. The research leads to a conclusion that the quasisynchronous X-ray bursts are of different natures: disturbances can propagate from one region to another through extended magnetic loops, or the bursts may be induced by processes going on in the lower solar layers.

3.5 High-Temperature Plasma Dynamics

The results of research described in the above chapters raise a question as to the varying configuration of magnetic structures where high-temperature plasma is localized. Due to a high spatial resolution of the Mg XII spectroheliometer, the recorded image series provide data for the analysis of not only photometric characteristics, but also of the dynamics of multiscale high-temperature plasma formations.

As mentioned above, the experimental data of observations lead to a conclusion that structures in the high-temperature plasma possess high-speed dynamics with characteristic time periods of several seconds [76, 78]. This fact has not been previously noticed because of the low temporal resolution (of more than 5 min) of all imaging devices that recorded the high-temperature plasma. The GOES X-ray monitor has high temporal resolution, but because of the lack of spatial resolution, it cannot detect a specific region undergoing changes or determine the way in which their fine structure varies.

Figure 3.26 presents the development of a process observed in the high-temperature coronal plasma. A series of images recorded on February 10, 2002,

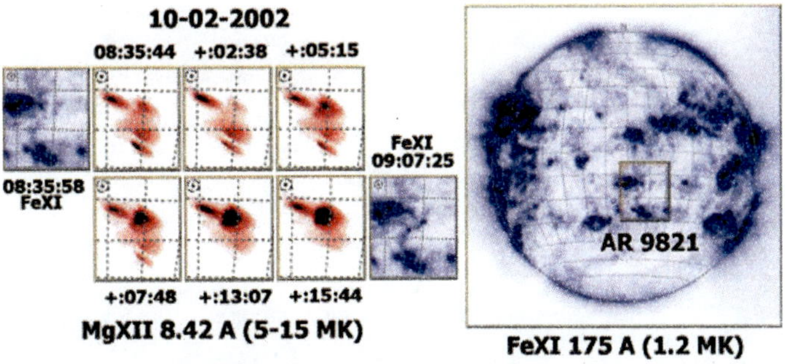

Fig. 3.26 Dynamics of high-temperature coronal plasma observed in Mg XII 8.42 Å line and 175 Å channel on February 10, 2002

at about 09UT shows the structural change in high-temperature plasma sources in the active region NOAA 9821 within a $6' \times 6'$ field during 15 min. According to the images, the number of individual sources observed in the Mg XII line substantially reduces in the course of evolution, while their integrated luminance increases [17]. This indicates that with the release of energy the configurations of the AR magnetic field becomes simpler. The most interesting feature of these observations is that the coronal plasma observed in the 175 Å telescopic channel (Fe XI–XII line, $T \sim 1.5$ MK) remained practically undisturbed during the observation period. This confirms the conclusion of the above chapter that the coronal and high-temperature plasma are localized in different magnetic structures.

Other examples of impulsive and long-lived soft X-ray sources characterized by high dynamics are given in Fig. 3.27. The figure is based on images recorded by the EIT/SOHO telescope in the Fe XII 195 Å line on which the positions of high-temperature sources of emission observed in each event by the Mg XII spectroheliometer are outlined. For the impulsive source recorded on February 20, 2002, two consecutive positions are shown that demonstrate its upward movement. The first position corresponds to 02:43:03 UT time, when the source center was at the altitude of ~66,000 km, and the second one—to 03:21:02 UT. The source had moved during this period of time upward in the corona to the altitude of 106,000 km. The average rate of rise thus equaled 17 km/s. The visible size of the high-temperature region had increased meanwhile from 1.5 to 2.4 angular minutes.

Figure 3.28 presents time profiles for emission sources in the Mg XII line and within the 1–8 Å wavelength range that were constructed based on observation data from the GOES-10 satellite. Both time profiles have a similar form characterized by a rather fast increase of the emission flux at the onset of the event and by a long-term decay phase of 2 h for the impulsive source and about 14 h for the long-lived one. At the same time, the emission decreases in the first case by three orders of the value, from 10^{-5} to 10^{-8} W/m^2, and in the second case by only 50 times, from 10^{-6} to 2×10^{-8} W/m^2. If the decay phase is expressed by a characteristic time τ during which the emissive power reduces e times, then $\tau = 17.4$ min for the

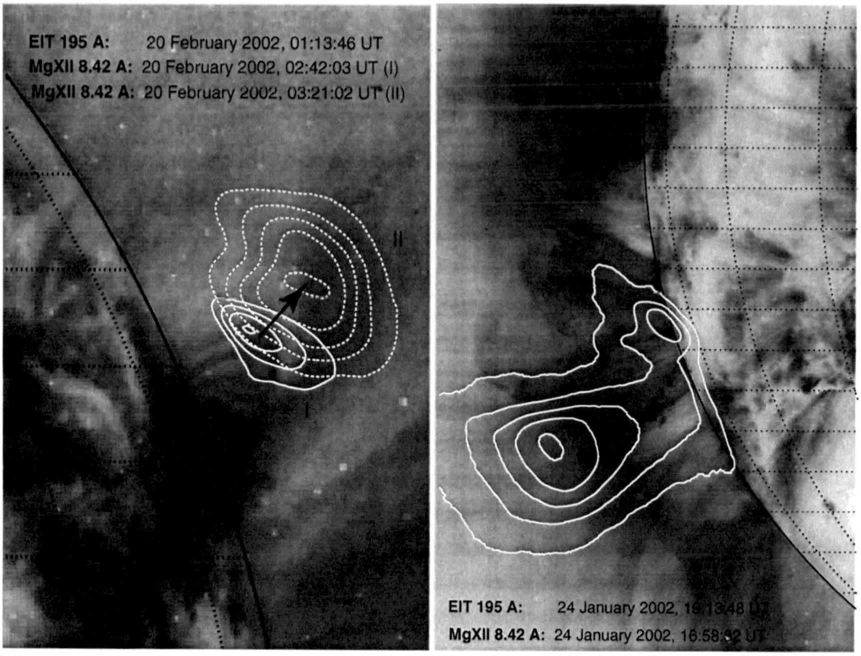

Fig. 3.27 Dynamics of high-temperature long-lived X-ray source observed in Mg XII 8.42 Å line (CORONAS-F) and 195 Å channel (EIT/SOHO) on February 20, 2002

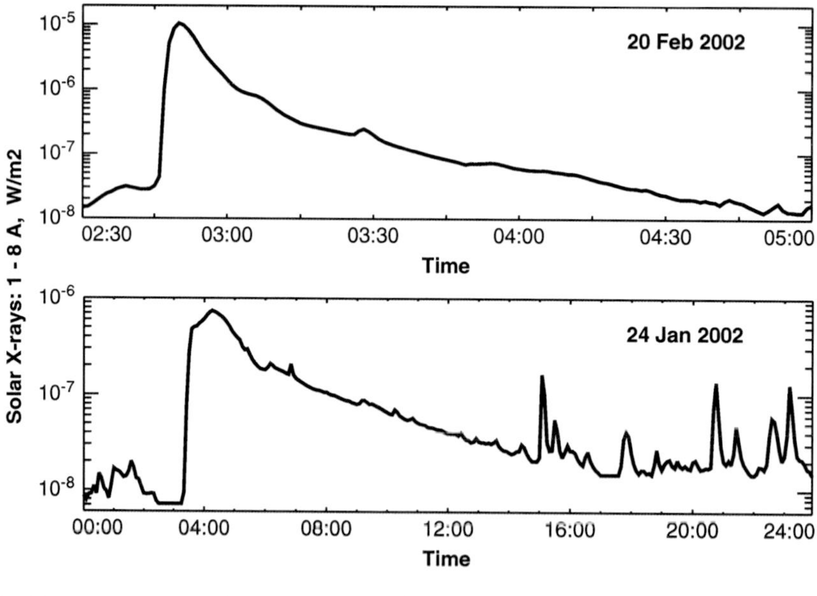

Fig. 3.28 Time profiles of emission in 8.42 Å spectral line (Mg XII) and within the 1–8 Å wavelength (GOES-10)

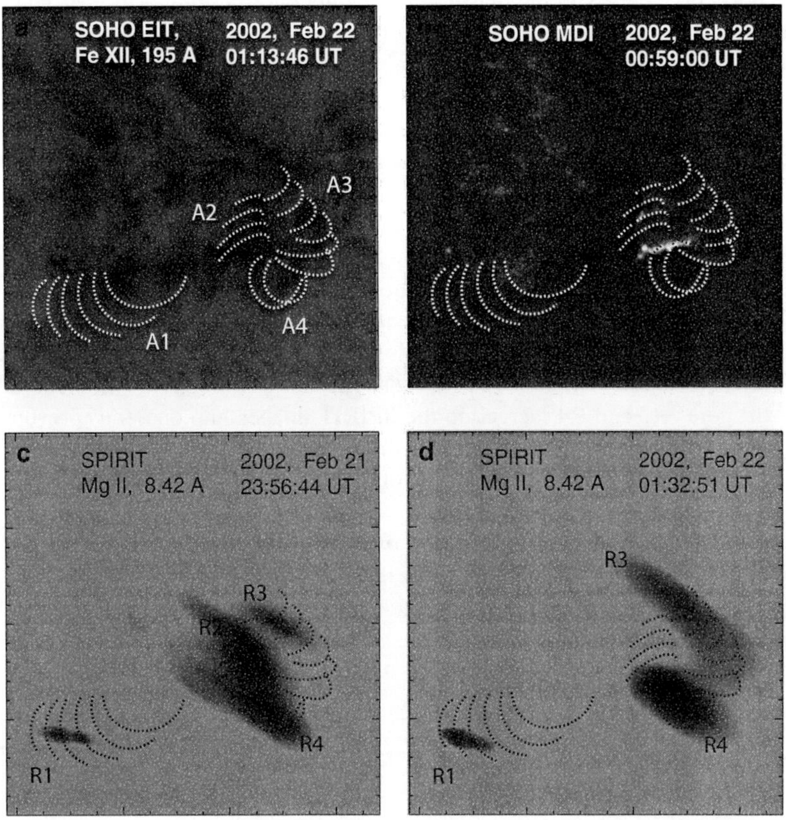

Fig. 3.29 Structure of active region in the photosphere and in the corona in different spectral ranges: (**a**) 195 Å by EIT/SOHO; (**b**) MDI/SOHO magnetogram; (**c**)–(**d**) Mg XII 8.42 Å line before and after the flare

impulsive source, while for the long-lived source $\tau = 214.7$ min, that is, is more than 3.5 h. The plasma-cooling modes in different types of high-temperature sources thus differ substantially from each other, at least by their characteristic times.

Figure 3.29 presents the structure of an active region in the photosphere and in the corona in the form of images of different emission ranges: (a) image of the active region in the 195 Å range of EIT/SOHO; (b) a MDI/SOHO magnetogram overlaid by the neutral line of the photospheric magnetic field; (c–d) SPIRIT images in the Mg XII line before and after the flare. All images are reduced to the same scale, and their relative displacement caused by the Sun rotation is compensated for.

Four groups of magnetic loops are identified in the analyzed EIT/SOHO and TRACE images and marked in the figure by A1–A4 symbols. These loop systems in Fig. 3.4c,d are aligned with high-temperature emission sources detected in the Mg XII spectroheliograph images. The comparison shows that the arrangement of the high-temperature plasma regions observed in the Mg XII line corresponds to coronal loop systems, and the plasma exhibits the highest luminosity near the tips of the loops. To determine the lifetime of the observed sources, SPIRIT images taken in several previous days were analyzed, and it was detected that all four sources had existed in the corona for at least 3 days. This substantially exceeds the characteristic time of heat-conductive cooling of the plasma of 10 MK temperature, which takes from one to several hours (depending on the electron concentration). It may therefore be assumed that energy release occurs in the "quiet" solar corona, heating the high-temperature sources even without flares. Such energy release most probably relates to magnetic reconnection high up in the corona at separators of a large-scale magnetic field of an active region.

Temporal dynamics of X-ray sources caused by an M4.4 class flare on February 21–22, 2002, was most interesting. The onset of the flare in Mg XII line was at about 23:45 UT on February 21, 2002 in core R2 shown in Fig. 3.29. At the moment of the flare, the cores R2 and R4 merged in the SPIRIT images into one, thus preventing the analysis of dynamics for either of them. The core R1 was located farther away from the flare center and even at the time of the flare maximum was observed in the images as an individual emission source.

The time profiles of brightness in the flare region R2–R4 (dotted curve) and in the remote core R1 (solid line) are given in Fig. 3.30. It is worthy of note that the time profiles demonstrate mutual correlation visible in many sections. In particular, a burst, though of a smaller extent, was also observed in the remote core R1 in a

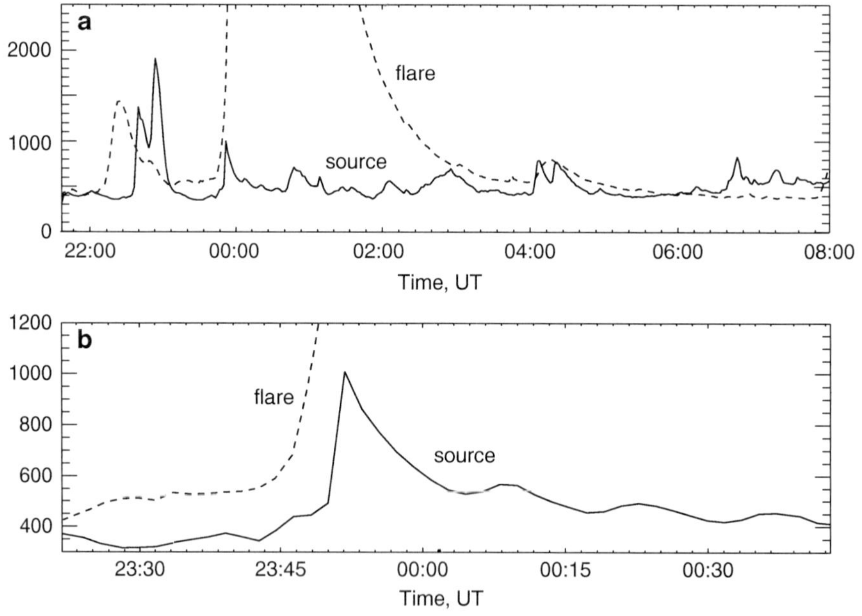

Fig. 3.30 Time profiles of brightness in the flare region R2–R4 (*dotted curve*) and in the remote core R1 (*solid line*)

while after the flare onset (dotted curve). The delay time for the second burst being $\sim$5 min and the distance between the cores $\sim 2 \cdot 10^5$ km, the propagation velocity of the disturbance from the flare center amounts to $\geq$650 km/s. The velocity may be higher if the flare disturbance propagates not straight but along an arc (for example, along magnetic lines). However, there are few presumptions of this kind because magnetosonic MHD disturbances can propagate in any direction with respect to the magnetic field, as opposed to the plasma that can flow only along the magnetic lines in an approaching strong field. The propagation velocity of the flare disturbance in the Sun corona estimated in this way agrees with the result achieved in [10], which analyzes coronal loop oscillations and demonstrates that a disturbance triggering such oscillations propagated radially from the flare center at a velocity of 700 km/s.

The propagating flare disturbance interacting with loop systems can cause loop oscillations at different periods [109]. Quasiperiodic intensity pulsations at an amplitude of 10–20% and a period of about 10 min were detected during a decay phase of a burst generated in the source R1 after a flare event. Natural plasma MHD oscillations excited after the flare inside the high-temperature sources are most probably observed in this case. Conditions necessary for such oscillations are specified in [81]. The observational manifestation of MHD oscillations of the plasma inside a source should include: (a) emission oscillations in the source and (b) lateral oscillations of the source that may be detected from the Doppler shift of spectral lines.

3.6 Possible Mechanisms of Heating Compact Hot Sources in the Corona

Two energy sources are most significant for heating the flare plasma in the solar corona: accelerated particles produced during the impulsive flare phase and shock waves effectively compressing and heating the plasma downstream of the shock front. The major mechanism is generally considered to be the heating by energetic particles, as a result of which the plasma temperature can increase to anomalously high values of 30 and even 100 million K. Nevertheless, the observed variety of flares might eliminate the possibility of their interpretation by a single mechanism. In addition, minor flares do not actually contain a nonthermal emission component related to energetic particles. Such events should obviously be interpreted using other heating mechanisms.

Both heating sources, energetic particles and shock waves, provide similar results of X-ray observations: compact emission sources hotter than the surrounding plasma are formed in the corona. A peculiarity of heating by fast particles may be an intensive nonthermal emission generated by particles decelerating in the plasma. However, the presence of this component in the spectrum should not be considered as a deciding argument, because, although the compressed plasma downstream of the shock front does not generate nonthermal emission, it can function as a target for the energetic particles producing such emission.

Additional information on a probable heating mechanism for the flare plasma in a specific event can be obtained by analyzing the temporal behavior of the plasma temperature and emission measure.

The analysis based on a developed multitemperature flare model shows that in the majority of events the temperature rises rapidly at the burst onset; the emission measure of the hottest component is the first to reach a maximum (Fig. 3.4). The emission measure of the hot component then begins to decrease, while the total emission measure mainly contributed to by the low-temperature plasma increases for some more time. The emission measure of the low-temperature component reaches a maximum with a delay of from 3–5 to 25–30 min for impulsive and long-decay flares, respectively. This resembles cooling of the hot plasma through heating of the adjacent cold regions. A similar conclusion was drawn in an article [87], which analyzes the DEM at different flare phases and shows that the amount of the hot (~10 MK) and cold (~1–1.5 MK) plasma reduces in the course of development, while that of the medium-temperature plasma (~2 MK) increases (Fig. 3.31). The phenomena of heat-conductive cooling therefore play a significant role in flares and are important for correct interpretation of the high-temperature coronal plasma dynamics.

According to the study, the dynamics of the emission measure for the components with different temperatures cannot be explained merely by heating and heat exchange processes [42]. In particular, as the density of the hot coronal plasma in IEs exceeds 10^{11} cm^{-3}, causes of the increased density of the matter in the emission region should also be searched for. Chromospheric evaporation, that is, a gas-dynamic rise of the plasma from the chromosphere heated by flare electrons, is

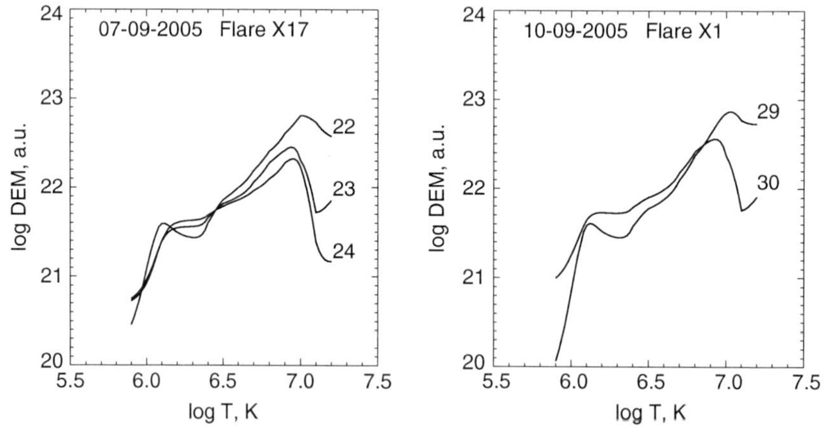

Fig. 3.31 Dynamics of DEM of two X-class flares, observation were made with $\Delta t = 1.5$ h. The DEM was calculated based on EUV spectra in the 280–330 Å spectral range observed by SPIRIT/CORONAS-F

often considered as a mechanism of supplying the matter to the corona. At the same time, the evaporated matter should be located *inside* flare loops, filling them rather uniformly; this contradicts observations of the present paper and other observations suggesting that high-temperature emission sources are formed *above* flare loops and have a dense core. From this viewpoint, fast shock waves as a heating source for the high-temperature plasma in flares have an important advantage over energetic particles: during heating, they substantially compress the gas downstream of the shock front and in addition continuously supply new matter to the high-temperature emission source [117].

An important argument in favor of shock waves heating the flare coronal plasma is the rise of a high-temperature source at a velocity of about 10 km/s during a flare observed in the Mg XII images for IEs [66]. This motion can be naturally interpreted in terms of the shock model as an upward displacement of the discontinuity surface that separates the hot region from the "undisturbed" plasma flow. The continuity condition $nv = \text{const}$ must be satisfied for the discontinuity surface, so the density of the matter compressed downstream of the shock front can be estimated from the ratio of the velocity of the supersonic plasma flowing from the reconnection region ($\sim$1,000 km/s) and the velocity of the front ($\sim$10 km/s). We thus determine that the plasma of a normal coronal density of $n{\sim}10^9\,\text{cm}^{-3}$ will be compressed down to $\sim 10^{11}\,\text{cm}^{-3}$, which agrees with the above densities estimated on the basis of observations.

The conclusion about the significance of heat-conductive cooling for flares agrees with theoretical results obtained in [14, 93] when solving a system of continuity equations for a shock front under coronal conditions. In adiabatic representation, the plasma heating by a shock wave generates hot regions in the corona with a temperature of more than 100 MK that propagate upward at a velocity of 1,000 km/s, which disagrees significantly both with the observations and with conclusions related to the flare plasma temperature content made in Sect. 3.2.

Theoretical results may be brought into agreement with the observations if heat-conductive cooling is taken into account. The difference between the shock front velocities in the adiabatic and nonadiabatic approximations is particularly convincing. In the former case, it is about 1,000 km/s, that is, two orders of magnitude higher than the observed one, while after the heat conduction effects in the plasma have been taken into account, it decreases to the observed values. Note that the plasma heating by electrons may play a significant role in forming large-scale high-temperature regions having long lifetimes and coronal densities of $\sim 2 \cdot 10^9\,\text{cm}^{-3}$ described in this paper. Such heating mode, however, should considerably differ from that of the dense chromospheric plasma, which is a thick target for accelerated electrons of any energy. With respect to fast electrons, coronal emission sources are a thin target whose heating efficiency depends to a great extent on its density and size, as well as on the electron energy.

A conclusion can thus be made about the volumetric plasma heating by shock waves during flare events.

3.7 Conclusions

We observed new type of long-life high-temperature plasma sources in the solar corona which are placed at height in the corona up to $0.3R_o$ and have the temperature of ~10 MK. For this sources as well as for flares, we determined DEM, spatial distribution of electron temperature and density which show substantial input of plasma with temperature 2–10 MK in energy budget of flares. 10-MK plasma which is observed at high altitudes in corona is concentrated above the coronal loops and shows cooling time from hours to days. Probably, it is heated permanently by magnetic reconnection in lower layers of corona. In these hot plasma sources, we observed oscillations with periods from 5 to 150 min. The differences in the power spectra of these oscillations indicate on the different mechanisms of plasma heating in these processes and allow to determine the type of the observed plasma source. Long-period oscillations could indicate own resonance frequencies of active region as a system of magnetic loops. We measured the speed of distribution of disturbance during flare events in the hot plasma source and made a conclusion about possible mechanism volumetric plasma heating by shock waves.

Acknowledgements We are grateful to F. F. Goryaev, S. N. Oparin from LPI, and V. V. Grechnev for helpful remarks and discussions. This work was supported in part by the Russian Foundation for Basic Research (project 11–02–01079-a), the Basic Research Program of the Presidium of the Russian Academy of Sciences (Program no. 16, Part 3), the "Plasma Processes in the Solar System," by grants of the President of Russian Federation (MK-3875.2011.2, MD-5510.2011.2), and grants no. 218816 (the SOTERIA project, http://soteria-space.eu) and no. 284461 (eHEROES) of the Seventh Framework Program of the European Union (FP07/2007–2013).

References

1. L.W. Acton, M.L. Finch, C.W. Gilbreth, J.L. Culhane, R.D. Bentley, J.A. Bowles, P. Guttridge, A.H. Gabriel, J.G. Firth, R.W. Hayes, Sol. Phys. **65**, 53 (1980)
2. L.A. Akimov, S.A. Beletskii, I.L. Belkina, O.I. Bugaenko, Y.I. Velikodskii, I.A. Zhitnik, A.P. Ignat'ev, V.V. Korokhin, S.V. Kuzin, G.P. Marchenko, A.A. Pertsov, Astron. Rep. **49**, 579 (2005)
3. L.A. Akimov, C.A. Beletski, I.L. Belkina, O.I. Bugaenko, Yu.I. Velikodski, I.A. Zhitnik, A.P. Ignat'ev, V.V. Korohin, S.V. Kuzin et al., Astronomicheskij zhurnal **49**, 579 (2005), *in russian*
4. L.A. Akimov, I.L. Belkina, S.V. Kuzin, A.A. Pertsov, I.A. Zhitnik, Astron. Lett. **34**, 851 (2008)
5. L.A. Akimov, I.L. Belikina, S.V. Kuzin et al., Pis'ma v Astronomicheskij zhurnal **34**, 1 (2008), *in russian*
6. S.S. Andreev, S.Yu. Zuev, E.B. Kluenkov, A.Ya. Lopatin, V.I. Luchin, K.A. Prohorov, N.N. Suslov, L.A. Salawenko, Poverkhnost'. Rentgenovskie, Sinkhrotronnye i Neitronnye Issledovaniya **2**, 8 (2003), *in russian*
7. M.J. Aschwanden, in *NATO Adv and ed Research Workshops*, 851 (2008)
8. M.J. Aschwanden, Springer (2008)
9. M.J. Aschwanden, J. Lim, D.E. Gary, J.A. Klimchuk, Astrophys. J. **454**, 512 (1995)

10. M.J. Aschwanden, L. Fletcher, C. Schrijver, D. Alexander, The Astrophys. J. **520**, 880 (1999)
11. W.E. Behring, Leonard Cohen, U. Feldman, G. A. Doschek, Astrophys. J. **203**, 521 (1976)
12. I.L. Bejgman, S.A. Bozhenkov, I.A. Zhitnik, S.V. Kuzin, I.Yu. Tolstihina, A.M. Urnov, Pis'ma v astron. zhurn. **31**, 39 (2005), *in russian*
13. A.M. Binello, H.E. Mason, P.J. Storey, Astron. Astrophys Suppl. **127**, 545 (1998)
14. S.A. Bogachev, B.V. Somov, S.O. Masuda, Astron. Lett. **24**, 543 (1998)
15. S.A. Bogachev, B.V. Somov, S.O. Masuda, Pis'ma v Astronomicheskij zhurnal **24**, 631 (1998), *in russian*
16. S.A. Bogachev, S.V. Kuzin, I.A. Zhitnik et al., Astronomicheskij Vestnik **39**, 571 (2005), *in russian*
17. S.A. Bogachev, S.V. Kuzin, I.A. Zhitnik, A.M. Urnov, V.V. Grechnev, Sol. Syst. Res. **39**, 508 (2005)
18. S.A. Bogachev, S.V. Kuzin, A.A. Pertsov, M.S. Zykov, Sol. Syst. Res. **44**, 166 (2010)
19. V.N. Borovik, V.V. Grechnev, O.I.etal. Bugaenko, in *Proceedings of IAU Symposium 226 on Coronal and Stellar Mass Ejections*, 13–17 September, Beijing, China, 508 (2005)
20. P. Brekke, W.T. Thompson, T.N. Woods, F.G. Eparvier, Astrophys. J. **536**, 959 (2000)
21. N.S. Brickhouse, J.C. Raymond, Astrophys. J. Suppl. **97**, 551 (1995)
22. J.W. Brosius, J.M. Davila, R.J. Thomas, B.C. Monsignori-Fossi, Astrophys. J. Suppl. **106**, 143 (1996)
23. J.W. Brosius, J.M. Davila, R.J. Thomas, Astrophys. J. Suppl. **119**, 255 (1998)
24. J.W. Brosius, R.J. Thomas, J.M. Davila, E. Landi, Astrophys. J. **543**, 1016 (2000)
25. I.J.D. Craig, J.C. Brown, Astron. Astrophys. **49**, 239 (1976)
26. G. Del Zanna, Astron. Astrophys. **481**, L69 (2008)
27. G. Del Zanna, H.E. Mason, Astron. Astrophys. **433**, 731 (2005)
28. J.-P. Delaboudinière, G.E. Artzner, J. Brunaud, A.H. Gabriel, J.F. Hochedez, F. Millier, X.Y. Song, B. Au, K.P. Dere, R.A. Howard, R. Kreplin, D.J. Michels, J.D. Moses, J.M. Defise, C. Jamar, P. Rochus, J.P. Chauvineau, J.P. Marioge, R.C. Catura, J.R. Lemen, L. Shing, R.A. Stern, J.B. Gurman, W.M. Neupert, A. Maucherat, F. Clette, P. Cugnon, E.L. van Dessel, Sol. Phys. **162**, 291 (1995)
29. K.P. Dere, Astrophys. J. **221**, 1062 (1978)
30. K.P. Dere, K.G. Widing, H.E. Mason, A.K. Bhatia, Astrophys. J. Suppl. **40**, 341 (1979)
31. K.P. Dere, E. Landi, H.E. Mason, B.C.M. Fossi, P.R. Young, Astron. Astrophys. Suppl. **125**, 149 (1997)
32. G.A. Doschek, in *Extreme Ultraviolet Astronomy*, 341 (1979)
33. G.A. Doschek, Astrophys. J. **527**, 426 (1999)
34. P.M. Edwin, B. Roberts, Sol. Phys. **76**, 239 (1982)
35. Yu.I. Ermolaev, L.M. Zelenyj, G.N. Zastenker, A.A. Petrukovich, I.G. Mitrofanov, M.L. Litvak, I.S. Veselovskij, M.I. Panasuk, L.L. Lazutin, A.V. Dmitriev, A.V. Zhukov, S.N. Kuznetsov, I.N. Magkova, B.Yu. Yushkov, V.G. Kurt, A.A. Gnezdilov, R.V. Gorgutsa, A.K. Markeev, D.E. Sobolev, V.V. Fomichev, V.D. Kuznetsov, S.I. Boldyrev, I.M. Chertok, K.A. Boyarchuk, I.V. Krasheninnikov, O.P. Kolomijtsev, L.N. Lestchenko, A.V. Belov, S.P. Gajdash, H.D. Kanonidi, S.A. Bogachev, I.A. Zhitnik, A.P. Ignat'ev, S.V. Kuzin, S.N. Oparin, A.A. Pertsov, V.A. Slemzin, N.K. Suhodrev, S.V. Shestov, V.I. Vlasov, I.V. Chashey, E.V. Vashenyuk, Ya.A. Saharov, A.N. Danilin, V.M. Bogod, S.H. Tohchukova, A.V. Mihalev, A.B. Beletski, N.V. Kostyleva, M.A. Chernigovskaya, V.V. Grechnev, K. Kudela, Geomagnetizm i Ajeronomija **1**, 23 (2005), *in russian*
36. U. Feldman, P. Mandelbaum, J.F. Seely, G.A. Doschek, H. Gursky, Astrophys. J. Suppl. **81**, 387 (1992)
37. A. Fludra, J. Sylwester, Sol. Phys. **105**, 323 (1986)
38. K.V. Getman, M.A. Livshits, Astron. Rep. **44**, 255 (2000)
39. E. Gibson, M.: Mir (1998), *in russian*
40. L. Golub, J.M. Pasachoff, Cambridge University Press (2000)
41. V.V. Grechnev, I.M. Chertok, V.A. Slemzin, S.V.etal. Kuzin, J. Geophys. Res. **110**, A9 (2005)

42. V.V. Grechnev, A.M. Uralov, V.G. Zandanov, G.V. Rudenko, V.N. Borovik, I.Y. Grigorieva, V.A. Slemzin, V.A. Bogachev, S.V. Kuzin, Publ. Astron. Soc. Jpn. **58**, 55 (2006)
43. C.J. Greer, Ir. Astron. J. **22**, 197 (1995)
44. Y.I. Grineva, V.I. Karev, V.V. Korneev, V.V. Krutov, S.L. Mandelstam, L.A. Vainstein, B.N. Vasilyev, I.A. Zhitnik, Sol. Phys. **29**, 441 (1973)
45. M.L. Kaiser, Adv. Space Res. **36**, 1483 (2005)
46. T. Kato, Physica Scripta Volume T **73**, 98 (1997)
47. F.P. Keenan, R.J. Thomas, W.M. Neupert, V.J. Foster, P.J.F. Brown, S.S. Tayal, Mon. Not. R. Astron. Soc. **278**, 773 (1996)
48. F.P. Keenan, K.M. Aggarwal, R.S.I. Ryans, R.O. Milligan, D.S. Bloomfield, J.W. Brosius, J.M. Davila, R.J. Thomas, Astrophys. J. **624**, 428 (2005)
49. F.P. Keenan, D.B. Jess, K.M. Aggarwal, R.J. Thomas, J.W. Brosius, J.M. Davila, Mon. Not. R. Astron. Soc. **376**, 205 (2007)
50. A. Kepa, B. Sylwester, M. Siarkowski, J. Sylwester, Adv. Space Res. **42**, 828 (2008)
51. J.A. Klimchuk, Sol. Phys. **193**, 53 (2000)
52. N.N. Kolachevskij, A.S. Pirozhkov, E.N. Ragozin, Kvantovaja Jelektronika **25**, 843 (1998), *in russian*
53. T. Kosugi, K. Matsuzaki, T. Sakao, T. Shimizu, Y. Sone, S. Tachikawa, T. Hashimoto, K. Minesugi, A. Ohnishi, T. Yamada, S. Tsuneta, H. Hara, K. Ichimoto, Y. Suematsu, M. Shimojo, T. Watanabe, S. Shimada, J.M. Davis, L.D. Hill, J.K. Owens, A.M. Title, J.L. Culhane, L.K. Harra, G.A. Doschek, L. Golub, Sol. Phys. **243**, 3 (2007)
54. S.V. Kuzin, S.V. Shestov, A.A. Pertsov, A.A. Reva, S.Yu. Zuev, A.Ya. Lopatin, V.I. Luchin, H. Zhou, T. Huo, Poverhnost'. Rentgenovskie, sinhrotronnye i nejtronnye issledovanija **7**, 1 (2008), *in russian*
55. S.V. Kuzin, S.A. Bogachev, I.A. Zhitnik, A.A. Pertsov, A.P. Ignatiev, A.M. Mitrofanov, V.A. Slemzin, S.V. Shestov, N.K. Sukhodrev, O.I. Bugaenko, Adv. Space Res. **43**, 1001 (2009)
56. S.V. Kuzin, S.A. Bogachev, I.A. Zhitnik, S.V. Shestov, V.A. Slemzin, A.V. Mitrofanov, N.K. Suhodrev, A.A. Pertsov, A.P. Ignat'ev, O.I. Bugaenko, Yu.S. Ivanov, A.A. Reva, M.S. Zykov, A.S. Ul'janov, S.N. Oparin, A.L. Goncharov, T.A. Shergina, A.M. Urnov, V.A. Solov'ev, S.G. Popova, Izvestnija RAN. Serija fizicheskaja **1**, v pechati (2009), *in russian*
57. E. Landi, Astron. Astrophys. **382**, 1106 (2002)
58. E. Landi, U. Feldman, Astrophys. J. **672**, 674 (2008)
59. E. Landi, M. Landini, Astron. Astrophys. **340**, 265 (1998)
60. E. Landi, G. Del Zanna, P.R. Young, K.P. Dere, H.E. Mason, M. Landini, Astrophys. J. Suppl. **162**, 261 (2006)
61. J. Lang, D.H. Brooks, A.C. Lanzafame, R. Martin, C.D. Pike, W.T. Thompson, Astron. Astrophys. **463**, 339 (2007)
62. R.P. Lin, B.R. Dennis, A.O. Benz, Dordrecht, Kluwer (2009)
63. M. Malinovsky, L. Heroux, Astrophys. J. **181**, 1009 (1973)
64. S.L. Mandel'shtam, A.I. Efimov, Uspehi fizicheskih nauk **LXIII**, 163 (1957), *in russian*
65. H.E. Mason, B.C.M. Fossi, Astron. Astrophys. Rev. **6**, 123 (1994)
66. S. Masuda, in *Multi-Wavelength Observations of Coronal Structure and Dynamics*, 123 (1994)
67. P. Mazzotta, G. Mazzitelli, S. Colafrancesco, N. Vittorio, Astron. Astrophys. Supplement **133**, 403 (1998)
68. S.W. McIntosh, J.C. Brown, P.G. Judge, Astron. Astrophys. **333**, 333 (1998)
69. J.M. McTiernan, G.H. Fisher, P. Li, Astrophys. J. **514**, 472 (1999)
70. B.C. Monsignori-Fossi, M. Landini, Adv. Space Res. **11**, 281 (1991)
71. W.M. Neupert, S.O. Kastner, Astron. Astrophys. **128**, 181 (1983)
72. W.M. Neupert, G.L. Epstein, R.J. Thomas, W.T. Thompson, Sol. Phys. **137**, 87 (1992)
73. Y. Ogawara, L.W. Acton, R.D. Bentley, M.E. Bruner, J.L. Culhane, E. Hiei, T. Hirayama, H.S. Hudson, T. Kosugi, J.R. Lemen, K.T. Strong, S. Tsuneta, Y. Uchida, T. Watanabe, M. Yoshimori, Publ. Astron. Soc. Jpn. **44**, L41 (1992)
74. V.N. Oraevskij, I.I. Sobel'man, Pis'ma v astron. zhurn. **28**, 457 (2002), *in russian*

75. B.E. Patchett, R.A. Harrison, E.C. Sawyer, J. Br. Interplanet. Soc. **43**, 181 (1990)
76. A.A. Pertsov, I.A. Zhitnik, in *Proc. of "Modern problems of Physics of Solar and Stellar activity", N.Novgorod, 2–7 June 2003*, 181 (1990)
77. A.A. Pertsov, I.A. Zhitnik, in *Trudy konferencii "Aktual'nye problemy fiziki solnechnoj i zvezdnoj aktivnosti, N. Novgorod*, 457 (2002), *in russian*
78. A. Pertsov, I. Zhitnik, O. Bougaenko, A. Ignatiev, V. Krutov, S. Kuzin, A. Mitrofanov, S. Oparin, V. Slemzin, in *ESA Special Publication*, 181 (1990)
79. K.J.H. Phillips, Cambridge University Press (1964)
80. S.R. Pottasch, Space Sci. Rev. **3**, 816 (1964)
81. B. Robert, P.M. Edwin, A.O. Benz, Astrophys. J. **279**, 857 (1984)
82. P. Saint-Hilaire, A.O. Benz, Sol. Phys. **210**, 287 (2002)
83. J.T. Schmelz, J.L.R. Saba, K.T. Strong, H.D. Winter, J.W. Brosius, Astrophys. J. **523**, 432 (1999)
84. J.T. Schmelz, J. Scott, L.A. Rightmire, Astrophys. J. Lett. **684**, L115 (2008)
85. S.V. Shestov, S.A. Bozhenkov, I.A. Zhitnik, S.V. Kuzin, A.M. Urnov, I.L. Bejgman, F.F. Gorjaev, I.Yu. Tolstihina, Pis'ma v Astron. zhurn. **34**, 38 (2008), *in russian*
86. S.V. Shestov, A.M. Urnov, S.V. Kuzin, I.A. Zhitnik, S.A. Bogachev, Pis'ma v Astron. zhurn. **35**, 50 (2009), *in russian*
87. S.V. Shestov, S.V. Kuzin, A.M. Urnov, A.S. Ulyanov, S.A. Bogachev, Astron. Lett. **36**, 1 (2010)
88. S.V. Shestov, S.V. Kuzin, A.M. Urnov, A.S. Ul'janov, S.A. Bogachev, Pis'ma v Astron. zhurn. **36**, 1 (2010), *in russian*
89. K. Shibasaki, Astrophys. J. **556**, 326 (2001)
90. K. Shibasaki, in *Yohkoh 8th Anniv. Symposium*, 326 (2001)
91. I.C. Shklovski, M.: Fizmatgiz (2009), *in russian*
92. I.I. Sobel'Man, I.A. Zhitnik, in *Society of Photo-Optical Instrumentation Engineers (SPIE) Conference Series*, 326 (2001)
93. B.V. Somov, T. Kosugi, S.A. Bogachev, S. Masuda, T. Sakao, Adv. Space Res. **35**, 1690 (2005)
94. X. Song, S. Jing-hong, L. Qi-peng, W. Zhao-lan, X. Zhong, Optic. Precis. Eng. **12**, 480 (2004)
95. P. Song, H.J. Singer, G.L. Siscoe, Am. Geophys. Union (2009)
96. A.V. Stepanov, Y.G. Kopylova, Y.T. Tsap, Cosmic Res. **46**, 294 (2008)
97. A.V. Stepanov, Yu.G. Kopylova, Yu.T. Cap, Kosmicheskie issledovanija **46**, 303 (2008), *in russian*
98. P.J. Storey, G. Del Zanna, H.E. Mason, C.J. Zeippen, Astron. Astrophys. **433**, 717 (2005)
99. K. Strong, M. Bruner, T. Tarbell, A. Title, C.J. Wolfson, Space Sci. Rev. **70**, 119 (1994)
100. J. Sylwester, I. Gaicki, Z. Kordylewski, M. Nowak, S. Kowalinski, M. Sjarkowski, W. Bentley, R.D. Trzebinski, M.W. Whyndham, P.R. Guttridge, J.L. Culhane, J. Lang, K.J.H. Phillips, C.M. Brown, G.A. Doschek, V.N. Oraevsky, S.I. Boldyrev, I.M. Kopaev, A.I. Stepanov, V.Y. Klepikov, in *Crossroads for European Solar and Heliospheric Physics. Recent Achievements and Future Mission Possibilities*, 119 (1994)
101. R.J. Thomas, W.M. Neupert, Astrophys. J. Suppl. **91**, 461 (1994)
102. R. Tousey, J.-D.F. Bartoe, G.E. Brueckner, J.D. Purcell, Appl. Optic. **16**, 870 (1977)
103. S. Tsuneta, L. Acton, M. Bruner, J. Lemen, W. Brown, R. Caravalho, R. Catura, S. Freeland, B. Jurcevich, J. Owens, Sol. Phys. **136**, 37 (1991)
104. A.M. Urnov, S.V. Shestov, S.A. Bogachev, F.F. Goryaev, I.A. Zhitnik, S.V. Kuzin, Astron. Lett. **33**, 396 (2007)
105. A.M. Urnov, S.V. Shestov, S.A. Bogachev, F.F. Goryaev, I.A. Zhitnik, S.V. Kuzin, Pis'ma v Astron. zhurn. **33**, 446 (2007), *in russian*
106. Y.I. Vitinskij, M. Kopetski, G.V. Kuklin, M.: Nauka (1998), *in russian*
107. Y.I. Vitinskij, M. Kopetskij, G.V. Kuklin, Moscow: Nauka (2007)
108. T.J. Wang, in *Proceedings of Chromospheric and Coronal Magnetic Fields, Katlenburg-Lindau, Germany*, 396 (2007)

109. T.J. Wang, S.K. Solanki, W. Curdt, D.E. Innes, I.E. Dammasch, Astrophys. J. 574, L101 (2002)
110. T.J. Wang, S.K. Solanki, W. Curdt, D.E. Innes, I.E. Dammasch, in *Proceedings European Conference and IAU Colloquia, Santorini, Greece*, L101 (2002)
111. T.J. Wang, S.K. Solanki, D.E. Innes et al., Astron. Astrophys. **402**, L17 (2003)
112. H.P. Warren, A.D. Warshall, Astrophys. J. **571**, 999 (2002)
113. S.M. White, R.J. Thomas, R.A. Schwartz, Sol. Phys. **227**, 231 (2005)
114. I. Zhitnik, S. Kuzin, A. Afanas'ev, O. Bugaenko, A. Ignat'ev, V. Krutov, A. Mitrofanov, S. Oparin, A. Pertsov, V. Slemzin, N. Sukhodrev, A. Umov, Adv. Space Res. **32**, 473 (2003)
115. I. Zhitnik, S. Kuzin, O. Bugaenko, A. Ignat'ev, V. Krutov, D. Lisin, A. Mitrofanov, S. Oparin, A. Pertsov, V. Slemzin, A. Urnov, Adv. Space Res. **32**, 2573 (2003)
116. I.A. Zhitnik, O.I. Bugaenko, A.P. Ignat'ev, V.V. Krutov, S.V. Kuzin, A.V. Mitrofanov, S.N. Oparin, A.A. Pertsov, V.A. Slemzin, A.I. Stepanov, A.M. Urnov, Mon. Not. R. Astron. Soc. **338**, 67 (2003)
117. I.A. Zhitnik, S.V. Kuzin, S.A. Bogachev, owner=sshestov et al., (Fizmatlit, Moscow, 2005)
118. I.A. Zhitnik, S.V. Kuzin, I.I. Sobel'man, O.I. Bugaenko, A.P. Ignat'ev, A.V. Mitrofanov, S.N. Oparin, A.A. Pertsov, V.A. Slemzin, N.K. Suhodrev, A.M. Urnov, Astronomicheskij Vestnik **39**, 495 (2005), *in russian*
119. I.A. Zhitnik, O.I. Bougaenko, J.-P. Delaboudiniere, A.P. Ignatiev, V.V. Korneev, V.V. Krutov, S.V. Kuzin, D.V. Lisin, A.V. Mitrofanov, S.N. Oparin, V.N. Oraevsky, A.A. Pertsov, V.A. Slemzin, I.I. Sobelman, A.I. Stepanov, J. Schwarz, in *ESA Special Publication*, 231 (2005)
120. I.A. Zhitnik, A.M. Urnov, in *Lecture Notes in Physics* (Springer, Berlin, 2006), p. 272
121. I.A. Zhitnik, S.V. Kuzin, A.M. Urnov, S.A. Bogachev, F.F. Goryaev, S.V. Shestov, Sol. Syst. Res. **40**, 272 (2006)
122. I.A. Zhitnik, S.V. Kuzin, A.M. Urnov, S.A. Bogachev, F.F. Goryaev, S.V. Shestov, Astronomicheskij Vestnik **40**, 299 (2006), *in russian*
123. I.A. Zhitnik, S.V. Kuzin, S.A. Bogachev et al., Fizmatlit (2006), *in russian*
124. H. Zhou, J. Zheng, T. Huo, G. Zhang, Z. Qi, P. Zhong, Optic. Precis. Eng. **12**, 480 (2004)
125. S.Yu. Zuev, A.V. Mitrofanov, Poverhnost'. Rentgenovskie, sinhrotronnye i nejtronnye issledovanija **1**, 81 (2002), *in russian*
126. S.Yu. Zuev, S.V. Kuzin, A.Ya. Lopatin, V.I. Luchin, V.N. Polkovnikov, N.N. Salastchenko, L.A. Suslov, N.N. Tsybin, S.V. Shestov, in *Materialy XIII Mezhdunarodnogo simpoziuma "Nanofizika i nanojelektronika"*, 299 (2006), *in russian*

Chapter 4
Populations of Excited Parabolic States of Hydrogen Beam in Fusion Plasmas

O. Marchuk and Yu. Ralchenko

Abstract The injection of a neutral beam into the core of magnetically confined plasmas is a foundation for various plasma diagnostics, including charge-exchange recombination spectroscopy, beam emission spectroscopy, and motional Stark effect diagnostics. We review the current status of statistical and nonstatistical collisional-radiative models used for calculation of populations of hydrogen beam excited states. The recently developed collisional-radiative (CR) model in parabolic states, which utilizes collisional data calculated in the Glauber approximation, is discussed in detail. CR simulations with this model show nonstatistical emission for the σ and π components of the H_α Stark multiplet and provide m-resolved populations. The results of simulations are in excellent agreement with the systematic motional Stark effect (MSE) measurements from the JET tokamak.

4.1 Neutral Beam Diagnostics for Fusion Plasmas

Efficient heating of plasma in fusion devices is one of the major problems in plasma physics. The injection of high-energy neutral beam (hydrogen or deuterium) into the confined plasma volume was found to be one of the most powerful tools to solve this problem [1]. There are practically no fusion devices, such as tokamaks and stellarators, which are not equipped with neutralbeam injectors. The beam particles penetrating the plasma volume are quickly ionized, and the resulting fast ions provide effective transfer of energy and momentum to the plasma ions and electrons through binary collisions.

During the last few decades, important advances in atomic physics resulted in development of various spectroscopic diagnostics based on neutral beam injection (NBI) [2]. In the modern fusion devices, the most important diagnostic methods include charge-exchange recombination spectroscopy (CXRS) [3,4], motional Stark effect (MSE) [5], beam emission spectroscopy (BES) [6], and fast-ion diagnostics (FIDA) [7, 8]. These techniques are used to infer such important parameters as space-resolved ion temperature and plasma rotation, concentration of impurities,

V. Shevelko and H. Tawara (eds.), *Atomic Processes in Basic and Applied Physics*, Springer Series on Atomic, Optical, and Plasma Physics 68,
DOI 10.1007/978-3-642-25569-4_4,

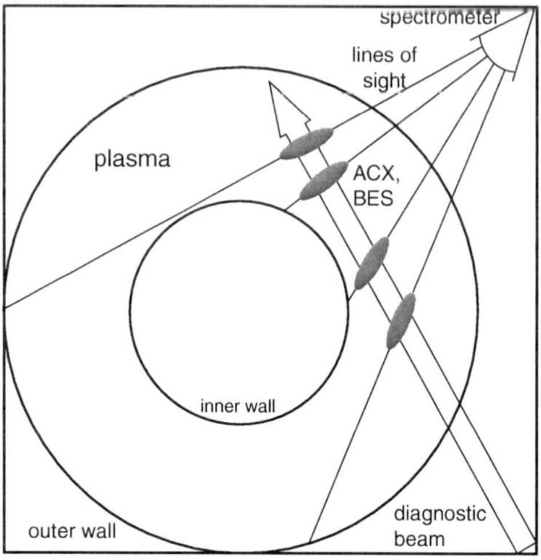

Fig. 4.1 Top view of injection of diagnostic beam into a tokamak plasma. Zones of active charge-exchange (ACX) and beam emission spectroscopy (BES) are shown as *gray spots*

effective charge of plasma, fuel ion ratio, intensity of electric field, and distribution function of fast ions along the plasma radius.

A schematic picture of NBI into a tokamak plasma is presented in Fig. 4.1. The diagnostic beam passes through the plasma volume which is magnetically confined between the outer and inner walls. The injection schemes vary between the radial and tangential types. The energy of the beam atoms entering plasma is typically in the range of 30–100 keV/u, and in many cases the heating beam is used for diagnostics as well. Plasma density along the beam path may vary between 10^{12}–$10^{14}\,\mathrm{cm}^{-3}$ and plasma temperature changes between a few eV at the last closed magnetic flux surface to a few keV in the plasma core. The applied magnetic field is mostly in the range of 1–5 T.

The beam penetration into a fusion plasma is accompanied by binary collisions with electrons, main plasma ions, and impurities. These processes are primarily responsible for emission of the photons observed by spectrometers at different lines of sight. The most important atomic processes leading to the emission of spectral lines used in a beam-based spectroscopy are listed below:

$$\mathrm{H}_0 + \{e, p, X^z\} \rightarrow \mathrm{H}^* + \{e, p, X^z\}, \tag{4.1}$$

$$\mathrm{H}_0 + X^z \rightarrow X^{(z-1)*} + p, \tag{4.2}$$

$$p + \mathrm{H}_0 \rightarrow \mathrm{H}^* + p. \tag{4.3}$$

Reaction (4.1) is the excitation of hydrogen atoms H_0 by electrons e, impurity ions X^z, and plasma protons p. The ensuing radiative stabilization of excited hydrogen atoms H^* results in emission of spectral lines, for example, H_α ($n = 2 - n = 3$) and H_β ($n = 2 - n = 4$), which are used for MSE and BES diagnostics. The charge-exchange reaction (4.2) is the process of electron capture to excited states of an impurity ion which is followed by emission of photons in visible and ultraviolet ranges of spectra. This process is used in the active CXRS [3]. The FIDA diagnostic is based on the capture of a beam atom electron by the fast protons in plasma as shown in reaction (4.3). As in the case of BES diagnostics, the usually observed spectrum is the H_α line emission. In this reaction, the fast protons are produced by ionization of beam atoms by collisions with plasma particles.

A quantitative description of emission of the H_α line in NBI diagnostics is rather complicated. In the rest frame of the beam atom, the bound electron experiences the translational electric field $\mathbf{F} = \mathbf{v} \times \mathbf{B}$, where $\mathbf{v}$ is the beam velocity and $\mathbf{B}$ is the magnetic field. The induced electric field on the order of 10^5 V/cm splits the zero-field spherical states into a multiplet of parabolic states characterized by the principal quantum number n, electric quantum number $k = n_1 - n_2$ and magnetic number m, with $n = n_1 + n_2 + m + 1$ [5, 9]. Below, we will often use notation (n, k, m) for parabolic states. The emitted spectral components of H_α are polarized according to the well-established selection rules [10], and spectral radiance depends also on the angle of observation. In addition, the existing neutral beam injectors utilizing positive ion sources produce three different energy components of atoms (H, H_2, H_3) simultaneously [11]. While the resulting H_α profile becomes rather complex, nonetheless unique information on ion components, plasma electric field, pitch angle, and other parameters can be derived from H_α analysis.

An example of a D_α line observed in the JET tokamak [12] is shown in Fig. 4.2. The nonshifted components of H_α and D_α lines originate at outer and inner walls

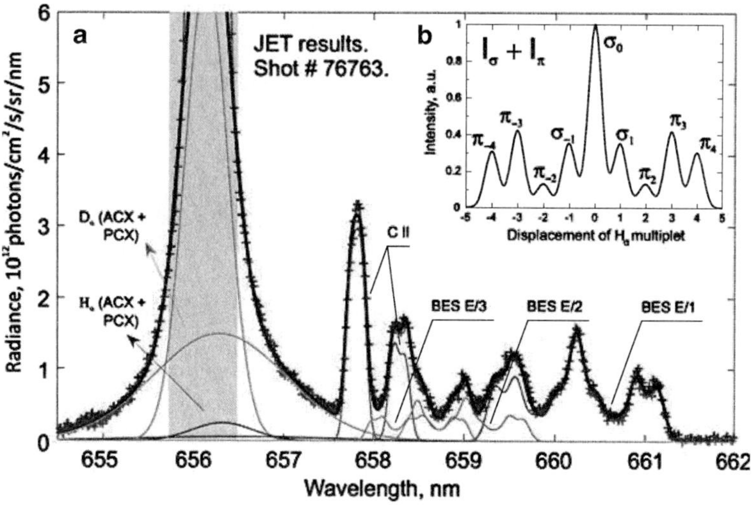

Fig. 4.2 Experimental data and fit of beam emission spectra measured from the tokamak JET [12]. The passive (PCX) and active (ACX) charge-exchange components of D_α and H_α lines are shown as *green* and *blue lines*, respectively. The Zeeman-split multiplet of C II edge lines (*red line*) is observed at 658 nm. The Doppler-shifted spectra of the first (*blue line*), second (*brown line*), and third (*green line*) energy beam components are located above 658 nm. The inset (**b**) shows the theoretical statistical intensities of σ and π components of H_α line multiplet

due to such processes as excitation of recycling neutrals by electrons and charge-exchange recombination of the main plasma ions on the same recycling neutrals and impurities. The Doppler-shifted H_α and D_α lines are the results of charge exchange between the beam atoms and ions in the plasma core. The H-to-D ion ratio in the plasma is obtained from comparing the intensities of H_α and D_α lines. The actual value of the Doppler shift of both lines depends on the plasma rotation speed and viewing geometry of the spectrometer. The widths of the lines are determined by the ion temperature that can be easily extracted from the CXRS spectra of impurities. The Zeeman-splitted C II transitions $1s^2 2s^2 3s - 1s^2 2s^2 3p$ appear as sharp peaks at 658 nm. These lines originate close to the plasma boundary in the region of maximum abundance of C II.

The Doppler-shifted H_α multiplets are also used for measurements of the electric field strength. The BES and MSE spectra generally show a very complex structure due to the presence of several beam components and numerous overlapping π ($\Delta m = 0$) and σ ($\Delta m = \pm 1$) radiative transitions [10]. It is not surprising therefore that the accurate description of MSE spectra is currently considered to be one of the most challenging tasks in active plasma spectroscopy. According to a general theory of radiation emission, the intensity of an observed multiplet can be represented as the sum of intensities of π and σ components I_π and I_σ:

$$I(\theta) = I_\pi \sin^2(\theta) + I_\sigma \left(1 + \cos^2(\theta)\right), \tag{4.4}$$

where θ is the angle between the direction of the electric field and the line of sight. The experimental spectrum between 659 and 661.5 nm (Fig. 4.2a) represents the MSE $\theta = 90°$ emission for the H_α multiplet. The calculated MSE spectrum with statistically distribution of state populations is shown in Fig. 4.2b. For statistical populations of magnetic levels of $n = 3$ states, the ratio between the intensities of σ and π components is well known and tabulated in the literature [10, 13]. For instance, the σ component ratio is $\sigma_1 : \sigma_0 = 0.1025 : 0.291$, and the π component ratios are $\pi_4 : \pi_3 : \pi_2 = 0.089 : 0.122 : 0.0385$. The measurements in Fig. 4.2a reveal considerable deviation from these values for all beam components. For instance, the ratio between the π_4 and π_3 lines is close to unity while the intensity of σ_1 is increased relative to the unshifted σ_0 component.

Such differences between statistical model predictions and experimental spectral patterns are observed practically on all existing fusion devices with NBI diagnostics thereby pointing out to a number of potential complications. Thus, derivation of plasma parameters would become less reliable for spectrometers with low resolution. Furthermore, unknown intensities of magnetic lines would significantly affect both the complexity of spectrum-fitting routines and the errors in derivation of the electric field strength. It is also important that nonstatistical emission of spectral lines would modify the overall emission of H_α which is expected to provide a calibration source for CXRS diagnostics in ITER plasma [14]. Finally, nonstatistical effects in calculated effective rate coefficients would strongly affect determination of such important parameters as, for example, concentration of fusion products, the knowledge of which is crucial for the control of ignition condition in a fusion plasma. All these issues call for development of nonstatistical models for description of magnetic line intensities emitted by the hydrogen beam in a strong electric field.

4.2 Statistical Models for Excited States of the Hydrogen Beam

The statistical approach to excited states of hydrogen implies that the populations of fine-structure states with the same principal quantum number n are proportional to their statistical weights. This is the most straightforward method to determine populations of excited states, beam emission rates, beam attenuation lengths, and some other parameters. The major advantage of statistical approach is the possibility to use widely available data for collisional processes connecting different n shells [15–19]. The first atomic models for excited states of hydrogen beam, which were developed with a primary interest in the beam decay length, unambiguously showed that the account of excited states is important for beam attenuation [20] and CXRS spectroscopy [21].

In spite of a relative simplicity of statistical approaches, a satisfactory agreement between various models was achieved only recently (see, e.g., [12]). One of the first comparisons of populations of the $n = 2$ and $n = 3$ excited states in a hydrogen beam was performed in [22], where considerable differences with the ADAS data [23] were found. For instance, populations of the $n = 3$ states differed by more than a factor of two. Later, a newly developed statistical model [24] confirmed the ADAS results based on the same set of elementary atomic data used. The reexamination of the ionization cross sections [25–28] for the excited $n = 2$ and $n = 3$ states completely resolved this outstanding discrepancy.

As shown in Fig. 4.3, the most frequently used statistical models [22–24] show excellent agreement for the density range of 10^{13}–$10^{14}\,cm^{-3}$. For low plasma densities populations of excited states increase linearly (coronal case). Then collisional ionization and (de)excitation prevent further growth of populations. The 20% agreement between statistical simulations in Fig. 4.3 is quite remarkable, and these models that were used as test beds for the available atomic data provided a solid foundation for development of nonstatistical simulations.

4.3 Non-statistical Description of Excited States of the Hydrogen Beam in Plasma

The statistical assumption for atomic levels is directly related to the local thermodynamic equilibrium (LTE) discussed in numerous papers and textbooks [29–32]. The LTE, or Boltzmann, population distribution is reached for a group of levels

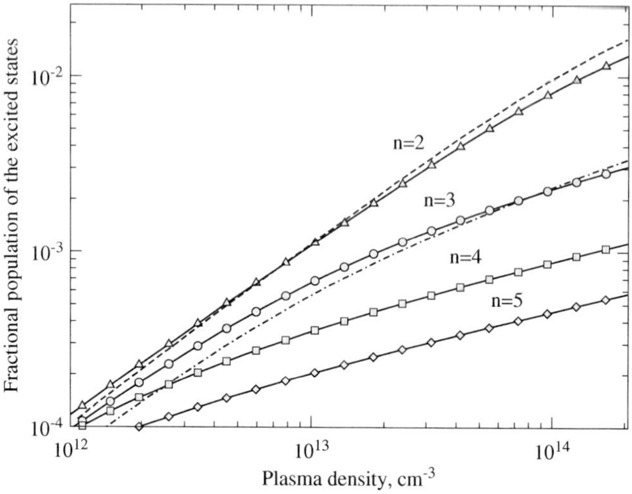

Fig. 4.3 Populations of excited states of hydrogen beam calculated under statistical assumption: $n = 2$ state [24] (*solid line with triangles*), [22] (*dashed line*), $n = 3$ state [24] (*solid line with circles*) [22] (*dashed point line*), $n = 4$ state [24] (*solid line with squares*), and $n = 5$ state [24] (*solid line with diamonds*). Energy of the beam is 40 keV/u, and plasma temperature is 2 keV (plasma temperature of 1 keV is taken in calculations of $n = 2$ excited state population in [22])

when their populations (say, N_j and N_k for levels j and k) are related as follows:

$$\frac{N_j}{N_k} = \frac{g_j}{g_k} \mathrm{e}^{-\Delta E_{jk}/T}, \tag{4.5}$$

where g_j and g_k are the corresponding statistical weights, ΔE_{jk} is the energy difference, and T is the particle temperature. For $\Delta E_{jk} \ll T$, which is the case for the hydrogen fine-structure states with the same n in hot fusion plasmas, the level populations are proportional to their g values. The LTE condition is established when collisional processes are much stronger than the radiative ones, that is

$$N_\mathrm{e} \cdot C_j^\mathrm{e} + \sum_i N_{Z_i} C_j^i \gg A_j, \tag{4.6}$$

and each direct collisional process is equilibrated by its inverse (e.g., excitation and deexcitation). In (4.6), N_e and C_j^e are the density of electrons and the total rate coefficient for electron collisions, N_{Z_i} and C_j^i are the same parameters for ions i with charge Z_i, and A_j is the total radiative decay rate for level j. It is well known [30] that due to the small threshold energies for transitions between fine structure levels within the same n the statistical distribution is established primarily through the $\Delta n = 0$ collisions with ions while electrons play a secondary role. It is clear that for a sufficiently low plasma density condition (4.6) cannot be fulfilled and statistical approach is not valid.

The LTE conditions for hydrogen atom in an induced electric field of tokamaks are more difficult to analyze. The true eigenstates for hydrogen in an electric field are the parabolic states [9] rather than the common spherical nl states. Their energies depend on the field strength and thus have to be recalculated for each simulation. As a result, the collisional cross sections between parabolic states are also to be generated for each set of plasma/beam parameters, and such data are not available in the literature. Besides, Stark splitting for high-n states can be comparable with the energy difference between the states with different n, and some of the levels may even plunge into continuum and thus disappear from the model.

The energy diagram for the $n = 5 - 8$ states of hydrogen as a function of the induced electric field at $B = 5\,\mathrm{T}$ is shown in Fig. 4.4. The calculations were performed up to the second-order Stark effect. The ranges of electric field for TEXTOR,

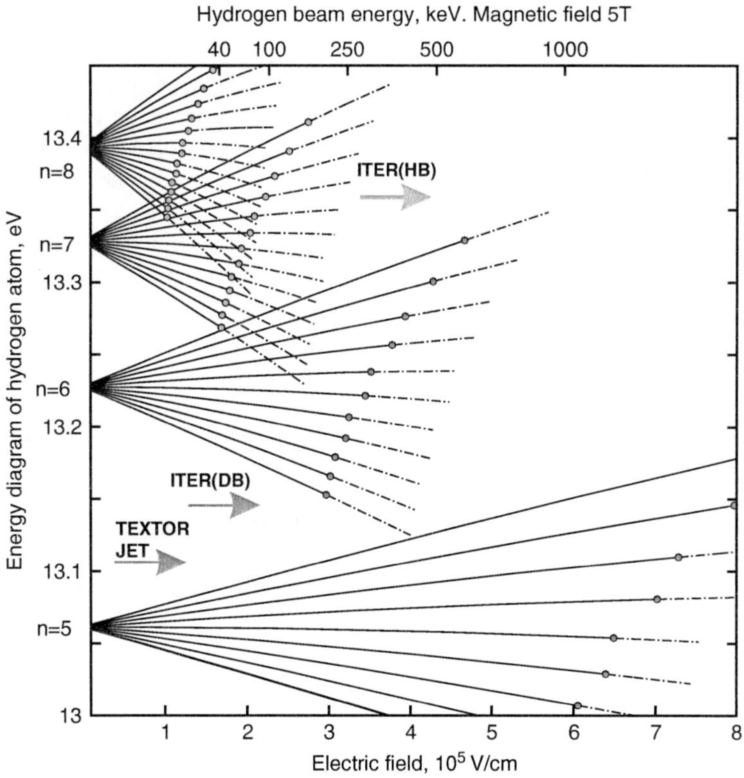

Fig. 4.4 Energy diagram for hydrogen $n = 5$–8 levels in an electric field (see also [33]). *Solid line* shows the result of second-order perturbation theory; points denote the field strength at which radiative rate equals to ionization rate [36]; *dashed-point line* denotes ionizing states. *Arrows* denote the conditions for TEXTOR, JET, and ITER diagnostic (DB) and heating beams (HB)

JET, and ITER (heating and diagnostic beams) tokamaks are shown by arrows. Even for the existing fusion devices such as TEXTOR and JET, the Stark effect results in level crossing for $n = 7$ and 8 which drastically affects populations of the corresponding states due to significant increase of the corresponding collisional cross sections. For the ITER heating beam, the level crossing occurs already for the states with $n = 5$ and $n = 6$. Another effect that strongly modifies the LTE criteria is the ionization of high-n states by the strong electric field [33–35]. Due to the exponential dependence on the width of the potential barrier, the ionization rates can differ by orders of magnitude for the blue and red components of the n-multiplet [36]. Since the ionized atoms immediately drift out of the beam volume, the unbalanced field ionization plays the same role as the radiative decay in (4.6), and therefore, stronger collisions are required to reach LTE. This in fact means that higher plasma densities are needed for LTE conditions under field ionization. These and other considerations highlight major problems in application of statistical methods to description of a hydrogen beam in fusion plasmas.

In response to systematic deviations of the observed MSE spectra from the statistical predictions, a few models for hydrogen atom in electric field were developed. The major objective here was a nonstatistical description of the relative intensities between the magnetic σ and π components. Currently, there exist three collisional-radiative (CR) models resolved over the Stark parabolic states. The first two m-models are based on the first Born approximation (FBA) collisional data [37, 38]. These two models again use statistical assumption for the high-n states. Although both models are based on the same approximations and include practically the same number of the m-resolved states, the results differ considerably. The latest model described in [39] was developed for m-resolved parabolic states using collisional data calculated in the Glauber approximation (GA) [40] for an arbitrary angle α between the electric field and incoming projectiles. This model showed a good agreement with the first measurements from the JET tokamak [6]. Below, we discuss this model in detail.

4.3.1 Collisional-Radiative Model in Parabolic States

The CR model for parabolic magnetic levels is rather similar to the common models for spherical states [29, 31]. The system of time-dependent rate equations is written as:

$$\frac{\mathrm{d}\hat{N}}{\mathrm{d}t} = \hat{A}\hat{N} \tag{4.7}$$

with $\hat{N}$ and $\hat{A}$ being the vector of parabolic state populations and the rate matrix, respectively. For one magnetic level, $j = n_j k_j m_j$ with population N_j the corresponding rate equation is:

$$\frac{dN_j}{dt} = \sum_i W_{ij} N_i - N_j \sum_i W_{ji} - N_j \left(S_j + \Gamma_j\right),$$
$$W_{ij} = N_e \left(C^e_{ij} + C^p_{ij}\right) + A_{ij}\Theta_{ij},$$
$$S_j = N_e \left(C^e_j + C^p_j\right), \tag{4.8}$$

where W_{ij} is the total rate of collisional and radiative processes connecting levels i and j, S_j is the collisional ionization rate including charge-exchange losses, Γ_j is the probability of ionization due to the electric field, C^e_{ij} and C^p_{ij} are the (de)excitation collisional rate coefficients due to electrons and protons, C^e_j and C^p_j are the corresponding ionization rate coefficients, and A_{ij} is the radiative decay rate. Θ_{ij} is the Heaviside step function which is 1 if $E_i > E_j$ (E_j is the energy of the level j) and 0 otherwise. Here, one assumes also a one component hydrogen plasma. The energy levels and oscillator strengths in the weak field approximation for parabolic states are well known [10, 41]. A good agreement was also obtained for energy levels and ionization rates in strong electric fields [33, 34], and a number of asymptotic approximations for ionization rates, valid for the weak field limits, were developed [36].

The major difficulty in application of such CR models stems from insufficient data for collisions. The model requires data on excitation of parabolic states by plasma electrons and ions. The cross sections for electron collisions are about an order of magnitude smaller than the corresponding cross sections for protons [26] in spherical basis, and therefore lower precision is sufficient for these data. The electron-impact excitation cross sections can be easily calculated from oscillator strengths using the asymptotic van Regermorter formula [42] since the projectile velocity is very high even for excitations from the ground state: for a typical plasma temperature in excess of 1 keV, the electron velocity v in the rest frame of the hydrogen atom is about ten times larger than the atomic unit of speed v_0. The situation is completely different for collisions involving main plasma ions and impurities. Here, for excitation from the ground state, the relative velocity $u = v/v_0$ varies between $\sqrt{2}$ and 4.5 for beam energies of 50 keV/u and 500 keV/u, respectively. For the beam energy below 100 keV/u, application of FBA is only justified either for excitations with $\Delta n = 0$ or for excitations between the states with rather high n's, above $n = 3$. This is why the m-models based on the FBA cross sections have a limited range of applicability.

A collisional-radiative model based on the Glauber approximation collisional cross sections was developed recently [39]. The GA data for collisions between the nl spherical states of hydrogen are known to be accurate above 50 keV/u, and at high energies these cross sections converge to the Born values. The theory of GA cross sections for parabolic states was formulated in [39]. A standard scattering problem described in numerous textbooks on collision theory has only one preferred direction, that is, axis z' along the velocity of the projectile. However, the collisions between the beam atom and plasma particles for MSE conditions also

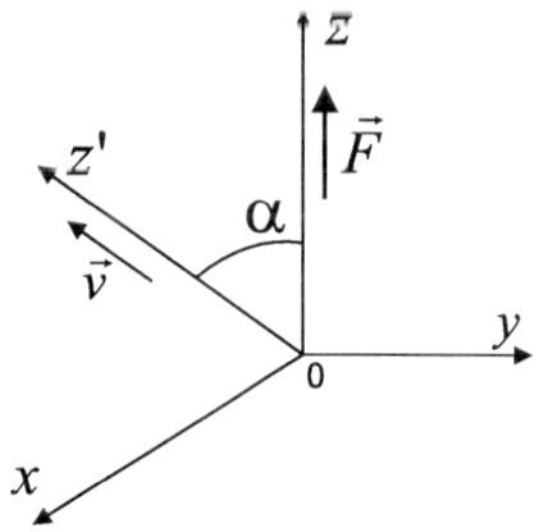

Fig. 4.5 Transformation between the spherical and parabolic wavefunctions having different quantization axes. The electric field **F** defines the quantization axes of the parabolic states in the plasma. The quantization axis z' chosen along the projectile velocity **v** is common in projectile-target collisional problems

involve another axis z in the direction of the induced electric field **F**, which is the quantization axis for parabolic states. A parabolic wavefunction quantized along z can be represented as a linear combination of spherical wavefunctions quantized along z' using the following expression [39]:

$$\psi_{nkm} = \sum_{l=|m|}^{n-1} C_{nk}^{lm} \sum_{m'=-l}^{l} d_{lm}^{m'}(\alpha)\varphi_{nlm'}, \tag{4.9}$$

where α is the angle between axes z and z', C_{nk}^{lm} is the Clebsch–Gordan coefficient [9], and $d_{lm}^{m'}$ is the rotation matrix [43]. This transformation was also utilized in the analysis of Rydberg atoms in the crossed electric and magnetic fields [44].

Equation (4.9) represents a double transformation which is explained in Fig. 4.5. First, note that the x- and y-axes can always be selected so that the z' axis is in the (xz) plane. The first transformation aligning axes z and z' corresponds to the rotation by angle α around the axis y and then, the second transformation between the spherical and parabolic bases is applied. Of course, for MSE conditions, $\alpha = 90°$. The theory, however, was formulated for an arbitrary angle α, and in this case all spherical states contribute to the sum in (4.9), with a weight depending on C_{nk}^{lm} and $d_{lm}^{m'}$. The scattering amplitude $F_{ab}(\mathbf{q})$ and cross section σ_{ab} for a transition between parabolic states $a = n_a k_a m_a$ and $b = n_b k_b m_b$ can be then written as

$$F_{ab}(\mathbf{q}) = \sum_{l_a=|m_a|}^{n_a-1} C_{n_a k_a}^{l_a m_a} \sum_{m'_a=-l_a}^{l_a} d_{l_a m_a}^{m'_a}(\alpha) \sum_{l_b=|m_b|}^{n_b-1} C_{n_b k_b}^{l_b m_b} \sum_{m'_b=-l_b}^{l_b} d_{l_b m_b}^{m'_b}(\alpha) F_{a'b'}(\mathbf{q}), \tag{4.10}$$

$$\sigma_{ab} = \frac{k_f}{k_i} \int |F_{ab}(\mathbf{q})|^2 \mathrm{d}\mathbf{q}, \tag{4.11}$$

where $\mathbf{q} \equiv \mathbf{k_i} - \mathbf{k_f}$ is the momentum transfer and $F_{a'b'}(\mathbf{q})$ is the "standard" scattering amplitude between spherical states a' and b'.

Substituting (4.10) into (4.11), one derives the following expansion for the cross section:

$$\sigma_{ab} = \sum_{pqrs} c_{ab}^{pq,rs} \rho_{pq,rs}, \tag{4.12}$$

where the weights $c_{ab,rs}^{pq}$ are the products of Clebsch–Gordan coefficients and rotation matrix elements, and the elements of the density matrix ρ [45, 46] are defined as:

$$\rho_{pq,rs} = \frac{k_f}{k_i} \int F_{pq}(\mathbf{q}) F_{rs}^*(\mathbf{q}) d\mathbf{q}. \tag{4.13}$$

The diagonal matrix elements with $p = r$ and $q = s$ are the usual cross sections. The states p, r and q, s belong to the multiplets with the principal quantum numbers n_a and n_b, respectively. The off-diagonal elements of the density matrix (4.13) are nonzero if

$$m_q - m_p = m_s - m_r. \tag{4.14}$$

If the initial state is the ground state, the density matrix is rather simple as it contains only four elements for excitations into $n = 2$ and 14 nonzero elements for excitations into $n = 3$ [47, 48]. For the states with higher n, the number of nonzero off-diagonal elements increases rapidly.

Equations (4.10) and (4.11) provide a complete recipe for calculation of collisional cross sections between parabolic states. The fact that it also includes off-diagonal matrix elements makes it more complex as compared to the standard techniques for spherical states. The calculated cross sections can be then used in the system of coupled equations (4.8). The cross sections of excitation and deexcitation are connected via the principle of the detailed balance.

The time-dependent rate equation system (4.8) should be accompanied by a proper set of initial conditions. For a beam entering plasma at time $t = 0$, it is customary to assume that the initial population is only in the ground state, that is, $N_0(t = 0) = 1$ and $N_j(t = 0) = 0$ for other j's. For simplicity, it is also assumed that the following conditions are fulfilled:

$$v_\mathrm{i} \ll v_\mathrm{b} \ll v_\mathrm{e}, \tag{4.15}$$

where v_i, v_b, and v_e is the velocity of plasma ions, beam particles, and plasma electrons, respectively. Correspondingly, the rate coefficients for atom–electron collisions are calculated using the usual Maxwellian averaging, and in the case of atom-ion collisions the rate coefficients are calculated as a product between the excitation cross section and the beam velocity.

4.3.2 Time-Dependent Solutions for Excited States

The time-dependent populations for m-resolved levels are shown in this section. Here, we expand our previous calculations by extending the manifold of included parabolic states up to $n = 10$ from $n = 5$ in [49]. Importantly, all excited states in this model are treated without any assumption on statistical equilibrium. This is done in order to compare not only the line ratios of Stark components but also the absolute populations for nonstatistical and statistical models. The calculation of populations of excited states was performed using the collisional-radiative code NOMAD [50] using the GA cross sections discussed above. The details on the atomic data used for ionization and charge exchange can be found in [39, 49].

The main features of a time-dependent simulation can be exemplified for a typical fusion plasma with electron and proton densities of $10^{13}\,\mathrm{cm}^{-3}$, plasma temperature of 3 keV, and magnetic field of 3 T. The beam energy is 50 keV/u, and at $t = 0$ all population was in the ground state of hydrogen. Figure 4.6 presents the results of calculations for the magnetic levels of $n = 2$ and $n = 3$ states. The shown quantities are the relative populations of excited levels with respect to the ground state $n_j = N_j/(g_j \cdot N_g)$.

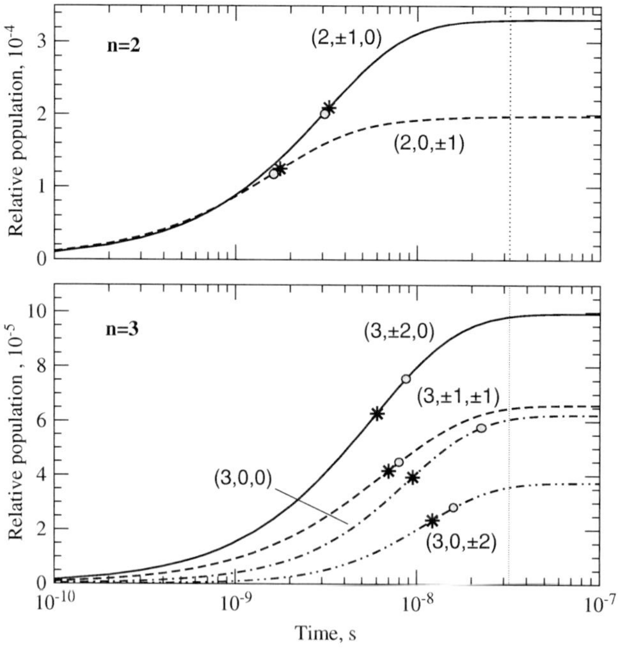

Fig. 4.6 Time-dependent solutions for populations of excited levels. Plasma density is $10^{13}\,\mathrm{cm}^{-3}$; plasma temperature is 3 keV; beam energy is 50 keV/u, and magnetic field is 3 T. *Top*: populations of the $n = 2$ parabolic states; *bottom*: populations of the $n = 3$ parabolic states. *Open circles* denote the radiative lifetime τ_r of the corresponding level, and the *asterisks* show the $0.63 \times n_i^{QSS}$ point. *Vertical dotted line* denotes the time for the beam particles to travel 0.1 m inside the plasma volume

The excited levels are primarily populated via excitations by plasma ions (mainly protons) while contribution of plasma electrons, although included in the CR modeling, is small. As is seen in Fig. 4.6, after several nanoseconds, the excited state populations start to approach the saturation phase which is usually called a quasi-steady state (QSS) [31]. In the case of a complete statistical distribution among the levels, the relative populations n_i within a specific n should be the same. It is clearly not the case for either $n = 2$ or $n = 3$ states: the relative populations of magnetic levels deviate from the Boltzmann statistics by more than 30% at the density of $10^{13}\,\text{cm}^{-3}$, and therefore, the assumption of a statistical distribution among the $n = 2$ and $n = 3$ magnetic levels is not valid in this plasma.

The QSS condition is achieved for different levels j at different characteristic times τ_j^{qs} which are determined by the total depopulation rate for this level:

$$\tau_j^{qs} \approx \frac{1}{A_j + S_j + N_\text{e} \sum_{i \neq j} (C_{ji}^\text{p} + C_{ji}^\text{e})}. \tag{4.16}$$

For a low-density coronal plasma, the characteristic time is close to the radiative lifetime $\tau_j^\text{r} = 1/A_j$, and the level population at this time reaches approximately 0.63 (=$1 - 1/e$) of the QSS population. The open circles in Fig. 4.6 show the radiative lifetimes, while the asterisks indicate the times where $n_i = 0.63 \times n_i^\text{QSS}$. Clearly, the coronal approximation is approximately valid for the $n = 2$ levels and, to some degree, for the $(3, \pm 1, \pm 1)$ levels in $n = 3$. The other $n = 3$ levels are strongly coupled by collisions due to a small energy difference, and therefore their populations deviate from coronal values. Nevertheless, the major conclusion of inadequacy of the statistical approach to QSS populations of $n = 2$ and $n = 3$ states remains valid. It should also be pointed out that the QSS regime is achieved for the excited levels only after penetrating some distance which is about 10 cm (vertical dashed line in Fig. 4.6) for this particular simulation. It is expected therefore that the beam emission intensity close to the plasma edge would be reduced with respect to the steady-state model predictions. This effect is indeed frequently observed in experiments.

The populations of the excited $n = 2$ and $n = 3$ states can also be compared with the results of the statistical model. The total QSS populations calculated in our model for the beam energy of 50 keV/u are approximately 10^{-3} and 6×10^{-4} for $n = 2$ and $n = 3$, respectively, which very well agrees with the statistical populations of Fig. 4.3 for 40 keV/u. One should, however, be cautious when applying Glauber approximation for calculation of atomic data below 50 keV/u since GA is essentially a high-energy approximation.

4.3.3 *Quasi-Steady-State Calculations*

It is very useful to study the dependence of H_α component intensities on the plasma density. The normalized ($\Sigma I_\sigma = \Sigma I_\pi = 1/2$) intensities of the σ- and π-components calculated for quasi-steady-state conditions in a wide range of

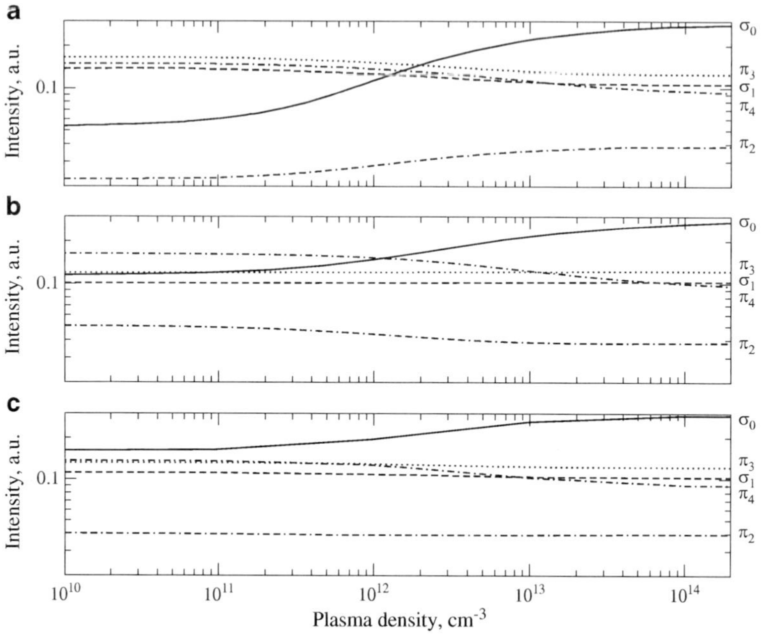

Fig. 4.7 Line intensity ratios for the σ and π components as a function of plasma density: (**a**) FBA model at the energy of 40 keV/u [37], (**b**) Glauber approximation at the energy of 50 keV/u [39], (**c**) FBA approximation at the energy of 80 keV/u [38]. The intensity of σ_0 component is shown as *solid line*; σ_1 component is shown as *dashed line*, π_2 component is shown as *dot-double-dash line*; π_3 component is shown as *dotted line*, and π_4 component is shown as *dashed-dot line*. The following normalization is assumed: $\sum I_\sigma = \sum I_\pi = 1/2$

plasma densities between $10^{10}\,\mathrm{cm}^{-3}$ and $2\times10^{14}\,\mathrm{cm}^{-3}$ are presented in Fig. 4.7. The three panels show results calculated using different sets of data: top for FBA with $E_B = 40$ keV/u, middle for GA and 50 keV/u, and bottom for FBA and 80 keV/u. It is obvious that the statistical model which is valid at high densities significantly deviates from the present QSS calculations even in the range of densities typical of fusion plasmas.

The H_α components can vary drastically with plasma density. For instance, the relative intensity of σ_0 increases by more than a factor of 5 in Fig. 4.7a. It would certainly be very interesting to design an experiment with relatively low plasma density; unfortunately, it would take a rather long distance for hydrogen atoms to reach quasy-steady state for characteristic beam energies in fusion devices.

The relative intensity ratios within the same polarization plane (π or σ) represents the most sensitive benchmark experiment to test theoretical predictions. A spectrometer to be used for this analysis is not required to be absolutely calibrated and the polarization properties of the front optics remain the same for either σ or π spectral lines.

The σ and π intensity ratios calculated with the GA collisional data are shown in Fig. 4.8 along with the experimental data from JET [6, 12, 39, 49]. The statistical

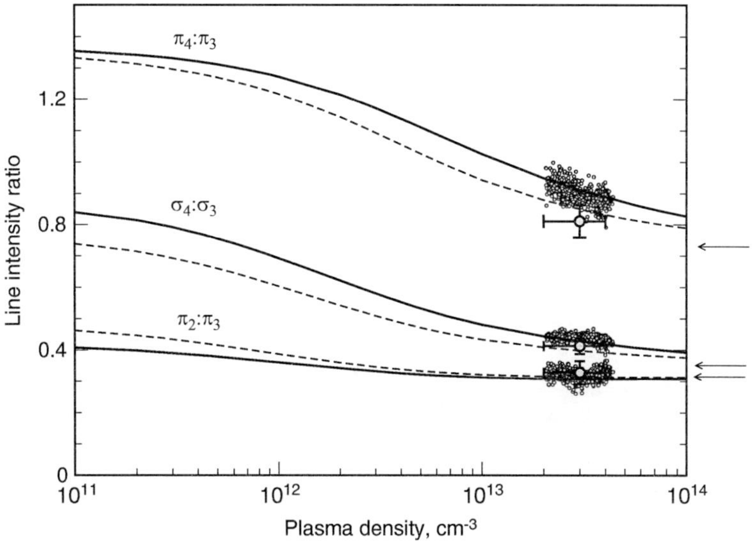

Fig. 4.8 Line intensity ratios for the σ and π components of H_α line. The recent experimental data from JET [12] are shown as *orange points* for π_4/π_3, σ_1/σ_0, and π_2/π_3 components. Experimental data from JET [6] for the same components are presented as *open circles* with errors. Theoretical results using the Glauber approximation are shown as *solid lines* [39]. The *dashed lines* present the results of calculations taking into account the C^{6+} impurity in the plasma at the level of 3% ($Z_{\rm eff} = 2$). The *arrows* denote the statistical values

limits are shown by arrows. The experimental data are obtained from several MSE spectra similar to Fig. 4.2, while the average plasma density is about 3×10^{13} cm^{-3}. The agreement between theoretical results and experimental data is seen to be very good. Clearly, both experimental data and calculations show a significant deviation from the statistical limits. For instance, the π_4/π_3 ratio exceeds unity at low plasma densities and approaches the statistical values of 0.73 only at the plasma density above 10^{14}. The σ_1/σ_0 ratio varies between 0.8 and 0.4 in the same density range. Finally, the ratio of π_2 and π_3 components shows a weaker dependence on density but still decreases by about 25% between coronal and statistical limits.

The old experimental data from JET [6] shown by large circles with error bars are somewhat closer to the statistical values than the results of [12]. This difference can be attributed to presence of impurities in the JET plasma. During the early measurements in 1990s, the concentration of carbon in JET was higher. Since collisions with the fully ionized carbon are rather strong, they are expected to shift the intensity ratios closer to their statistical values; indeed, this effect is observed in Fig. 4.8. In order to take impurity collisions into account, the H–C^{6+} cross sections between excited states were also calculated in the GA. For excitation from the ground state, the atomic-orbitals close-coupling nl-resolved data [51] were scaled appropriately. The calculations with account of collisions with C^{6+} shown by the dashed lines in Fig. 4.8 clearly decrease the π_4/π_3 ratio. It has to be mentioned that

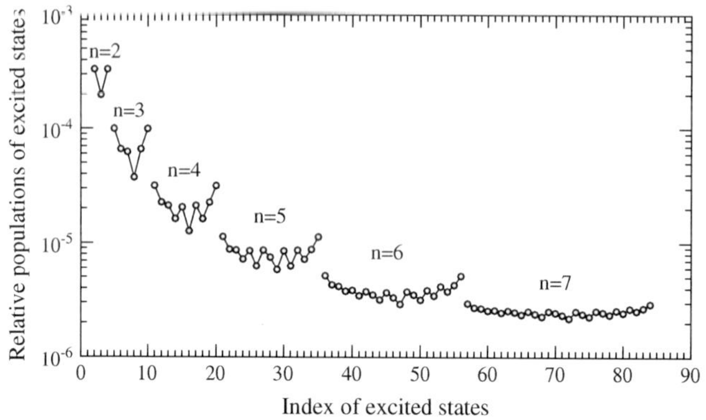

Fig. 4.9 Relative populations of excited parabolic states of hydrogen beam in plasma. Plasma density is $10^{13}\,\text{cm}^{-3}$; plasma temperature is 3 keV, and beam energy is 50 keV/u. A state index denotes the nondegenerated $(n, k, \pm m)$ state which are arranged according to their energies. The states of the same multiplet are connected by *solid lines*. The ionization induced by electric field was for simplicity excluded from calculations

the results obtained in [39] were the first calculations that explain these interesting measurements at JET [12].

As was already mentioned above, the relative populations n_i within the same principal quantum number n should be equal under LTE conditions. As shown in Fig. 4.9 for $(T, E_B, B) = (3\,\text{keV}, 50\,\text{keV/u}, 3\,\text{T})$, this condition is approximately fulfilled only for the $n = 7$ levels, while for n as high as 5 the variation between the populations of different m-levels reaches 40%. Note first that the relative populations of the $(n, \pm k, |m|)$ states, which are most separated from the central state $(n, 0, |m|)$, are very close since the corresponding energy difference is very small and the levels are easily equilibrated by ion collisions. It is also noteworthy that the populations of the levels degenerated in the first-order Stark effect are not necessarily close to each other. The transitions between these levels have $\Delta m = 2$, and therefore the cross sections are not as large. These examples show that the collisional-radiative model in parabolic states provides a straightforward way to analyze level populations and contributions of different physical processes.

4.4 Conclusion

Since the early days of fusion science, the NBI is among the most important tools not only for plasma heating but also for a localized diagnostics of plasma in tokamaks, stellarators, and other devices. The existing theoretical models for beam emission are mostly based on two methods: a statistical approach to calculation of level populations within atomic shells and a nonstatistical modeling for parabolic states of hydrogen. The statistical method utilizes the state-of-the-art atomic data for collisions between different n as well as accurate results for ionization and charge exchange. A significant progress has been achieved in improving consistency of statistical models. The recent comparisons show that different codes agree within to 20% for the range of interest for fusion parameters and beam energy above 40 keV/u.

Unfortunately, the statistical approach fails to provide an accurate description of some important effects. This deficiency is well known for the MSE in fusion devices where an extensive analysis of accurate experimental data has shown that the H_α component ratios are very different from statistical predictions. This problem motivated development of a new m-resolved model in parabolic states. This method is more complex as it requires calculation of a large amount of atomic data for transitions between the m states. Moreover, unlike the standard collision theory for spherical states, the off-diagonal elements of the collisional density matrix are also important.

The recently developed collisional-radiative model for parabolic states in a hydrogen beam with the new data obtained in the Glauber approximation was the first theory capable of accurately explaining the observed H_α σ and π ratios from the JET tokamak. Also, the effect of impurities on MSE profiles was clearly observed in simulations. The model shows that, as expected from general considerations, the deviations from the statistical values are most pronounced in the low-density limit where excitations from the ground state play the most important role. Even for typical fusion plasma conditions, the time-dependent simulations indicate a signification difference for the $n = 2$ and $n = 3$ parabolic states from statistical values. It was also found that the QSS phase for beam populations is reached on the time scale of about 3×10^{-8} s for plasma density of 10^{13} cm^{-3} which corresponds to the penetration distance of 10 cm. Although this feature has practically no effect on plasma diagnostics on such large fusion machines as ITER or JET, nonetheless it would be important for medium- and especially small-size devices.

The new CR model can also be applied to calculation of effective rate coefficients for H_α and H_β as well as to determination of beam emission characteristics. It would certainly be important to include more accurate data for heavy particle collisions calculated with close-coupling and other non-perturbative methods [52]. Even with the existing high-energy Glauber data, this approach can be considered as an important first step toward a unified NBI model which can incorporate all major needs of the beam-plasma diagnostics.

Acknowledgements O.M is thankful to the members of Atomic Spectroscopy Group at NIST for their hospitality. Work of Yu.R. supported in part by the Office of the Fusion Energy Sciences of the US Department of Energy.

References

1. R. Hemsworth et al., Nucl. Fusion **49**, 045006 (2009)
2. A. Malaquias et al., Rev. Sci. Instrum. **75**, 3393 (2004)

3. R.C. Isler, Phys. Rev. Lett. **38**, 1359 (1977)
4. M.G. von Hellermann et al., Rev. Sci. Instrum. **T120**, 19 (2005)
5. F.M. Levinton et al., Phys. Rev. Lett. **63**, 2060 (1989)
6. W. Mandl, R.C. Wolf, M.G. von Hellermann, H.P. Summers, Plasma Phys. Contr. Fusion **35**, 1373 (1993)
7. W.W. Heidbrink, K.H. Burrell, Y. Luo, N.A. Pablant, E. Ruskov, Plasma Phys. Contr. Fusion **46**, 1 (2004)
8. E. Delabie et al., Rev. Sci. Instrum. **79**, 10E552 (2008)
9. L.D. Landau, E.M. Lifshitz, *Quantum Mechanics: Non-Relativistic Theory* (Pergamon, Oxford, 1976), p. 344
10. H.A. Bethe, E.E. Salpeter, *Quantum Mechanics of One- and Two-Electron Atoms* (Plenum, New York, 1977), p. 234
11. R. Uhlemann, R.S. Hemsworth, G. Wang, H. Euringer, Rev. Sci. Instrum. **64**, 974 (1992)
12. E. Delabie et al., Plasma Phys. Contr. Fusion **52**, 125008 (2010)
13. N. Ryde, *Atoms and Molecules in Electric Fields* (Almqvist Wiksell International, Stockholm, 1976), p. 204
14. M.G. von Hellermann et al., Nucl. Instrum. Phys. Res. A **623**, 720 (2010)
15. T.G. Winter, Phys. Rev. A **80**, 032701 (2009)
16. D.R. Schultz, M.R. Strayer, J.C. Wells, Phys. Rev. Lett. **82**, 3976 (1999)
17. E.Y. Sidky, C.D. Lin, Phys. Rev. A **65**, 012711 (2001)
18. N. Toshima, Phys. Rev. A **50**, 3940 (1994)
19. D.R. Schultz, T.-G. Lee, S.D. Loch, J. Phys. B: At. Mol. Opt. Phys. **43**, 144002 (2010)
20. C.D. Boley, R.K. Janev, D.E. Post, Phys. Rev. Lett. **52**, 534 (1984)
21. R.C. Isler, R.E. Olson, Phys. Rev. A **37**, 3399 (1988)
22. I.H. Hutchinson, Plasma Phys. Contr. Fusion **44**, 71 (2002)
23. H.P. Summers, The ADAS User Manual, version 2.6 (2004), http://adas.phys.strath.ac.uk
24. O. Marchuk et al., Rev. Sci. Instrum. **79**, 10F532 (2008)
25. P.D. Fainstein, V.H. Ponce, R.D. Rivarola J. Phys. B: At. Mol. Opt. Phys. **23**, 1481 (1990)
26. R.K. Janev, J.J. Smith, Suppl. J. Nucl. Fusion **4** (1993)
27. G.H. Olivera, R.D. Rivarola, P.D. Fainstein, Phys. Rev. A **51**, 847 (1995)
28. International Atomic Energy Agency, http://www-amdis.iaea.org/ALADDIN/collision.html
29. H.R. Griem, *Plasma Spectroscopy* (McGraw-Hill Book Company, New York, 1964), p. 150
30. D.H. Sampson, J. Phys. B: At. Mol. Phys. **10**, 749 (1977)
31. T. Fujimoto, *Plasma Spectroscopy* (Clarendon Press, Oxford, 2004), p. 136
32. H.J. Kunze, *Introduction to Plasma Spectroscopy* (Springer, Berlin, 2009), p. 141
33. E. Luc-Koenig, A. Bachelier, J. Phys B.: Atom. Molec. Phys. **13**, 1743 (1980)
34. R.J. Damburg, V.V. Kolosov, J. Phys B.: Atom. Molec. Phys. **9**, 3149 (1976)
35. P. Lotte et al., in *Proceedings of 29th EPS Conference on Plasma Physics and Control Fusion*, http://crppwww.epfl.ch/~duval/O2_01.pdf (2002)
36. R.J. Damburg, V.V. Kolosov, J. Phys B.: Atom. Molec. Phys. **11**, 1921 (1978)
37. A. Boileau et al., J. Phys. B: At. Mol. Opt. Phys. **22**, L145 (1989)
38. M.F. Gu, C.T. Holcomb, R.J. Jayakuma, S.L. Allen, J. Phys. B: At. Mol. Opt. Phys. **41**, 095701 (2008)
39. O. Marchuk et al., J. Phys. B: At. Mol. Opt. Phys. **43**, 011002 (2010)
40. V. Franco, B.K. Thomas, Phys. Rev. A **4**, 945 (1971)
41. K. Omidvar, At. Data Nucl. Data Tables **28**, 215 (1983)
42. V. Fisher, V. Bernshtam, H. Goelten, Y. Maron, Phys. Rev. A **53**, 2425 (1996)
43. A.R. Edmonds, *Angular Momentum in Quantum Mechanics* (Princeton University Press, Princeton, 1957), p. 53
44. P.A. Braun, J. Phys. B: At. Mol. Opt. Phys. **24**, 2313 (1991)
45. K. Blum, *Density Matrix Theory and Applications* (Plenum Press, New York, 1981), p. 77
46. O. Schöller, J.S. Briggs, R.M. Dreizler, J. Phys. B.: At. Mol. Phys. **19**, 2505 (1986)
47. M.J. Alguard, C.W. Drake, Phys. Rev. A **8**, 27 (1973)
48. J.R. Ashburn et al., Phys. Rev. A **40**, 4885 (1989)

49. O. Marchuk et al., Nucl. Instrum. Phys. Res. A **623**, 738 (2010)
50. Yu.V. Ralchenko, Y. Maron, Quant. Spectr. Rad. Transf. **71**, 609 (2001)
51. N. Toshima, H. Tawara, NIFS Report NIFS-DATA **26**, 32 (1995)
52. O. Marchuk, Yu. Ralchenko, D.R. Schultz, Plasma Phys. Contr. Fusion (to be submitted)

Chapter 5
Atomic Processes in Dusty Plasmas

D.-H. Ki and Y.-D. Jung

Abstract Various atomic processes are investigated in dusty plasmas including strong collective interactions. The effects of ion temperature on the electron–dust collision process are investigated in complex dusty plasmas. It is found that the eikonal scattering phase and electron–dust grain collision cross section decrease with decreasing ion temperature in dusty plasmas. The ion wake effects on the Coulomb drag force are investigated in complex dusty plasmas. It is found that the ion wake effects enhance the Coulomb ion drag force. The effects of electron temperature and density on the ion–dust grain bremsstrahlung process are investigated in dusty plasmas. It is found that the ion–dust bremsstrahlung radiation cross section decreases with increasing electron density in dusty plasmas.

5.1 Introduction

The atomic processes in plasmas have received considerable attention since the collision and radiation processes has been widely used as the plasma diagnostic tool in plasma spectroscopy. Recent years, there has been a considerable interest in various physical processes in complex plasmas composed of plasma particles and highly charged dust grains exhibiting the strong electrostatic interactions. Hence, in this chapter, we have discussed several atomic processes in complex dusty plasmas containing the strong collective effects.

At first, we have discussed the effects of ion temperature on the electron–dust collision process in complex dusty plasmas. The second-order eikonal method is employed to obtain the scattering phase and cross section for the electron–dust grain collision as functions of the impact parameter, collision energy, ion temperature, density, and Debye length. The results show that the eikonal scattering phase and eikonal electron–dust grain collision cross section decrease with decreasing ion temperature in dusty plasmas. It is also found that the effect of ion temperature on the electron–dust grain collision process is more significant than the effect of electron density in dusty plasmas. In addition, it is shown that the density

V. Shevelko and H. Tawara (eds.), *Atomic Processes in Basic and Applied Physics*, 103
Springer Series on Atomic, Optical, and Plasma Physics 68,
DOI 10.1007/978-3-642-25569-4_5, © Springer-Verlag Berlin Heidelberg 2012

effect is more significant when the electron temperature is comparable to the ion temperature.

In second, we have investigated the ion wake effects on the Coulomb drag force in complex dusty plasmas. It is shown that the ion wake effects significantly enhance the Coulomb ion drag force. It is also found that the ion wake effects on the Coulomb drag force increase with an increase of the Debye length. In addition, the ion wake effects on the momentum transfer cross section and Coulomb drag force are found to be increased with increasing thermal Mach number, i.e., decreasing plasma temperature. It is also found that the Coulomb ion drag force would be stronger for smaller dust grains.

In third, we have discussed the effects of electron temperature and density on the ion–dust grain bremsstrahlung process in dusty plasmas. The ion–dust bremsstrahlung radiation cross section is obtained as a function of the dust charge, dust radius, Debye length, collision energy, radiation energy, electron density, and electron temperature by using the Born approximation. It is shown that the ion–dust bremsstrahlung radiation cross section decreases with an increase of the electron density in dusty plasmas. It is also shown that the electron temperature suppresses the bremsstrahlung radiation cross section. In addition, the effect of electron temperature on the ion–dust bremsstrahlung process is found to be more significant than the effect of electron density in dusty plasmas.

5.2 Electron–Dust Collisions in Complex Dusty Plasmas

The elastic electron–ion collision in plasmas has received much attention since this process has been widely used as the plasma diagnostic tool, and the process is also one of the major atomic processes in various plasmas. Recent years, there has been a considerable interest in collision and radiation processes in complex plasmas composed of plasma particles and highly charged dust grains exhibiting the strong electrostatic interactions [1–5]. Moreover, it has been known that the collective plasma–dust interactions are ubiquitous in various astrophysical and laboratory plasmas. Hence, the numerous physical processes have been extensively explored in order to obtain the information on the plasma parameters such as the density and temperature of plasma particles in complex dusty plasmas [6]. It has been shown that the most of dusty plasmas would be mainly constituted of thermal electrons, ions, and negatively charged dusty grains, i.e., three-component dusty plasmas. Hence, it would be expected that the electron–dust grain collision in dusty plasmas would be different from that in only two-component electron–dust plasmas due to the influence of the plasma ions in dusty plasmas. However, the effect of the ion temperature on the electron–dust grain collision has not been investigates as yet. Thus, in this section, we have investigated the influence of the ion temperature on the elastic collision process due to the interaction between the electron and negatively charged dust grain in three-component dusty plasmas. The scattering phase and cross section for the electron–dust grain collision are obtained as functions of the

impact parameter, collision energy, ion temperature, density, and Debye length by using the second-order eikonal method with the impact parameter analysis.

For an interaction potential $V(\mathbf{r})$, the solution $\psi_k(\mathbf{r})$ of the nonrelativistic Schrödinger equation would be expressed by the following integral form of the Lippmann–Schwinger equation [7]:

$$\psi_k(\mathbf{r}) = \varphi_k(\mathbf{r}) + \frac{2\mu}{\hbar^2}\int \mathrm{d}^3\mathbf{r}' V(\mathbf{r}')\psi_k(\mathbf{r}')G(\mathbf{r},\mathbf{r}'). \tag{5.1}$$

Here, $\varphi_k(\mathbf{r})$ and $G(\mathbf{r},\mathbf{r}')$ are, respectively, the solution of the homogeneous equation and the Green's function:

$$\left(\nabla^2 + k^2\right)\varphi_k(\mathbf{r}) = 0, \tag{5.2}$$

$$\left(\nabla^2 + k^2\right)G(\mathbf{r},\mathbf{r}') = \delta(\mathbf{r},\mathbf{r}'), \tag{5.3}$$

where $k[=(2\mu E/\hbar^2)^{1/2}]$ is the wave number, μ is reduced mass of the collision system, $E(=\mu v^2/2)$ is the collision energy, v is the relative collision velocity, and $\delta(\mathbf{r},\mathbf{r}')$ is the Dirac delta function. By using the cylindrical coordinate system such as $\mathbf{r} = \mathbf{b} + z\hat{\mathbf{n}}$, where $\mathbf{b}$ is the impact parameter, $\hat{\mathbf{n}}$ is the unit vector normal to the momentum transfer $\Delta\mathbf{k}(\equiv \mathbf{k}_i - \mathbf{k}_f)$, $\mathbf{k}_i$ and $\mathbf{k}_f$ are, respectively, the incident and final wave vectors, the eikonal scattering amplitude $f_E(\Delta\mathbf{k})$ would then be obtained as the following integral form:

$$f_{\mathrm{E}}(\Delta\mathbf{k}) = -\frac{\mu}{2\pi\hbar^2}\int \mathrm{d}^3\mathbf{r}\,\exp\left[-\frac{\mathrm{i}\mu}{\hbar^2 k_i}\int_{-\infty}^{z}\mathrm{d}z' V(\mathbf{b},z')\right]\mathrm{e}^{\mathrm{i}\Delta\mathbf{k}\cdot\mathbf{r}}V(\mathbf{r}). \tag{5.4}$$

Since the differential eikonal collision cross section is determined by the relation $\mathrm{d}\sigma_{\mathrm{E}}/\mathrm{d}\Omega = |f_{\mathrm{E}}(\Delta\mathbf{k})|^2$, the total elastic eikonal collision cross section (σ_{E}) would be expressed as

$$\sigma_{\mathrm{E}}(k) = \int \mathrm{d}^2\mathbf{b}|\exp[\mathrm{i}\chi_{\mathrm{E}}(k,b)] - 1|^2, \tag{5.5}$$

where $\mathrm{d}\Omega$ is the differential solid angle and $|\mathbf{k}_i| = |\mathbf{k}_f| = k$ for the elastic collision process. Here, the total eikonal scattering phase $\chi_{\mathrm{E}}(k,b)$ would be represented by the following series expansion technique [8]:

$$\chi_{\mathrm{E}}(k,b) = \sum_l \left(-\frac{\mu}{\hbar^2}\right)^{l+1}\frac{1}{(l+1)!}\frac{1}{k}\left(\frac{\partial}{\partial k}\frac{1}{k} - \frac{b}{k^2}\frac{\partial}{\partial b}\right)^l\int_{-\infty}^{\infty}\mathrm{d}z V^{l+1}(z,b). \tag{5.6}$$

In dusty plasmas composed of electrons, ions, and negatively charged dusty grains, the quasineutral condition [9] at equilibrium would be represented by

$$n_{\mathrm{e}0} + Zn_{\mathrm{d}0} = n_{\mathrm{i}0} = n_0, \tag{5.7}$$

where $n_{\alpha 0}(\alpha = e, i, d)$ are, respectively, the equilibrium densities for electrons (e), ions (i), and dust grains (d), Z is the charge number of the dust grain, and n_0 is the total plasma density. In the case of $|e\phi/k_{\mathrm{B}}T_\alpha| \ll 1 \quad (\alpha = e, i)$, the solution of the Poisson equation [10] in three-component dusty plasmas would be obtained by

$$\begin{aligned} \nabla^2\phi &= 4\pi e(n_{\mathrm{e}} + Zn_{\mathrm{d}} - n_{\mathrm{i}}) \\ &\approx 4\pi e n_{\mathrm{e}0}\frac{e\phi}{k_{\mathrm{B}}T_{\mathrm{e}}}\left(1 + \frac{n_0}{n_{\mathrm{e}0}}\frac{T_{\mathrm{e}}}{T_{\mathrm{i}}}\right), \end{aligned} \tag{5.8}$$

where ϕ is the electrostatic scalar potential in dusty plasmas, k_{B} is the Boltzmann constant, and T_α is the temperature of species α. Then, the effective Debye shielding distance [9] λ_{eff} in three-component dusty plasmas is given by

$$\lambda_{\mathrm{eff}} = \lambda_{\mathrm{D}}\left(\frac{n_{\mathrm{e}0}}{n_0} + \frac{T_{\mathrm{e}}}{T_{\mathrm{i}}}\right)^{-1/2}, \tag{5.9}$$

where $\lambda_{\mathrm{D}}(\equiv k_{\mathrm{B}}T_{\mathrm{e}}/4\pi n_0 e^2)^{1/2}$ is the standard Debye length in electron–ion plasmas and the correction factor $(n_{\mathrm{e}0}/n_0 + T_{\mathrm{e}}/T_{\mathrm{i}})^{-1/2}$ represents for the influence of the density and temperature ratios on the Debye length. Since the correction factor is usually greater than unity, the effective Debye length in dusty plasmas is expected to be greater than that in conventional electron–ion plasmas. In spherical polar coordinates with their origin at the center of the dust grain, the interaction potential $V_{\mathrm{i-d}}(r)$ between the electron and negatively charged dust grain with charge $-Ze$ in dusty plasmas is then represented by the Yukawa potential with the effective λ_{eff}:

$$V_{\mathrm{i-d}}(r) = \frac{Ze^2}{r}\exp\left[-\left(\frac{n_{\mathrm{e}0}}{n_0} + \frac{T_{\mathrm{e}}}{T_{\mathrm{i}}}\right)^{1/2} r/\lambda_{\mathrm{D}}\right]. \tag{5.10}$$

By using (5.6) and (5.10) with the identity of the zeroth-order of the MacDonald function [10], the total eikonal scattering phase χ_{E} including the first- and second-order contributions for the elastic election–dust grain collision in dusty plasmas would be given by

$$\chi_{\mathrm{E}}(\bar{b}, \bar{E}, \xi, \bar{\lambda}_{\mathrm{D}}) = -\frac{2}{\bar{E}^{1/2}}K_0(\xi\bar{b}/\bar{\lambda}_{\mathrm{D}}) - \frac{\xi}{2\bar{E}^{3/2}\bar{\lambda}_{\mathrm{D}}}\frac{a_Z}{a}K_0(2\xi\bar{b}/\bar{\lambda}_{\mathrm{D}}), \tag{5.11}$$

where $\bar{b}(\equiv b/a)$ is the scaled impact parameter, a is the radius of the spherical dust grain, $a_Z \equiv a_0/Z$, $a_0(=\hbar^2/m_{\mathrm{e}}e^2)$ is the first Bohr radius of the hydrogen atom, m_{e} is the mass of the electron, $\bar{E}(\equiv E/Z^2Ry)$ is the scaled collision energy, $Ry(=m_{\mathrm{e}}e^4/2\hbar^2 \approx 13.6\,\mathrm{eV})$ is the Rydberg constant, $\xi \equiv (n_{\mathrm{e}0}/n_0 + T_{\mathrm{e}}/T)^{1/2}$, and $\bar{\lambda}_{\mathrm{D}}(\equiv \lambda_{\mathrm{D}}/a)$ is the scaled effective Debye length. Hence, the scaled differential collision cross section $\partial\bar{\sigma}_{\mathrm{E}}[\equiv (\mathrm{d}\sigma_{\mathrm{E}}/\mathrm{d}\bar{b})/\pi a^2]$ in units of πa^2 within the framework

of the second-order eikonal analysis for the elastic electron–dust grain collision in dusty plasmas is then found to be

$$\begin{aligned}\partial\bar{\sigma}_{\mathrm{E}} &= \frac{\mathrm{d}\sigma_{\mathrm{E}}}{\mathrm{d}\bar{b}}/\pi a^2 = 2\bar{b}\left|\exp\left[\mathrm{i}\chi_{\mathrm{E}}(\bar{b},\bar{E},\xi,\bar{\lambda}D)\right]-1\right|^2 \\ &= 2\bar{b}\left|\exp\left[-\mathrm{i}\frac{2}{\bar{E}^{1/2}}K_0\left[\left(\frac{n_{\mathrm{e}0}}{n_0}+\frac{T_{\mathrm{e}}}{T}\right)^{1/2}\bar{b}/\bar{\lambda}_{\mathrm{D}}\right]\right.\right. \\ &\quad \left.\left.-\mathrm{i}\frac{\left(\frac{n_{\mathrm{e}0}}{n_0}+\frac{T_{\mathrm{e}}}{T}\right)^{1/2}}{2\bar{E}^{3/2}\bar{\lambda}_D}\frac{a_Z}{a}K_0\left[2\left(\frac{n_{\mathrm{e}0}}{n_0}+\frac{T_{\mathrm{e}}}{T}\right)^{1/2}\bar{b}/\bar{\lambda}_{\mathrm{D}}\right]\right]-1\right|^2. \qquad (5.12)\end{aligned}$$

This eikonal approximation would be quite reliable to investigate the electron–dust grain collision in dusty plasmas since the Sommerfeld parameter $(Ze^2/r_{\mathrm{i}}E)\mathrm{e}^{-r_{\mathrm{i}}/\lambda_{\mathrm{eff}}}$, where r_{i} is the interparticle distance between the electron and dust grain, for the electron–dust interaction in typical circumstances of dusty plasmas is usually less than unity. For typical circumstances of dusty plasmas, it has been shown that $Z \approx 100$–1,000, $a \approx 0.01$–1 μm, and $\lambda_{\mathrm{D}}/a \approx 5$–100 [1].

Figure 5.1 shows the three-dimensional plot of the total eikonal phase χ_{E} as a function of the temperature ratio $\bar{T}(\equiv T_{\mathrm{e}}/T_{\mathrm{i}})$ and density ratio $\bar{n}(\equiv n_{\mathrm{e}0}/n_0)$. As shown, the total eikonal scattering phase decreases with increasing temperature ratio $\bar{T}$ in dusty plasmas. It is found that the diminution of the ion temperature strongly suppresses the eikonal scattering phase for the electron–dust grain collision. Hence, we can understand that the eikonal scattering phase decreases when

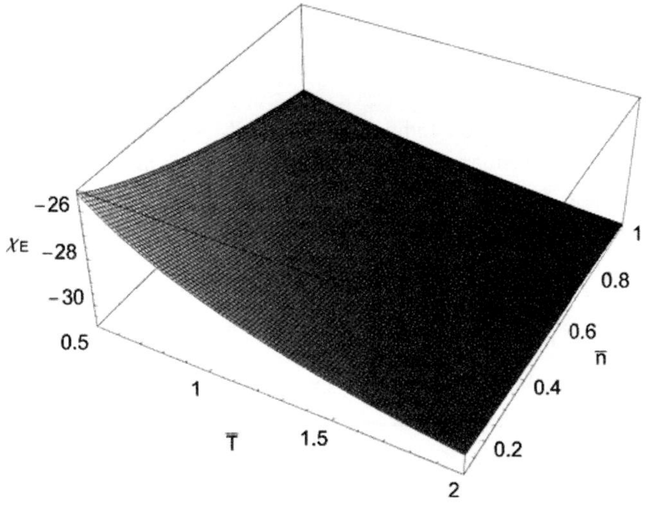

Fig. 5.1 The three-dimensional plot of the total eikonal scattering phase χ_{E} as a function of the temperature ratio $\bar{T}(\equiv T_{\mathrm{e}}/T_{\mathrm{i}})$ and density ratio $\bar{n}(\equiv n_{\mathrm{e}0}/n_0)$ when $\bar{b} = 7$, $\bar{\lambda}_{\mathrm{D}} = 50$, $a_Z/a = 5 \times 10^{-5}$, and $\bar{E} = 5$. From [11]

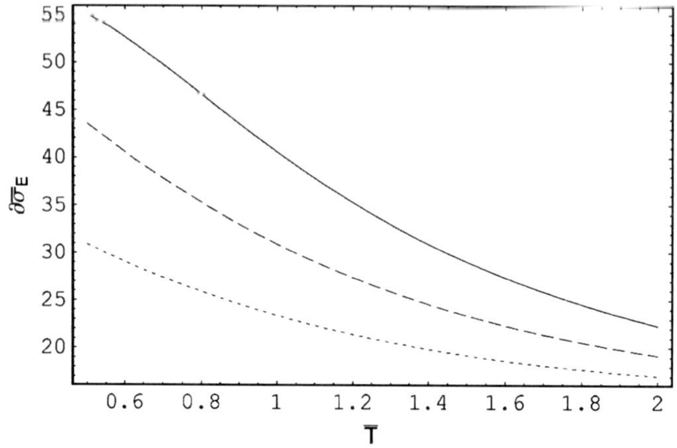

Fig. 5.2 The scaled differential collision cross section $\partial\bar{\sigma}_E$ as a function of the temperature ratio $\bar{T}(\equiv T_e/T_i)$ for various values of the density ratio $\bar{n}(\equiv n_{e0}/n_0)$ when $\bar{b} = 7$, $\bar{\lambda}_D = 50$, $a_Z/a = 5 \times 10^{-5}$, and $\bar{E} = 10$. The *solid line* represents the case of $\bar{n} = 0.1$. The *dashed line* represents the case of $\bar{n} = 0.5$. The *dotted line* represents the case of $\bar{n} = 1$. From [11]

the ion temperature is smaller than the electron temperature in three-component dusty plasmas. It is also shown that the eikonal scattering phase decreases with an increase of the density ratio $\bar{n}$ in dusty plasmas. We also understand that the eikonal scattering phase decreases the electron equilibrium density. Thus, it is found that the ion temperature effect plays an important role in the scattering phase. In addition, the electron equilibrium density is quite important to determine the eikonal scattering phase in dusty plasmas. Figure 5.2 represents the scaled differential electron–dust grain collision cross section $\partial\bar{\sigma}_E$ as a function of the temperature ratio $\bar{T}$ for various values of the density ratio $\bar{n}$. As it is seen, the eikonal electron–dust grain collision cross section decreases with a decrease of the ion temperature. Then, it is found that the influence of the ion temperature strongly suppresses the electron–dust grain collision cross section in three-component dusty plasmas. Hence, it would be expected that the electron–dust grain collision cross sections in the one-temperature plasma ($T_e/T_i \approx 1$) such as the dust combustion would be smaller than those in conventional dusty plasmas ($T_e/T_i \gg 1$) due to the effect of the ion temperature. It is also found that the density effect on the scattering cross section decreases with an increase of the temperature ratio $\bar{T}$. Thus, it would be also expected that the density effects the electron–dust grain collision cross sections are more significant in the one-temperature plasma since the density dependence decreases with decreasing ion temperature. From this work, we have found that the ion temperature and density effects play important roles in the elastic electron–dust grain collision in complex dusty plasmas. These results would provide useful information on the collision processes in three-component complex dusty plasmas.

5.3 Ion Drag in Complex Dusty Plasmas

Recently, there has been a considerable interest in dynamics of plasmas containing charged dust grains including nonlinear collective effects and strong electrostatic interactions between the charged components in laboratory and space dusty plasmas [1, 2, 12]. Hence, the physical processes including collisions and radiations have been extensively explored in order to obtain the information on relevant plasma parameters of dusty plasmas [2, 3]. In addition, the momentum transfer collision [13–16] in dusty plasmas has been of a great interest since the momentum transfer process is related to the Coulomb ion drag force due to the ion–dust interaction. It has been known that the particle interaction in plasmas would be described by the Debye–Hückel model obtained by the linearization of the Poisson equation with the Boltzmann distribution. However, recently, it is shown that the ion–dust interaction potential including the ion flow in dusty plasmas would not be properly described by only the standard Debye–Hückel model so that the importance of the effect of the wake field has been proposed on the potential of the charged dust grain. Very recently, it is found that the influence of the wake field due to the ion flow produces the additional far-field terms [17, 18] in the ion–dust interaction potential. In addition, it has been found that the plasma wake generates the anisotropy [17, 18] of the plasma density around the charged dust grain. Hence, it would be expected that the Coulomb ion drag force including the influence of the wake field is quite different from that without the ion wake effect. Thus, in this section, we investigate the effects of the ion wake on the Coulomb drag force in dusty plasmas. The effective interaction potential model [17, 18] including the additional far-field terms due to the ion wake apart from the standard Debye–Hückel shielding term is employed in order to obtain the momentum transfer cross section and Coulomb ion drag force in dusty plasmas. In addition, the Born analysis [19] is applied to obtain the scattering amplitude as a function of the Debye length, dust charge, ion charge, plasma temperature, thermal Mach number, and collision wave number.

For the potential scattering, the differential momentum transfer cross section [20, 21] $\mathrm{d}\sigma_{\mathrm{M}}$ would be written as

$$\mathrm{d}\sigma_{\mathrm{M}} = \left|f(\mathbf{k}_i, \mathbf{k}_f)\right|^2 (1 - \cos\theta)\mathrm{d}\Omega, \tag{5.13}$$

where $f(\mathbf{k}_i, \mathbf{k}_f)$ is the scattering amplitude and $\mathbf{k}_i$ and $\mathbf{k}_f$ are, respectively, the wave vectors of the incident and final scattered waves, θ is the scattering angle, the factor $(1 - \cos\theta)$ denotes the fraction of the momentum transfer of the incident projectile, and $\mathrm{d}\Omega$ is the differential solid angle. It has been known that the momentum transfer cross section is relevant for investigating the Coulomb drag force for the particle–dust grain interaction in dusty plasmas. In the Born analysis [19], the scattering amplitude $f(\mathbf{k}_i, \mathbf{k}_f)$ would be expressed as

$$f(\mathbf{k}_i, \mathbf{k}_f) = -\frac{\mu}{2\pi\hbar^2} \int \mathrm{d}^3\mathbf{r}\, V(\mathbf{r}) \mathrm{e}^{\mathrm{i}(\mathbf{k}_i - \mathbf{k}_f)\cdot\mathbf{r}}, \tag{5.14}$$

where μ is the reduced mass of the collision system, $\hbar$ is the rationalized Planck constant, and $V(\mathbf{r})$ is the interaction potential. It is well known that the interaction potential in plasmas can be obtained by linearizing the Boltzmann–Poisson equation. However, it is found that the ion–dust interaction potential including the influence of the plasma wake field in dusty plasmas would not be properly described by only the ordinary Debye–Hückel term [13]. Very recently, an excellent work by Morfill and Ivlev [17] has provided the useful analytic form of the effective potential for the charged dust including the effects of the ion wake in dusty plasmas has been obtained by the superposition of the Debye–Hückel term and additional far-field terms including the Legendre polynomials. Using the effective potential model [17], the interaction potential energy $V_{\text{eff}}(\mathbf{k})$ between the projectile ion and target dust grain in dusty plasmas including the influence of the ion wake on the plasma density would be given by

$$V_{\text{eff}}(r,\varphi) = q_{\text{i}}Q\left[\frac{\mathrm{e}^{-r/\lambda_{\text{D}}}}{r} - \sqrt{\frac{8}{\pi}}\frac{M_{\text{T}}\lambda_{\text{D}}^2}{r^3}\cos\varphi - \left(2-\frac{\pi}{2}\right)\frac{M_{\text{T}}^2\lambda_{\text{D}}^2}{r^3}(3\cos^2\varphi - 1)\right], \tag{5.15}$$

where $q_{\text{i}}(=ze)$ is the charge of the ion, $Q(=-Ze)$ is the charge of the dust grain, λ_{D} is the Debye length, λ_{D} is the value of the thermal Mach number, u_{i} is the ion velocity, $v_{T_{\text{i}}}$ is the ion thermal velocity, and φ is the angle between the projectile velocity and position vector. As it is seen, the plasma wakes in dusty plasmas produce the nonlinear interactions which are similar to the Cherenkov radiations [22] in the moving medium.

It has been also shown that the dipole dependence of the effective potential in plasmas would be related to the Landau damping effect [23]. According to the contributions from the real and imaginary parts of the complex scattering amplitude, the total momentum transfer cross section for the interaction between the ion and dust grain including the influence of the ion wake field is then found to be

$$\sigma_{\text{M}}(k, M_{\text{T}}) = \sigma_0 \int_0^{\pi} \mathrm{d}\theta \sin\theta(1-\cos\theta)\left\{\left[\frac{Ka\cos(Ka) + (\frac{a}{\lambda_{\text{D}}})\sin(Ka)}{Ka\left[(Ka)^2 + (\frac{a}{\lambda_{\text{D}}})^2\right]} + 4\left(2-\frac{\pi}{2}\right)M_{\text{T}}^2\left(\frac{\lambda_{\text{D}}}{a}\right)^2[-3S_4(Ka) + 3C_3(Ka) + S_2(Ka)]\right]^2 + \left(2\sqrt{\frac{2}{\pi}}M_{\text{T}}\left(\frac{\lambda_{\text{D}}}{a}\right)^2\right)^2[2S_3(Ka) - C_2(Ka)]^2\right\}, \tag{5.16}$$

where $\sigma_0 \equiv 2\pi[zZ(m_{\text{i}}/m_{\text{e}})(a^2/a_0)]^2$, m_{i} is the ion mass, m_{e} is the electron mass, $a_0(=\hbar^2/m_{\text{e}}e^2)$ is the Bohr radius of the hydrogen atom, $K \equiv 2k\sin(\theta/2)$,

$k \equiv |\mathbf{k}_i| = |\mathbf{k}_f|$ for the elastic scattering, and $S_n(Ka)$ and $C_n(Ka)$ are, respectively, given by

$$S_n(Ka) = \int_{Ka}^{\infty} \mathrm{d}\xi \xi^{-n} \sin \xi, \tag{5.17}$$

$$C_n(Ka) = \int_{Ka}^{\infty} \mathrm{d}\xi \xi^{-n} \cos \xi. \tag{5.18}$$

As shown in (5.18), the momentum transfer cross section has the strong dependence of the Mach number due to the ion wake field. It is shown that the Coulomb ion drag force $F_{\mathrm{CD}}(= n_{\mathrm{i}} m_{\mathrm{i}} \sigma_{\mathrm{M}} v_{\mathrm{i,tot}} u_{\mathrm{i}})$ would be represented by the time rate of the momentum transfer from the ion projectile to the target dust grain, where n_{i} is the ion density and $v_{\mathrm{i,tot}}(= u_{\mathrm{i}}^2 + 8v_{T_{\mathrm{i}}}^2/\pi)^{1/2}$ is the total ion velocity [3]. Hence, the Coulomb ion drag force including the influence of the ion wake field in dusty plasmas is then obtained in the following form:

$$\begin{aligned} F_{\mathrm{CD}}(\bar{k}, M_{\mathrm{T}}) = {} & F_0 \bar{k}^2 M_{\mathrm{T}}^{-1} (M_{\mathrm{T}}^2 + 8/\pi)^{1/2} \\ & \times \int_{-1}^{1} \mathrm{d}y (1-y) \left\{ \left[\frac{A \cos A + (a/\lambda_D) \sin A}{A[2\bar{k}^2(1-y) + (a/\lambda_{\mathrm{D}})^2]} \right. \right. \\ & \left. + 4\left(2 - \frac{\pi}{2}\right) M_{\mathrm{T}}^2 \left(\frac{\lambda_{\mathrm{D}}}{a}\right)^2 [-3S_4(A) + 3C_3(A) + S_2(A)] \right]^2 \\ & \left. + \left(2\sqrt{\frac{2}{\pi}} M_{\mathrm{T}} \left(\frac{\lambda_{\mathrm{D}}}{a}\right)^2\right)^2 [2S_3(A) - C_2(A)]^2 \right\}, \end{aligned} \tag{5.19}$$

where $F_0 \equiv 2\pi(n_{\mathrm{i}}/m_{\mathrm{i}})[zZ(m_{\mathrm{i}}/m_{\mathrm{e}})(a/a_0)]^2$, $y \equiv \cos\varphi$, $\bar{k}(\equiv ka)$, and $A \equiv \bar{k}\sqrt{2(1-y)}$ is the scaled collision wave number. The contribution from the Coulomb electron drag would be neglected in dusty plasmas since the electron drag is quite smaller than the ion drag by the mass factor $(m_{\mathrm{e}}/m_{\mathrm{i}})^{1/2}$ [24].

Figure 5.3 shows the scaled total momentum transfer cross section $\bar{\sigma}_{\mathrm{M}}(\equiv \sigma_{\mathrm{M}}/\sigma_0)$ as a function of the scaled Debye length $\bar{\lambda}_{\mathrm{D}}(\equiv \lambda_{\mathrm{D}}/a)$. In addition, Fig. 5.4 represents the scaled total momentum transfer cross section $\bar{\sigma}_{\mathrm{M}}$ as a function of the thermal Mach number M_{T}. As shown in these figures, the momentum transfer cross section increases with an increase of the scaled Debye length. Hence, it can be expected that the momentum transfer cross sections would be greater for smaller dust grains. It is also shown that the momentum transfer cross section drastically increases with increasing Mach number. Then, it is understood that the wake field effects significantly enhance the cross section. In addition, it is found that the momentum transfer cross section decreases with an increase of the scaled collision wave number $\bar{k}$. Figure 5.5 represents the scaled Coulomb ion drag force

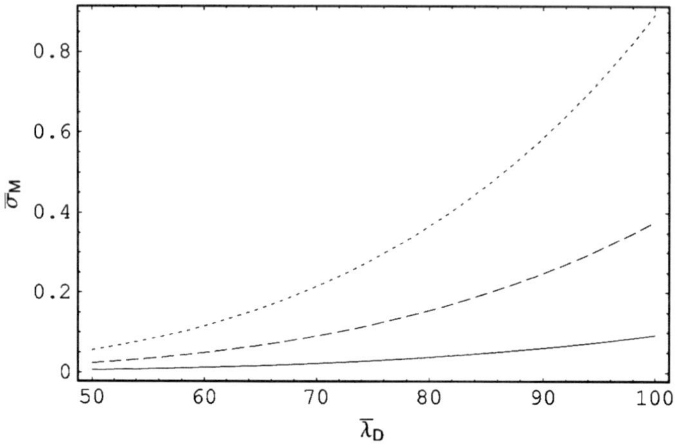

Fig. 5.3 The scaled total momentum transfer cross section $\bar{\sigma}_M$ as a function of the scaled Debye length $\bar{\lambda}_D$ for $\bar{k} = 1{,}000$. The *solid line* represents the case of $M_T = 0.1$. The *dashed line* represents the case of $M_T = 0.2$. The *dotted line* represents the case of $M_T = 0.3$. From [25]

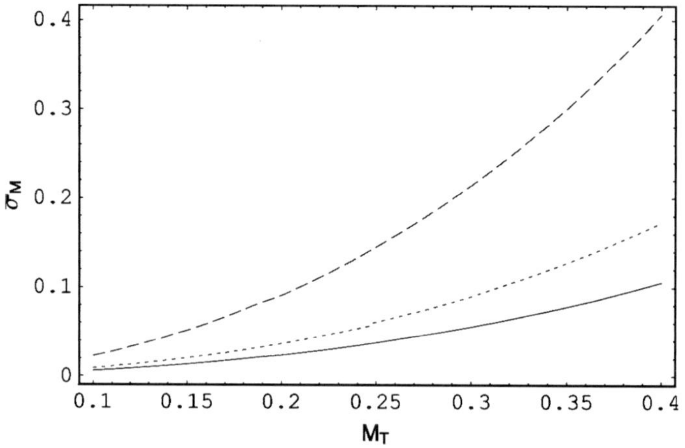

Fig. 5.4 The scaled total momentum transfer cross section $\bar{\sigma}_M$ as a function of the thermal Mach number M_T. The *solid line* is the case of $\bar{\lambda}_D = 50$ and $\bar{k} = 1{,}000$. The *dashed line* is the case of $\bar{\lambda}_D = 70$ and $\bar{k} = 1{,}000$. The *dotted line* is the case of $\bar{\lambda}_D = 70$ and $\bar{k} = 1{,}400$. From [25]

$\bar{F}_{CD}(\equiv F_{CD}/F_0)$ as a function of the thermal Mach number M_T for various values of the scaled Debye length $\bar{\lambda}_D$. As it is seen, the Coulomb ion drag force also increases with increasing Mach number and scaled Debye length. Hence, it can be also expected that the Coulomb ion drag force would be stronger for smaller dust grains. It is also found that the dependence of the thermal Mach number on the momentum transfer cross section is stronger than that on the Coulomb ion drag force due to the factor $(1 + 8/\pi M_T)^{-1/2}$. Therefore, in this work, we have found

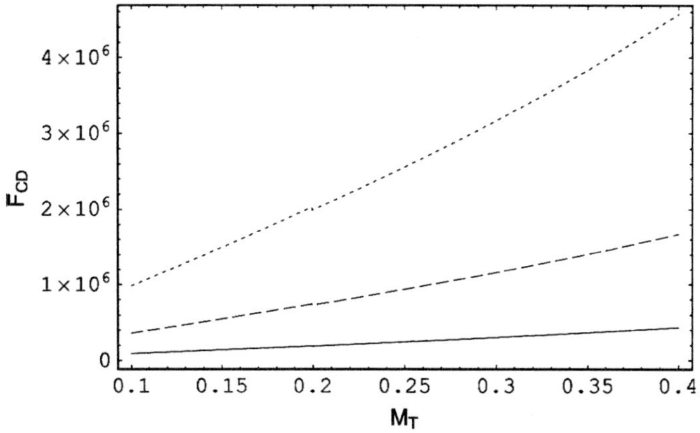

Fig. 5.5 The scaled Coulomb ion drag force $\bar{F}_{CD}$ as a function of the thermal Mach number M_T for $\bar{k} = 1{,}000$. The *solid line* represents the case of $\bar{\lambda}_D = 50$. The *dashed line* represents the case of $\bar{\lambda}_D = 70$. The *dotted line* represents the case of $\bar{\lambda}_D = 90$. From [25]

that the influence of the ion wake field plays an important role in the Coulomb drag force and momentum transfer cross section for the ion–dust grain interaction in dusty plasmas. These results provide useful information on the effects of the plasma wake field on the Coulomb drag in dusty plasmas.

5.4 Ion–Dust Bremsstrahlung Spectrum in Dusty Plasmas

The bremsstrahlung [16, 26–31] in plasmas has received much attention since this process is known to be one of the most fundamental processes, and the continuum radiation due to the bremsstrahlung process has been a diagnostic tool for investigating the plasma parameters. Recently, there has been a considerable interest in physical processes in plasmas encompassing highly charged dust grains [1–5]. In addition, it has been shown that the collective dust–plasma interactions are ubiquitous in various astrophysical and laboratory plasmas [3]. Hence, the collision and radiation processes have been extensively investigated in order to obtain the information on the plasma parameters such as the plasma density and temperature in dusty plasmas. In most cases, the dusty plasmas are consisting of thermal electrons and ions, and negatively charged dusty grains, i.e., three-component dusty plasmas. Hence, the ion–dust grain interaction in dusty plasmas would be different from that in only two-component ion–dust plasmas due to the influence of the plasma electrons in dusty plasmas. Thus, in this section, we have investigated the effects of electron density and temperature on the bremsstrahlung process due to the interaction of ion and negatively charged dust grain in three-component dusty plasmas. The ion–dust grain bremsstrahlung radiation cross section is obtained as

a function of the dust charge, dust radius, Debye length, collision energy, radiation energy, electron density, and electron temperature by using the Born approximation for the initial and final states of the projectile ion.

In the Born approximation, the differential ion–dust grain bremsstrahlung cross section [19] $d^2\sigma_b$ can be obtained by the second-order nonrelativistic perturbation method:

$$d^2\sigma_b = d\sigma_C \cdot dW_\omega, \tag{5.20}$$

where the differential elastic scattering cross section $d\sigma_C$ can be represented by

$$d\sigma_C = \frac{1}{2\pi\hbar v_0^2}\left|\bar{V}(\mathbf{q})\right|^2 q dq, \tag{5.21}$$

where $\hbar$ is the rationalized Planck constant, v_0 is the initial relative collision velocity, and $\bar{V}(\mathbf{q})$ is the Fourier transformation of the ion–dust grain interaction potential $V(\mathbf{r})$:

$$\bar{V}(\mathbf{q}) = \int d^3\mathbf{r}\, e^{-i\mathbf{q}\cdot\mathbf{r}}\, V(\mathbf{r}), \tag{5.22}$$

$\mathbf{q}(=\mathbf{k}_0 - \mathbf{k}_f)$ is the momentum transfer, and $\mathbf{k}_0$ and $\mathbf{k}_f$ are, respectively, the wave vectors of the initial and final states of the projectile ion. Here, dW_ω represents the differential photon emission probability within the frequency between ω and $\omega + d\omega$:

$$dW_\omega = \frac{\alpha}{4\pi^2}\Lambda_i^2 \sum_{\hat{\mathbf{e}}} |\hat{\mathbf{e}} \cdot \mathbf{q}|^2 \frac{d\omega}{\omega} d\Omega, \tag{5.23}$$

where $\alpha(=e^2/\hbar c \cong 1/137)$ is the fine structure constant, $\Lambda_i(\equiv \hbar/m_i c)$ is the Compton wavelength of the ion, m_i is the mass of the ion, c is the velocity of the light, $\hat{\mathbf{e}}$ is the unit photon polarization vector, and $d\Omega$ is the differential sold angle. In dusty plasmas consisting of electrons, ions, and negatively charged dusty grains, the quasineutral condition [9] at equilibrium is represented by

$$n_{e0} + Z_d n_{d0} = n_{i0} = n_0, \tag{5.24}$$

where n_{j0} $(j = e, i, d)$ are, respectively, the equilibrium densities for electrons (e), ions (i), and dust grains (d), $-Z_d$ is the charge number of the dusty grains, and n_0 is the total plasma density. In the case of $\left|e\phi/k_B T_j\right| \ll 1$ $(j = e, i)$, the Poisson's equation [5] in three-component dusty plasmas can be written as

$$\begin{aligned} \nabla^2\phi &= 4\pi e(n_e + Z_d n_d - n_i) \\ &\approx 4\pi e n_{e0}\frac{e\phi}{k_B T_e}\left(1 + \frac{n_0}{n_{e0}}\frac{T_e}{T_i}\right), \end{aligned} \tag{5.25}$$

where ϕ is the electrostatic potential in dusty plasmas, k_B is the Boltzmann constant, and T_j is the temperature of species j. The effective Debye length [9] λ_{eff} in three-component dusty plasmas is then given by

$$\lambda_{\text{eff}} = \lambda_D \left(\frac{n_{e0}}{n_0} + \frac{T_e}{T_i} \right)^{-1}, \tag{5.26}$$

where $\lambda_D (\equiv k_B T_e / 4\pi n_0 e^2)^{1/2}$ is the standard Debye length in electron–ion plasmas and the correction factor $(n_{e0}/n_0 + T_e/T_i)$ stands for the influence of the dusty plasma on the Debye length. Since the correction factor is usually greater than unity, the effective Debye length in dusty plasmas is found to be greater than that in customary electron–ion plasmas. Hence, in spherical polar coordinates with their origin at the center of the dust grain, the interaction potential $V_{i-d}(r)$ between the ion with charge ze and negatively charged dust grain with charge $-Z_d e$ in dusty plasmas is represented by the Yukawa form with the effective Debye length λ_{eff}:

$$V_{i-d}(r) = -\frac{Z_d z e^2}{r} \exp(-r/\lambda_{\text{eff}}). \tag{5.27}$$

For the sake of simplicity, the dust grains are assumed to be spherical shapes throughout in this work. After some algebra, the Fourier transformation $\bar{V}_{i-d}(q)$ of the ion–dust grain interaction potential is then given by

$$\bar{V}_{i-d}(q) = -\frac{4\pi Z_d z e^2 a}{q} \frac{\exp\left(-\frac{a}{\lambda_{\text{eff}}}\right)}{\left(\frac{a}{\lambda_{\text{eff}}}\right)^2 + (qa)^2} \left[\left(\frac{a}{\lambda_{\text{eff}}} \right) \sin(qa) + (qa)\cos(qa) \right], \tag{5.28}$$

where a is the radius of the spherical dust grain. Hence, the ion–dust bremsstrahlung cross section in dusty plasmas is then given by

$$d^2\sigma_b = \frac{16}{3} \frac{\alpha^3 a_0^2}{\bar{E}_0} \frac{m_e}{m_i} Z_d^2 z^2 \exp(-2\bar{\lambda}_{\text{eff}}^{-1}) \left(\frac{\bar{\lambda}_{\text{eff}}^{-1} \sin\bar{q} + \bar{q}\cos\bar{q}}{\bar{\lambda}_{\text{eff}}^{-2} + \bar{q}^2} \right)^2 \bar{q} d\bar{q} \frac{d\omega}{\omega}, \tag{5.29}$$

where $a_0 (= \hbar^2/m_e e^2)$ is the first Bohr radius of the hydrogen atom, m_e is the mass of the electron, $\bar{E}_0 (\equiv m_i v_0^2 / 2Ry)$ is the scaled initial collision energy, $Ry (= m_e e^4 / 2\hbar^2 \approx 13.6\,\text{eV})$ is the Rydberg constant, $\bar{q} (\equiv qa)$ is the scaled momentum transfer, and $\bar{\lambda}_{\text{eff}} (\equiv \lambda_{\text{eff}}/a)$ is the scaled effective Debye length. This expression of the bremsstrahlung cross section is reliable when the kinetic energy of the projectile ion is greater than the interaction energy between the ion and dust grain due to the prerequisite of the Born approximation. It has been also shown that the continuum spectrum due to the bremsstrahlung process would be investigated through the bremsstrahlung radiation cross section [32] defined as $d^2\chi_b / d\bar{\varepsilon} d\bar{q} \equiv \hbar\omega (d\sigma_b / \hbar d\omega d\bar{q})$, where $\bar{\varepsilon} (\equiv \varepsilon / Ry)$ is the scaled photon energy and $\varepsilon (\equiv \hbar\omega)$ is the photon energy. After some mathematical manipulations, the ion–dust grain bremsstrahlung radiation cross section $\mathrm{d}^2\chi_\mathrm{b}/\mathrm{d}\bar{\varepsilon}$ in three-component dusty plasmas is then obtained by the following analytic form:

$$\begin{aligned}\frac{\mathrm{d}^2\chi_\mathrm{b}}{\mathrm{d}\bar{\varepsilon}} &= \frac{4}{3}\frac{\alpha^3 a_0^2}{\bar{E}_0}(\eta Z_\mathrm{d} z)^2 \exp\left(-2\xi^{-1}\bar{\lambda}_\mathrm{D}^{-1}\right)\left|\ln\left[\frac{1+\left(\bar{a}\eta\xi\bar{\lambda}_\mathrm{D}\left(\sqrt{\bar{E}_0}+\sqrt{\bar{E}_f}\right)\right)^2}{1+\left(\bar{a}\eta\xi\bar{\lambda}_\mathrm{D}\left(\sqrt{\bar{E}_0}-\sqrt{\bar{E}_f}\right)\right)^2}\right]\right.\\
&+\left(1+2\xi^{-1}\bar{\lambda}_\mathrm{D}^{-1}\right)\left[\cosh\left(2\xi^{-1}\bar{\lambda}_\mathrm{D}^{-1}\right)+\sinh\left(2\xi^{-1}\bar{\lambda}_\mathrm{D}^{-1}\right)\right]\\
&\times\left\{\sum_{l=1}^{2}\left[\mathrm{Ci}\left((-1)^l 2\mathrm{i}\xi^{-1}\bar{\lambda}_\mathrm{D}^{-1}+2\bar{a}\eta\left(\sqrt{\bar{E}_0}+\sqrt{\bar{E}_f}\right)\right)\right.\right.\\
&-\left.\mathrm{Ci}\left((-,1)^l 2\mathrm{i}\xi^{-1}\bar{\lambda}_\mathrm{D}^{-1}+2\bar{a}\eta\left(\sqrt{\bar{E}_0}-\sqrt{\bar{E}_f}\right)\right)\right]\\
&+i\sum_{l=1}^{2}\left[\mathrm{Si}\left(2\mathrm{i}\xi^{-1}\bar{\lambda}_\mathrm{D}^{-1}+(-1)^l 2\bar{a}\eta\left(\sqrt{\bar{E}_0}+\sqrt{\bar{E}_f}\right)\right)\right.\\
&-\left.\mathrm{Si}\left(2i\xi^{-1}\bar{\lambda}_\mathrm{D}^{-1}+(-1)^l 2\bar{a}\eta\left(\sqrt{\bar{E}_0}-\sqrt{\bar{E}_f}\right)\right)\right]\\
&+\frac{2\cos\left(\bar{a}\eta\left(\sqrt{\bar{E}_0}+\sqrt{\bar{E}_f}\right)\right)}{1+\left(\bar{a}\eta\xi\bar{\lambda}_\mathrm{D}\left(\sqrt{\bar{E}_0}+\sqrt{\bar{E}_f}\right)\right)^2}-\frac{2\cos\left(\bar{a}\eta\left(\sqrt{\bar{E}_0}-\sqrt{\bar{E}_f}\right)\right)}{1+\left(\bar{a}\eta\xi\bar{\lambda}_\mathrm{D}\left(\sqrt{\bar{E}_0}-\sqrt{\bar{E}_f}\right)\right)^2}\\
&-\frac{2\bar{a}\eta\xi\bar{\lambda}_\mathrm{D}\left(\sqrt{\bar{E}_0}+\sqrt{\bar{E}_f}\right)\sin\left(\bar{a}\eta\left(\sqrt{\bar{E}_0}+\sqrt{\bar{E}_f}\right)\right)}{1+\left(\bar{a}\eta\xi\bar{\lambda}_\mathrm{D}\left(\sqrt{\bar{E}_0}+\sqrt{\bar{E}_f}\right)\right)^2}\\
&+\left.\frac{2\bar{a}\eta\xi\bar{\lambda}_\mathrm{D}\left(\sqrt{\bar{E}_0}-\sqrt{\bar{E}_f}\right)\sin\left(\bar{a}\eta\left(\sqrt{\bar{E}_0}-\sqrt{\bar{E}_f}\right)\right)}{1+\left(\bar{a}\eta\xi\bar{\lambda}_\mathrm{D}\left(\sqrt{\bar{E}_0}-\sqrt{\bar{E}_f}\right)\right)^2}\right|^2,\end{aligned}\tag{5.30}$$

where $\bar{E}_f(\equiv\bar{E}_0-\bar{\varepsilon})$ is the scaled final projectile energy, $\bar{a}(\equiv a/a_0)$ is the scaled radius of the dust grain, $\bar{\lambda}_\mathrm{D}\equiv\lambda_\mathrm{D}/a$, $\xi\equiv(n_{\mathrm{e}0}/n_0+T_\mathrm{e}/T_\mathrm{i})^{-1}$, $\mathrm{Ci}(x)(=-\int_x^\infty \mathrm{d}t\cos t/t)$, and $\mathrm{Si}(x)(=\int_0^x \mathrm{d}t\sin t/t)$ are the cosine and sine integrals [33], respectively. Very recently, an excellent discussion [34] on the additional part of the electrostatic potential in dusty plasmas is given due to the ion absorption by the dust grain and ion-neutral collisions. It has been found that the electrostatic potential would be represented only by the standard Debye–Hückel form if there is no ion flux on the surface of the dust grain, i.e., nonabsorbing dust grains. In this work, we just retain only the Debye–Hückel form of the interaction potential since we consider the bremsstrahlung process due to the scattering of the ion by the nonabsorbing dust grain. However, if the ion flux is existed on the dust grain, the attractive part of the potential has to be included in the ion–dust bremsstrhlung process. Hence, it would be expected that the additional part of the interaction potential suppresses the ion–dust grain bremsstrahlung spectrum obtained by the Yukawa-type Debye–Hückel potential. The validity of the Born approximation can be considered by using the Born parameter [19], $|V|/E$, for the potential scattering, where $|V|$ a typical strength of the interaction potential. Since the Born parameter $\frac{Z_d z e^2}{r_i E} e^{-r_i/\lambda_{eff}}$, where r_i is the interparticle distance between the ion and dust grain, for the ion–dust interaction in typical circumstances of dusty plasmas is usually less than unity, the Born approximation would be quite reliable to investigate the high-energy ion–dust bremsstrahlung process in dusty plasmas. In typical dusty plasmas, the range of the Coulomb coupling parameter for the ion–dust interaction in typical dusty plasmas is found to be $\Gamma_{i\text{–}d}(= \frac{Z_d z e^2}{r_i k_B T_i}) < 1$. However, the Coulomb coupling parameter for the dust–dust interaction [4] $\Gamma_{d\text{–}d}(= \frac{Z_d^2 e^2}{r_d k_B T_d})$, where r_d is the interparticle distance between the dust grains and T_d is the dust temperature, can be greater than the unity. Hence, for $\bar{E}_0, \bar{E}_f > 1$, the ion–dust grain bremsstrahlung radiation cross sections would be reliable since the interaction energy between the ion and charged dust grain is usually much less than the collision energy due to the considerable size of the dust grain. The ion–atom bremsstrahlung emission is expected to be quite different from the electron–ion bremsstrahlung emission due to the polarization of the target atom. Hence, the electron–atom and ion–atom polarization bremsstrahlung processes have to be considered in the partially ionized plasmas. However, the ion–atom polarization bremsstrahlung process is neglected in this work since the dust plasma is assumed to be completely ionized. For typical circumstances of dusty plasmas, it has been shown that $Z_d \approx$ 100–1,000, $a \approx$ 0.01–1 μm, and $\lambda_D/a \approx$ 5–100 [1].

In order to specifically investigate the effects of electron temperature and density on the ion–dust bremsstrahlung process in a dusty plasma, we set $Z_d = 200$, $a = 0.1$ μm, $\lambda_D/a = 50$, $m_i/m_e = 1{,}840$, and $z = 1$. Figure 5.6 shows the scaled bremsstrahlung radiation cross section $\partial^2_{\bar{\varepsilon}} \bar{\chi}_b [\equiv (d^2\chi_b/d\bar{\varepsilon})/\pi a_0^2]$ in units of πa_0^2 for the interaction of the ion with the negatively charged dust grain in dusty plasmas as a function of the temperature ratio T_e/T for various values of the density ratio n_{e0}/n_0. In addition, Fig. 5.7 represents the scaled bremsstrahlung radiation cross section $\partial^2_{\bar{\varepsilon}} \bar{\chi}_b$ as a function of the density ratio n_{e0}/n_0 for various values of the temperature ratio T_e/T_i. As shown in these figures, it found that the ion–dust bremsstrahlung radiation cross decreases with an increase of the ratio of the electron temperature to the ion temperature in a dusty plasma. It is also shown that the electron density suppresses the ion–dust bremsstrahlung radiation cross section. It is interesting to note that the effect of electron density on the bremsstrahlung radiation cross section diminishes with an increase of the electron temperature since the high-energy electrons can be readily repelled by the negatively charged dusty grains in dusty plasmas [5]. Figure 5.8 represents the surface plot of the effects of electron temperature and density on the ion–dust bremsstrahlung process $F(\bar{T}, \bar{n})$

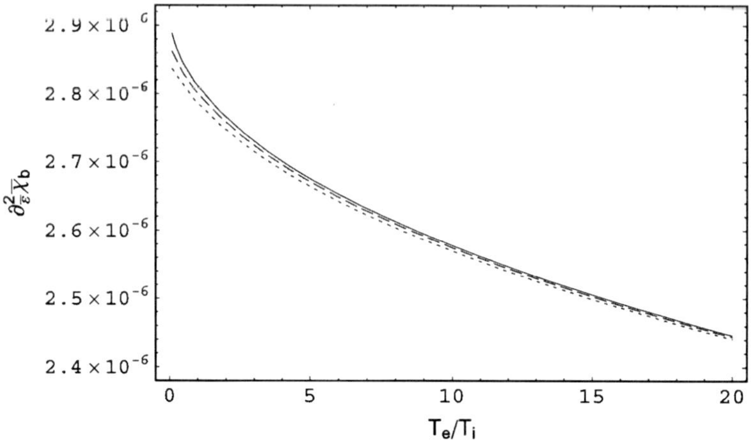

Fig. 5.6 The scaled ion–dust grain bremsstrahlung radiation cross section $\partial^2_{\bar{\varepsilon}}\bar{\chi}_b$ as a function of the temperature ratio T_e/T_i for $\bar{E}_0 = 5$, $\bar{\varepsilon} = 2$, $T_e = 5 \times 10^3$ K, $n_0 = 10^{10}\,\mathrm{cm}^{-3}$, and $a = 9.7 \times 10^{-5}$ cm. The *solid line* represents the case of $n_{e0}/n_0 = 0$. The *dashed line* represents the case of $n_{e0}/n_0 = 0.5$. The *dotted line* represents the case of $n_{e0}/n_0 = 0.9$. From [35]

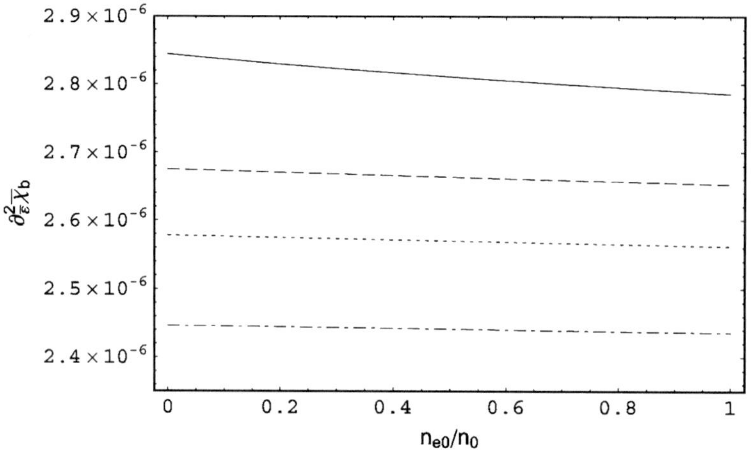

Fig. 5.7 The scaled ion–dust grain bremsstrahlung radiation cross section $\partial^2_{\bar{\varepsilon}}\bar{\chi}_b$ as a function of the temperature ratio n_{e0}/n_0 for $\bar{E}_0 = 5$, $\bar{\varepsilon} = 2$, $T_e = 5 \times 10^3$ K, $n_0 = 10^{10}\,\mathrm{cm}^{-3}$, and $a = 9.7 \times 10^{-5}$ cm. The *solid line* represents the case of $T_e/T_i = 0.5$. The *dashed line* represents the case of $T_e/T_i = 5$. The *dotted line* represents the case of $T_e/T_i = 10$. The *dot-dashed line* represents the case of $T_e/T_i = 20$. From [35]

as a function of the temperature ratio $\bar{T}(\equiv T_e/T_i)$ and density ratio $\bar{n}(\equiv n_{e0}/n_0)$. As it is seen, the effects of electron temperature and density significantly suppress the ion–dust grain bremsstrahlung radiation cross section. The effect of electron density is found to be important for low-temperature ratios. It is also found that the effect of electron temperature on the bremsstrahlung radiation cross section is

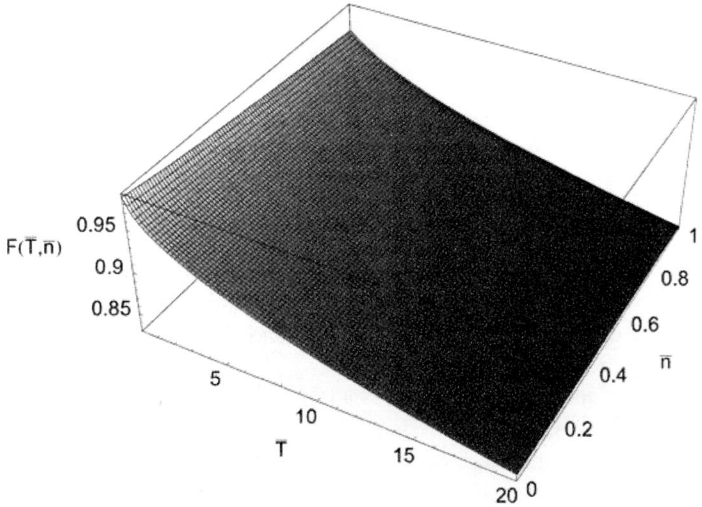

Fig. 5.8 The surface plot of the effects of electron temperature and density on the ion–dust grain bremsstrahlung process $F(\bar{T}, \bar{n})$ as a function of the temperature ratio $\bar{T}$ and density ratio $\bar{n}$ for $\bar{E}_0 = 5, \bar{\varepsilon} = 2, T_e = 5 \times 10^3$ K, $n_0 = 10^{10}\,\text{cm}^{-3}$, and $a = 9.7 \times 10^{-5}$ cm. From [35]

more significant than the effect of electron density in dusty plasmas. Hence, we have found that the effects of electron temperature and density play important roles in the ion–dust grain bremsstrahlung process in dusty plasmas containing electrons, ions, and negatively charged dust grains. These results would provide useful information on the ion–dust bremsstrahlung emission spectrum and also the radiation due to the interaction between dust particle chains and streaming ions in the plasma sheath in dusty plasmas.

5.5 Conclusions

In this chapter, we have discussed various atomic processes in dusty plasmas including the strong collective interactions.

Firstly, the effects of ion temperature on the electron–dust collision process are investigated in complex dusty plasmas by using the second-order eikonal analysis. We have found that the eikonal scattering phase and eikonal electron–dust grain collision cross section decrease with decreasing ion temperature in dusty plasmas. In addition, we found that the effect of ion temperature on the electron–dust grain collision process is more significant than the effect of electron density in dusty plasmas. We also found that the density effect is more significant when the electron temperature is comparable to the ion temperature. Hence, we have found that the ion temperature and density effects play important roles in the elastic electron–dust grain collision in complex dusty plasmas.

Secondly, the ion wake effects on the Coulomb drag force are investigated in complex dusty plasmas. We have found that the ion wake effects significantly enhance the Coulomb ion drag force. We also found that the ion wake effects on the Coulomb drag force increase with an increase of the Debye length. In addition, we found that the ion wake effects on the momentum transfer cross section and Coulomb drag force increase with increasing thermal Mach number, i.e., decreasing plasma temperature. We also found that the Coulomb ion drag force would be stronger for smaller dust grains. Hence, we have found that the influence of the ion wake field plays an important role in the Coulomb drag force and momentum transfer cross section for the ion–dust grain interaction in dusty plasmas.

Thirdly, the effects of electron temperature and density on the ion–dust grain bremsstrahlung process are investigated in dusty plasmas. We obtained the ion–dust bremsstrahlung radiation cross section as a function of the dust charge, dust radius, Debye length, collision energy, radiation energy, electron density, and electron temperature by using the Born approximation. We have found that the ion–dust bremsstrahlung radiation cross section decreases with increasing electron density in dusty plasmas. We also found that the electron temperature suppresses the bremsstrahlung radiation cross section. In addition, we found that the effect of electron temperature on the ion–dust bremsstrahlung process is more significant than the effect of electron density in dusty plasmas. Hence, we have found that the effects of electron temperature and density play important roles in the ion–dust grain bremsstrahlung process in dusty plasmas containing electrons, ions, and negatively charged dust grains.

These results on atomic processes in dusty plasmas would provide useful information on the plasma parameters and physical properties of complex dusty plasmas.

Acknowledgements One of the authors (Y.-D. J.) gratefully acknowledges Dr. M. Rosenberg for the useful discussions and warm hospitality while visiting the Department of Electrical and Computer Engineering at the University of California, San Diego. He would also like to thank Prof. H. Tawara, Prof. T. Kato, Prof. M. Sato, Prof. Y. Hirooka, Prof. I. Murakami, and Prof. D. Kato for their warm hospitality and support while visiting the National Institute for Fusion Science (NIFS) in Japan as a long-term visiting professor. This research was supported by Basic Science Research Program through the National Research Foundation of Korea (NRF) funded by the Ministry of Education, Science, and Technology (Grant No. 2011–0003099).

Reference

1. D.A. Mendis, M. Rosenberg, Ann. Rev. Astron. Astrophys. **32**, 419 (1994)
2. A. Bouchoule, *Dusty Plasmas: Physics, Chemistry and Technological Impacts in Plasma Processing* (Wiley, Chichester, 1999)
3. P.K. Shukla, A.A. Mamum, *Introduction to Dusty Plasma Physics* (Institute of Physics Publishing, Bristol, 2002)
4. T.S. Ramazanov, K.N. Dzhumagulova, A.N. Jumabekov, M.K. Dosbolayev, Phys. Plasmas **15**, 053704 (2008)

5. S.A. Maiorov, T.S. Ramazanov, K.N. Dzhumagulova, A.N. Jumabekov, M.K. Dosbolayev, Phys. Plasmas **15**, 093701 (2008)
6. V. Fortov, I. Iakubov, A. Khrapak, *Physics of Strongly Coupled Plasma* (Oxford University Press, Oxford, 2006)
7. S.P. Khare: *Introduction to the Theory of Collisions of Electrons with Atoms and Molecules* (Plenum, New York, 2002)
8. Z. Metawei, Acta Phys. Polonica B **31**, 713 (2000)
9. S. Vidhya Lakshmi, R. Bharuthram, P.K. Shukla, Astrophys. Space Sci. **209**, 213 (1993)
10. E.T. Whittaker, G.N. Watson, *A Course of Modern Analysis*, 4th edn. (Cambridge University Press, Cambridge, 1952)
11. D.-H. Ki, Y.-D. Jung, J. Appl. Phys. **108**, 086101 (2010)
12. P. Bliokh, V. Sinitsin, V. Yaroshenko, *Dusty and Self-Gravitational Plasma in Space* (Kluwer, Dordrecht, 1995)
13. V.N. Tsytovich, G.E. Morfill, S.V. Vladimirov, H. Thomas, *Elementary Physics of Complex Plasmas* (Springer, Berlin, 2008)
14. J. Perrin, P. Molinas-Mata, P. Belenguer, J. Phys. D **27**, 2499 (1994)
15. J.E. Daugherty, D.B. Graves, J. Appl. Phys. **78**, 2279 (1995)
16. H.F. Beyer, V.P. Shevelko, *Introduction to the Physics of Highly Charged Ions* (Institute of Physics Publishing, Bristol, 2003)
17. G. Morfill, A.V. Ivlev, Rev. Mod. Phys. **81**, 1353 (2009)
18. V.E. Fortov, G.E. Morfill, *Complex and Dusty Plasma* (CRC Press, Boca Raton, 2010)
19. R.J. Gould, *Electromagnetic Processes* (Princeton University Press, Princeton, 2006)
20. G.A. Kobzev, I.T. Iakubov, M.M. Popovich, *Transport and Optical Properties of Nonideal Plasmas* (Plenum, New York, 1995)
21. S. Geltman, *Topics in Atomic Collision* (Krieger, Malabar, 1997)
22. V.L. Ginzburg, *Application of Electrodynamics in Theoretical Physics and Astrophysics* (Gordon and Breach, New York, 1989)
23. P.K. Shukla, L. Stenflo, R. Bingham, Phys. Lett. A **359**, 218 (2006)
24. L. Spitzer Jr., *Physical Processes in the Interstellar Medium* (Wiley, New York, 1978)
25. D.-H. Ki, Y.-D. Jung, Appl. Phys. Lett. **97**, 101502 (2010)
26. H.A. Bethe, E.E. Salpeter, *Quantum Mechanics of One- and Two-Electron Atoms* (Springer, Berlin, 1957)
27. G. Bekefi, *Radiation Processes in Plasmas* (Wiley, New York, 1966)
28. H. Totsuji, Phys. Rev. A **32**, 3005 (1985)
29. V.P. Shevelko, *Atoms and Their Spectroscopic Properties* (Springer, Berlin, 1997)
30. H.F. Beyer, H.-J. Kluge, V.P. Shevelko, *X-Ray Radiation of Highly Charged Ions* (Springer, Berlin, 1997)
31. V. Shevelko, H. Tawara, *Atomic Multielectron Processes* (Springer, Berlin, 1998)
32. J.D. Jackson, *Classical Electrodynamics* (Wiley, New York, 1962)
33. G. Arfken, *Mathematical Methods for Physicists* (Academic, New York 1966)
34. S.A. Khrapak, G.E. Morfill, Phys. Plasmas **15**, 084502 (2008)
35. Y.-D. Jung, I. Murakami, J. Appl. Phys. **105**, 106106 (2009)

Part II
Atomic Heavy-Particle Collisions

Chapter 5
Atomic Processes in Dusty Plasmas

D.-H. Ki and Y.-D. Jung

Abstract Various atomic processes are investigated in dusty plasmas including strong collective interactions. The effects of ion temperature on the electron–dust collision process are investigated in complex dusty plasmas. It is found that the eikonal scattering phase and electron–dust grain collision cross section decrease with decreasing ion temperature in dusty plasmas. The ion wake effects on the Coulomb drag force are investigated in complex dusty plasmas. It is found that the ion wake effects enhance the Coulomb ion drag force. The effects of electron temperature and density on the ion–dust grain bremsstrahlung process are investigated in dusty plasmas. It is found that the ion–dust bremsstrahlung radiation cross section decreases with increasing electron density in dusty plasmas.

5.1 Introduction

The atomic processes in plasmas have received considerable attention since the collision and radiation processes has been widely used as the plasma diagnostic tool in plasma spectroscopy. Recent years, there has been a considerable interest in various physical processes in complex plasmas composed of plasma particles and highly charged dust grains exhibiting the strong electrostatic interactions. Hence, in this chapter, we have discussed several atomic processes in complex dusty plasmas containing the strong collective effects.

At first, we have discussed the effects of ion temperature on the electron–dust collision process in complex dusty plasmas. The second-order eikonal method is employed to obtain the scattering phase and cross section for the electron–dust grain collision as functions of the impact parameter, collision energy, ion temperature, density, and Debye length. The results show that the eikonal scattering phase and eikonal electron–dust grain collision cross section decrease with decreasing ion temperature in dusty plasmas. It is also found that the effect of ion temperature on the electron–dust grain collision process is more significant than the effect of electron density in dusty plasmas. In addition, it is shown that the density

V. Shevelko and H. Tawara (eds.), *Atomic Processes in Basic and Applied Physics*, Springer Series on Atomic, Optical, and Plasma Physics 68, DOI 10.1007/978-3-642-25569-4_5, © Springer-Verlag Berlin Heidelberg 2012

effect is more significant when the electron temperature is comparable to the ion temperature.

In second, we have investigated the ion wake effects on the Coulomb drag force in complex dusty plasmas. It is shown that the ion wake effects significantly enhance the Coulomb ion drag force. It is also found that the ion wake effects on the Coulomb drag force increase with an increase of the Debye length. In addition, the ion wake effects on the momentum transfer cross section and Coulomb drag force are found to be increased with increasing thermal Mach number, i.e., decreasing plasma temperature. It is also found that the Coulomb ion drag force would be stronger for smaller dust grains.

In third, we have discussed the effects of electron temperature and density on the ion–dust grain bremsstrahlung process in dusty plasmas. The ion–dust bremsstrahlung radiation cross section is obtained as a function of the dust charge, dust radius, Debye length, collision energy, radiation energy, electron density, and electron temperature by using the Born approximation. It is shown that the ion–dust bremsstrahlung radiation cross section decreases with an increase of the electron density in dusty plasmas. It is also shown that the electron temperature suppresses the bremsstrahlung radiation cross section. In addition, the effect of electron temperature on the ion–dust bremsstrahlung process is found to be more significant than the effect of electron density in dusty plasmas.

5.2 Electron–Dust Collisions in Complex Dusty Plasmas

The elastic electron–ion collision in plasmas has received much attention since this process has been widely used as the plasma diagnostic tool, and the process is also one of the major atomic processes in various plasmas. Recent years, there has been a considerable interest in collision and radiation processes in complex plasmas composed of plasma particles and highly charged dust grains exhibiting the strong electrostatic interactions [1–5]. Moreover, it has been known that the collective plasma–dust interactions are ubiquitous in various astrophysical and laboratory plasmas. Hence, the numerous physical processes have been extensively explored in order to obtain the information on the plasma parameters such as the density and temperature of plasma particles in complex dusty plasmas [6]. It has been shown that the most of dusty plasmas would be mainly constituted of thermal electrons, ions, and negatively charged dusty grains, i.e., three-component dusty plasmas. Hence, it would be expected that the electron–dust grain collision in dusty plasmas would be different from that in only two-component electron–dust plasmas due to the influence of the plasma ions in dusty plasmas. However, the effect of the ion temperature on the electron–dust grain collision has not been investigates as yet. Thus, in this section, we have investigated the influence of the ion temperature on the elastic collision process due to the interaction between the electron and negatively charged dust grain in three-component dusty plasmas. The scattering phase and cross section for the electron–dust grain collision are obtained as functions of the

impact parameter, collision energy, ion temperature, density, and Debye length by using the second-order eikonal method with the impact parameter analysis.

For an interaction potential $V(\mathbf{r})$, the solution $\psi_k(\mathbf{r})$ of the nonrelativistic Schrödinger equation would be expressed by the following integral form of the Lippmann–Schwinger equation [7]:

$$\psi_k(\mathbf{r}) = \varphi_k(\mathbf{r}) + \frac{2\mu}{\hbar^2}\int \mathrm{d}^3\mathbf{r}' V(\mathbf{r}')\psi_k(\mathbf{r}')G(\mathbf{r},\mathbf{r}'). \tag{5.1}$$

Here, $\varphi_k(\mathbf{r})$ and $G(\mathbf{r},\mathbf{r}')$ are, respectively, the solution of the homogeneous equation and the Green's function:

$$\left(\nabla^2 + k^2\right)\varphi_k(\mathbf{r}) = 0, \tag{5.2}$$

$$\left(\nabla^2 + k^2\right)G(\mathbf{r},\mathbf{r}') = \delta(\mathbf{r},\mathbf{r}'), \tag{5.3}$$

where $k[= (2\mu E/\hbar^2)^{1/2}]$ is the wave number, μ is reduced mass of the collision system, $E(= \mu v^2/2)$ is the collision energy, v is the relative collision velocity, and $\delta(\mathbf{r},\mathbf{r}')$ is the Dirac delta function. By using the cylindrical coordinate system such as $\mathbf{r} = \mathbf{b} + z\hat{\mathbf{n}}$, where $\mathbf{b}$ is the impact parameter, $\hat{\mathbf{n}}$ is the unit vector normal to the momentum transfer $\Delta\mathbf{k}(\equiv \mathbf{k}_i - \mathbf{k}_f)$, $\mathbf{k}_i$ and $\mathbf{k}_f$ are, respectively, the incident and final wave vectors, the eikonal scattering amplitude $f_E(\Delta\mathbf{k})$ would then be obtained as the following integral form:

$$f_{\mathrm{E}}(\Delta\mathbf{k}) = -\frac{\mu}{2\pi\hbar^2}\int \mathrm{d}^3\mathbf{r}\,\exp\left[-\frac{\mathrm{i}\mu}{\hbar^2 k_i}\int_{-\infty}^{z}\mathrm{d}z' V(\mathbf{b},z')\right]\mathrm{e}^{\mathrm{i}\Delta\mathbf{k}\cdot\mathbf{r}}V(\mathbf{r}). \tag{5.4}$$

Since the differential eikonal collision cross section is determined by the relation $\mathrm{d}\sigma_{\mathrm{E}}/\mathrm{d}\Omega = |f_{\mathrm{E}}(\Delta\mathbf{k})|^2$, the total elastic eikonal collision cross section (σ_{E}) would be expressed as

$$\sigma_{\mathrm{E}}(k) = \int \mathrm{d}^2\mathbf{b}\left|\exp\left[\mathrm{i}\chi_{\mathrm{E}}(k,b)\right] - 1\right|^2, \tag{5.5}$$

where $\mathrm{d}\Omega$ is the differential solid angle and $|\mathbf{k}_i| = |\mathbf{k}_f| = k$ for the elastic collision process. Here, the total eikonal scattering phase $\chi_{\mathrm{E}}(k,b)$ would be represented by the following series expansion technique [8]:

$$\chi_{\mathrm{E}}(k,b) = \sum_l \left(-\frac{\mu}{\hbar^2}\right)^{l+1}\frac{1}{(l+1)!}\frac{1}{k}\left(\frac{\partial}{\partial k}\frac{1}{k} - \frac{b}{k^2}\frac{\partial}{\partial b}\right)^l \int_{-\infty}^{\infty}\mathrm{d}z V^{l+1}(z,b). \tag{5.6}$$

In dusty plasmas composed of electrons, ions, and negatively charged dusty grains, the quasineutral condition [9] at equilibrium would be represented by

$$n_{\mathrm{e}0} + Zn_{\mathrm{d}0} = n_{\mathrm{i}0} = n_0, \tag{5.7}$$

where $n_{\alpha 0}(\alpha = e, i, d)$ are, respectively, the equilibrium densities for electrons (e), ions (i), and dust grains (d), Z is the charge number of the dust grain, and n_0 is the total plasma density. In the case of $|e\phi/k_B T_\alpha| \ll 1 \quad (\alpha = e, i)$, the solution of the Poisson equation [10] in three-component dusty plasmas would be obtained by

$$\begin{aligned}\nabla^2\phi &= 4\pi e(n_e + Zn_d - n_i)\\ &\approx 4\pi e n_{e0}\frac{e\phi}{k_B T_e}\left(1 + \frac{n_0}{n_{e0}}\frac{T_e}{T_i}\right),\end{aligned} \tag{5.8}$$

where ϕ is the electrostatic scalar potential in dusty plasmas, k_B is the Boltzmann constant, and T_α is the temperature of species α. Then, the effective Debye shielding distance [9] λ_{eff} in three-component dusty plasmas is given by

$$\lambda_{eff} = \lambda_D\left(\frac{n_{e0}}{n_0} + \frac{T_e}{T_i}\right)^{-1/2}, \tag{5.9}$$

where $\lambda_D(\equiv k_B T_e/4\pi n_0 e^2)^{1/2}$ is the standard Debye length in electron–ion plasmas and the correction factor $(n_{e0}/n_0 + T_e/T_i)^{-1/2}$ represents for the influence of the density and temperature ratios on the Debye length. Since the correction factor is usually greater than unity, the effective Debye length in dusty plasmas is expected to be greater than that in conventional electron–ion plasmas. In spherical polar coordinates with their origin at the center of the dust grain, the interaction potential $V_{i-d}(r)$ between the electron and negatively charged dust grain with charge $-Ze$ in dusty plasmas is then represented by the Yukawa potential with the effective λ_{eff}:

$$V_{i-d}(r) = \frac{Ze^2}{r}\exp\left[-\left(\frac{n_{e0}}{n_0} + \frac{T_e}{T_i}\right)^{1/2} r/\lambda_D\right]. \tag{5.10}$$

By using (5.6) and (5.10) with the identity of the zeroth-order of the MacDonald function [10], the total eikonal scattering phase χ_E including the first- and second-order contributions for the elastic election–dust grain collision in dusty plasmas would be given by

$$\chi_E(\bar{b}, \bar{E}, \xi, \bar{\lambda}_D) = -\frac{2}{\bar{E}^{1/2}}K_0(\xi\bar{b}/\bar{\lambda}_D) - \frac{\xi}{2\bar{E}^{3/2}\bar{\lambda}_D}\frac{a_Z}{a}K_0(2\xi\bar{b}/\bar{\lambda}_D), \tag{5.11}$$

where $\bar{b}(\equiv b/a)$ is the scaled impact parameter, a is the radius of the spherical dust grain, $a_Z \equiv a_0/Z$, $a_0(=\hbar^2/m_e e^2)$ is the first Bohr radius of the hydrogen atom, m_e is the mass of the electron, $\bar{E}(\equiv E/Z^2 Ry)$ is the scaled collision energy, $Ry(= m_e e^4/2\hbar^2 \approx 13.6\,\text{eV})$ is the Rydberg constant, $\xi \equiv (n_{e0}/n_0 + T_e/T)^{1/2}$, and $\bar{\lambda}_D(\equiv \lambda_D/a)$ is the scaled effective Debye length. Hence, the scaled differential collision cross section $\partial\bar{\sigma}_E[\equiv (d\sigma_E/d\bar{b})/\pi a^2]$ in units of πa^2 within the framework

of the second-order eikonal analysis for the elastic electron–dust grain collision in dusty plasmas is then found to be

$$\partial\bar{\sigma}_{\mathrm{E}} = \frac{\mathrm{d}\sigma_{\mathrm{E}}}{\mathrm{d}\bar{b}}/\pi a^2 = 2\bar{b}\left|\exp\left[\mathrm{i}\chi_{\mathrm{E}}(\bar{b},\bar{E},\xi,\bar{\lambda}D)\right]-1\right|^2$$
$$= 2\bar{b}\left|\exp\left[-\mathrm{i}\frac{2}{\bar{E}^{1/2}}K_0\left[\left(\frac{n_{\mathrm{e}0}}{n_0}+\frac{T_{\mathrm{e}}}{T}\right)^{1/2}\bar{b}/\bar{\lambda}_{\mathrm{D}}\right]\right.\right.$$
$$\left.\left.-\mathrm{i}\frac{\left(\frac{n_{\mathrm{e}0}}{n_0}+\frac{T_{\mathrm{e}}}{T}\right)^{1/2}}{2\bar{E}^{3/2}\bar{\lambda}_D}\frac{a_Z}{a}K_0\left[2\left(\frac{n_{\mathrm{e}0}}{n_0}+\frac{T_{\mathrm{e}}}{T}\right)^{1/2}\bar{b}/\bar{\lambda}_{\mathrm{D}}\right]\right]-1\right|^2. \quad (5.12)$$

This eikonal approximation would be quite reliable to investigate the electron–dust grain collision in dusty plasmas since the Sommerfeld parameter $(Ze^2/r_{\mathrm{i}}E)\mathrm{e}^{-r_{\mathrm{i}}/\lambda_{\mathrm{eff}}}$, where r_{i} is the interparticle distance between the electron and dust grain, for the electron–dust interaction in typical circumstances of dusty plasmas is usually less than unity. For typical circumstances of dusty plasmas, it has been shown that $Z \approx 100$–1,000, $a \approx 0.01$–1 μm, and $\lambda_{\mathrm{D}}/a \approx 5$–100 [1].

Figure 5.1 shows the three-dimensional plot of the total eikonal phase χ_{E} as a function of the temperature ratio $\bar{T}(\equiv T_{\mathrm{e}}/T_{\mathrm{i}})$ and density ratio $\bar{n}(\equiv n_{\mathrm{e}0}/n_0)$. As shown, the total eikonal scattering phase decreases with increasing temperature ratio $\bar{T}$ in dusty plasmas. It is found that the diminution of the ion temperature strongly suppresses the eikonal scattering phase for the electron–dust grain collision. Hence, we can understand that the eikonal scattering phase decreases when

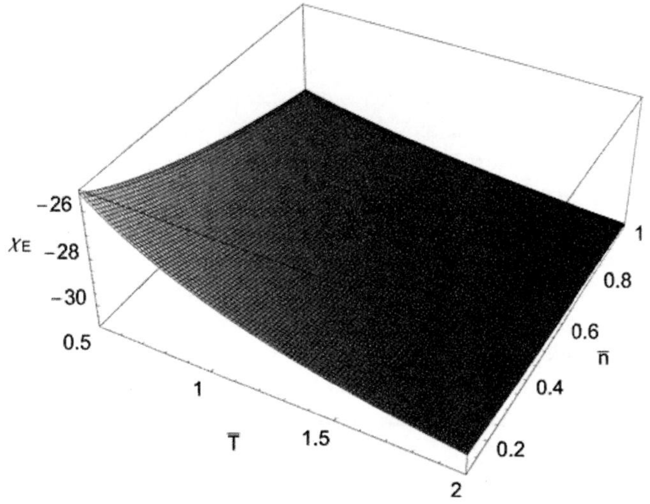

Fig. 5.1 The three-dimensional plot of the total eikonal scattering phase χ_{E} as a function of the temperature ratio $\bar{T}(\equiv T_{\mathrm{e}}/T_{\mathrm{i}})$ and density ratio $\bar{n}(\equiv n_{\mathrm{e}0}/n_0)$ when $\bar{b}=7$, $\bar{\lambda}_{\mathrm{D}}=50$, $a_Z/a = 5\times10^{-5}$, and $\bar{E}=5$. From [11]

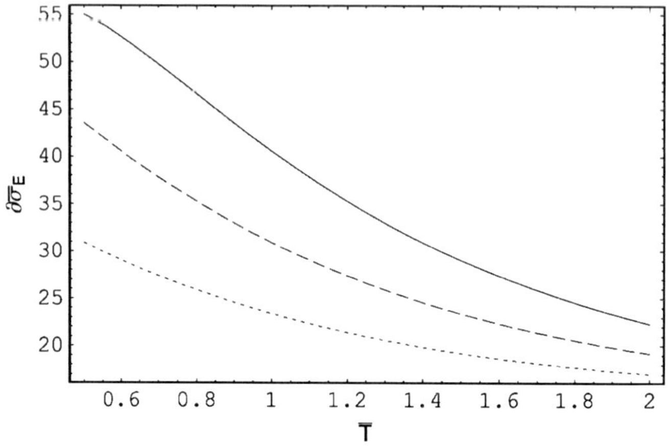

Fig. 5.2 The scaled differential collision cross section $\partial\bar{\sigma}_E$ as a function of the temperature ratio $\bar{T}(\equiv T_e/T_i)$ for various values of the density ratio $\bar{n}(\equiv n_{e0}/n_0)$ when $\bar{b}=7$, $\bar{\lambda}_D=50$, $a_Z/a=5\times10^{-5}$, and $\bar{E}=10$. The *solid line* represents the case of $\bar{n}=0.1$. The *dashed line* represents the case of $\bar{n}=0.5$. The *dotted line* represents the case of $\bar{n}=1$. From [11]

the ion temperature is smaller than the electron temperature in three-component dusty plasmas. It is also shown that the eikonal scattering phase decreases with an increase of the density ratio $\bar{n}$ in dusty plasmas. We also understand that the eikonal scattering phase decreases the electron equilibrium density. Thus, it is found that the ion temperature effect plays an important role in the scattering phase. In addition, the electron equilibrium density is quite important to determine the eikonal scattering phase in dusty plasmas. Figure 5.2 represents the scaled differential electron–dust grain collision cross section $\partial\bar{\sigma}_E$ as a function of the temperature ratio $\bar{T}$ for various values of the density ratio $\bar{n}$. As it is seen, the eikonal electron–dust grain collision cross section decreases with a decrease of the ion temperature. Then, it is found that the influence of the ion temperature strongly suppresses the electron–dust grain collision cross section in three-component dusty plasmas. Hence, it would be expected that the electron–dust grain collision cross sections in the one-temperature plasma ($T_e/T_i \approx 1$) such as the dust combustion would be smaller than those in conventional dusty plasmas ($T_e/T_i \gg 1$) due to the effect of the ion temperature. It is also found that the density effect on the scattering cross section decreases with an increase of the temperature ratio $\bar{T}$. Thus, it would be also expected that the density effects the electron–dust grain collision cross sections are more significant in the one-temperature plasma since the density dependence decreases with decreasing ion temperature. From this work, we have found that the ion temperature and density effects play important roles in the elastic electron–dust grain collision in complex dusty plasmas. These results would provide useful information on the collision processes in three-component complex dusty plasmas.

5.3 Ion Drag in Complex Dusty Plasmas

Recently, there has been a considerable interest in dynamics of plasmas containing charged dust grains including nonlinear collective effects and strong electrostatic interactions between the charged components in laboratory and space dusty plasmas [1, 2, 12]. Hence, the physical processes including collisions and radiations have been extensively explored in order to obtain the information on relevant plasma parameters of dusty plasmas [2, 3]. In addition, the momentum transfer collision [13–16] in dusty plasmas has been of a great interest since the momentum transfer process is related to the Coulomb ion drag force due to the ion–dust interaction. It has been known that the particle interaction in plasmas would be described by the Debye–Hückel model obtained by the linearization of the Poisson equation with the Boltzmann distribution. However, recently, it is shown that the ion–dust interaction potential including the ion flow in dusty plasmas would not be properly described by only the standard Debye–Hückel model so that the importance of the effect of the wake field has been proposed on the potential of the charged dust grain. Very recently, it is found that the influence of the wake field due to the ion flow produces the additional far-field terms [17, 18] in the ion–dust interaction potential. In addition, it has been found that the plasma wake generates the anisotropy [17, 18] of the plasma density around the charged dust grain. Hence, it would be expected that the Coulomb ion drag force including the influence of the wake field is quite different from that without the ion wake effect. Thus, in this section, we investigate the effects of the ion wake on the Coulomb drag force in dusty plasmas. The effective interaction potential model [17, 18] including the additional far-field terms due to the ion wake apart from the standard Debye–Hückel shielding term is employed in order to obtain the momentum transfer cross section and Coulomb ion drag force in dusty plasmas. In addition, the Born analysis [19] is applied to obtain the scattering amplitude as a function of the Debye length, dust charge, ion charge, plasma temperature, thermal Mach number, and collision wave number.

For the potential scattering, the differential momentum transfer cross section [20, 21] $\mathrm{d}\sigma_{\mathrm{M}}$ would be written as

$$\mathrm{d}\sigma_{\mathrm{M}} = \left|f(\mathbf{k}_i, \mathbf{k}_f)\right|^2 (1 - \cos\theta)\mathrm{d}\Omega, \tag{5.13}$$

where $f(\mathbf{k}_i, \mathbf{k}_f)$ is the scattering amplitude and $\mathbf{k}_i$ and $\mathbf{k}_f$ are, respectively, the wave vectors of the incident and final scattered waves, θ is the scattering angle, the factor $(1 - \cos\theta)$ denotes the fraction of the momentum transfer of the incident projectile, and $\mathrm{d}\Omega$ is the differential solid angle. It has been known that the momentum transfer cross section is relevant for investigating the Coulomb drag force for the particle–dust grain interaction in dusty plasmas. In the Born analysis [19], the scattering amplitude $f(\mathbf{k}_i, \mathbf{k}_f)$ would be expressed as

$$f(\mathbf{k}_i, \mathbf{k}_f) = -\frac{\mu}{2\pi\hbar^2} \int \mathrm{d}^3\mathbf{r}\, V(\mathbf{r}) \mathrm{e}^{\mathrm{i}(\mathbf{k}_i - \mathbf{k}_f)\cdot\mathbf{r}}, \tag{5.14}$$

where μ is the reduced mass of the collision system, $\hbar$ is the rationalized Planck constant, and $V(\mathbf{r})$ is the interaction potential. It is well known that the interaction potential in plasmas can be obtained by linearizing the Boltzmann–Poisson equation. However, it is found that the ion–dust interaction potential including the influence of the plasma wake field in dusty plasmas would not be properly described by only the ordinary Debye–Hückel term [13]. Very recently, an excellent work by Morfill and Ivlev [17] has provided the useful analytic form of the effective potential for the charged dust including the effects of the ion wake in dusty plasmas has been obtained by the superposition of the Debye–Hückel term and additional far-field terms including the Legendre polynomials. Using the effective potential model [17], the interaction potential energy $V_{\text{eff}}(\mathbf{k})$ between the projectile ion and target dust grain in dusty plasmas including the influence of the ion wake on the plasma density would be given by

$$V_{\text{eff}}(r,\varphi) = q_{\mathrm{i}} Q \left[\frac{\mathrm{e}^{-r/\lambda_{\mathrm{D}}}}{r} - \sqrt{\frac{8}{\pi}} \frac{M_{\mathrm{T}} \lambda_{\mathrm{D}}^2}{r^3} \cos\varphi - \left(2 - \frac{\pi}{2}\right) \frac{M_{\mathrm{T}}^2 \lambda_{\mathrm{D}}^2}{r^3} (3\cos^2\varphi - 1) \right], \tag{5.15}$$

where $q_{\mathrm{i}}(=ze)$ is the charge of the ion, $Q(=-Ze)$ is the charge of the dust grain, λ_{D} is the Debye length, λ_{D} is the value of the thermal Mach number, u_{i} is the ion velocity, $v_{T_{\mathrm{i}}}$ is the ion thermal velocity, and φ is the angle between the projectile velocity and position vector. As it is seen, the plasma wakes in dusty plasmas produce the nonlinear interactions which are similar to the Cherenkov radiations [22] in the moving medium.

It has been also shown that the dipole dependence of the effective potential in plasmas would be related to the Landau damping effect [23]. According to the contributions from the real and imaginary parts of the complex scattering amplitude, the total momentum transfer cross section for the interaction between the ion and dust grain including the influence of the ion wake field is then found to be

$$\begin{aligned}\sigma_{\mathrm{M}}(k, M_{\mathrm{T}}) = \sigma_0 \int_0^{\pi} \mathrm{d}\theta \sin\theta (1-\cos\theta) \Bigg\{ \Bigg[&\frac{Ka\cos(Ka) + (\frac{a}{\lambda_{\mathrm{D}}})\sin(Ka)}{Ka\left[(Ka)^2 + (\frac{a}{\lambda_{\mathrm{D}}})^2\right]} \\ &+ 4\left(2-\frac{\pi}{2}\right) M_{\mathrm{T}}^2 \left(\frac{\lambda_{\mathrm{D}}}{a}\right)^2 [-3S_4(Ka) + 3C_3(Ka) + S_2(Ka)] \Bigg]^2 \\ &+ \left(2\sqrt{\frac{2}{\pi}} M_{\mathrm{T}} \left(\frac{\lambda_{\mathrm{D}}}{a}\right)^2\right)^2 [2S_3(Ka) - C_2(Ka)]^2 \Bigg\},\end{aligned} \tag{5.16}$$

where $\sigma_0 \equiv 2\pi[zZ(m_{\mathrm{i}}/m_{\mathrm{e}})(a^2/a_0)]^2$, m_{i} is the ion mass, m_{e} is the electron mass, $a_0(=\hbar^2/m_{\mathrm{e}}e^2)$ is the Bohr radius of the hydrogen atom, $K \equiv 2k\sin(\theta/2)$,

$k \equiv |\mathbf{k}_i| = |\mathbf{k}_f|$ for the elastic scattering, and $S_n(Ka)$ and $C_n(Ka)$ are, respectively, given by

$$S_n(Ka) = \int_{Ka}^{\infty} d\xi \xi^{-n} \sin \xi, \tag{5.17}$$

$$C_n(Ka) = \int_{Ka}^{\infty} d\xi \xi^{-n} \cos \xi. \tag{5.18}$$

As shown in (5.18), the momentum transfer cross section has the strong dependence of the Mach number due to the ion wake field. It is shown that the Coulomb ion drag force $F_{\mathrm{CD}}(= n_{\mathrm{i}} m_{\mathrm{i}} \sigma_{\mathrm{M}} v_{\mathrm{i,tot}} u_{\mathrm{i}})$ would be represented by the time rate of the momentum transfer from the ion projectile to the target dust grain, where n_{i} is the ion density and $v_{\mathrm{i,tot}}(= u_{\mathrm{i}}^2 + 8v_{T_{\mathrm{i}}}^2/\pi)^{1/2}$ is the total ion velocity [3]. Hence, the Coulomb ion drag force including the influence of the ion wake field in dusty plasmas is then obtained in the following form:

$$\begin{aligned} F_{\mathrm{CD}}(\bar{k}, M_{\mathrm{T}}) = {} & F_0 \bar{k}^2 M_{\mathrm{T}}^{-1} (M_{\mathrm{T}}^2 + 8/\pi)^{1/2} \\ & \times \int_{-1}^{1} dy (1-y) \left\{ \left[\frac{A \cos A + (a/\lambda_D) \sin A}{A[2\bar{k}^2(1-y) + (a/\lambda_{\mathrm{D}})^2]} \right. \right. \\ & \left. + 4\left(2 - \frac{\pi}{2}\right) M_{\mathrm{T}}^2 \left(\frac{\lambda_{\mathrm{D}}}{a}\right)^2 [-3S_4(A) + 3C_3(A) + S_2(A)] \right]^2 \\ & \left. + \left(2\sqrt{\frac{2}{\pi}} M_{\mathrm{T}} \left(\frac{\lambda_{\mathrm{D}}}{a}\right)^2\right)^2 [2S_3(A) - C_2(A)]^2 \right\}, \end{aligned} \tag{5.19}$$

where $F_0 \equiv 2\pi(n_{\mathrm{i}}/m_{\mathrm{i}})[zZ(m_{\mathrm{i}}/m_{\mathrm{e}})(a/a_0)]^2$, $y \equiv \cos\varphi$, $\bar{k}(\equiv ka)$, and $A \equiv \bar{k}\sqrt{2(1-y)}$ is the scaled collision wave number. The contribution from the Coulomb electron drag would be neglected in dusty plasmas since the electron drag is quite smaller than the ion drag by the mass factor $(m_{\mathrm{e}}/m_{\mathrm{i}})^{1/2}$ [24].

Figure 5.3 shows the scaled total momentum transfer cross section $\bar{\sigma}_{\mathrm{M}}(\equiv \sigma_{\mathrm{M}}/\sigma_0)$ as a function of the scaled Debye length $\bar{\lambda}_{\mathrm{D}}(\equiv \lambda_{\mathrm{D}}/a)$. In addition, Fig. 5.4 represents the scaled total momentum transfer cross section $\bar{\sigma}_{\mathrm{M}}$ as a function of the thermal Mach number M_{T}. As shown in these figures, the momentum transfer cross section increases with an increase of the scaled Debye length. Hence, it can be expected that the momentum transfer cross sections would be greater for smaller dust grains. It is also shown that the momentum transfer cross section drastically increases with increasing Mach number. Then, it is understood that the wake field effects significantly enhance the cross section. In addition, it is found that the momentum transfer cross section decreases with an increase of the scaled collision wave number $\bar{k}$. Figure 5.5 represents the scaled Coulomb ion drag force

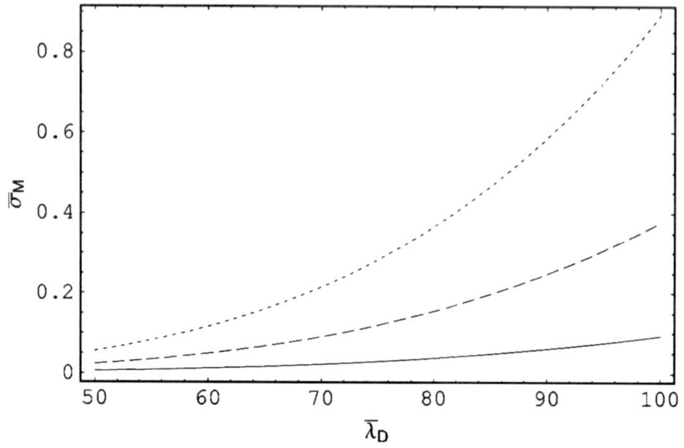

Fig. 5.3 The scaled total momentum transfer cross section $\bar{\sigma}_M$ as a function of the scaled Debye length $\bar{\lambda}_D$ for $\bar{k} = 1{,}000$. The *solid line* represents the case of $M_T = 0.1$. The *dashed line* represents the case of $M_T = 0.2$. The *dotted line* represents the case of $M_T = 0.3$. From [25]

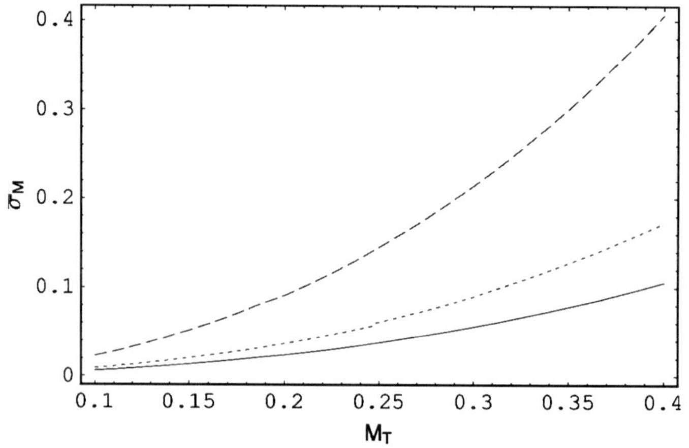

Fig. 5.4 The scaled total momentum transfer cross section $\bar{\sigma}_M$ as a function of the thermal Mach number M_T. The *solid line* is the case of $\bar{\lambda}_D = 50$ and $\bar{k} = 1{,}000$. The *dashed line* is the case of $\bar{\lambda}_D = 70$ and $\bar{k} = 1{,}000$. The *dotted line* is the case of $\bar{\lambda}_D = 70$ and $\bar{k} = 1{,}400$. From [25]

$\bar{F}_{CD}(\equiv F_{CD}/F_0)$ as a function of the thermal Mach number M_T for various values of the scaled Debye length $\bar{\lambda}_D$. As it is seen, the Coulomb ion drag force also increases with increasing Mach number and scaled Debye length. Hence, it can be also expected that the Coulomb ion drag force would be stronger for smaller dust grains. It is also found that the dependence of the thermal Mach number on the momentum transfer cross section is stronger than that on the Coulomb ion drag force due to the factor $(1 + 8/\pi M_T)^{-1/2}$. Therefore, in this work, we have found

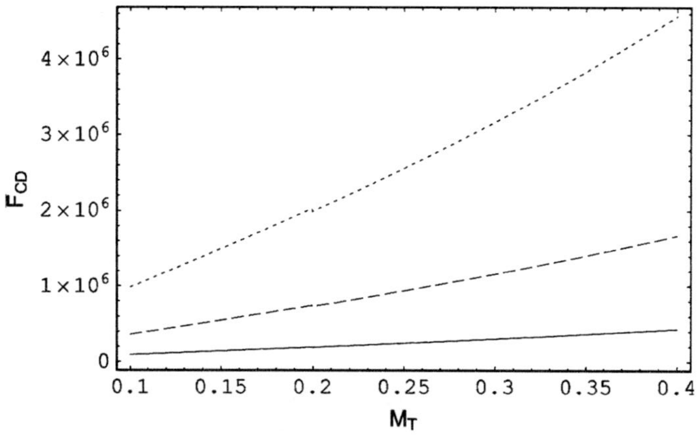

Fig. 5.5 The scaled Coulomb ion drag force $\bar{F}_{\mathrm{CD}}$ as a function of the thermal Mach number M_{T} for $\bar{k} = 1{,}000$. The *solid line* represents the case of $\bar{\lambda}_{\mathrm{D}} = 50$. The *dashed line* represents the case of $\bar{\lambda}_{\mathrm{D}} = 70$. The *dotted line* represents the case of $\bar{\lambda}_{\mathrm{D}} = 90$. From [25]

that the influence of the ion wake field plays an important role in the Coulomb drag force and momentum transfer cross section for the ion–dust grain interaction in dusty plasmas. These results provide useful information on the effects of the plasma wake field on the Coulomb drag in dusty plasmas.

5.4 Ion–Dust Bremsstrahlung Spectrum in Dusty Plasmas

The bremsstrahlung [16, 26–31] in plasmas has received much attention since this process is known to be one of the most fundamental processes, and the continuum radiation due to the bremsstrahlung process has been a diagnostic tool for investigating the plasma parameters. Recently, there has been a considerable interest in physical processes in plasmas encompassing highly charged dust grains [1–5]. In addition, it has been shown that the collective dust–plasma interactions are ubiquitous in various astrophysical and laboratory plasmas [3]. Hence, the collision and radiation processes have been extensively investigated in order to obtain the information on the plasma parameters such as the plasma density and temperature in dusty plasmas. In most cases, the dusty plasmas are consisting of thermal electrons and ions, and negatively charged dusty grains, i.e., three-component dusty plasmas. Hence, the ion–dust grain interaction in dusty plasmas would be different from that in only two-component ion–dust plasmas due to the influence of the plasma electrons in dusty plasmas. Thus, in this section, we have investigated the effects of electron density and temperature on the bremsstrahlung process due to the interaction of ion and negatively charged dust grain in three-component dusty plasmas. The ion–dust grain bremsstrahlung radiation cross section is obtained as

a function of the dust charge, dust radius, Debye length, collision energy, radiation energy, electron density, and electron temperature by using the Born approximation for the initial and final states of the projectile ion.

In the Born approximation, the differential ion–dust grain bremsstrahlung cross section [19] $\mathrm{d}^2\sigma_\mathrm{b}$ can be obtained by the second-order nonrelativistic perturbation method:

$$\mathrm{d}^2\sigma_\mathrm{b} = \mathrm{d}\sigma_\mathrm{C} \cdot \mathrm{d}W_\omega, \tag{5.20}$$

where the differential elastic scattering cross section $\mathrm{d}\sigma_\mathrm{C}$ can be represented by

$$\mathrm{d}\sigma_\mathrm{C} = \frac{1}{2\pi\hbar v_0^2}\left|\bar{V}(\mathbf{q})\right|^2 q\mathrm{d}q, \tag{5.21}$$

where $\hbar$ is the rationalized Planck constant, v_0 is the initial relative collision velocity, and $\bar{V}(\mathbf{q})$ is the Fourier transformation of the ion–dust grain interaction potential $V(\mathbf{r})$:

$$\bar{V}(\mathbf{q}) = \int \mathrm{d}^3\mathbf{r}\,\mathrm{e}^{-\mathrm{i}\mathbf{q}\cdot\mathbf{r}}\,V(\mathbf{r}), \tag{5.22}$$

$\mathbf{q}(=\mathbf{k}_0 - \mathbf{k}_f)$ is the momentum transfer, and $\mathbf{k}_0$ and $\mathbf{k}_f$ are, respectively, the wave vectors of the initial and final states of the projectile ion. Here, $\mathrm{d}W_\omega$ represents the differential photon emission probability within the frequency between ω and $\omega + \mathrm{d}\omega$:

$$\mathrm{d}W_\omega = \frac{\alpha}{4\pi^2}\Lambda_\mathrm{i}^2 \sum_{\hat{\mathbf{e}}} |\hat{\mathbf{e}} \cdot \mathbf{q}|^2 \frac{\mathrm{d}\omega}{\omega}\mathrm{d}\Omega, \tag{5.23}$$

where $\alpha(= e^2/\hbar c \cong 1/137)$ is the fine structure constant, $\Lambda_\mathrm{i}(\equiv \hbar/m_\mathrm{i}c)$ is the Compton wavelength of the ion, m_i is the mass of the ion, c is the velocity of the light, $\hat{\mathbf{e}}$ is the unit photon polarization vector, and $\mathrm{d}\Omega$ is the differential sold angle. In dusty plasmas consisting of electrons, ions, and negatively charged dusty grains, the quasineutral condition [9] at equilibrium is represented by

$$n_{\mathrm{e}0} + Z_\mathrm{d} n_{\mathrm{d}0} = n_{\mathrm{i}0} = n_0, \tag{5.24}$$

where n_{j0} $(j = e, i, d)$ are, respectively, the equilibrium densities for electrons (e), ions (i), and dust grains (d), $-Z_\mathrm{d}$ is the charge number of the dusty grains, and n_0 is the total plasma density. In the case of $\left|e\phi/k_\mathrm{B}T_j\right| \ll 1$ $(j = e, i)$, the Poisson's equation [5] in three-component dusty plasmas can be written as

$$\begin{aligned} \nabla^2\phi &= 4\pi e(n_\mathrm{e} + Z_\mathrm{d}n_\mathrm{d} - n_\mathrm{i}) \\ &\approx 4\pi e n_{\mathrm{e}0}\frac{e\phi}{k_\mathrm{B}T_\mathrm{e}}\left(1 + \frac{n_0}{n_{\mathrm{e}0}}\frac{T_\mathrm{e}}{T_\mathrm{i}}\right), \end{aligned} \tag{5.25}$$

where ϕ is the electrostatic potential in dusty plasmas, k_B is the Boltzmann constant, and T_j is the temperature of species j. The effective Debye length [9] λ_{eff} in three-component dusty plasmas is then given by

$$\lambda_{\text{eff}} = \lambda_D \left(\frac{n_{e0}}{n_0} + \frac{T_e}{T_i} \right)^{-1}, \tag{5.26}$$

where $\lambda_D (\equiv k_B T_e / 4\pi n_0 e^2)^{1/2}$ is the standard Debye length in electron–ion plasmas and the correction factor $(n_{e0}/n_0 + T_e/T_i)$ stands for the influence of the dusty plasma on the Debye length. Since the correction factor is usually greater than unity, the effective Debye length in dusty plasmas is found to be greater than that in customary electron–ion plasmas. Hence, in spherical polar coordinates with their origin at the center of the dust grain, the interaction potential $V_{i-d}(r)$ between the ion with charge ze and negatively charged dust grain with charge $-Z_d e$ in dusty plasmas is represented by the Yukawa form with the effective Debye length λ_{eff}:

$$V_{i-d}(r) = -\frac{Z_d z e^2}{r} \exp(-r/\lambda_{\text{eff}}). \tag{5.27}$$

For the sake of simplicity, the dust grains are assumed to be spherical shapes throughout in this work. After some algebra, the Fourier transformation $\bar{V}_{i-d}(q)$ of the ion–dust grain interaction potential is then given by

$$\bar{V}_{i-d}(q) = -\frac{4\pi Z_d z e^2 a}{q} \frac{\exp\left(-\frac{a}{\lambda_{\text{eff}}}\right)}{\left(\frac{a}{\lambda_{\text{eff}}}\right)^2 + (qa)^2} \left[\left(\frac{a}{\lambda_{\text{eff}}} \right) \sin(qa) + (qa)\cos(qa) \right], \tag{5.28}$$

where a is the radius of the spherical dust grain. Hence, the ion–dust bremsstrahlung cross section in dusty plasmas is then given by

$$d^2\sigma_b = \frac{16}{3} \frac{\alpha^3 a_0^2}{\bar{E}_0} \frac{m_e}{m_i} Z_d^2 z^2 \exp(-2\bar{\lambda}_{\text{eff}}^{-1}) \left(\frac{\bar{\lambda}_{\text{eff}}^{-1} \sin\bar{q} + \bar{q}\cos\bar{q}}{\bar{\lambda}_{\text{eff}}^{-2} + \bar{q}^2} \right)^2 \bar{q}\, d\bar{q} \frac{d\omega}{\omega}, \tag{5.29}$$

where $a_0 (= \hbar^2/m_e e^2)$ is the first Bohr radius of the hydrogen atom, m_e is the mass of the electron, $\bar{E}_0 (\equiv m_i v_0^2 / 2Ry)$ is the scaled initial collision energy, $Ry (= m_e e^4 / 2\hbar^2 \approx 13.6\,\text{eV})$ is the Rydberg constant, $\bar{q} (\equiv qa)$ is the scaled momentum transfer, and $\bar{\lambda}_{\text{eff}} (\equiv \lambda_{\text{eff}}/a)$ is the scaled effective Debye length. This expression of the bremsstrahlung cross section is reliable when the kinetic energy of the projectile ion is greater than the interaction energy between the ion and dust grain due to the prerequisite of the Born approximation. It has been also shown that the continuum spectrum due to the bremsstrahlung process would be investigated through the bremsstrahlung radiation cross section [32] defined as $d^2\chi_b / d\bar{\varepsilon} d\bar{q} \equiv \hbar\omega (d\sigma_b / \hbar d\omega d\bar{q})$, where $\bar{\varepsilon} (\equiv \varepsilon / Ry)$ is the scaled photon energy and $\varepsilon (\equiv \hbar\omega)$ is the photon energy. After some mathematical manipulations, the ion–dust grain bremsstrahlung radiation cross section $\mathrm{d}^2\chi_\mathrm{b}/\mathrm{d}\bar{\varepsilon}$ in three-component dusty plasmas is then obtained by the following analytic form:

$$
\begin{aligned}
\frac{\mathrm{d}^2\chi_\mathrm{b}}{\mathrm{d}\bar{\varepsilon}} = \frac{4}{3}\frac{\alpha^3 a_0^2}{\bar{E}_0}(\eta Z_\mathrm{d} z)^2 \exp\left(-2\xi^{-1}\bar{\lambda}_\mathrm{D}^{-1}\right) \Bigg| \ln\left[\frac{1+\left(\bar{a}\eta\xi\bar{\lambda}_\mathrm{D}\left(\sqrt{\bar{E}_0}+\sqrt{\bar{E}_f}\right)\right)^2}{1+\left(\bar{a}\eta\xi\bar{\lambda}_\mathrm{D}\left(\sqrt{\bar{E}_0}-\sqrt{\bar{E}_f}\right)\right)^2}\right] \\
+\left(1+2\xi^{-1}\bar{\lambda}_\mathrm{D}^{-1}\right)\left[\cosh\left(2\xi^{-1}\bar{\lambda}_\mathrm{D}^{-1}\right)+\sinh\left(2\xi^{-1}\bar{\lambda}_\mathrm{D}^{-1}\right)\right] \\
\times\Bigg\{\sum_{l=1}^{2}\Bigg[\mathrm{Ci}\left((-1)^l 2\mathrm{i}\xi^{-1}\bar{\lambda}_\mathrm{D}^{-1}+2\bar{a}\eta\left(\sqrt{\bar{E}_0}+\sqrt{\bar{E}_f}\right)\right) \\
-\mathrm{Ci}\left((-,1)^l 2\mathrm{i}\xi^{-1}\bar{\lambda}_\mathrm{D}^{-1}+2\bar{a}\eta\left(\sqrt{\bar{E}_0}-\sqrt{\bar{E}_f}\right)\right)\Bigg] \\
+i\sum_{l=1}^{2}\Bigg[\mathrm{Si}\left(2\mathrm{i}\xi^{-1}\bar{\lambda}_\mathrm{D}^{-1}+(-1)^l 2\bar{a}\eta\left(\sqrt{\bar{E}_0}+\sqrt{\bar{E}_f}\right)\right) \\
-\mathrm{Si}\left(2i\xi^{-1}\bar{\lambda}_\mathrm{D}^{-1}+(-1)^l 2\bar{a}\eta\left(\sqrt{\bar{E}_0}-\sqrt{\bar{E}_f}\right)\right)\Bigg] \\
+\frac{2\cos\left(\bar{a}\eta\left(\sqrt{\bar{E}_0}+\sqrt{\bar{E}_f}\right)\right)}{1+\left(\bar{a}\eta\xi\bar{\lambda}_\mathrm{D}\left(\sqrt{\bar{E}_0}+\sqrt{\bar{E}_f}\right)\right)^2}-\frac{2\cos\left(\bar{a}\eta\left(\sqrt{\bar{E}_0}-\sqrt{\bar{E}_f}\right)\right)}{1+\left(\bar{a}\eta\xi\bar{\lambda}_\mathrm{D}\left(\sqrt{\bar{E}_0}-\sqrt{\bar{E}_f}\right)\right)^2} \\
-\frac{2\bar{a}\eta\xi\bar{\lambda}_\mathrm{D}\left(\sqrt{\bar{E}_0}+\sqrt{\bar{E}_f}\right)\sin\left(\bar{a}\eta\left(\sqrt{\bar{E}_0}+\sqrt{\bar{E}_f}\right)\right)}{1+\left(\bar{a}\eta\xi\bar{\lambda}_\mathrm{D}\left(\sqrt{\bar{E}_0}+\sqrt{\bar{E}_f}\right)\right)^2} \\
+\frac{2\bar{a}\eta\xi\bar{\lambda}_\mathrm{D}\left(\sqrt{\bar{E}_0}-\sqrt{\bar{E}_f}\right)\sin\left(\bar{a}\eta\left(\sqrt{\bar{E}_0}-\sqrt{\bar{E}_f}\right)\right)}{1+\left(\bar{a}\eta\xi\bar{\lambda}_\mathrm{D}\left(\sqrt{\bar{E}_0}-\sqrt{\bar{E}_f}\right)\right)^2}\Bigg|^2,
\end{aligned}
\tag{5.30}
$$

where $\bar{E}_f(\equiv\bar{E}_0-\bar{\varepsilon})$ is the scaled final projectile energy, $\bar{a}(\equiv a/a_0)$ is the scaled radius of the dust grain, $\bar{\lambda}_\mathrm{D}\equiv\lambda_\mathrm{D}/a$, $\xi\equiv(n_{\mathrm{e}0}/n_0+T_\mathrm{e}/T_\mathrm{i})^{-1}$, $\mathrm{Ci}(x)(=-\int_x^\infty \mathrm{d}t\cos t/t)$, and $\mathrm{Si}(x)(=\int_0^x \mathrm{d}t\sin t/t)$ are the cosine and sine integrals [33], respectively. Very recently, an excellent discussion [34] on the additional part of the electrostatic potential in dusty plasmas is given due to the ion absorption by the dust grain and ion-neutral collisions. It has been found that the electrostatic potential would be represented only by the standard Debye–Hückel form if there is no ion flux on the surface of the dust grain, i.e., nonabsorbing dust grains. In this work, we just retain only the Debye–Hückel form of the interaction potential since we consider the bremsstrahlung process due to the scattering of the ion by the nonabsorbing dust grain. However, if the ion flux is existed on the dust grain, the attractive part of the potential has to be included in the ion–dust bremsstrhlung process. Hence, it would be expected that the additional part of the interaction potential suppresses the ion–dust grain bremsstrahlung spectrum obtained by the Yukawa-type Debye–Hückel potential. The validity of the Born approximation can be considered by using the Born parameter [19], $|V|/E$, for the potential scattering, where $|V|$ a typical strength of the interaction potential. Since the Born parameter $\frac{Z_d z e^2}{r_i E} e^{-r_i/\lambda_{\text{eff}}}$, where r_i is the interparticle distance between the ion and dust grain, for the ion–dust interaction in typical circumstances of dusty plasmas is usually less than unity, the Born approximation would be quite reliable to investigate the high-energy ion–dust bremsstrahlung process in dusty plasmas. In typical dusty plasmas, the range of the Coulomb coupling parameter for the ion–dust interaction in typical dusty plasmas is found to be $\Gamma_{\text{i–d}}(=\frac{Z_d z e^2}{r_i k_B T_i}) < 1$. However, the Coulomb coupling parameter for the dust–dust interaction [4] $\Gamma_{\text{d–d}}(=\frac{Z_d^2 e^2}{r_d k_B T_d})$, where r_d is the interparticle distance between the dust grains and T_d is the dust temperature, can be greater than the unity. Hence, for $\bar{E}_0, \bar{E}_f > 1$, the ion–dust grain bremsstrahlung radiation cross sections would be reliable since the interaction energy between the ion and charged dust grain is usually much less than the collision energy due to the considerable size of the dust grain. The ion–atom bremsstrahlung emission is expected to be quite different from the electron–ion bremsstrahlung emission due to the polarization of the target atom. Hence, the electron–atom and ion–atom polarization bremsstrahlung processes have to be considered in the partially ionized plasmas. However, the ion–atom polarization bremsstrahlung process is neglected in this work since the dust plasma is assumed to be completely ionized. For typical circumstances of dusty plasmas, it has been shown that $Z_d \approx$ 100–1,000, $a \approx$ 0.01–1 μm, and $\lambda_D/a \approx$ 5–100 [1].

In order to specifically investigate the effects of electron temperature and density on the ion–dust bremsstrahlung process in a dusty plasma, we set $Z_d = 200$, $a = 0.1$ μm, $\lambda_D/a = 50$, $m_i/m_e = 1{,}840$, and $z = 1$. Figure 5.6 shows the scaled bremsstrahlung radiation cross section $\partial^2_{\bar{\varepsilon}} \bar{\chi}_b [\equiv (d^2\chi_b/d\bar{\varepsilon})/\pi a_0^2]$ in units of πa_0^2 for the interaction of the ion with the negatively charged dust grain in dusty plasmas as a function of the temperature ratio T_e/T for various values of the density ratio n_{e0}/n_0. In addition, Fig. 5.7 represents the scaled bremsstrahlung radiation cross section $\partial^2_{\bar{\varepsilon}} \bar{\chi}_b$ as a function of the density ratio n_{e0}/n_0 for various values of the temperature ratio T_e/T_i. As shown in these figures, it found that the ion–dust bremsstrahlung radiation cross decreases with an increase of the ratio of the electron temperature to the ion temperature in a dusty plasma. It is also shown that the electron density suppresses the ion–dust bremsstrahlung radiation cross section. It is interesting to note that the effect of electron density on the bremsstrahlung radiation cross section diminishes with an increase of the electron temperature since the high-energy electrons can be readily repelled by the negatively charged dusty grains in dusty plasmas [5]. Figure 5.8 represents the surface plot of the effects of electron temperature and density on the ion–dust bremsstrahlung process $F(\bar{T}, \bar{n})$

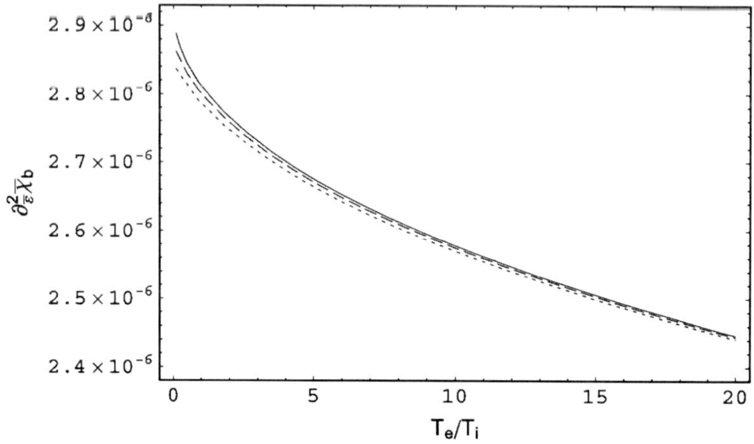

Fig. 5.6 The scaled ion–dust grain bremsstrahlung radiation cross section $\partial^2_{\bar{\varepsilon}}\bar{\chi}_{\mathrm{b}}$ as a function of the temperature ratio $T_{\mathrm{e}}/T_{\mathrm{i}}$ for $\bar{E}_0 = 5$, $\bar{\varepsilon} = 2$, $T_{\mathrm{e}} = 5 \times 10^3$ K, $n_0 = 10^{10}\,\mathrm{cm}^{-3}$, and $a = 9.7 \times 10^{-5}$ cm. The *solid line* represents the case of $n_{\mathrm{e}0}/n_0 = 0$. The *dashed line* represents the case of $n_{\mathrm{e}0}/n_0 = 0.5$. The *dotted line* represents the case of $n_{\mathrm{e}0}/n_0 = 0.9$. From [35]

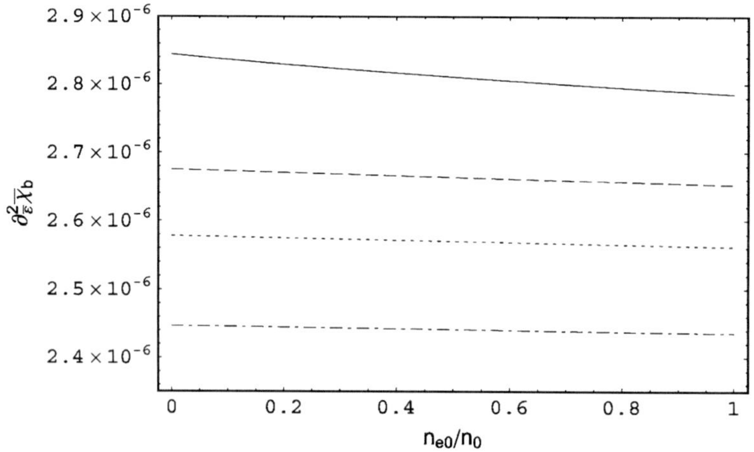

Fig. 5.7 The scaled ion–dust grain bremsstrahlung radiation cross section $\partial^2_{\bar{\varepsilon}}\bar{\chi}_{\mathrm{b}}$ as a function of the temperature ratio $n_{\mathrm{e}0}/n_0$ for $\bar{E}_0 = 5$, $\bar{\varepsilon} = 2$, $T_{\mathrm{e}} = 5 \times 10^3$ K, $n_0 = 10^{10}\,\mathrm{cm}^{-3}$, and $a = 9.7 \times 10^{-5}$ cm. The *solid line* represents the case of $T_{\mathrm{e}}/T_{\mathrm{i}} = 0.5$. The *dashed line* represents the case of $T_{\mathrm{e}}/T_{\mathrm{i}} = 5$. The *dotted line* represents the case of $T_{\mathrm{e}}/T_{\mathrm{i}} = 10$. The *dot-dashed line* represents the case of $T_{\mathrm{e}}/T_{\mathrm{i}} = 20$. From [35]

as a function of the temperature ratio $\bar{T}(\equiv T_{\mathrm{e}}/T_{\mathrm{i}})$ and density ratio $\bar{n}(\equiv n_{\mathrm{e}0}/n_0)$. As it is seen, the effects of electron temperature and density significantly suppress the ion–dust grain bremsstrahlung radiation cross section. The effect of electron density is found to be important for low-temperature ratios. It is also found that the effect of electron temperature on the bremsstrahlung radiation cross section is

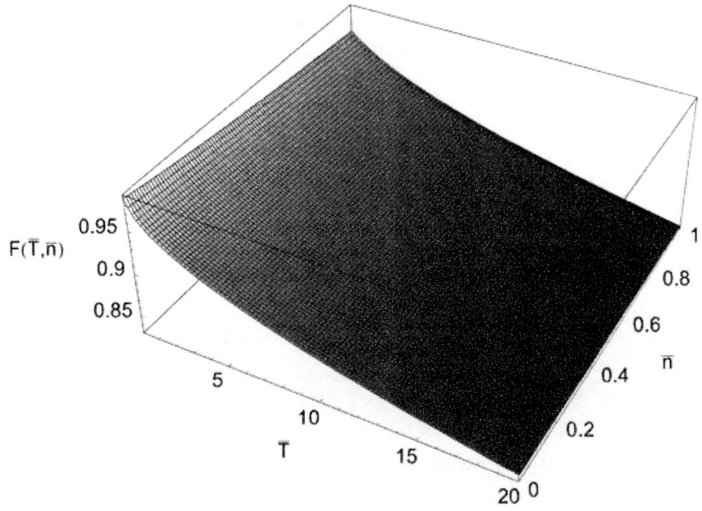

Fig. 5.8 The surface plot of the effects of electron temperature and density on the ion–dust grain bremsstrahlung process $F(\bar{T},\bar{n})$ as a function of the temperature ratio $\bar{T}$ and density ratio $\bar{n}$ for $\bar{E}_0 = 5$, $\bar{\varepsilon} = 2$, $T_e = 5 \times 10^3$ K, $n_0 = 10^{10}\,\mathrm{cm}^{-3}$, and $a = 9.7 \times 10^{-5}$ cm. From [35]

more significant than the effect of electron density in dusty plasmas. Hence, we have found that the effects of electron temperature and density play important roles in the ion–dust grain bremsstrahlung process in dusty plasmas containing electrons, ions, and negatively charged dust grains. These results would provide useful information on the ion–dust bremsstrahlung emission spectrum and also the radiation due to the interaction between dust particle chains and streaming ions in the plasma sheath in dusty plasmas.

5.5 Conclusions

In this chapter, we have discussed various atomic processes in dusty plasmas including the strong collective interactions.

Firstly, the effects of ion temperature on the electron–dust collision process are investigated in complex dusty plasmas by using the second-order eikonal analysis. We have found that the eikonal scattering phase and eikonal electron–dust grain collision cross section decrease with decreasing ion temperature in dusty plasmas. In addition, we found that the effect of ion temperature on the electron–dust grain collision process is more significant than the effect of electron density in dusty plasmas. We also found that the density effect is more significant when the electron temperature is comparable to the ion temperature. Hence, we have found that the ion temperature and density effects play important roles in the elastic electron–dust grain collision in complex dusty plasmas.

Secondly, the ion wake effects on the Coulomb drag force are investigated in complex dusty plasmas. We have found that the ion wake effects significantly enhance the Coulomb ion drag force. We also found that the ion wake effects on the Coulomb drag force increase with an increase of the Debye length. In addition, we found that the ion wake effects on the momentum transfer cross section and Coulomb drag force increase with increasing thermal Mach number, i.e., decreasing plasma temperature. We also found that the Coulomb ion drag force would be stronger for smaller dust grains. Hence, we have found that the influence of the ion wake field plays an important role in the Coulomb drag force and momentum transfer cross section for the ion–dust grain interaction in dusty plasmas.

Thirdly, the effects of electron temperature and density on the ion–dust grain bremsstrahlung process are investigated in dusty plasmas. We obtained the ion–dust bremsstrahlung radiation cross section as a function of the dust charge, dust radius, Debye length, collision energy, radiation energy, electron density, and electron temperature by using the Born approximation. We have found that the ion–dust bremsstrahlung radiation cross section decreases with increasing electron density in dusty plasmas. We also found that the electron temperature suppresses the bremsstrahlung radiation cross section. In addition, we found that the effect of electron temperature on the ion–dust bremsstrahlung process is more significant than the effect of electron density in dusty plasmas. Hence, we have found that the effects of electron temperature and density play important roles in the ion–dust grain bremsstrahlung process in dusty plasmas containing electrons, ions, and negatively charged dust grains.

These results on atomic processes in dusty plasmas would provide useful information on the plasma parameters and physical properties of complex dusty plasmas.

Acknowledgements One of the authors (Y.-D. J.) gratefully acknowledges Dr. M. Rosenberg for the useful discussions and warm hospitality while visiting the Department of Electrical and Computer Engineering at the University of California, San Diego. He would also like to thank Prof. H. Tawara, Prof. T. Kato, Prof. M. Sato, Prof. Y. Hirooka, Prof. I. Murakami, and Prof. D. Kato for their warm hospitality and support while visiting the National Institute for Fusion Science (NIFS) in Japan as a long-term visiting professor. This research was supported by Basic Science Research Program through the National Research Foundation of Korea (NRF) funded by the Ministry of Education, Science, and Technology (Grant No. 2011–0003099).

Reference

1. D.A. Mendis, M. Rosenberg, Ann. Rev. Astron. Astrophys. **32**, 419 (1994)
2. A. Bouchoule, *Dusty Plasmas: Physics, Chemistry and Technological Impacts in Plasma Processing* (Wiley, Chichester, 1999)
3. P.K. Shukla, A.A. Mamum, *Introduction to Dusty Plasma Physics* (Institute of Physics Publishing, Bristol, 2002)
4. T.S. Ramazanov, K.N. Dzhumagulova, A.N. Jumabekov, M.K. Dosbolayev, Phys. Plasmas **15**, 053704 (2008)

5. S.A. Maiorov, T.S. Ramazanov, K.N. Dzhumagulova, A.N. Jumabekov, M.K. Dosbolayev, Phys. Plasmas **15**, 093701 (2008)
6. V. Fortov, I. Iakubov, A. Khrapak, *Physics of Strongly Coupled Plasma* (Oxford University Press, Oxford, 2006)
7. S.P. Khare: *Introduction to the Theory of Collisions of Electrons with Atoms and Molecules* (Plenum, New York, 2002)
8. Z. Metawei, Acta Phys. Polonica B **31**, 713 (2000)
9. S. Vidhya Lakshmi, R. Bharuthram, P.K. Shukla, Astrophys. Space Sci. **209**, 213 (1993)
10. E.T. Whittaker, G.N. Watson, *A Course of Modern Analysis*, 4th edn. (Cambridge University Press, Cambridge, 1952)
11. D.-H. Ki, Y.-D. Jung, J. Appl. Phys. **108**, 086101 (2010)
12. P. Bliokh, V. Sinitsin, V. Yaroshenko, *Dusty and Self-Gravitational Plasma in Space* (Kluwer, Dordrecht, 1995)
13. V.N. Tsytovich, G.E. Morfill, S.V. Vladimirov, H. Thomas, *Elementary Physics of Complex Plasmas* (Springer, Berlin, 2008)
14. J. Perrin, P. Molinas-Mata, P. Belenguer, J. Phys. D **27**, 2499 (1994)
15. J.E. Daugherty, D.B. Graves, J. Appl. Phys. **78**, 2279 (1995)
16. H.F. Beyer, V.P. Shevelko, *Introduction to the Physics of Highly Charged Ions* (Institute of Physics Publishing, Bristol, 2003)
17. G. Morfill, A.V. Ivlev, Rev. Mod. Phys. **81**, 1353 (2009)
18. V.E. Fortov, G.E. Morfill, *Complex and Dusty Plasma* (CRC Press, Boca Raton, 2010)
19. R.J. Gould, *Electromagnetic Processes* (Princeton University Press, Princeton, 2006)
20. G.A. Kobzev, I.T. Iakubov, M.M. Popovich, *Transport and Optical Properties of Nonideal Plasmas* (Plenum, New York, 1995)
21. S. Geltman, *Topics in Atomic Collision* (Krieger, Malabar, 1997)
22. V.L. Ginzburg, *Application of Electrodynamics in Theoretical Physics and Astrophysics* (Gordon and Breach, New York, 1989)
23. P.K. Shukla, L. Stenflo, R. Bingham, Phys. Lett. A **359**, 218 (2006)
24. L. Spitzer Jr., *Physical Processes in the Interstellar Medium* (Wiley, New York, 1978)
25. D.-H. Ki, Y.-D. Jung, Appl. Phys. Lett. **97**, 101502 (2010)
26. H.A. Bethe, E.E. Salpeter, *Quantum Mechanics of One- and Two-Electron Atoms* (Springer, Berlin, 1957)
27. G. Bekefi, *Radiation Processes in Plasmas* (Wiley, New York, 1966)
28. H. Totsuji, Phys. Rev. A **32**, 3005 (1985)
29. V.P. Shevelko, *Atoms and Their Spectroscopic Properties* (Springer, Berlin, 1997)
30. H.F. Beyer, H.-J. Kluge, V.P. Shevelko, *X-Ray Radiation of Highly Charged Ions* (Springer, Berlin, 1997)
31. V. Shevelko, H. Tawara, *Atomic Multielectron Processes* (Springer, Berlin, 1998)
32. J.D. Jackson, *Classical Electrodynamics* (Wiley, New York, 1962)
33. G. Arfken, *Mathematical Methods for Physicists* (Academic, New York 1966)
34. S.A. Khrapak, G.E. Morfill, Phys. Plasmas **15**, 084502 (2008)
35. Y.-D. Jung, I. Murakami, J. Appl. Phys. **105**, 106106 (2009)

Chapter 5
Atomic Processes in Dusty Plasmas

D.-H. Ki and Y.-D. Jung

Abstract Various atomic processes are investigated in dusty plasmas including strong collective interactions. The effects of ion temperature on the electron–dust collision process are investigated in complex dusty plasmas. It is found that the eikonal scattering phase and electron–dust grain collision cross section decrease with decreasing ion temperature in dusty plasmas. The ion wake effects on the Coulomb drag force are investigated in complex dusty plasmas. It is found that the ion wake effects enhance the Coulomb ion drag force. The effects of electron temperature and density on the ion–dust grain bremsstrahlung process are investigated in dusty plasmas. It is found that the ion–dust bremsstrahlung radiation cross section decreases with increasing electron density in dusty plasmas.

5.1 Introduction

The atomic processes in plasmas have received considerable attention since the collision and radiation processes has been widely used as the plasma diagnostic tool in plasma spectroscopy. Recent years, there has been a considerable interest in various physical processes in complex plasmas composed of plasma particles and highly charged dust grains exhibiting the strong electrostatic interactions. Hence, in this chapter, we have discussed several atomic processes in complex dusty plasmas containing the strong collective effects.

At first, we have discussed the effects of ion temperature on the electron–dust collision process in complex dusty plasmas. The second-order eikonal method is employed to obtain the scattering phase and cross section for the electron–dust grain collision as functions of the impact parameter, collision energy, ion temperature, density, and Debye length. The results show that the eikonal scattering phase and eikonal electron–dust grain collision cross section decrease with decreasing ion temperature in dusty plasmas. It is also found that the effect of ion temperature on the electron–dust grain collision process is more significant than the effect of electron density in dusty plasmas. In addition, it is shown that the density

V. Shevelko and H. Tawara (eds.), *Atomic Processes in Basic and Applied Physics*, Springer Series on Atomic, Optical, and Plasma Physics 68,
DOI 10.1007/978-3-642-25569-4_5,

effect is more significant when the electron temperature is comparable to the ion temperature.

In second, we have investigated the ion wake effects on the Coulomb drag force in complex dusty plasmas. It is shown that the ion wake effects significantly enhance the Coulomb ion drag force. It is also found that the ion wake effects on the Coulomb drag force increase with an increase of the Debye length. In addition, the ion wake effects on the momentum transfer cross section and Coulomb drag force are found to be increased with increasing thermal Mach number, i.e., decreasing plasma temperature. It is also found that the Coulomb ion drag force would be stronger for smaller dust grains.

In third, we have discussed the effects of electron temperature and density on the ion–dust grain bremsstrahlung process in dusty plasmas. The ion–dust bremsstrahlung radiation cross section is obtained as a function of the dust charge, dust radius, Debye length, collision energy, radiation energy, electron density, and electron temperature by using the Born approximation. It is shown that the ion–dust bremsstrahlung radiation cross section decreases with an increase of the electron density in dusty plasmas. It is also shown that the electron temperature suppresses the bremsstrahlung radiation cross section. In addition, the effect of electron temperature on the ion–dust bremsstrahlung process is found to be more significant than the effect of electron density in dusty plasmas.

5.2 Electron–Dust Collisions in Complex Dusty Plasmas

The elastic electron–ion collision in plasmas has received much attention since this process has been widely used as the plasma diagnostic tool, and the process is also one of the major atomic processes in various plasmas. Recent years, there has been a considerable interest in collision and radiation processes in complex plasmas composed of plasma particles and highly charged dust grains exhibiting the strong electrostatic interactions [1–5]. Moreover, it has been known that the collective plasma–dust interactions are ubiquitous in various astrophysical and laboratory plasmas. Hence, the numerous physical processes have been extensively explored in order to obtain the information on the plasma parameters such as the density and temperature of plasma particles in complex dusty plasmas [6]. It has been shown that the most of dusty plasmas would be mainly constituted of thermal electrons, ions, and negatively charged dusty grains, i.e., three-component dusty plasmas. Hence, it would be expected that the electron–dust grain collision in dusty plasmas would be different from that in only two-component electron–dust plasmas due to the influence of the plasma ions in dusty plasmas. However, the effect of the ion temperature on the electron–dust grain collision has not been investigates as yet. Thus, in this section, we have investigated the influence of the ion temperature on the elastic collision process due to the interaction between the electron and negatively charged dust grain in three-component dusty plasmas. The scattering phase and cross section for the electron–dust grain collision are obtained as functions of the

impact parameter, collision energy, ion temperature, density, and Debye length by using the second-order eikonal method with the impact parameter analysis.

For an interaction potential $V(\mathbf{r})$, the solution $\psi_k(\mathbf{r})$ of the nonrelativistic Schrödinger equation would be expressed by the following integral form of the Lippmann–Schwinger equation [7]:

$$\psi_k(\mathbf{r}) = \varphi_k(\mathbf{r}) + \frac{2\mu}{\hbar^2}\int \mathrm{d}^3\mathbf{r}' V(\mathbf{r}')\psi_k(\mathbf{r}')G(\mathbf{r},\mathbf{r}'). \tag{5.1}$$

Here, $\varphi_k(\mathbf{r})$ and $G(\mathbf{r},\mathbf{r}')$ are, respectively, the solution of the homogeneous equation and the Green's function:

$$\left(\nabla^2 + k^2\right)\varphi_k(\mathbf{r}) = 0, \tag{5.2}$$

$$\left(\nabla^2 + k^2\right)G(\mathbf{r},\mathbf{r}') = \delta(\mathbf{r},\mathbf{r}'), \tag{5.3}$$

where $k[= (2\mu E/\hbar^2)^{1/2}]$ is the wave number, μ is reduced mass of the collision system, $E(= \mu v^2/2)$ is the collision energy, v is the relative collision velocity, and $\delta(\mathbf{r},\mathbf{r}')$ is the Dirac delta function. By using the cylindrical coordinate system such as $\mathbf{r} = \mathbf{b} + z\hat{\mathbf{n}}$, where $\mathbf{b}$ is the impact parameter, $\hat{\mathbf{n}}$ is the unit vector normal to the momentum transfer $\Delta\mathbf{k}(\equiv \mathbf{k}_i - \mathbf{k}_f)$, $\mathbf{k}_i$ and $\mathbf{k}_f$ are, respectively, the incident and final wave vectors, the eikonal scattering amplitude $f_E(\Delta\mathbf{k})$ would then be obtained as the following integral form:

$$f_{\mathrm{E}}(\Delta\mathbf{k}) = -\frac{\mu}{2\pi\hbar^2}\int \mathrm{d}^3\mathbf{r}\,\exp\left[-\frac{\mathrm{i}\mu}{\hbar^2 k_i}\int_{-\infty}^{z}\mathrm{d}z' V(\mathbf{b},z')\right]\mathrm{e}^{\mathrm{i}\Delta\mathbf{k}\cdot\mathbf{r}}V(\mathbf{r}). \tag{5.4}$$

Since the differential eikonal collision cross section is determined by the relation $\mathrm{d}\sigma_{\mathrm{E}}/\mathrm{d}\Omega = |f_{\mathrm{E}}(\Delta\mathbf{k})|^2$, the total elastic eikonal collision cross section (σ_{E}) would be expressed as

$$\sigma_{\mathrm{E}}(k) = \int \mathrm{d}^2\mathbf{b}\left|\exp\left[\mathrm{i}\chi_{\mathrm{E}}(k,b)\right] - 1\right|^2, \tag{5.5}$$

where $\mathrm{d}\Omega$ is the differential solid angle and $|\mathbf{k}_i| = |\mathbf{k}_f| = k$ for the elastic collision process. Here, the total eikonal scattering phase $\chi_{\mathrm{E}}(k,b)$ would be represented by the following series expansion technique [8]:

$$\chi_{\mathrm{E}}(k,b) = \sum_l \left(-\frac{\mu}{\hbar^2}\right)^{l+1}\frac{1}{(l+1)!}\frac{1}{k}\left(\frac{\partial}{\partial k}\frac{1}{k} - \frac{b}{k^2}\frac{\partial}{\partial b}\right)^l \int_{-\infty}^{\infty}\mathrm{d}z V^{l+1}(z,b). \tag{5.6}$$

In dusty plasmas composed of electrons, ions, and negatively charged dusty grains, the quasineutral condition [9] at equilibrium would be represented by

$$n_{\mathrm{e}0} + Zn_{\mathrm{d}0} = n_{\mathrm{i}0} = n_0, \tag{5.7}$$

where $n_{\alpha 0}(\alpha = e, i, d)$ are, respectively, the equilibrium densities for electrons (e), ions (i), and dust grains (d), Z is the charge number of the dust grain, and n_0 is the total plasma density. In the case of $|e\phi/k_B T_\alpha| \ll 1$ $(\alpha = e, i)$, the solution of the Poisson equation [10] in three-component dusty plasmas would be obtained by

$$\nabla^2\phi = 4\pi e(n_e + Zn_d - n_i)$$
$$\approx 4\pi e n_{e0}\frac{e\phi}{k_B T_e}\left(1 + \frac{n_0}{n_{e0}}\frac{T_e}{T_i}\right), \tag{5.8}$$

where ϕ is the electrostatic scalar potential in dusty plasmas, k_B is the Boltzmann constant, and T_α is the temperature of species α. Then, the effective Debye shielding distance [9] λ_{eff} in three-component dusty plasmas is given by

$$\lambda_{\text{eff}} = \lambda_D\left(\frac{n_{e0}}{n_0} + \frac{T_e}{T_i}\right)^{-1/2}, \tag{5.9}$$

where $\lambda_D(\equiv k_B T_e/4\pi n_0 e^2)^{1/2}$ is the standard Debye length in electron–ion plasmas and the correction factor $(n_{e0}/n_0 + T_e/T_i)^{-1/2}$ represents for the influence of the density and temperature ratios on the Debye length. Since the correction factor is usually greater than unity, the effective Debye length in dusty plasmas is expected to be greater than that in conventional electron–ion plasmas. In spherical polar coordinates with their origin at the center of the dust grain, the interaction potential $V_{i-d}(r)$ between the electron and negatively charged dust grain with charge $-Ze$ in dusty plasmas is then represented by the Yukawa potential with the effective λ_{eff}:

$$V_{i-d}(r) = \frac{Ze^2}{r}\exp\left[-\left(\frac{n_{e0}}{n_0} + \frac{T_e}{T_i}\right)^{1/2} r/\lambda_D\right]. \tag{5.10}$$

By using (5.6) and (5.10) with the identity of the zeroth-order of the MacDonald function [10], the total eikonal scattering phase χ_E including the first- and second-order contributions for the elastic election–dust grain collision in dusty plasmas would be given by

$$\chi_E(\bar{b}, \bar{E}, \xi, \bar{\lambda}_D) = -\frac{2}{\bar{E}^{1/2}}K_0(\xi\bar{b}/\bar{\lambda}_D) - \frac{\xi}{2\bar{E}^{3/2}\bar{\lambda}_D}\frac{a_Z}{a}K_0(2\xi\bar{b}/\bar{\lambda}_D), \tag{5.11}$$

where $\bar{b}(\equiv b/a)$ is the scaled impact parameter, a is the radius of the spherical dust grain, $a_Z \equiv a_0/Z$, $a_0(=\hbar^2/m_e e^2)$ is the first Bohr radius of the hydrogen atom, m_e is the mass of the electron, $\bar{E}(\equiv E/Z^2 Ry)$ is the scaled collision energy, $Ry(=m_e e^4/2\hbar^2 \approx 13.6\,\text{eV})$ is the Rydberg constant, $\xi \equiv (n_{e0}/n_0 + T_e/T)^{1/2}$, and $\bar{\lambda}_D(\equiv \lambda_D/a)$ is the scaled effective Debye length. Hence, the scaled differential collision cross section $\partial\bar{\sigma}_E[\equiv (d\sigma_E/d\bar{b})/\pi a^2]$ in units of πa^2 within the framework

of the second-order eikonal analysis for the elastic electron–dust grain collision in dusty plasmas is then found to be

$$\begin{aligned}\partial\bar{\sigma}_{\mathrm{E}} &= \frac{\mathrm{d}\sigma_{\mathrm{E}}}{\mathrm{d}\bar{b}}/\pi a^2 = 2\bar{b}\left|\exp\left[\mathrm{i}\chi_{\mathrm{E}}(\bar{b},\bar{E},\xi,\bar{\lambda}\,D)\right]-1\right|^2 \\ &= 2\bar{b}\left|\exp\left[-\mathrm{i}\frac{2}{\bar{E}^{1/2}}K_0\left[\left(\frac{n_{\mathrm{e}0}}{n_0}+\frac{T_{\mathrm{e}}}{T}\right)^{1/2}\bar{b}/\bar{\lambda}_{\mathrm{D}}\right]\right.\right. \\ &\quad \left.\left. -\mathrm{i}\frac{\left(\frac{n_{\mathrm{e}0}}{n_0}+\frac{T_{\mathrm{e}}}{T}\right)^{1/2}}{2\bar{E}^{3/2}\bar{\lambda}_D}\frac{a_Z}{a}K_0\left[2\left(\frac{n_{\mathrm{e}0}}{n_0}+\frac{T_{\mathrm{e}}}{T}\right)^{1/2}\bar{b}/\bar{\lambda}_{\mathrm{D}}\right]\right]-1\right|^2. \end{aligned} \tag{5.12}$$

This eikonal approximation would be quite reliable to investigate the electron–dust grain collision in dusty plasmas since the Sommerfeld parameter $(Ze^2/r_{\mathrm{i}}E)\mathrm{e}^{-r_{\mathrm{i}}/\lambda_{\mathrm{eff}}}$, where r_{i} is the interparticle distance between the electron and dust grain, for the electron–dust interaction in typical circumstances of dusty plasmas is usually less than unity. For typical circumstances of dusty plasmas, it has been shown that $Z \approx 100$–$1{,}000$, $a \approx 0.01$–$1\,\mu$m, and $\lambda_{\mathrm{D}}/a \approx 5$–$100$ [1].

Figure 5.1 shows the three-dimensional plot of the total eikonal phase χ_{E} as a function of the temperature ratio $\bar{T}(\equiv T_{\mathrm{e}}/T_{\mathrm{i}})$ and density ratio $\bar{n}(\equiv n_{\mathrm{e}0}/n_0)$. As shown, the total eikonal scattering phase decreases with increasing temperature ratio $\bar{T}$ in dusty plasmas. It is found that the diminution of the ion temperature strongly suppresses the eikonal scattering phase for the electron–dust grain collision. Hence, we can understand that the eikonal scattering phase decreases when

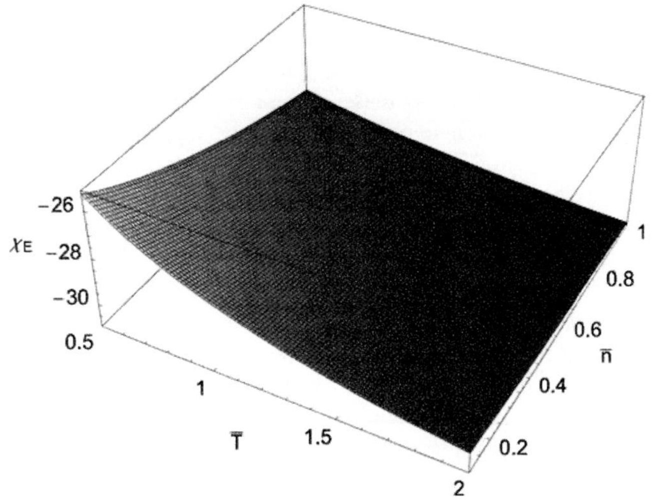

Fig. 5.1 The three-dimensional plot of the total eikonal scattering phase χ_{E} as a function of the temperature ratio $\bar{T}(\equiv T_{\mathrm{e}}/T_{\mathrm{i}})$ and density ratio $\bar{n}(\equiv n_{\mathrm{e}0}/n_0)$ when $\bar{b}=7$, $\bar{\lambda}_{\mathrm{D}}=50$, $a_Z/a = 5\times10^{-5}$, and $\bar{E}=5$. From [11]

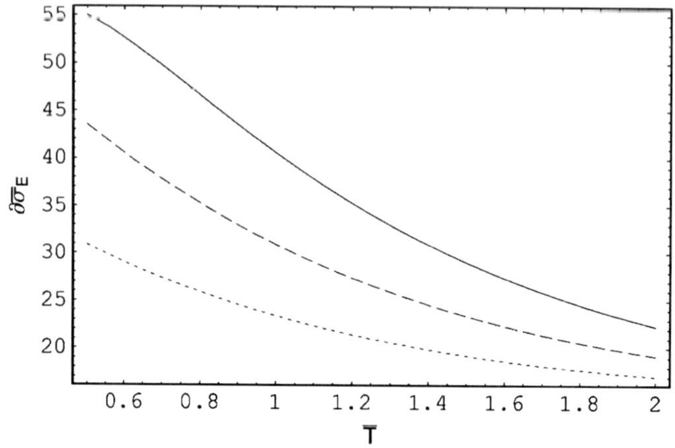

Fig. 5.2 The scaled differential collision cross section $\partial\bar{\sigma}_{\mathrm{E}}$ as a function of the temperature ratio $\bar{T}(\equiv T_{\mathrm{e}}/T_{\mathrm{i}})$ for various values of the density ratio $\bar{n}(\equiv n_{\mathrm{e}0}/n_0)$ when $\bar{b} = 7$, $\bar{\lambda}_{\mathrm{D}} = 50$, $a_Z/a = 5 \times 10^{-5}$, and $\bar{E} = 10$. The *solid line* represents the case of $\bar{n} = 0.1$. The *dashed line* represents the case of $\bar{n} = 0.5$. The *dotted line* represents the case of $\bar{n} = 1$. From [11]

the ion temperature is smaller than the electron temperature in three-component dusty plasmas. It is also shown that the eikonal scattering phase decreases with an increase of the density ratio $\bar{n}$ in dusty plasmas. We also understand that the eikonal scattering phase decreases the electron equilibrium density. Thus, it is found that the ion temperature effect plays an important role in the scattering phase. In addition, the electron equilibrium density is quite important to determine the eikonal scattering phase in dusty plasmas. Figure 5.2 represents the scaled differential electron–dust grain collision cross section $\partial\bar{\sigma}_{\mathrm{E}}$ as a function of the temperature ratio $\bar{T}$ for various values of the density ratio $\bar{n}$. As it is seen, the eikonal electron–dust grain collision cross section decreases with a decrease of the ion temperature. Then, it is found that the influence of the ion temperature strongly suppresses the electron–dust grain collision cross section in three-component dusty plasmas. Hence, it would be expected that the electron–dust grain collision cross sections in the one-temperature plasma ($T_{\mathrm{e}}/T_{\mathrm{i}} \approx 1$) such as the dust combustion would be smaller than those in conventional dusty plasmas ($T_{\mathrm{e}}/T_{\mathrm{i}} \gg 1$) due to the effect of the ion temperature. It is also found that the density effect on the scattering cross section decreases with an increase of the temperature ratio $\bar{T}$. Thus, it would be also expected that the density effects the electron–dust grain collision cross sections are more significant in the one-temperature plasma since the density dependence decreases with decreasing ion temperature. From this work, we have found that the ion temperature and density effects play important roles in the elastic electron–dust grain collision in complex dusty plasmas. These results would provide useful information on the collision processes in three-component complex dusty plasmas.

5.3 Ion Drag in Complex Dusty Plasmas

Recently, there has been a considerable interest in dynamics of plasmas containing charged dust grains including nonlinear collective effects and strong electrostatic interactions between the charged components in laboratory and space dusty plasmas [1, 2, 12]. Hence, the physical processes including collisions and radiations have been extensively explored in order to obtain the information on relevant plasma parameters of dusty plasmas [2, 3]. In addition, the momentum transfer collision [13–16] in dusty plasmas has been of a great interest since the momentum transfer process is related to the Coulomb ion drag force due to the ion–dust interaction. It has been known that the particle interaction in plasmas would be described by the Debye–Hückel model obtained by the linearization of the Poisson equation with the Boltzmann distribution. However, recently, it is shown that the ion–dust interaction potential including the ion flow in dusty plasmas would not be properly described by only the standard Debye–Hückel model so that the importance of the effect of the wake field has been proposed on the potential of the charged dust grain. Very recently, it is found that the influence of the wake field due to the ion flow produces the additional far-field terms [17, 18] in the ion–dust interaction potential. In addition, it has been found that the plasma wake generates the anisotropy [17, 18] of the plasma density around the charged dust grain. Hence, it would be expected that the Coulomb ion drag force including the influence of the wake field is quite different from that without the ion wake effect. Thus, in this section, we investigate the effects of the ion wake on the Coulomb drag force in dusty plasmas. The effective interaction potential model [17, 18] including the additional far-field terms due to the ion wake apart from the standard Debye–Hückel shielding term is employed in order to obtain the momentum transfer cross section and Coulomb ion drag force in dusty plasmas. In addition, the Born analysis [19] is applied to obtain the scattering amplitude as a function of the Debye length, dust charge, ion charge, plasma temperature, thermal Mach number, and collision wave number.

For the potential scattering, the differential momentum transfer cross section [20, 21] $\mathrm{d}\sigma_{\mathrm{M}}$ would be written as

$$\mathrm{d}\sigma_{\mathrm{M}} = \left|f(\mathbf{k}_i, \mathbf{k}_f)\right|^2 (1 - \cos\theta)\mathrm{d}\Omega, \tag{5.13}$$

where $f(\mathbf{k}_i, \mathbf{k}_f)$ is the scattering amplitude and $\mathbf{k}_i$ and $\mathbf{k}_f$ are, respectively, the wave vectors of the incident and final scattered waves, θ is the scattering angle, the factor $(1 - \cos\theta)$ denotes the fraction of the momentum transfer of the incident projectile, and $\mathrm{d}\Omega$ is the differential solid angle. It has been known that the momentum transfer cross section is relevant for investigating the Coulomb drag force for the particle–dust grain interaction in dusty plasmas. In the Born analysis [19], the scattering amplitude $f(\mathbf{k}_i, \mathbf{k}_f)$ would be expressed as

$$f(\mathbf{k}_i, \mathbf{k}_f) = -\frac{\mu}{2\pi\hbar^2}\int \mathrm{d}^3\mathbf{r}\, V(\mathbf{r})\mathrm{e}^{\mathrm{i}(\mathbf{k}_i - \mathbf{k}_f)\cdot\mathbf{r}}, \tag{5.14}$$

where μ is the reduced mass of the collision system, $\hbar$ is the rationalized Planck constant, and $V(\mathbf{r})$ is the interaction potential. It is well known that the interaction potential in plasmas can be obtained by linearizing the Boltzmann–Poisson equation. However, it is found that the ion–dust interaction potential including the influence of the plasma wake field in dusty plasmas would not be properly described by only the ordinary Debye–Hückel term [13]. Very recently, an excellent work by Morfill and Ivlev [17] has provided the useful analytic form of the effective potential for the charged dust including the effects of the ion wake in dusty plasmas has been obtained by the superposition of the Debye–Hückel term and additional far-field terms including the Legendre polynomials. Using the effective potential model [17], the interaction potential energy $V_{\mathrm{eff}}(\mathbf{k})$ between the projectile ion and target dust grain in dusty plasmas including the influence of the ion wake on the plasma density would be given by

$$V_{\mathrm{eff}}(r,\varphi) = q_{\mathrm{i}}Q\left[\frac{\mathrm{e}^{-r/\lambda_{\mathrm{D}}}}{r} - \sqrt{\frac{8}{\pi}}\frac{M_{\mathrm{T}}\lambda_{\mathrm{D}}^{2}}{r^{3}}\cos\varphi - \left(2-\frac{\pi}{2}\right)\frac{M_{\mathrm{T}}^{2}\lambda_{\mathrm{D}}^{2}}{r^{3}}(3\cos^{2}\varphi - 1)\right], \tag{5.15}$$

where $q_{\mathrm{i}}(=ze)$ is the charge of the ion, $Q(=-Ze)$ is the charge of the dust grain, λ_{D} is the Debye length, λ_{D} is the value of the thermal Mach number, u_{i} is the ion velocity, $v_{T_{\mathrm{i}}}$ is the ion thermal velocity, and φ is the angle between the projectile velocity and position vector. As it is seen, the plasma wakes in dusty plasmas produce the nonlinear interactions which are similar to the Cherenkov radiations [22] in the moving medium.

It has been also shown that the dipole dependence of the effective potential in plasmas would be related to the Landau damping effect [23]. According to the contributions from the real and imaginary parts of the complex scattering amplitude, the total momentum transfer cross section for the interaction between the ion and dust grain including the influence of the ion wake field is then found to be

$$\begin{aligned}\sigma_{\mathrm{M}}(k,M_{\mathrm{T}}) = \sigma_0\int_0^{\pi}\mathrm{d}\theta\,\sin\theta(1-\cos\theta)\Bigg\{\Bigg[&\frac{Ka\cos(Ka)+(\frac{a}{\lambda_{\mathrm{D}}})\sin(Ka)}{Ka\left[(Ka)^2+(\frac{a}{\lambda_{\mathrm{D}}})^2\right]}\\ &+4\left(2-\frac{\pi}{2}\right)M_{\mathrm{T}}^{2}\left(\frac{\lambda_{\mathrm{D}}}{a}\right)^{2}[-3S_4(Ka)+3C_3(Ka)+S_2(Ka)]\Bigg]^{2}\\ &+\left(2\sqrt{\frac{2}{\pi}}M_{\mathrm{T}}\left(\frac{\lambda_{\mathrm{D}}}{a}\right)^{2}\right)^{2}[2S_3(Ka)-C_2(Ka)]^{2}\Bigg\},\end{aligned} \tag{5.16}$$

where $\sigma_0 \equiv 2\pi[zZ(m_{\mathrm{i}}/m_{\mathrm{e}})(a^2/a_0)]^2$, m_{i} is the ion mass, m_{e} is the electron mass, $a_0(=\hbar^2/m_{\mathrm{e}}e^2)$ is the Bohr radius of the hydrogen atom, $K \equiv 2k\sin(\theta/2)$,

$k \equiv |\mathbf{k}_i| = |\mathbf{k}_f|$ for the elastic scattering, and $S_n(Ka)$ and $C_n(Ka)$ are, respectively, given by

$$S_n(Ka) = \int_{Ka}^{\infty} \mathrm{d}\xi \xi^{-n} \sin \xi, \tag{5.17}$$

$$C_n(Ka) = \int_{Ka}^{\infty} \mathrm{d}\xi \xi^{-n} \cos \xi. \tag{5.18}$$

As shown in (5.18), the momentum transfer cross section has the strong dependence of the Mach number due to the ion wake field. It is shown that the Coulomb ion drag force $F_{\mathrm{CD}}(= n_{\mathrm{i}} m_{\mathrm{i}} \sigma_{\mathrm{M}} v_{\mathrm{i,tot}} u_{\mathrm{i}})$ would be represented by the time rate of the momentum transfer from the ion projectile to the target dust grain, where n_{i} is the ion density and $v_{\mathrm{i,tot}}(= u_{\mathrm{i}}^2 + 8v_{T_{\mathrm{i}}}^2/\pi)^{1/2}$ is the total ion velocity [3]. Hence, the Coulomb ion drag force including the influence of the ion wake field in dusty plasmas is then obtained in the following form:

$$\begin{aligned} F_{\mathrm{CD}}(\bar{k}, M_{\mathrm{T}}) = {} & F_0 \bar{k}^{2} M_{\mathrm{T}}^{-1} (M_{\mathrm{T}}^2 + 8/\pi)^{1/2} \\ & \times \int_{-1}^{1} \mathrm{d}y (1-y) \left\{ \left[\frac{A \cos A + (a/\lambda_D) \sin A}{A[2\bar{k}^2(1-y) + (a/\lambda_{\mathrm{D}})^2]} \right. \right. \\ & \left. + 4\left(2 - \frac{\pi}{2}\right) M_{\mathrm{T}}^2 \left(\frac{\lambda_{\mathrm{D}}}{a}\right)^2 [-3S_4(A) + 3C_3(A) + S_2(A)] \right]^2 \\ & \left. + \left(2\sqrt{\frac{2}{\pi}} M_{\mathrm{T}} \left(\frac{\lambda_{\mathrm{D}}}{a}\right)^2\right)^2 [2S_3(A) - C_2(A)]^2 \right\}, \end{aligned} \tag{5.19}$$

where $F_0 \equiv 2\pi(n_{\mathrm{i}}/m_{\mathrm{i}})[zZ(m_{\mathrm{i}}/m_{\mathrm{e}})(a/a_0)]^2$, $y \equiv \cos\varphi$, $\bar{k}(\equiv ka)$, and $A \equiv \bar{k}\sqrt{2(1-y)}$ is the scaled collision wave number. The contribution from the Coulomb electron drag would be neglected in dusty plasmas since the electron drag is quite smaller than the ion drag by the mass factor $(m_{\mathrm{e}}/m_{\mathrm{i}})^{1/2}$ [24].

Figure 5.3 shows the scaled total momentum transfer cross section $\bar{\sigma}_{\mathrm{M}}(\equiv \sigma_{\mathrm{M}}/\sigma_0)$ as a function of the scaled Debye length $\bar{\lambda}_{\mathrm{D}}(\equiv \lambda_{\mathrm{D}}/a)$. In addition, Fig. 5.4 represents the scaled total momentum transfer cross section $\bar{\sigma}_{\mathrm{M}}$ as a function of the thermal Mach number M_{T}. As shown in these figures, the momentum transfer cross section increases with an increase of the scaled Debye length. Hence, it can be expected that the momentum transfer cross sections would be greater for smaller dust grains. It is also shown that the momentum transfer cross section drastically increases with increasing Mach number. Then, it is understood that the wake field effects significantly enhance the cross section. In addition, it is found that the momentum transfer cross section decreases with an increase of the scaled collision wave number $\bar{k}$. Figure 5.5 represents the scaled Coulomb ion drag force

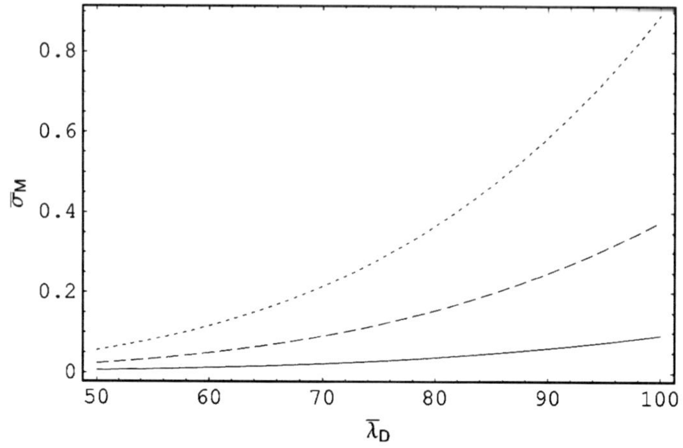

Fig. 5.3 The scaled total momentum transfer cross section $\bar{\sigma}_M$ as a function of the scaled Debye length $\bar{\lambda}_D$ for $\bar{k} = 1{,}000$. The *solid line* represents the case of $M_T = 0.1$. The *dashed line* represents the case of $M_T = 0.2$. The *dotted line* represents the case of $M_T = 0.3$. From [25]

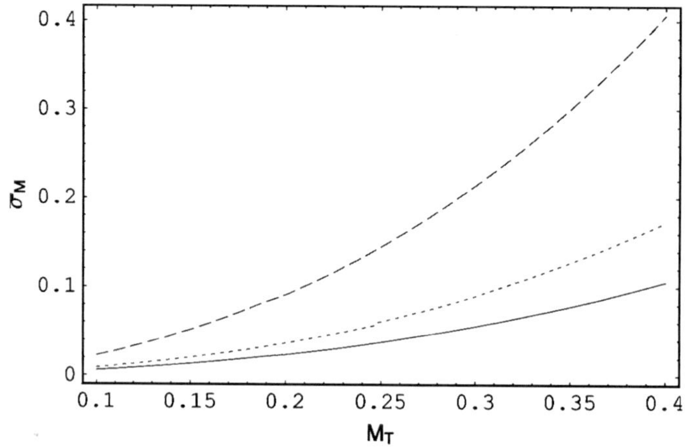

Fig. 5.4 The scaled total momentum transfer cross section $\bar{\sigma}_M$ as a function of the thermal Mach number M_T. The *solid line* is the case of $\bar{\lambda}_D = 50$ and $\bar{k} = 1{,}000$. The *dashed line* is the case of $\bar{\lambda}_D = 70$ and $\bar{k} = 1{,}000$. The *dotted line* is the case of $\bar{\lambda}_D = 70$ and $\bar{k} = 1{,}400$. From [25]

$\bar{F}_{CD}(\equiv F_{CD}/F_0)$ as a function of the thermal Mach number M_T for various values of the scaled Debye length $\bar{\lambda}_D$. As it is seen, the Coulomb ion drag force also increases with increasing Mach number and scaled Debye length. Hence, it can be also expected that the Coulomb ion drag force would be stronger for smaller dust grains. It is also found that the dependence of the thermal Mach number on the momentum transfer cross section is stronger than that on the Coulomb ion drag force due to the factor $(1 + 8/\pi M_T)^{-1/2}$. Therefore, in this work, we have found

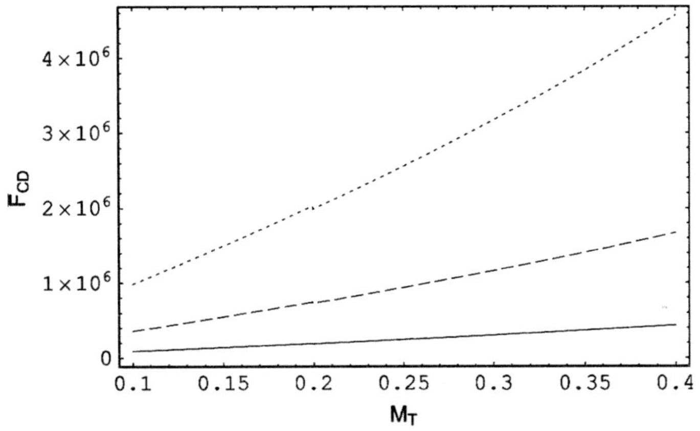

Fig. 5.5 The scaled Coulomb ion drag force $\bar{F}_{\rm CD}$ as a function of the thermal Mach number $M_{\rm T}$ for $\bar{k} = 1{,}000$. The *solid line* represents the case of $\bar{\lambda}_{\rm D} = 50$. The *dashed line* represents the case of $\bar{\lambda}_{\rm D} = 70$. The *dotted line* represents the case of $\bar{\lambda}_{\rm D} = 90$. From [25]

that the influence of the ion wake field plays an important role in the Coulomb drag force and momentum transfer cross section for the ion–dust grain interaction in dusty plasmas. These results provide useful information on the effects of the plasma wake field on the Coulomb drag in dusty plasmas.

5.4 Ion–Dust Bremsstrahlung Spectrum in Dusty Plasmas

The bremsstrahlung [16, 26–31] in plasmas has received much attention since this process is known to be one of the most fundamental processes, and the continuum radiation due to the bremsstrahlung process has been a diagnostic tool for investigating the plasma parameters. Recently, there has been a considerable interest in physical processes in plasmas encompassing highly charged dust grains [1–5]. In addition, it has been shown that the collective dust–plasma interactions are ubiquitous in various astrophysical and laboratory plasmas [3]. Hence, the collision and radiation processes have been extensively investigated in order to obtain the information on the plasma parameters such as the plasma density and temperature in dusty plasmas. In most cases, the dusty plasmas are consisting of thermal electrons and ions, and negatively charged dusty grains, i.e., three-component dusty plasmas. Hence, the ion–dust grain interaction in dusty plasmas would be different from that in only two-component ion–dust plasmas due to the influence of the plasma electrons in dusty plasmas. Thus, in this section, we have investigated the effects of electron density and temperature on the bremsstrahlung process due to the interaction of ion and negatively charged dust grain in three-component dusty plasmas. The ion–dust grain bremsstrahlung radiation cross section is obtained as

a function of the dust charge, dust radius, Debye length, collision energy, radiation energy, electron density, and electron temperature by using the Born approximation for the initial and final states of the projectile ion.

In the Born approximation, the differential ion–dust grain bremsstrahlung cross section [19] $\mathrm{d}^2\sigma_\mathrm{b}$ can be obtained by the second-order nonrelativistic perturbation method:

$$\mathrm{d}^2\sigma_\mathrm{b} = \mathrm{d}\sigma_\mathrm{C} \cdot \mathrm{d}W_\omega, \tag{5.20}$$

where the differential elastic scattering cross section $\mathrm{d}\sigma_\mathrm{C}$ can be represented by

$$\mathrm{d}\sigma_\mathrm{C} = \frac{1}{2\pi\hbar v_0^2}\left|\bar{V}(\mathbf{q})\right|^2 q\mathrm{d}q, \tag{5.21}$$

where $\hbar$ is the rationalized Planck constant, v_0 is the initial relative collision velocity, and $\bar{V}(\mathbf{q})$ is the Fourier transformation of the ion–dust grain interaction potential $V(\mathbf{r})$:

$$\bar{V}(\mathbf{q}) = \int \mathrm{d}^3\mathbf{r}\, \mathrm{e}^{-\mathrm{i}\mathbf{q}\cdot\mathbf{r}}\, V(\mathbf{r}), \tag{5.22}$$

$\mathbf{q}(=\mathbf{k}_0 - \mathbf{k}_f)$ is the momentum transfer, and $\mathbf{k}_0$ and $\mathbf{k}_f$ are, respectively, the wave vectors of the initial and final states of the projectile ion. Here, $\mathrm{d}W_\omega$ represents the differential photon emission probability within the frequency between ω and $\omega + \mathrm{d}\omega$:

$$\mathrm{d}W_\omega = \frac{\alpha}{4\pi^2}\Lambda_\mathrm{i}^2 \sum_{\hat{\mathbf{e}}} |\hat{\mathbf{e}}\cdot\mathbf{q}|^2 \frac{\mathrm{d}\omega}{\omega}\mathrm{d}\Omega, \tag{5.23}$$

where $\alpha(=e^2/\hbar c \cong 1/137)$ is the fine structure constant, $\Lambda_\mathrm{i}(\equiv \hbar/m_\mathrm{i}c)$ is the Compton wavelength of the ion, m_i is the mass of the ion, c is the velocity of the light, $\hat{\mathbf{e}}$ is the unit photon polarization vector, and $\mathrm{d}\Omega$ is the differential sold angle. In dusty plasmas consisting of electrons, ions, and negatively charged dusty grains, the quasineutral condition [9] at equilibrium is represented by

$$n_{\mathrm{e}0} + Z_\mathrm{d} n_{\mathrm{d}0} = n_{\mathrm{i}0} = n_0, \tag{5.24}$$

where n_{j0} $(j = e, i, d)$ are, respectively, the equilibrium densities for electrons (e), ions (i), and dust grains (d), $-Z_\mathrm{d}$ is the charge number of the dusty grains, and n_0 is the total plasma density. In the case of $\left|e\phi/k_\mathrm{B}T_j\right| \ll 1$ $(j = e, i)$, the Poisson's equation [5] in three-component dusty plasmas can be written as

$$\begin{aligned} \nabla^2\phi &= 4\pi e(n_\mathrm{e} + Z_\mathrm{d}n_\mathrm{d} - n_\mathrm{i}) \\ &\approx 4\pi e n_{\mathrm{e}0}\frac{e\phi}{k_\mathrm{B}T_\mathrm{e}}\left(1 + \frac{n_0}{n_{\mathrm{e}0}}\frac{T_\mathrm{e}}{T_\mathrm{i}}\right), \end{aligned} \tag{5.25}$$

where ϕ is the electrostatic potential in dusty plasmas, k_B is the Boltzmann constant, and T_j is the temperature of species j. The effective Debye length [9] λ_{eff} in three-component dusty plasmas is then given by

$$\lambda_{\text{eff}} = \lambda_D \left(\frac{n_{e0}}{n_0} + \frac{T_e}{T_i} \right)^{-1}, \tag{5.26}$$

where $\lambda_D (\equiv k_B T_e / 4\pi n_0 e^2)^{1/2}$ is the standard Debye length in electron–ion plasmas and the correction factor $(n_{e0}/n_0 + T_e/T_i)$ stands for the influence of the dusty plasma on the Debye length. Since the correction factor is usually greater than unity, the effective Debye length in dusty plasmas is found to be greater than that in customary electron–ion plasmas. Hence, in spherical polar coordinates with their origin at the center of the dust grain, the interaction potential $V_{i-d}(r)$ between the ion with charge ze and negatively charged dust grain with charge $-Z_d e$ in dusty plasmas is represented by the Yukawa form with the effective Debye length λ_{eff}:

$$V_{i-d}(r) = -\frac{Z_d z e^2}{r} \exp(-r/\lambda_{\text{eff}}). \tag{5.27}$$

For the sake of simplicity, the dust grains are assumed to be spherical shapes throughout in this work. After some algebra, the Fourier transformation $\bar{V}_{i-d}(q)$ of the ion–dust grain interaction potential is then given by

$$\bar{V}_{i-d}(q) = -\frac{4\pi Z_d z e^2 a}{q} \frac{\exp\left(-\frac{a}{\lambda_{\text{eff}}}\right)}{\left(\frac{a}{\lambda_{\text{eff}}}\right)^2 + (qa)^2} \left[\left(\frac{a}{\lambda_{\text{eff}}} \right) \sin(qa) + (qa)\cos(qa) \right], \tag{5.28}$$

where a is the radius of the spherical dust grain. Hence, the ion–dust bremsstrahlung cross section in dusty plasmas is then given by

$$d^2\sigma_b = \frac{16}{3} \frac{\alpha^3 a_0^2}{\bar{E}_0} \frac{m_e}{m_i} Z_d^2 z^2 \exp(-2\bar{\lambda}_{\text{eff}}^{-1}) \left(\frac{\bar{\lambda}_{\text{eff}}^{-1} \sin\bar{q} + \bar{q}\cos\bar{q}}{\bar{\lambda}_{\text{eff}}^{-2} + \bar{q}^2} \right)^2 \bar{q} d\bar{q} \frac{d\omega}{\omega}, \tag{5.29}$$

where $a_0 (= \hbar^2/m_e e^2)$ is the first Bohr radius of the hydrogen atom, m_e is the mass of the electron, $\bar{E}_0 (\equiv m_i v_0^2 / 2Ry)$ is the scaled initial collision energy, $Ry (= m_e e^4 / 2\hbar^2 \approx 13.6\,\text{eV})$ is the Rydberg constant, $\bar{q} (\equiv qa)$ is the scaled momentum transfer, and $\bar{\lambda}_{\text{eff}} (\equiv \lambda_{\text{eff}}/a)$ is the scaled effective Debye length. This expression of the bremsstrahlung cross section is reliable when the kinetic energy of the projectile ion is greater than the interaction energy between the ion and dust grain due to the prerequisite of the Born approximation. It has been also shown that the continuum spectrum due to the bremsstrahlung process would be investigated through the bremsstrahlung radiation cross section [32] defined as $d^2\chi_b/d\bar{\varepsilon}d\bar{q} \equiv \hbar\omega(d\sigma_b/\hbar d\omega d\bar{q})$, where $\bar{\varepsilon} (\equiv \varepsilon/Ry)$ is the scaled photon energy and $\varepsilon (\equiv \hbar\omega)$ is the photon energy. After some mathematical manipulations, the ion–dust grain bremsstrahlung radiation cross section $d^2\chi_b/d\bar{\varepsilon}$ in three-component dusty plasmas is then obtained by the following analytic form:

$$\frac{d^2\chi_b}{d\bar{\varepsilon}} = \frac{4}{3}\frac{\alpha^3 a_0^2}{\bar{E}_0}(\eta Z_d z)^2 \exp\left(-2\xi^{-1}\bar{\lambda}_D^{-1}\right)\Bigg| \ln\left[\frac{1+\left(\bar{a}\eta\xi\bar{\lambda}_D\left(\sqrt{\bar{E}_0}+\sqrt{\bar{E}_f}\right)\right)^2}{1+\left(\bar{a}\eta\xi\bar{\lambda}_D\left(\sqrt{\bar{E}_0}-\sqrt{\bar{E}_f}\right)\right)^2}\right]$$
$$+\left(1+2\xi^{-1}\bar{\lambda}_D^{-1}\right)\left[\cosh\left(2\xi^{-1}\bar{\lambda}_D^{-1}\right)+\sinh\left(2\xi^{-1}\bar{\lambda}_D^{-1}\right)\right]$$
$$\times\Bigg\{\sum_{l=1}^{2}\Bigg[\mathrm{Ci}\left((-1)^l 2i\xi^{-1}\bar{\lambda}_D^{-1}+2\bar{a}\eta\left(\sqrt{\bar{E}_0}+\sqrt{\bar{E}_f}\right)\right)$$
$$-\mathrm{Ci}\left((-,1)^l 2i\xi^{-1}\bar{\lambda}_D^{-1}+2\bar{a}\eta\left(\sqrt{\bar{E}_0}-\sqrt{\bar{E}_f}\right)\right)\Bigg]$$
$$+i\sum_{l=1}^{2}\Bigg[\mathrm{Si}\left(2i\xi^{-1}\bar{\lambda}_D^{-1}+(-1)^l 2\bar{a}\eta\left(\sqrt{\bar{E}_0}+\sqrt{\bar{E}_f}\right)\right)$$
$$-\mathrm{Si}\left(2i\xi^{-1}\bar{\lambda}_D^{-1}+(-1)^l 2\bar{a}\eta\left(\sqrt{\bar{E}_0}-\sqrt{\bar{E}_f}\right)\right)\Bigg]$$
$$+\frac{2\cos\left(\bar{a}\eta\left(\sqrt{\bar{E}_0}+\sqrt{\bar{E}_f}\right)\right)}{1+\left(\bar{a}\eta\xi\bar{\lambda}_D\left(\sqrt{\bar{E}_0}+\sqrt{\bar{E}_f}\right)\right)^2}-\frac{2\cos\left(\bar{a}\eta\left(\sqrt{\bar{E}_0}-\sqrt{\bar{E}_f}\right)\right)}{1+\left(\bar{a}\eta\xi\bar{\lambda}_D\left(\sqrt{\bar{E}_0}-\sqrt{\bar{E}_f}\right)\right)^2}$$
$$-\frac{2\bar{a}\eta\xi\bar{\lambda}_D\left(\sqrt{\bar{E}_0}+\sqrt{\bar{E}_f}\right)\sin\left(\bar{a}\eta\left(\sqrt{\bar{E}_0}+\sqrt{\bar{E}_f}\right)\right)}{1+\left(\bar{a}\eta\xi\bar{\lambda}_D\left(\sqrt{\bar{E}_0}+\sqrt{\bar{E}_f}\right)\right)^2}$$
$$+\frac{2\bar{a}\eta\xi\bar{\lambda}_D\left(\sqrt{\bar{E}_0}-\sqrt{\bar{E}_f}\right)\sin\left(\bar{a}\eta\left(\sqrt{\bar{E}_0}-\sqrt{\bar{E}_f}\right)\right)}{1+\left(\bar{a}\eta\xi\bar{\lambda}_D\left(\sqrt{\bar{E}_0}-\sqrt{\bar{E}_f}\right)\right)^2}\Bigg|^2, \tag{5.30}$$

where $\bar{E}_f(\equiv \bar{E}_0-\bar{\varepsilon})$ is the scaled final projectile energy, $\bar{a}(\equiv a/a_0)$ is the scaled radius of the dust grain, $\bar{\lambda}_D \equiv \lambda_D/a$, $\xi \equiv (n_{e0}/n_0 + T_e/T_i)^{-1}$, $\mathrm{Ci}(x)(= -\int_x^\infty dt \cos t/t)$, and $\mathrm{Si}(x)(= \int_0^x dt \sin t/t)$ are the cosine and sine integrals [33], respectively. Very recently, an excellent discussion [34] on the additional part of the electrostatic potential in dusty plasmas is given due to the ion absorption by the dust grain and ion neutral collisions. It has been found that the electrostatic potential would be represented only by the standard Debye–Hückel form if there is no ion flux on the surface of the dust grain, i.e., nonabsorbing dust grains. In this work, we just retain only the Debye–Hückel form of the interaction potential since we consider the bremsstrahlung process due to the scattering of the ion by the nonabsorbing dust grain. However, if the ion flux is existed on the dust grain, the attractive part of the potential has to be included in the ion–dust bremsstrhlung process. Hence, it would be expected that the additional part of the interaction potential suppresses the ion–dust grain bremsstrahlung spectrum obtained by the Yukawa-type Debye–Hückel potential. The validity of the Born approximation can be considered by using the Born parameter [19], $|V|/E$, for the potential scattering, where $|V|$ a typical strength of the interaction potential. Since the Born parameter $\frac{Z_d z e^2}{r_i E} e^{-r_i/\lambda_{eff}}$, where r_i is the interparticle distance between the ion and dust grain, for the ion–dust interaction in typical circumstances of dusty plasmas is usually less than unity, the Born approximation would be quite reliable to investigate the high-energy ion–dust bremsstrahlung process in dusty plasmas. In typical dusty plasmas, the range of the Coulomb coupling parameter for the ion–dust interaction in typical dusty plasmas is found to be $\Gamma_{i-d}(=\frac{Z_d z e^2}{r_i k_B T_i}) < 1$. However, the Coulomb coupling parameter for the dust–dust interaction [4] $\Gamma_{d-d}(=\frac{Z_d^2 e^2}{r_d k_B T_d})$, where r_d is the interparticle distance between the dust grains and T_d is the dust temperature, can be greater than the unity. Hence, for $\bar{E}_0, \bar{E}_f > 1$, the ion–dust grain bremsstrahlung radiation cross sections would be reliable since the interaction energy between the ion and charged dust grain is usually much less than the collision energy due to the considerable size of the dust grain. The ion–atom bremsstrahlung emission is expected to be quite different from the electron–ion bremsstrahlung emission due to the polarization of the target atom. Hence, the electron–atom and ion–atom polarization bremsstrahlung processes have to be considered in the partially ionized plasmas. However, the ion–atom polarization bremsstrahlung process is neglected in this work since the dust plasma is assumed to be completely ionized. For typical circumstances of dusty plasmas, it has been shown that $Z_d \approx 100$–1,000, $a \approx 0.01$–1 μm, and $\lambda_D/a \approx 5$–100 [1].

In order to specifically investigate the effects of electron temperature and density on the ion–dust bremsstrahlung process in a dusty plasma, we set $Z_d = 200$, $a = 0.1$ μm, $\lambda_D/a = 50$, $m_i/m_e = 1{,}840$, and $z = 1$. Figure 5.6 shows the scaled bremsstrahlung radiation cross section $\partial^2_{\bar{\varepsilon}} \bar{\chi}_b [\equiv (d^2\chi_b/d\bar{\varepsilon})/\pi a_0^2]$ in units of πa_0^2 for the interaction of the ion with the negatively charged dust grain in dusty plasmas as a function of the temperature ratio T_e/T for various values of the density ratio n_{e0}/n_0. In addition, Fig. 5.7 represents the scaled bremsstrahlung radiation cross section $\partial^2_{\bar{\varepsilon}} \bar{\chi}_b$ as a function of the density ratio n_{e0}/n_0 for various values of the temperature ratio T_e/T_i. As shown in these figures, it found that the ion–dust bremsstrahlung radiation cross decreases with an increase of the ratio of the electron temperature to the ion temperature in a dusty plasma. It is also shown that the electron density suppresses the ion–dust bremsstrahlung radiation cross section. It is interesting to note that the effect of electron density on the bremsstrahlung radiation cross section diminishes with an increase of the electron temperature since the high-energy electrons can be readily repelled by the negatively charged dusty grains in dusty plasmas [5]. Figure 5.8 represents the surface plot of the effects of electron temperature and density on the ion–dust bremsstrahlung process $F(\bar{T}, \bar{n})$

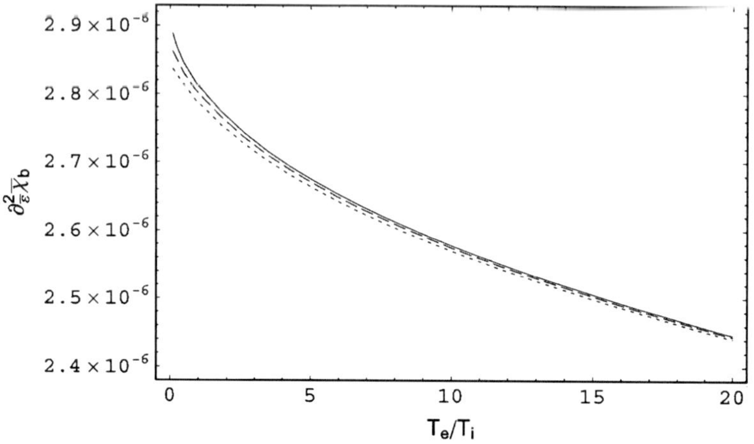

Fig. 5.6 The scaled ion–dust grain bremsstrahlung radiation cross section $\partial^2_{\bar{\varepsilon}}\bar{\chi}_b$ as a function of the temperature ratio T_e/T_i for $\bar{E}_0 = 5$, $\bar{\varepsilon} = 2$, $T_e = 5 \times 10^3$ K, $n_0 = 10^{10}\,\mathrm{cm}^{-3}$, and $a = 9.7 \times 10^{-5}$ cm. The *solid line* represents the case of $n_{e0}/n_0 = 0$. The *dashed line* represents the case of $n_{e0}/n_0 = 0.5$. The *dotted line* represents the case of $n_{e0}/n_0 = 0.9$. From [35]

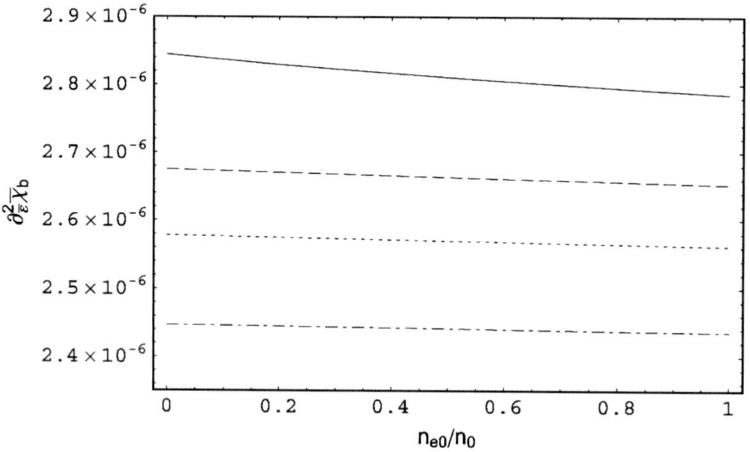

Fig. 5.7 The scaled ion–dust grain bremsstrahlung radiation cross section $\partial^2_{\bar{\varepsilon}}\bar{\chi}_b$ as a function of the temperature ratio n_{e0}/n_0 for $\bar{E}_0 = 5$, $\bar{\varepsilon} = 2$, $T_e = 5 \times 10^3$ K, $n_0 = 10^{10}\,\mathrm{cm}^{-3}$, and $a = 9.7 \times 10^{-5}$ cm. The *solid line* represents the case of $T_e/T_i = 0.5$. The *dashed line* represents the case of $T_e/T_i = 5$. The *dotted line* represents the case of $T_e/T_i = 10$. The *dot-dashed line* represents the case of $T_e/T_i = 20$. From [35]

as a function of the temperature ratio $\bar{T}(\equiv T_e/T_i)$ and density ratio $\bar{n}(\equiv n_{e0}/n_0)$. As it is seen, the effects of electron temperature and density significantly suppress the ion–dust grain bremsstrahlung radiation cross section. The effect of electron density is found to be important for low-temperature ratios. It is also found that the effect of electron temperature on the bremsstrahlung radiation cross section is

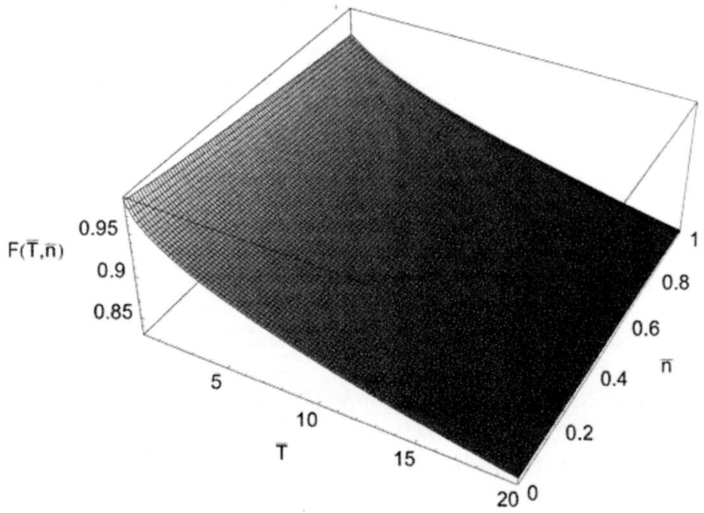

Fig. 5.8 The surface plot of the effects of electron temperature and density on the ion–dust grain bremsstrahlung process $F(\bar{T},\bar{n})$ as a function of the temperature ratio $\bar{T}$ and density ratio $\bar{n}$ for $\bar{E}_0 = 5$, $\bar{\varepsilon} = 2$, $T_e = 5 \times 10^3$ K, $n_0 = 10^{10}\,\text{cm}^{-3}$, and $a = 9.7 \times 10^{-5}$ cm. From [35]

more significant than the effect of electron density in dusty plasmas. Hence, we have found that the effects of electron temperature and density play important roles in the ion–dust grain bremsstrahlung process in dusty plasmas containing electrons, ions, and negatively charged dust grains. These results would provide useful information on the ion–dust bremsstrahlung emission spectrum and also the radiation due to the interaction between dust particle chains and streaming ions in the plasma sheath in dusty plasmas.

5.5 Conclusions

In this chapter, we have discussed various atomic processes in dusty plasmas including the strong collective interactions.

Firstly, the effects of ion temperature on the electron–dust collision process are investigated in complex dusty plasmas by using the second-order eikonal analysis. We have found that the eikonal scattering phase and eikonal electron–dust grain collision cross section decrease with decreasing ion temperature in dusty plasmas. In addition, we found that the effect of ion temperature on the electron–dust grain collision process is more significant than the effect of electron density in dusty plasmas. We also found that the density effect is more significant when the electron temperature is comparable to the ion temperature. Hence, we have found that the ion temperature and density effects play important roles in the elastic electron–dust grain collision in complex dusty plasmas.

Secondly, the ion wake effects on the Coulomb drag force are investigated in complex dusty plasmas. We have found that the ion wake effects significantly enhance the Coulomb ion drag force. We also found that the ion wake effects on the Coulomb drag force increase with an increase of the Debye length. In addition, we found that the ion wake effects on the momentum transfer cross section and Coulomb drag force increase with increasing thermal Mach number, i.e., decreasing plasma temperature. We also found that the Coulomb ion drag force would be stronger for smaller dust grains. Hence, we have found that the influence of the ion wake field plays an important role in the Coulomb drag force and momentum transfer cross section for the ion–dust grain interaction in dusty plasmas.

Thirdly, the effects of electron temperature and density on the ion–dust grain bremsstrahlung process are investigated in dusty plasmas. We obtained the ion–dust bremsstrahlung radiation cross section as a function of the dust charge, dust radius, Debye length, collision energy, radiation energy, electron density, and electron temperature by using the Born approximation. We have found that the ion–dust bremsstrahlung radiation cross section decreases with increasing electron density in dusty plasmas. We also found that the electron temperature suppresses the bremsstrahlung radiation cross section. In addition, we found that the effect of electron temperature on the ion–dust bremsstrahlung process is more significant than the effect of electron density in dusty plasmas. Hence, we have found that the effects of electron temperature and density play important roles in the ion–dust grain bremsstrahlung process in dusty plasmas containing electrons, ions, and negatively charged dust grains.

These results on atomic processes in dusty plasmas would provide useful information on the plasma parameters and physical properties of complex dusty plasmas.

Acknowledgements One of the authors (Y.-D. J.) gratefully acknowledges Dr. M. Rosenberg for the useful discussions and warm hospitality while visiting the Department of Electrical and Computer Engineering at the University of California, San Diego. He would also like to thank Prof. H. Tawara, Prof. T. Kato, Prof. M. Sato, Prof. Y. Hirooka, Prof. I. Murakami, and Prof. D. Kato for their warm hospitality and support while visiting the National Institute for Fusion Science (NIFS) in Japan as a long-term visiting professor. This research was supported by Basic Science Research Program through the National Research Foundation of Korea (NRF) funded by the Ministry of Education, Science, and Technology (Grant No. 2011–0003099).

Reference

1. D.A. Mendis, M. Rosenberg, Ann. Rev. Astron. Astrophys. **32**, 419 (1994)
2. A. Bouchoule, *Dusty Plasmas: Physics, Chemistry and Technological Impacts in Plasma Processing* (Wiley, Chichester, 1999)
3. P.K. Shukla, A.A. Mamum, *Introduction to Dusty Plasma Physics* (Institute of Physics Publishing, Bristol, 2002)
4. T.S. Ramazanov, K.N. Dzhumagulova, A.N. Jumabekov, M.K. Dosbolayev, Phys. Plasmas **15**, 053704 (2008)

5. S.A. Maiorov, T.S. Ramazanov, K.N. Dzhumagulova, A.N. Jumabekov, M.K. Dosbolayev, Phys. Plasmas **15**, 093701 (2008)
6. V. Fortov, I. Iakubov, A. Khrapak, *Physics of Strongly Coupled Plasma* (Oxford University Press, Oxford, 2006)
7. S.P. Khare: *Introduction to the Theory of Collisions of Electrons with Atoms and Molecules* (Plenum, New York, 2002)
8. Z. Metawei, Acta Phys. Polonica B **31**, 713 (2000)
9. S. Vidhya Lakshmi, R. Bharuthram, P.K. Shukla, Astrophys. Space Sci. **209**, 213 (1993)
10. E.T. Whittaker, G.N. Watson, *A Course of Modern Analysis*, 4th edn. (Cambridge University Press, Cambridge, 1952)
11. D.-H. Ki, Y.-D. Jung, J. Appl. Phys. **108**, 086101 (2010)
12. P. Bliokh, V. Sinitsin, V. Yaroshenko, *Dusty and Self-Gravitational Plasma in Space* (Kluwer, Dordrecht, 1995)
13. V.N. Tsytovich, G.E. Morfill, S.V. Vladimirov, H. Thomas, *Elementary Physics of Complex Plasmas* (Springer, Berlin, 2008)
14. J. Perrin, P. Molinas-Mata, P. Belenguer, J. Phys. D **27**, 2499 (1994)
15. J.E. Daugherty, D.B. Graves, J. Appl. Phys. **78**, 2279 (1995)
16. H.F. Beyer, V.P. Shevelko, *Introduction to the Physics of Highly Charged Ions* (Institute of Physics Publishing, Bristol, 2003)
17. G. Morfill, A.V. Ivlev, Rev. Mod. Phys. **81**, 1353 (2009)
18. V.E. Fortov, G.E. Morfill, *Complex and Dusty Plasma* (CRC Press, Boca Raton, 2010)
19. R.J. Gould, *Electromagnetic Processes* (Princeton University Press, Princeton, 2006)
20. G.A. Kobzev, I.T. Iakubov, M.M. Popovich, *Transport and Optical Properties of Nonideal Plasmas* (Plenum, New York, 1995)
21. S. Geltman, *Topics in Atomic Collision* (Krieger, Malabar, 1997)
22. V.L. Ginzburg, *Application of Electrodynamics in Theoretical Physics and Astrophysics* (Gordon and Breach, New York, 1989)
23. P.K. Shukla, L. Stenflo, R. Bingham, Phys. Lett. A **359**, 218 (2006)
24. L. Spitzer Jr., *Physical Processes in the Interstellar Medium* (Wiley, New York, 1978)
25. D.-H. Ki, Y.-D. Jung, Appl. Phys. Lett. **97**, 101502 (2010)
26. H.A. Bethe, E.E. Salpeter, *Quantum Mechanics of One- and Two-Electron Atoms* (Springer, Berlin, 1957)
27. G. Bekefi, *Radiation Processes in Plasmas* (Wiley, New York, 1966)
28. H. Totsuji, Phys. Rev. A **32**, 3005 (1985)
29. V.P. Shevelko, *Atoms and Their Spectroscopic Properties* (Springer, Berlin, 1997)
30. H.F. Beyer, H.-J. Kluge, V.P. Shevelko, *X-Ray Radiation of Highly Charged Ions* (Springer, Berlin, 1997)
31. V. Shevelko, H. Tawara, *Atomic Multielectron Processes* (Springer, Berlin, 1998)
32. J.D. Jackson, *Classical Electrodynamics* (Wiley, New York, 1962)
33. G. Arfken, *Mathematical Methods for Physicists* (Academic, New York 1966)
34. S.A. Khrapak, G.E. Morfill, Phys. Plasmas **15**, 084502 (2008)
35. Y.-D. Jung, I. Murakami, J. Appl. Phys. **105**, 106106 (2009)

Chapter 7
Electron Loss, Excitation, and Pair Production in Relativistic Collisions of Heavy Atomic Particles

A.B. Voitkiv and B. Najjari

Abstract We present a short review on some of the recent developments in the field of relativistic atomic collisions. In this review, we consider several processes which take place in collisions at high (relativistic) energies. They include: (a) projectile-electron excitation and loss in collisions between an ion, which initially carries electron(s), and a neutral atom; (b) bound-free electron-positron pair production in collisions between a bare nucleus and a neutral atom; and (c) bound–bound pair production in collisions between bare nuclei, which becomes possible only if the colliding nuclei possess electric charges of opposite sign.

7.1 Introduction

During the last few decades, collisions of singly and multiply charged ions X^{Z+} ($Z \sim$ 1–10) with atoms and molecules, occurring at impact velocities substantially exceeding the typical orbiting velocities of outer-shell atomic electrons, have been a subject of extensive research, both experimental and theoretical. The study of different processes, accompanying such collisions, are of great interest not only for the basic atomic physics research but it also has many applications in other fields of physics such as plasma physics, astrophysics, and radiation physics.

With the advent of accelerators of relativistic heavy ions, much higher impact energies ($\sim$0.1–200 GeV/u) and projectile charge states ($Z \sim$ 30–92) had become accessible for the explorations in experiments on ion–atom collisions.

Three basic atomic physics processes can occur in collisions between a bare projectile-nucleus and a target-atom (see, e.g., [1, 2]). (a) The atom can be excited or ionized by the interaction with the projectile. (b) One or more atomic electrons can be picked up by the projectile-nucleus and form bound or low-lying continuum states of the corresponding projectile-ion. The pick-up process can proceed with or without emission of radiation and is called radiative or nonradiative electron capture, respectively. A combination of (a) and (b) can also occur. Besides, in

V. Shevelko and H. Tawara (eds.), *Atomic Processes in Basic and Applied Physics*, Springer Series on Atomic, Optical, and Plasma Physics 68,
DOI 10.1007/978-3-642-25569-4_7,

relativistic collisions pair production becomes possible with cross sections reaching quite substantial values in the case of extreme relativistic impact energies when the collision velocity approaches very closely the speed of light in vacuum [1, 3].

Highly charged projectiles produced at accelerators of heavy ions are often not fully stripped and instead carry one or more very tightly bound electrons. If such projectiles collide with atomic targets, these electrons can be excited and/or lost. In the rest frame of the ion, this can be viewed as excitation or "ionization" of the ion by the impact of the incident atom. In addition to the nucleus, the atom has electrons which may influence the motion of the electrons of the ion in different ways. As a result, the physics of the ion excitation and "ionization" by the neutral atom impact will in general strongly differ from that for excitation and ionization in collisions with a bare atomic nucleus. Thus, in collisions of partially stripped ions with neutral atoms, a qualitatively new process—projectile-electron excitation and/or loss—becomes possible (for a recent review on the theory of these processes see [4]).

In the present chapter we concentrate our attention on projectile-electron transitions occurring in relativistic collisions between highly charged ions and atoms. In addition to excitation and loss of electrons carried initially by ions, which exist already before the collision, we shall also consider transitions involving negative energy states leading to electron–positron pair production. The latter can be viewed as a transition between electronic states with negative and positive total energy. If we assume that these states are strongly influenced only by the field of the ionic nucleus while the field of the atom acts merely as a collisional perturbation (which couples these states resulting in pair production) and neglect the interaction between the created electron and positron, then the analogy between pair production and projectile-electron excitation and loss processes becomes obvious.

In the case of free pair production, in which both created particles move in the continuum, the above assumption looks quite natural for asymmetric collisions where the charge of the ionic nucleus is larger than that of the neutral atom. This assumption is also valid for bound-free pair production where the electron is created in a bound state of the ion. Besides, since the difference in the velocities of the electron and positron is typically much larger than 1 a.u., the neglect of the electron–positron interaction clearly represents a good approximation.

As was mentioned, one can find common points between free and bound-free pair production, occurring in collisions between a bare nucleus and a neutral atom, and projectile-electron transitions caused by collisions with atoms. Such an analogy, however, does not hold for bound–bound pair production, which may occur in collisions of nuclei having charges of different sign and in which both electron and positron are created in bound states. Instead, the latter bears certain important similarities with electron capture process.

During the last 20 years a large number of experimental investigations has been performed on projectile-electron excitation and loss in collisions between relativistic highly charged ions and solid and gaseous targets. In particular, a variety of very heavy projectiles with the net charge 52–91 a.u. were used in the experiments. The experiments have also covered a very large interval of impact energies ranging from comparatively low relativistic energies of ~100–200 MeV/u [5–13] to the range of "intermediate" (10 GeV/u) [14, 15] and extreme relativistic energies (160 GeV/u) [16–18], where the projectile velocity already only fractionally differs from the speed of light $c = 137$ a.u.. Most of the data, however, have been collected for impact energies not exceeding a few hundreds of MeV/u. For impact energies above 1 GeV/u, there exists just a few experimental results. They include the data on electron loss from 10.8 GeV/u $Au^{78+}(1s)$ ions penetrating solid targets [14, 15] and on electron loss from 160 GeV/u $Pb^{81+}(1s)$ ions colliding with solid [16] and gaseous [17] targets (see also [18]).

Besides, one should also note that the experimental data on the elementary cross sections in collisions at sufficiently high impact energies, which were collected using solid targets, are not very accurate. For instance, loss cross sections, reported for 160 GeV/u $Pb^{81+}(1s)$ projectiles penetrating solid and gas targets differ between themselves roughly by a factor of 2 (see [16, 17]). Moreover, in the case of 10.8 GeV/u $Au^{78+}(1s)$ ions, experimental data of [14] and [15], which both were collected for collisions with solid-state targets, also differ by about of a factor of 2. Note that the reason for the former difference, as was recently explained in [19], lies in multiple collisions suffered by the projectiles when they move in solids which makes an accurate experimental determination of the values for elementary ion–atom cross sections very difficult.

The chapter is organized as follows. In the next section we consider electron loss and simultaneous loss–excitation (in which in the same collision event one of the electrons is lost, whereas the other one undergoes a transition into an excited state of the ion) occurring in collisions with atoms at comparatively low relativistic energies where the screening effect of atomic electrons is weak and can be neglected, but the role of higher-order effects in the interaction between the electrons of the projectile and the nucleus of the atomic target is very essential. In Sect. 7.3, we discuss electron loss at "intermediate" impact energies where the screening effect of atomic electrons becomes important, whereas the higher-order contributions still remain noticeable even for single-electron processes. In Sect. 7.4, we consider projectile-electron loss at very high collision energies, and finally in Sect. 7.5, we discuss electron–positron pair production including only its bound-free and bound–bound versions.

7.2 Electron Loss and Simultaneous Loss-Excitation at Comparatively Low Relativistic Impact Energies

7.2.1 Electron Loss

If the momentum transfer in the collision between a projectile-ion and many-electron target-atom is much larger than the typical momenta of atomic electrons, then the latter ones play just a minor role in the projectile-electron excitation and loss processes. Under such circumstances, the transitions of the electron of the ion in collisions with neutral atoms can be considered as caused solely by the interaction between this electron and the (unscreened) nucleus of the atom. Such a situation takes place, for instance, in the case of the electron loss from $U^{90+}(1s^2)$ ions occurring in collisions with atoms at an impact energy of 105 MeV/u. Loss cross sections for these ions were measured in [5–8] (for the discussion of electron loss cross sections measured in collisions with relatively light atoms, see [10, 11]).

Indeed, one can show that at this collision energy, the screening effect of atomic electrons is weak and can be ignored. Besides, since the effective threshold for the opening of the inelastic atomic mode in the case of electron loss from $U^{90+}(1s^2)$ is approximately equal to 240 MeV/u, the antiscreening effect of the atomic electrons is very weak as well and may be disregarded, both in the case of many-electron and very light atoms (including hydrogen). Thus, the process of electron loss from 105 MeV/u $U^{90+}(1s^2)$ can indeed be reduced to a three-body problem and considered accordingly.

In Fig. 7.1 we present results of calculations for the total cross section for single electron loss from 105 MeV/u $U^{90+}(1s^2)$ ions in collisions with different atomic targets ranging from beryllium to gold. The calculations were performed using the first-order perturbation theory of projectile-target collisions [20, 21] and the following three-body distorted-wave models: the symmetric eikonal approximation (SEA), the modified symmetric eikonal approximation, and the continuum distorted-wave–eikonal initial state model (CDW-EIS). In all of these models, which were considered in [22–24], the presence of atomic electrons is ignored.

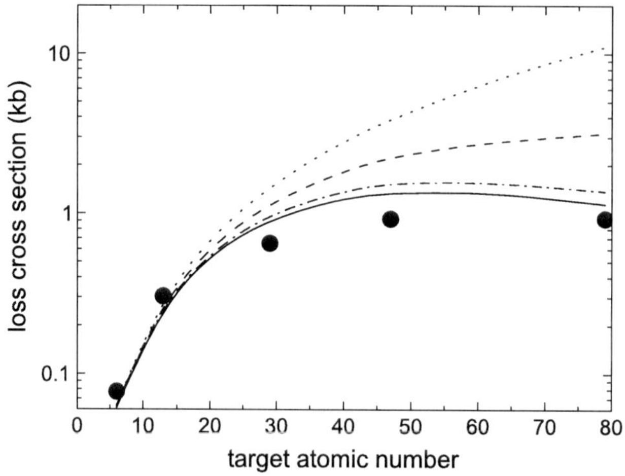

Fig. 7.1 The total cross section (per electron) for single electron loss from 105 MeV/u $U^{90+}(1s^2)$ colliding with various targets. *Circles* show experimental results obtained for collisions with solid-state targets of carbon, aluminum, copper, silver, and gold [5]. *Dot curve* shows results of the first order calculation. *Dash*, *dash-dot*, and *solid curves* display results of calculations using distorted-wave methods

In Fig. 7.1 the theoretical results are also compared with experimental data for the loss cross section reported in [5] for collisions of 105 MeV/u $U^{90+}(1s^2)$ with solid-state targets of beryllium, carbon, aluminum, copper, silver, and gold. In the calculation, the electron loss was considered as occurring from the ground state of a hydrogen-like ion whose effective nuclear charge was determined from the binding energy of the electrons in $U^{90+}(1s^2)$.

It is seen in Fig. 7.1 that, in collisions with targets having small atomic numbers ($Z_A/v \ll 1$), all the theoretical models yield very similar loss cross section values. However, when the ratio Z_A/v increases, the difference between the results of the first-order approximation on one side and the distorted-wave models on the other rapidly grows. Compared to the first order result, for collisions with atoms of gold, the SEA and CDW-EIS models yield the loss cross sections which are smaller by about a factor of 3.7 and 10, respectively.

A comparison with the experimental data shows a strong failure of the first-order approximation. This approximation, predicting the dependence $\sim Z_A^2$ for the loss cross section, does not reproduce the "saturation" of this cross section, which is clearly seen in the experimental data at $Z_A \gtrsim 40$, and overestimates these data by more than the order of magnitude in collisions with the gold target.

The results obtained using the symmetric eikonal approximation are also not in agreement with the experimental data. Although compared to the first-order approximation, this amplitude predicts a much weaker growth of the loss cross section with increasing Z_A; at very large Z_A, this cross section does substantially overestimate the experimental results.

In contrast, the application of the so-called modified eikonal approximation [23] leads to results much closer to the experimental data. The application of the CDW-EIA model [24] is also quite successful in this case yielding cross section values which are closest to the experimental data. Taking into account that the experimental data shown in Fig. 7.1 possess a possible systematic error of up to 20% (see [1]), one can say that the results for loss obtained using the modified eikonal and CDW-EIS approximations are in reasonable agreement with the experiment.

Figure 7.2 displays results for the total cross section for the electron loss from 220 MeV/u $U^{91+}(1s)$ ions. The impact energy of 220 MeV/u is not yet sufficiently high to make the screening effect of the atomic electrons important for the electron loss process. Besides, this impact energy is still below the effective threshold of 240 MeV/u for the antiscreening collision mode. Therefore, similarly to the case of electron loss from 105 MeV/u $U^{90+}(1s^2)$ ions, the presence of the atomic electrons can be neglected, and again the projectile-electron loss process can be considered using the three-body models.

Similarly to the electron loss from 105 MeV/u $U^{90+}(1s^2)$ ions, all the theoretical models again yield very close results for collisions with targets having low atomic numbers ($Z_A/v \ll 1$). When the atomic number of the target increases, the difference between the results of the first-order approximation, on one hand, and the distorted-wave amplitudes, on the other, rapidly increases and reaches about a factor of 2.5–3 for collisions with the gold target. Compared to the case at 105 MeV/u,

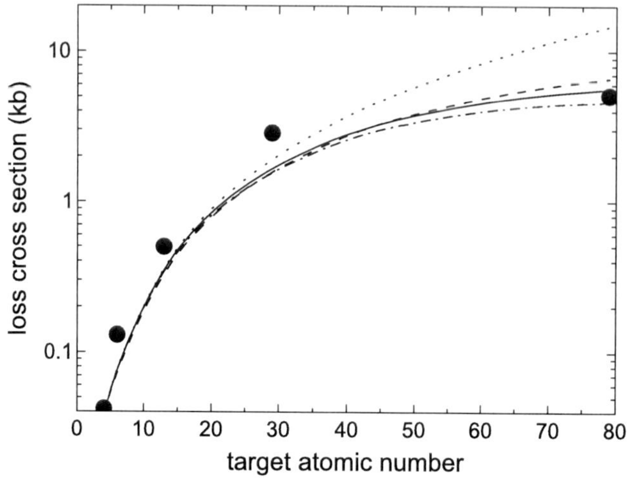

Fig. 7.2 The total cross section for the electron loss from 220 MeV/u $U^{91+}(1s)$ ions colliding with different targets. *Circles* show experimental results measured in [5] for the loss in collisions with solid-state targets of beryllium, carbon, aluminum, copper, and gold. *Dot curve* shows results of the first-order calculation. *Dash, dash-dot*, and *solid curves* display results of calculations using distorted-wave methods

the smaller differences between the first order and the other results are probably related to the fact that at an impact energy of 220 MeV/u, the effective strength of the interaction between the electron of the ion and the nucleus of the atom, is weaker. This point also may be responsible for the smaller difference between the predictions of the SEA and CDW-EIS models.

Taking again into account that the possible systematic error in the experimental data is of up to 20%, one may come to the conclusion that the first-order approximation strongly fails also in the case of the loss from 220 MeV/u $U^{91+}(1s)$ ions. However, now already all the distorted-wave methods yield results which are in reasonable agreement with the experiment.

In Fig. 7.3, we show results for the total cross sections of single electron loss from 105 MeV/u $U^{89+}(1s^2 2s)$. At this relatively low impact energy, there is a very large difference between the cross sections for the electron loss from the K and L shells. Therefore, one can assume that the loss occurs only from the L shell. The $2s$-electron in the initial and final states of the undistorted ion is described by considering this electron as moving in the Coulomb field of the ionic core (the nucleus plus the two K-shell electrons) whose effective charge is determined from the binding energy of the $2s$-electron in $U^{89+}(1s^2 2s)$.

Compared to the K-shell electrons, the $2s$-electron of the uranium ion is substantially less tightly bound. However, due to the relatively low value of the impact energy, the typical minimum momentum transfer to this electron, which is necessary to remove it from the ion, is still very large on the atomic scale. Therefore, although the screening effect is now substantially larger than in the case of the

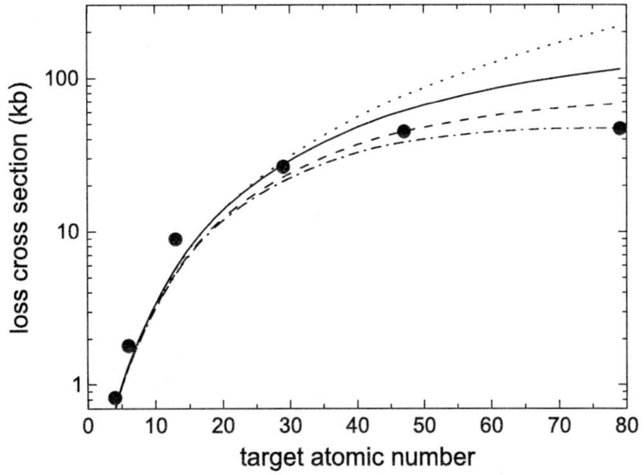

Fig. 7.3 The total cross section for the electron loss from 105 MeV/u $U^{89+}(1s^2 2s)$ ions colliding with different targets. *Circles* show experimental results measured in [5] for the loss in collisions with solid state targets of beryllium, carbon, aluminum, copper, silver and gold. *Dot curve* shows results of the first order calculation. *Dash*, *dash-dot* and *solid curves* display results of calculations using distorted-wave methods

electron loss from the K-shell, it nevertheless still remains quite modest and can be neglected. Indeed, according to the first order calculations, the reduction of the loss cross section caused by screening effect the reaches about 11%. This is to be compared with the difference of a factor of 2–4 between the first-order result and the predictions of the distorted-wave models which do not take the screening effect into account but instead attempt to describe the interaction between the electron of the ion and the nucleus of the atom in a better way.

The important difference between the previous cases and the electron loss from 105 $U^{89+}(1s^2 2s)$ ions is that the effective threshold for the antiscreening mode is about 60 MeV/u and, thus, this mode is now open.

In theoretical results shown in Fig. 7.3, the contribution of the antiscreening mode was estimated by treating this mode within the first-order perturbation theory. Note that, since the relative contribution of this mode to the loss cross section scales approximately as $1/Z_A$, the antiscreening effect has to be taken into account in collisions with light targets, like beryllium and carbon, but may be simply neglected in collisions with atoms of silver and gold.

It is seen in Fig. 7.3 that, similarly to the case with 105 MeV/u $U^{90+}(1s^2)$ ions, both the first-order and distorted-wave calculations yield very close cross section values for collisions with atomic targets having low atomic numbers. When the ratio Z_A/v increases, the difference between the results of these models rapidly starts to grow. Yet, compared to the case of the electron loss from 105 MeV/u $U^{90+}(1s^2)$, this difference remains substantially smaller reaching "merely" a factor of 2–4 for collisions with atoms of gold. The smaller difference could be attributed to the fact that the loss from the L-shell occurs in collisions with smaller momentum transfers corresponding to larger impact parameters, where the interaction between the electron of the ion and the nucleus of the atom is weaker.

An interesting peculiarity in the theoretical data shown in Fig. 7.3 is that now the CDW-EIS model overestimates the loss cross section by up to a factor of 2. Compared to the other distorted-wave calculations, this model now yields the worst agreement with the experiment. This is rather surprising, especially taking into account the good results obtained with the CDW-EIS amplitude for the loss from 105 MeV/u $U^{90}(1s^2)$ ions. The reasons for this failure are not clear.

In contrast, the application of the modified eikonal approximation yields good agreement with the experiment also in the case of the electron loss from the lithium-like uranium ion.

Compared to the simple first-order approach, the more elaborated distorted-wave models are supposed to improve the treatment of the interaction between the electron and the nucleus of the atom. As a result, these models are expected to yield better descriptions for the loss process. As we have just seen, this is indeed the case.

What, however, is important to keep in mind is that, similarly to the first-order theory, the distorted-wave models represent in essence high-energy (or high-velocity) approximations which in general have solid grounds only provided the impact energy is "sufficiently high."

Indeed, such models, like the CDW-EIS and SEA, were first proposed and turned out to be very successful to treat ionization and excitation of very light atomic systems (hydrogen and helium) occurring in collisions with fast bare nuclei whose impact velocities are much larger than the typical velocities of the "active" atomic electron [25, 26]. In such collisions, the range of the "sufficiently high" impact energies is of course reached.

When we consider excitation or loss of an electron, which is initially bound by a very strong field (like, e.g., in a few-electron uranium ions), even collision energies about $\sim$100 MeV/u can hardly be considered as "sufficiently high" since the typical collision time is not yet (much) smaller than the typical electron transition time. In such a case, results of the distorted-wave models should be taken with due caution.

7.2.2 Simultaneous Loss-Excitation

If a heavy ion initially carries several electrons, then more than one electron of the ion can be simultaneously excited and/or lost in a collision with a neutral atom. In [12], simultaneous excitation and loss of the projectile electrons was investigated experimentally for 223.2 MeV/u U^{90+} ions impinging on atomic targets of Ar, Kr, and Xe. In the collision between the projectile-ion and the target-atom, one of the two electrons of U^{90+} was ejected, and the other was simultaneously excited into the L-subshell states of U^{91+}. Note that such a process represents one of the simplest and basic processes which can occur with projectiles having initially more than one electron.

There are essentially two qualitatively different possibilities to get the simultaneous two-electron transitions. The first is that the atom in the collision effectively interacts with only one electron inducing its transition while the second electron undergoes a transition due to the electron-electron correlations in the ion and/or due to rearrangement in the final state of the ion: the second electron "tries" to adjust its wave function to the Hamiltonian which was "suddenly" changed because of the rapid removal of the first electron which leads to the population of excited states of the new Hamiltonian. These processes are often called the two-step-1 and shake-off, respectively.

The other possibility is that the field of the atom is strong enough to interact simultaneously with each of the two electrons and to become the main driving force for both electrons to undergo transitions. In case when the interaction with the atomic nucleus involves the exchange of only two virtual photons (one per electron), such a process is often referred to as the two-step-2 process.

It was shown in [27] that, provided the condition $\frac{Z_I Z_A}{v} > 0.4$ is fulfilled, where Z_I and Z_A are the charges of the ionic and atomic nuclei, respectively, and v the collision velocity, two-electron transitions in a heavy helium-like ion occurring in collisions with an atom are governed mainly by the "independent" interactions between the atom and each of the electrons of the ion. This condition, in particular, was very well fulfilled in the experiment [12].

When the electron transitions are governed by the independent interactions, the application of the independent electron model often yields reasonable results. According to this model, the cross section for the simultaneous loss–excitation is evaluated as

$$\sigma = 2\pi \int_0^\infty \mathrm{d}b b P(b), \tag{7.1}$$

where the probability $P(b)$ for the two-electron process is given by

$$P(b) = 2\, P_{\mathrm{exc}}(b)\, P_{\mathrm{loss}}(b). \tag{7.2}$$

Here, $P_{\mathrm{exc}}(b)$ and $P_{\mathrm{loss}}(b)$ are the probabilities for single-electron excitation and loss, respectively, in a collision with a given value of the impact parameter b.

Considering transitions of a single electron, which was initially very tightly bound by the field of a highly charged nucleus, we have already seen that in collisions with many-electron atoms, provided the collision energy is not too large, these transitions are caused mainly by the interaction between the electron of the ion and the nucleus of the atom. Compared to single-electron transitions, the two-electron processes are characterized by larger momentum and energy transfers and, thus, by even smaller impact parameters. Therefore, when considering the simultaneous loss–excitation one can also neglect the presence of the atomic electrons.

Thus, within the simplified picture of the ion–atom collision, sketched in the above paragraphs, the theoretical treatment of the process of the simultaneous loss and excitation is reduced to the finding of the single-electron transition probabilities within the three-body problem in which a relativistic electron is moving in the external electromagnetic fields generated by the nuclei of the colliding ion and atom.

In addition to the experimental data, the authors of [12] also reported results of their calculations for the simultaneous excitation–loss cross sections. These calculations were performed by using (7.1) and (7.2) and evaluating the single-electron transition probabilities $P_{\mathrm{exc}}(b)$ and $P_{\mathrm{loss}}(b)$ in the first-order interaction between the electron of the ion and the nucleus of the atom.

This two-electron process occurs at effectively very small impact parameters, where the field of the atomic nucleus acting on the electrons of the ion may be quite strong and where the first-order results for the transition probabilities may be not very reliable. Indeed, the calculations and the experimental data of [12] were in agreement only for collisions with Ar atoms, while for collisions with Kr and, especially, Xe target, the theoretical results of [12] very substantially overestimated the experimental data.

Recently, the problem of the simultaneous excitation and loss of the projectile electrons in collisions of 223.2 MeV/u U^{90+} ions with atoms was considered theoretically in [28]. The consideration of [28] was also based on the independent electron model. However, in addition to using the first-order theory, the transition probabilities were also evaluated in [28] by using the distorted-wave models. The results of [28] are shown in Fig. 7.4 where they are compared with the experimental data from [12]. It is clearly seen in the figure that, compared to the first-order calculations, the distorted-wave models are capable of a much better description of the experimental data.

In the situation considered, a rather good agreement with experiment was obtained by combining the distorted-wave methods, applied effectively to single-electron problem, with the independent electron approximation. One should note, however, that a more sophisticated development of distorted-wave models enables one to apply them to two-electron systems directly, that is, without using the independent electron approximation.

7.3 Electron Loss at "Intermediate" γ

It is known [4] that at asymptotically high impact energies, there occurs saturation, both for higher-order effects in the projectile-target interaction and for the screening and antiscreening effects of the atomic electrons, when their absolute and relative roles cease to vary with increasing collision energy. However, if this region of energies is not yet reached, then the increase in the impact energy decreases the role of higher-order effects but increases the influence of the screening.

In experiment of Claytor et al. [15], the loss cross section was measured for incident 10.8 GeV/u ($\gamma = 12.6$) $Au^{78+}(1s)$ ions penetrating different solid-state targets. At this collision energy, the typical impact parameters $b \lesssim b_{\max} = \gamma v/\omega_{\mathrm{eff}}$

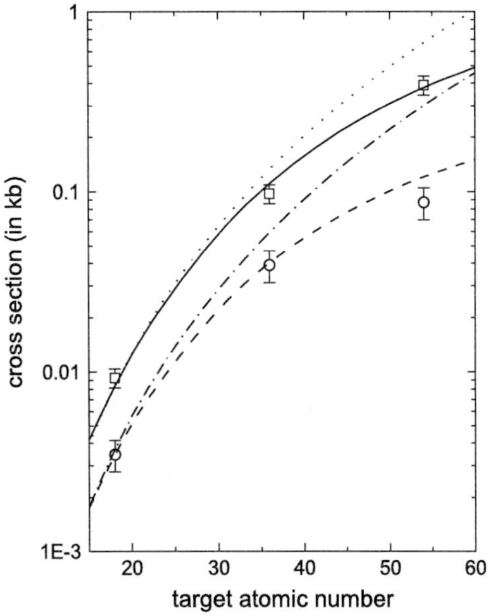

Fig. 7.4 Total cross sections for the reactions 223.2 $U^{90+}(1s^2)+Z_t \rightarrow U^{91+}(n=2,j)+e^-+\cdots$, where $j=1/2$ and $j=3/2$ are the angular momentum of the L-shell states of the hydrogen-like uranium ion. The cross sections are given as a function of the atomic number Z_t of the target atom. *Circles* and *squares* with the corresponding error bars are experimental data for the $j=3/2$ and $j=1/2$ cases, respectively, reported in [12] for collisions with argon, krypton, and xenon gas targets. *Dot* ($j=1/2$) and *dash-dot* ($j=3/2$) *curves* show the cross sections calculated with the single-electron transition probabilities obtained in the first order of perturbation theory in the interaction between the electron and the nucleus of the atom. *Solid* ($j=1/2$) and *dash* ($j=3/2$) *curves* display theoretical results obtained using the symmetric eikonal approximation to estimate the excitation probability and the CDW-EIS approximation to calculate the loss probability

(v is the collision velocity and $\omega_{\text{eff}} \sim Z_I^2$ is the averaged transition frequency), which would give the main contribution to the cross sections for the total electron loss in collisions with an unscreened atomic nucleus, are already much larger than the size of the inner shells in very heavy elements (although still substantially smaller than 1 a.u.). This means that in collisions with very heavy targets (like, e.g., gold), the influence of the screening effect of the atomic electrons on the electron loss process has to be taken into account, and the three-body models, used in the previous sections, may no longer be applied.

In addition to the necessity to account for the screening effects (and, in the case of collisions with light atoms, also for the antiscreening), one should pay attention to the fact that the targets in the experiment of [15] included carbon, aluminum, copper, silver and gold and that for collisions with the latter two one could already expect a noticeable deviation from the predictions of the first-order perturbation theory.

Taking all this into account, an appropriate candidate for calculating electron loss from 10.8 GeV/u $Au^{78+}(1s)$ could be the version of symmetric eikonal

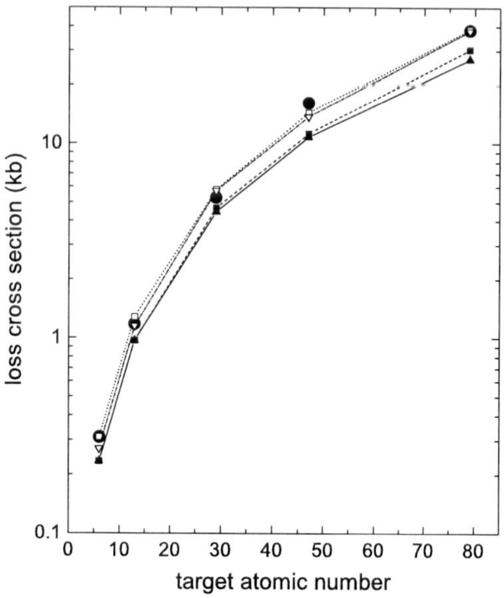

Fig. 7.5 The total cross section for electron loss from 10.8 GeV/u $Au^{78+}(1s)$ in collisions with atoms of carbon, aluminum, copper, silver, and gold. *Solid circles*: experimental data from [15]. *Open squares* connected by *dot lines*: theoretical results of [29]. *Open triangles*, connected by *dash-dot lines*: theoretical results from [30]. *Solid squares* connected by *dash lines*: our first-order calculations. *Solid triangles* connected by *solid lines*: results of our eikonal calculation. Note that all the *lines* just for guiding the eye

approximation which was considered in [31]. In this approximation, which shall below be labeled as SE, the electrons of the atom are taken into account regarding them as being "frozen" in space during the short collision time. The distortion effects, produced by a neutral atom on the initial and final states of electron of the projectile-ion, are taken into account via eikonal phases.

Besides, a calculation in the first-order approximation also very desirable since only by comparing its results with those of a more sophisticated calculation can yield an idea about the actual size of the higher order effects in this process.

Figure 7.5 shows a comparison between the experimental data of [15] and different theoretical results. The theoretical results include those of [29] and [30] and our first-order and eikonal results.

The results of Anholt and Becker [29] are based on the theory for projectile-electron loss, which they developed, and the semirelativistic description of the initial and final electronic states in the gold ion. The results of [30] are cross section estimates using a simple semiqualitative method. Our calculations are based on the theory first developed in [20, 21], in these calculations the ground and continuum states of the electron in Au^{78+} are described by the relativistic (Coulomb–Dirac) wave functions.

It is seen in the figure that the results of [29] and [30] are rather close to each other and agree well with the experimental data. At the same time, the results, obtained in the first-order and eikonal approximations, are noticeably lower than the other theoretical predictions and the experiment.

Concerning the comparison of the different calculations, it is quite expected that our turned out to be noticeably lower than those of [29]. Indeed, compared to the semirelativistic description of the electron, its fully relativistic treatment yields smaller loss cross sections. Besides, the theory of [29] predicting the dependence $\sigma_{\text{loss}} \sim A + B \ln \gamma$ largely ignores the relativistic peculiarities in the screening effect which also leads to an overestimation of the cross section.

The method of [30] involves a number of approximations whose accuracy is not very clear. Therefore, it is not easy to find out why results of this method are (or are not) in agreement with other calculations.

At first glance, it is surprising that the most rigorous set of theoretical data has the largest deviations from the experimental results. The reason for this is that the calculations have been done for the projectiles colliding with atoms whereas in the experiment the projectiles were stripped in solids.

As was already mentioned in the Introduction, there is a difference by roughly a factor of 2 between experimental cross sections reported for the loss from 160 GeV/u Pb^{81+} ions in collisions with solid and gas targets with the solid-state cross sections being larger. The origin of this difference, which was explained in [19], lies in multiple collisions between the projectile and atoms inside solids which effectively enhance the electron loss process. Since the magnitude of this difference depends on the impact energy and decreases when the energy decreases; similar reasons are probably responsible for the observed disagreement between our results for atomic targets and the experimental data from [15].

As was already mentioned, the range of the "intermediate" impact energies ($\sim$1–30 GeV/u) is characterized by the importance of both screening and higher-order effects. Therefore, into account that the SE model performs quite well at these impact energies [31], it is worth to consider some more examples of projectile-electron excitation and loss in this impact range.

7.3.1 *Single-Electron Loss*

We shall start with single electron loss.

In Fig. 7.6, we show results for the electron loss from incident $Au^{78+}(1s)$ projectiles in collisions with neutral Au atoms at impact energies 1–30 GeV/u. These results include our first Born and eikonal calculations as well as experimental data from [14] and [15] on the electron loss from 10.8 GeV/u $Au^{78+}(1s)$ projectiles.

In the case considered, our eikonal and first-order results differ by 15–35%. As expected, when the impact energy increases, the difference between them decreases. On overall, the difference is not very large but nevertheless should be taken into account.

Both the first Born and eikonal cross sections agree neither with the experimental data of [14] nor with that of [15]. The data from [14] are substantially smaller (by about of 30–50%), while the data reported in [15] are considerably larger (by about a factor of 1.3–1.4) than our results.

As was already mentioned in the introduction, there is a difference by roughly a factor of 2 between experimental cross sections reported for the loss from

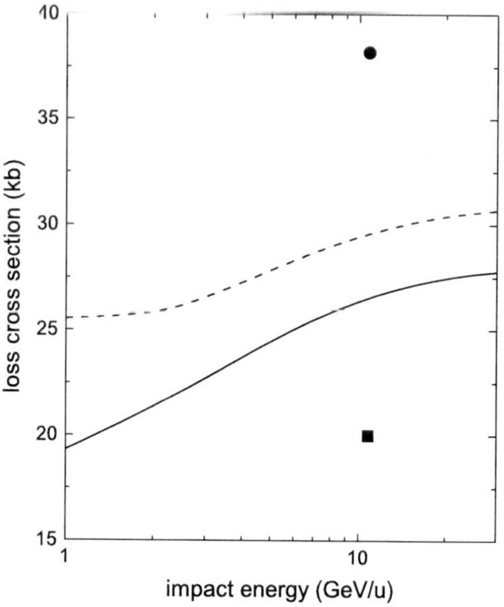

Fig. 7.6 Cross section for the electron loss from $Au^{78+}(1s)$ ions in collisions with neutral Au atoms. *Solid curve*: eikonal results. *Dash curve*: first Born results. The *circle* displays experimental data from [14], while the *square* shows the result of [15] scaled to the gold target

160 GeV/u Pb^{81+} ions in collisions with solid and gas targets with the solid-state cross sections being larger. The origin of this difference, which was explained in [19], lies in multiple collisions between the projectile and atoms inside solids which effectively enhance the electron loss process. Since the magnitude of this difference depends on the impact energy and decreases when the energy decreases, similar reasons are probably responsible for the observed disagreement between our results for atomic targets and the experimental data from [15].

7.3.2 *Two-Electron Transitions*

7.3.2.1 Double-Electron Loss

Figure 7.7 shows calculated cross sections for double-electron loss from $Au^{77+}(1s^2)$ projectiles incident on neutral Au atoms at impact energies 1–30 GeV/u. Compared to the single loss, now the difference between the eikonal and first Born results is much more pronounced. For impact energies $\sim$ 1–5 GeV/u, these calculations predict even qualitatively different dependencies of the cross sections on the collision energy. The absolute difference between the first Born and eikonal cross sections ranges between $\simeq$2.5 at 1 GeV/u and $\simeq$1.5 at 30 GeV/u. Moreover, as additional calculations show, for impact energies above 30 GeV/u, the ratio $\simeq$1.5 remains almost a constant, and thus even at asymptotically high collision energies, the first Born calculation still substantially overestimates the cross section values.

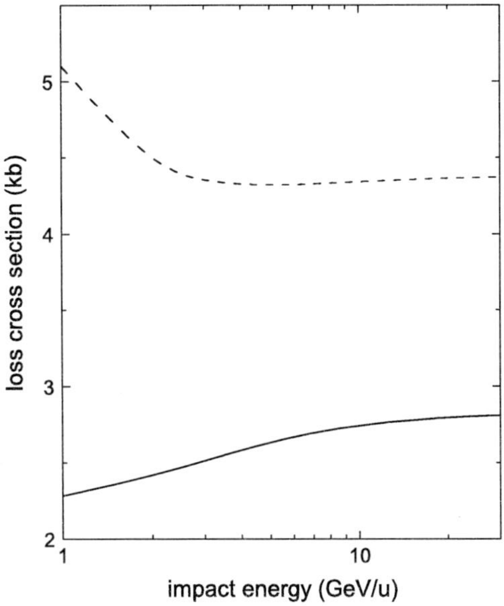

Fig. 7.7 Cross sections for double electron loss from $Au^{77+}(1s^2)$ ions in collisions with neutral Au atoms. Solid curve: eikonal results. *Dash curve*: first Born results

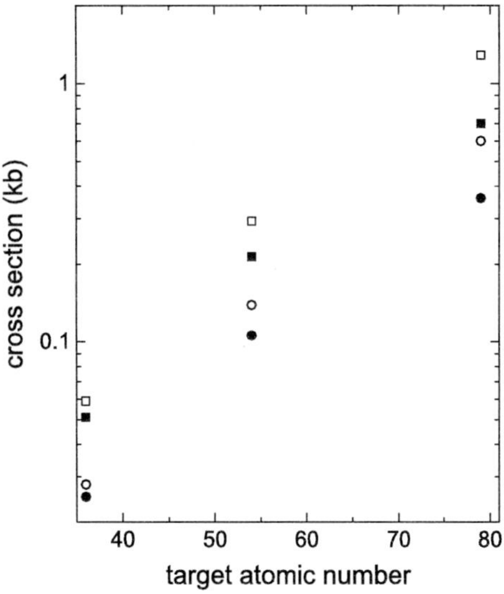

Fig. 7.8 Cross sections for the simultaneous electron loss–excitation, $U^{90+}(1s^2) \rightarrow U^{91+}(n = 2, j) + e^-$, occurring in collisions with Kr, Xe, and Au atomic targets at 20 GeV/u. $j = 1/2$ and $j = 3/2$ are the angular momentum of the states of the hydrogen-like uranium ion. *Squares* and *circles* show results for $j = 1/2$ and $j = 3/2$, respectively. *Open* and *solid symbols* denote the cross sections obtained using the first Born approximation and the SE model, respectively

7.3.2.2 Simultaneous Loss-Excitation

In Fig. 7.8, we present results of calculations for the simultaneous projectile-electron excitation and loss, $U^{90+}(1s^2) \rightarrow U^{91+}(n = 2, j) + e^-$, where n and j are, respectively, the principal quantum number and the total angular momentum of the excited state of the hydrogen-like ion. The projectile is assumed to collide with Kr,

Xe, and Au atomic targets at an impact energy of 20 GeV/u. Similarly to the case of double-electron loss, considered above, we observe that the differences between the cross sections, calculated with the first Born and eikonal probabilities, can be quite substantial if the atom is sufficiently heavy. For collisions leading to the population of the states with $j = 1/2$, the first Born results are larger by a factor of $\simeq$1.16 (Kr), $\simeq$1.38 (Xe) and $\simeq$1.84 (Au). When the states with $j = 3/2$ are populated this ratio is $\simeq$1.12, $\simeq$1.31 and $\simeq$1.67, respectively.

We see that the difference between the first Born and eikonal cross sections turns out to be somewhat smaller for transitions involving the states with $j = 3/2$. This can be attributed to the fact that, compared to the $j = 1/2$ case, these transitions are characterized on average by larger impact parameters. As a result, the field of the atom acting on the electrons of the ion is weaker, and therefore the first Born treatment becomes less inaccurate.

As additional calculations show, the differences between the results of the eikonal and first Born calculation does not substantially change when the impact energy increases further. Thus, like for double-electron loss, even at asymptotically high impact energies, the first Born calculation may considerably overestimate the cross section for simultaneous projectile-electron excitation and loss.

At these rather high energies, where the electric dipole transitions become already dominant, the pure excitation $U^{90+}(1s^2) \to U^{90+}(1s; n = 2, j)$ (or the excitation $U^{91+}(1s) \to U^{91+}(n = 2, j)$, see Fig. 7.9) proceeds more efficiently into the states with $j = 3/2$ (due to the most "powerful" dipole transition $1s_{1/2}(+1/2) \to 2p_{3/2}(+3/2)$). However, according to both the first Born and eikonal results, the cross sections for simultaneous loss–excitation are larger for transitions into the states with $j = 1/2$. The origin of this interesting peculiarity can be understood by considering the single-electron transition probabilities. At this impact energy, the

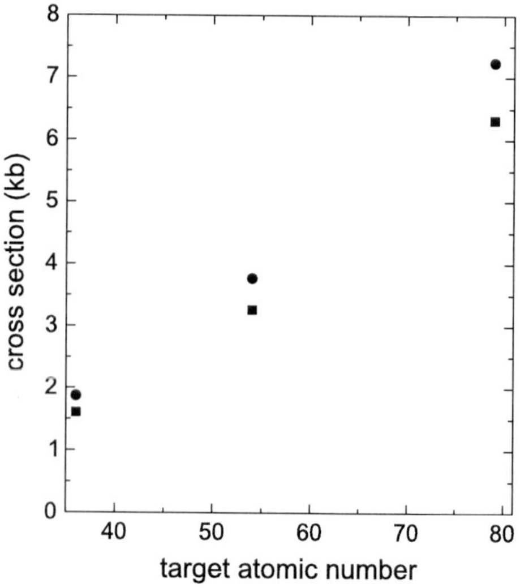

Fig. 7.9 Cross sections for the excitation of 20 GeV/u $U^{91+}(1s)$ projectiles into the states with $n = 2, j = 1/2$ (*squares*) and $n = 2$, $j = 3/2$ (*circles*) in collisions with Kr, Xe, and Au atomic targets. The cross sections were calculated using the SE model

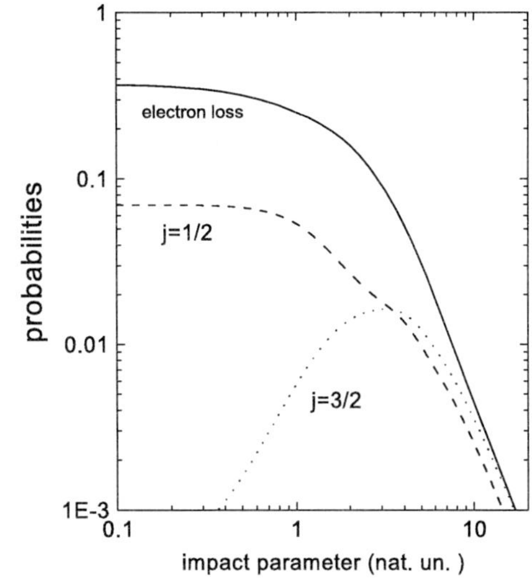

Fig. 7.10 Probabilities for projectile-electron loss and excitation in collisions of 20 GeV/u $U^{91+}(1s)$ ions with Au atoms given as a function of the impact parameter. The probabilities were calculated using the SE. *Solid curve*: the probability for electron loss. *Dash curve*: the probability for excitation into the states with $n = 2$, $j = 1/2$. *Dot curve*: the probability for excitation into the states with $n = 2$, $j = 3/2$

electron loss is also already dominated by the electric dipole transitions. However, compared to the excitation, this process involves larger energy-momentum transfer and, consequently, effectively occurs at smaller impact parameters. As a result, it turns out that the probability for electron loss has a larger overlap with the probability for excitation into the states with $j = 1/2$ (see for illustration Fig. 7.10).

7.4 Electron Loss in Collisions at Asymptotically High Impact Energies

Here, we consider the cases studied experimentally in [16] and [17], where a much higher collision energy was considered. In these experiments, the loss and capture cross sections were measured for Pb^{81+} and Pb^{82+} ions incident on solid and gas targets at an impact energy of 33 TeV, where the corresponding collisional Lorentz factor is very high ($\gamma = 168$).

Explorations of ion–atom collisions at such very high impact energies are of special interest because during the interaction between a projectile-ion and a target-atom, both these particles can be exposed to extremely intense and extraordinarily short pulses of the electromagnetic field.

For instance, in collisions of 33 TeV $Pb^{81+}(1s)$ ions Au atoms, the typical duration of the electromagnetic pulses, generated by the atoms in the rest frame of the ions, are $\lesssim 10^{-21}$ s. The peak intensities of these pulses reach $\sim 10^{31}$ W/cm^2. These field parameters may be compared with the parameters of state-of-the-art laser systems whose shortest pulse lengths are about $\sim 10^{-16}$–10^{-15} s and whose peak intensities do not exceed $\sim 10^{21}$–10^{22} W/cm^2. These intensities are so large that such pulses, despite the very short interaction time, not only enable one to induce

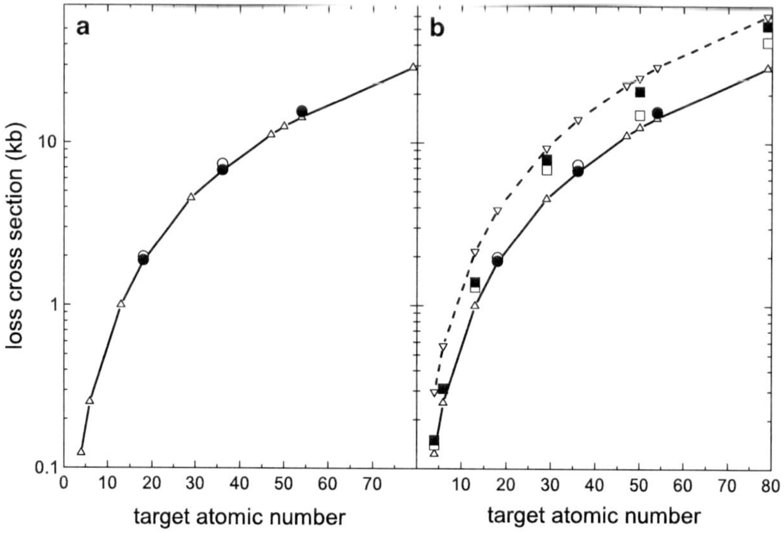

Fig. 7.11 The total section for electron loss from 33 TeV $Pb^{81+}(1s)$ projectiles. (**a**) *Circles* represent experimental data from [17] on the electron loss in gas targets (Ar ($Z_A = 18$), Kr ($Z_A = 36$) and Xe ($Z_A = 54$)) where the open and solid symbols refer to the "ionization" and "capture" experimental scenarios, respectively. *Up triangles* connected by guiding *solid line* are results of the first-order calculation for $Z_A = 4, 6, 13, 18, 29, 36, 47, 50, 54$, and 79. (**b**) *Circles* and *up triangles*: same as in the part (**a**) of the figure. *Squares* show the experimental data from [16] on the electron loss in solid-state targets (Be, C, Al, Cu, Sn and Au). *Down triangles* connected by guiding *dash line* display theoretical results of [29]. Note that both in (**a**) and (**b**) the *open* and *solid symbols* denoting the experimental data refer to the "ionization" and "capture" experimental scenarios, respectively

transitions between electron states with positive energies leading to the excitation of and the electron loss from very tightly bound $Pb^{81+}(1s)$ ions but also cause transitions between negative and positive states resulting in the electron–positron pair production with quite noticeable cross sections.

In Fig. 7.11a, shown are experimental data from [17] on the electron loss cross sections measured for 33 TeV Pb projectiles ($\gamma = 168$) which were penetrating three gas targets (Ar, Kr, and Xe). The loss cross sections were measured for the "ionization" scenario, in which the incident projectiles were $Pb^{81+}(1s)$ ions. These cross sections were also obtained in the "capture" scenario in which a beam of initially bare Pb^{82+} nuclei was traversing the same targets: in the latter case, the electron had to be captured into a bound state of the projectile before it could be lost in a consequent collision. The experimental loss data extracted in both scenarios are rather close to each other.

Figure 7.11a also shows results of calculations for the loss from 33 TeV $Pb^{81+}(1s)$ projectiles for a more extended set of atomic targets (Be, C, Al, Ar, Cu, Kr, Ag, Sn, Xe and Au). These calculations were performed using the first-order perturbation theory and describing the initial and final electron states by the Coulomb–Dirac wave functions.

In addition to the results, already shown in Fig. 7.11a, b contains also experimental data from [16] as well as theoretical results from [29] for the electron loss from the ground state of the hydrogen-like lead ions. In the experiment of [16], beams of incident 33 hydrogen-like ("ionization" scenario) and bare ("capture" scenario) lead ions were penetrating solid state targets. The figure presents results for the loss cross sections extracted from the measurements carried out in both scenarios.

Two main conclusions can be drawn from the Fig. 7.11. First, according to Fig. 7.11a, there is a good agreement between the experimental data obtained for the gaseous targets and the results of our first-order calculation. Second, the experimental loss cross sections measured in collisions with solid targets are substantially larger that those obtained for gas targets.

The reason for the latter was clarified in [19]. It turned out that it is excitations, suffered by the projectiles when they penetrate solids. Compared to the cross section for the loss from the ground state, the cross sections for the loss from excited states are larger. Therefore, if the beam of the hydrogen-like projectiles attains, due to collisions with atoms of the foil, a noticeable fraction of excited-state ions, the cross section for the electron loss from such a beam may effectively become larger. The use of a simplified two-state model in [16] had, in fact, introduced quite a substantial error into the values of the loss cross sections reported there. This also explains the very substantial deviations between the theoretical and experimental results for the loss cross sections observed in Fig. 7.11b.

7.4.1 Electron Loss from Heavy Many-Electron Ions

Electron loss processes from heavy many-electron ions (like Xe^{24+}, Pb^{25+}, U^{28+}) in collisions with atoms and molecules are of crucial importance in many applications in plasma and accelerator physics. In particular, the international FAIR project [32], where heavy many-electron ions (e.g., U^{28+} are planned to be accelerated up to energies of a few GeV/u, requires benchmarks for electron loss cross sections since these reactions may dominate among other beam loss processes.

Cross sections for electron loss from many-electrons ions at relativistic energies were recently considered in the Born approximation using the momentum-transfer representation (see [33, 34], and chapter by V. Shevelko et al. of this book). There, the Coulomb gauge for the atomic field and the nonrelativistic wave functions are used, and the electron-loss cross section is presented in terms of the usual Born form factor (arising from the scalar potential) and an additional term containing an integral over the derivative of the radial wave function of the active projectile electron in the initial bound $n\ell$ state (which appear due to the vector potential of the atomic field). The use of the nonrelativistic wave functions is justified by the fact that the main contribution to the loss cross sections at high collision energies is given by the loss of outermost projectile electrons which can be treated as nonrelativistic particles.

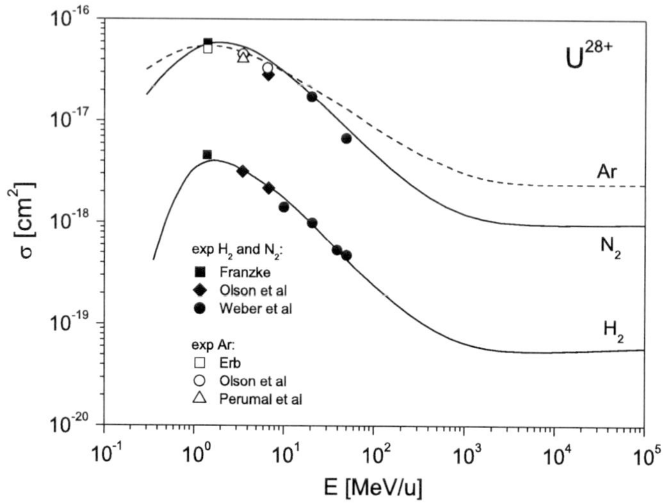

Fig. 7.12 Total electron loss cross sections in collisions of U^{28+} projectile-ions with H_2, N_2 and Ar targets as a function of the ion energy. The data for H_2 and N_2 targets are given in units of cm^2/mol. Symbols, experiment—H_2 and N_2 targets: *filled square* [35], *filled diamond* [36], *filled circle* [37]; Ar target: *open square* [38], *open circle* [36], *open triangle* [39]. Theory, curves: calculations from [34]

Figure 7.12 shows experimental data and theoretical calculations of the total electron loss cross sections for U^{28+} ions colliding with H_2, N_2, and Ar targets over a wide energy range. Theoretical results were obtained with taking into account multielectron loss processes which can contribute up to more than 50% to the total loss cross sections at low and intermediate energies (see [34] for details).

7.5 Capture of Leptons via Pair Production Mechanism

One of the most fascinating predictions of quantum electrodynamics is the possibility of converting energy into matter. Starting with the paper by Sauter [40], electron–positron pair production from vacuum in the presence of external electromagnetic fields has been attracting the attention of different physical communities.

Pair production has been studied theoretically in the presence of electromagnetic fields of various configurations (e. g., in the combination of Coulomb and high-energy photon fields [41], in high-energy collisions of charged particles [42–46], in constant and uniform fields [47], in slowly varying super-strong Coulomb fields [48], in colliding laser fields [49], and in crystals [50]), and also in the presence of gravitational fields [51].

Pair production can occur with a noticeable probability (a) if the external field is strong enough to provide an energy of the order of the electron rest energy mc^2 on a

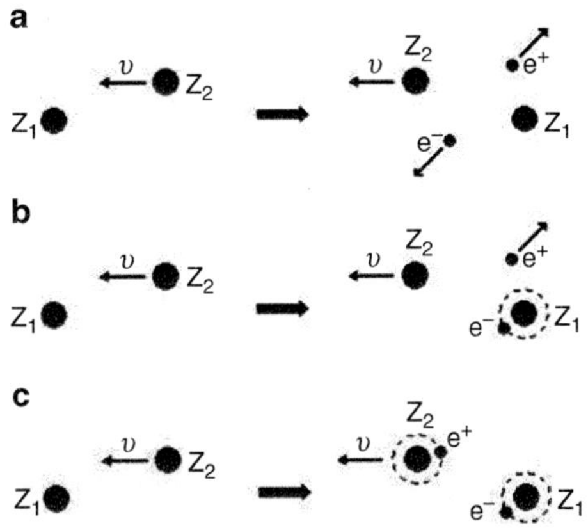

Fig. 7.13 A sketch of different pair production (sub-)processes: free (**a**), bound–free (**b**), and bound–bound (**c**) pair production

distance of the order of the electron Compton wavelength $\lambda_C = \hbar/mc$,[1] where $\hbar$ is the Planck's constant, (b) or/and if the field varies in time so rapidly that its typical frequencies multiplied by $\hbar$ are larger than $2mc^2$.

Experimentally, pair production has been explored only in the case of rapidly varying electromagnetic fields (for instance, in relativistic heavy-ion collisions [52–55], photon–laser collisions [56], in the collision of an intense laser beam with a solid target [57]). Note also that the observation of pair production in the collision of two light beams is one of the main goals of future intense-laser facilities [58].

Landau and Lifshitz [42] were the first to estimate the cross section for pair production in relativistic collisions of charged particles in which the created electron and positron freely move in space after the collision is over (see Fig. 7.13a). Such a process is termed free pair production, and it was studied in much detail in a vast amount of theoretical and experimental papers (see for recent reviews, e.g., [45,46], and also references therein).

7.5.1 Bound-Free Pair Production

During the last 2 decades, another kind of pair production process occurring in relativistic nuclear collisions has attracted much attention (see, e.g., [2, 46, 59], and references therein) in which the electron is created in a bound state with one of the colliding nuclei (see Fig. 7.13b).

There have been several experimental studies on bound–free pair production in relativistic ion–atom collisions, in which the total cross sections for this process were reported [16, 17, 52–55]. In these experiments, bare ions (La^{57+}, Au^{79+},

---

[1] A so strong field has an amplitude of the order of the so-called critical field $E_{cr} = m^2c^3/\hbar e = 1.3 \times 10^{16}$ V/cm, where e is the absolute value of the electron charge.

Pb^{82+}, U^{92+}) were incident on solid foils of different chemical elements ranging from beryllium to gold. In these experiments, also a very broad interval of impact energies was considered starting with relatively low energy collisions ($\sim$0.5 GeV/u) up to extreme relativistic collisions ($\sim$160 GeV/u Pb^{82+}).

Ions penetrating foils "see" atomic nuclei surrounded by electrons and the process of the pair production may be better regarded as occurring in collisions with neutral atoms rather than with the bare atomic nuclei. In ion–atom collisions, the bound–free pair production may be influenced by the presence of atomic electrons. Since this influence increases when the impact energy increases, below we shall restrict our attention to the discussion of the bound-free pair production by incident 33 TeV Pb^{82+} ions.

In Fig. 7.14, we show results for the total cross section for the electron capture via the pair production by 33 TeV Pb^{82+} projectiles incident on different targets. At this very high impact energy, the pair production represents the main capture mechanism: its contribution to the capture amounts from $\simeq$60% in collisions with Be atoms up to $\simeq$96% in collisions with Au atoms. Experimental data for these

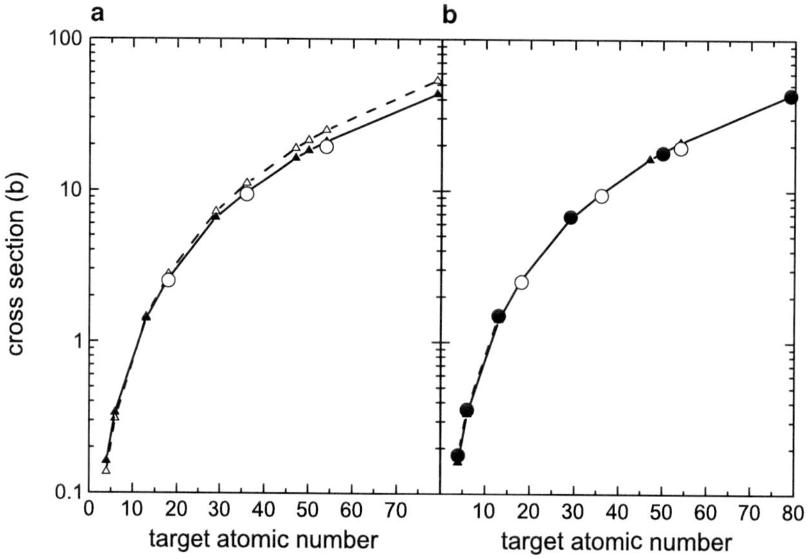

Fig. 7.14 Cross sections for the electron capture from the pair production in collisions of 33 TeV Pb^{82+} projectiles with gas- and solid-state targets given as a function of the target atomic number. (**a**) *Open circles* are experimental data from [17] for collisions with Ar, Kr, and Xe gas targets. *Solid triangles* connected by *solid curve* are results of the calculations for collisions with atoms having atomic numbers $Z_A = 4, 6, 13, 18, 29, 36, 47, 50, 54$, and 79. *Open triangles* connected by *dash curve* are results for the pair production in collisions with the bare atomic nuclei. The *curves* are just to guide the eye. (**b**) *Open circles* and *solid triangles* connected by *solid curve* represent the same results as in (**a**). *Solid circles* are data from [16], obtained for collisions with solid-state targets (Be, C, Al, Cu, Sn, and Au)

cross sections were reported in [16] and [17] for collisions with solid-state and gas targets, respectively.

The theoretical cross sections shown in the figure were calculated in the first-order perturbation theory taking into account the screening and antiscreening effects of atomic electrons. The electron and positron were described using the Coulomb–Dirac wave functions. In the rest frame of the ion, the energy spectrum of the positrons extends to tens of mc^2. Such positrons carry in general a big amount of angular momentum which makes it necessary to take into account continuum states with large values of the angular quantum number κ. Theoretical results shown in the figure were obtained by directly integrating over the interval of the total positron energies $[mc^2; 30mc^2]$ and summing over the positron angular momenta corresponding to $\kappa \leq \kappa_{\max} = 30$. The contributions from the positron states with higher energy and/or larger κ were evaluated by an extrapolation. The capture cross sections were calculated to all bound states with the principal quantum number $n \leq 6$. According to our estimates, the capture into the states with larger n is negligible and can safely be neglected. In the figure, only the total capture cross section is displayed.

Calculations in the light-cone approximation show that the pair production in collisions between 33 TeV Pb^{82+} and atoms ranging between Be and Au can be well described within the first-order theory in the interaction between the lepton transition current and the field of the atom. The reasons for this are the very high values of the impact energy and of the projectile charge so that even such a heavy atom like gold still represents an effectively weak perturbation.

The calculations for the capture cross section were done assuming the single-collision condition. Therefore, they should be first of all compared with the experimental results obtained for collisions with gas targets. Such a comparison is shown in Fig. 7.14a where a good agreement is seen between the experiment and theory. Moreover, in contrast to the experimental loss cross sections, the experimental capture cross sections in collisions with both solids and gases fall on the same curve. Therefore, in contrast to the calculated loss cross sections, the calculated cross sections are in agreement also with the experimental data for solid targets.

Comparing calculated results for the pair production with neutral atoms and the corresponding bare atomic nuclei, we see that the atoms are more effective at smaller values of Z_A. This reflects the relative importance of the contribution from the antiscreening mode. In the case of collisions with atoms having small Z_A, the antiscreening overcompensates a small decrease in the cross section caused by the screening of the atomic nucleus by the atomic electrons inherent to the elastic atomic mode. For heavy targets, the screening effect becomes stronger while the antiscreening contribution decreases (in relative terms) which leads to the reduction of the capture cross section compared to the case with the bare nuclei.

Compared to the electron loss, the pair production with capture involves much larger momentum transfers. Therefore, the screening effect of the atomic electrons is much weaker. For instance, for the pair production in collisions of 33 TeV Pb with

Be and Au atoms, this effect reduces the cross section for the capture to the ground state by about 5 and 24%, respectively.

Note also that the magnitude of the screening effect obtained in the calculations, which employ the exact Coulomb–Dirac wave functions, turns out to be somewhat larger compared to that obtained in calculations where semi-relativistic Darwin and Furry wave functions are used. This increase can be attributed to the fact that, compared to the semirelativistic, the fully relativistic treatment of the pair production predicts an enhancement in the emission of positrons with the intermediate energies ($\sim$1–20 mc^2 in the rest frame of the ion), whereas according to the semirelativistic description, there should be more positrons with higher energies. Since the creation of more energetic positrons implies larger values of the momentum transfers in the collision, the screening effect tends to be weaker in the semirelativistic consideration.

The magnitude of the screening effect for the pair production may be compared to the corresponding reduction by a factor of about 1.4 and 2 due to the screening effect of the atomic electrons in the elastic target mode in the case of the electron loss from the ground state of Pb^{81+} in collisions with the same atoms. It is obvious that such large differences in the magnitude of this effect are caused by the fact that, compared to the electron loss process, the pair production involves much larger momentum transfers (or, in other words, proceeds at much smaller impact parameters).

7.5.2 *Bound–Bound Pair Production*

When the colliding nuclei possess charges of different signs, yet another pair production process becomes possible in which not only the electron but also the positron is created in a bound state (see Fig. 7.13c). Following [60], we shall call this process bound–bound pair production. Compared to the free and bound–free cases, bound–bound pair production has a number of interesting features; in particular, it has an intrinsic nonperturbative dependence on charges of *both* colliding nuclei. This, as well as the fact that this process completes the picture of the basic (single-) pair production processes occurring in high-energy collisions of charged particles, makes its study of significant interest. Besides, bound–bound pair production which results in creating bound states of antimatter (in particular, antihydrogen) may also be relevant in connection with testing CPT invariance and the weak equivalence principle [61].

Let us consider the collision of two nuclei with charges Z_1 and Z_2 (say $Z_1 > 0$ and $Z_2 < 0$). Impact parameter values characteristic for this process are of the order of λ_C (see Fig. 7.17 below) and, thus, are much larger than the nuclear size. Therefore, one can treat the nuclei as point-like particles (our estimation using extended ("real") nuclei instead of point-like ones indeed shows very little change—well below 1%—in the cross section for bound–bound pair production). Since we deal with high impact energies, one can also assume that the initial velocities of these particles are not changed in the collision.

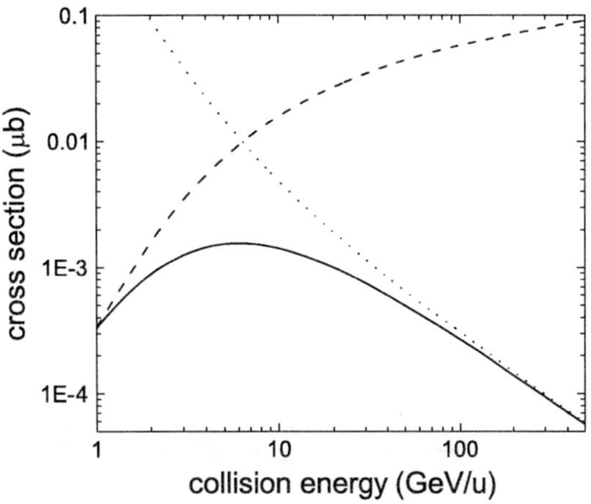

Fig. 7.15 Cross sections for pair production and electron capture given as a function to the collision energy. *Solid curve*: $p^- + \mathrm{U}^{92+} \rightarrow \overline{H}(1s) + \mathrm{U}^{91+}(1s)$. *Dash curve*: $p^- + \mathrm{U}^{92+} \rightarrow \overline{H}(1s) + \mathrm{e}^- + \mathrm{U}^{92+}$. *Dot curve* shows twice the cross section for the reaction $\mathrm{H}(1s) + \mathrm{U}^{92+} \rightarrow p^+ + \mathrm{U}^{91+}(1s)$

It can be shown [60] that the asymptotic form of the cross section for the bound–bound pair production is given by

$$\sigma \sim \frac{Z_1^5\,|Z_2^5|}{\gamma}. \tag{7.3}$$

This dependence is significantly different from the corresponding ones in the case of free and bound–free pair productions which read $\sigma_{\mathrm{f}} \sim Z_1^2 Z_2^2 \log^3(\gamma)$ and $\sigma_{\mathrm{bf}} \sim Z_1^5 Z_2^2 \log(\gamma)$, respectively (see, e.g., [1]). Note that in σ_{bf}, Z_1 is the charge of the nucleus carrying away the created electron.

In Fig. 7.15, we show the cross section for the reaction $p^- + \mathrm{U}^{92+} \rightarrow \overline{H}(1s) + \mathrm{U}^{91+}(1s)$ (solid curve). The dependence of the cross section on the impact energy is not monotonous. At the relatively low collision energies, the cross section increases with the energy reaching a maximum at about 5–7 GeV/u. With a further energy increase, the cross section starts to decrease with an increasing slope and reaches its asymptotic energy dependence $\sim 1/\gamma$ already within the energy interval displayed in the figure.

The cross section for bound–bound pair production can be compared with that for bound–free pair creation. We have calculated the cross section for the latter process (shown in Fig. 7.15 by dash curve), treating it as a transition between the negative- and positive-energy Coulomb states centered on the antiproton which is induced by the interaction with the charge Z_1 taken into account in lowest-order perturbation theory.

At relatively low impact energies, both cross sections increase and are rather close in magnitude to each other. However, at larger impact energies, the cross sections start to demonstrate qualitatively different behaviors, and the difference between them increases very rapidly.

Such an interrelation between these cross sections can be understood by noting the following. At very low collision energies, the spectrum of the electromagnetic field generated by the colliding particles does not have enough high-frequency components necessary to create an electron–positron pair. As a result, the cross sections for both pair production processes are very small. An increase in the impact energy leads to an increase of the high-frequency components of the field and both cross sections grow rather rapidly. However, when the impact energy increases further, the conditions for the bound–bound pair production begin to deteriorate. Indeed, the electron and positron are created on different nuclei, and therefore the difference between their momenta increases with the impact energy. This effectively reduces the overlap between the states ψ_i and ψ_f, making bound–bound pair production more difficult to occur.

This, of course, does not occur in bound–free pair production since both leptons are created on/around the same nucleus. In this case, when the impact energy grows, the range of the impact parameters efficiently contributing to the process grows as well ($\sim\gamma$) leading to the logarithmic increase in the cross section.

Bound–bound pair production can be viewed as a collision-induced transition between states of the electron with negative and positive total energies bound in the field of the charge $Z_2 < 0$ and charge $Z_1 > 0$, respectively. This is reminiscent of the atomic collision process of nonradiative electron capture (for a review see, e.g., [1, 2]) in which an electron initially bound in the atom undergoes a transition into a bound state in the ion: $(Z_\mathrm{a}+\mathrm{e}^-) + Z_\mathrm{i} \to Z_\mathrm{a}+(Z_\mathrm{i}+\mathrm{e}^-)$, where Z_a and Z_i are the charges of the atomic and ionic nuclei. Therefore, it is of interest to compare the cross sections for these two processes.

Such a comparison is presented in Fig. 7.15 where dot curve shows twice the cross section for the reaction $H(1s)+\mathrm{U}^{92+} \to p^+ + \mathrm{U}^{91+}(1s)$, calculated using the simplest description mentioned above. At relatively low and intermediate collision energies, where the electron capture is much more probable than the bound–bound pair production, the two cross sections show a qualitatively different behavior. However, at higher impact energies the cross sections approach each other, cross, and when the energy increases further, demonstrate exactly the same energy dependence with the bound–bound pair production cross section being a factor of 2 larger.

The factor of 2 is of statistical origin reflecting the difference between the averaging over the spin projections in the initial state in the case of electron capture and the corresponding summation in the case of pair production. Apart from this, the cross sections in the limit $\gamma \gg 1$ are identical. This circumstance can be understood by taking into account the symmetry between these two processes and observing that at $\gamma \gg 1$, the absolute values of the momentum transfers in both processes become essentially the same. Thus, at asymptotically high impact energies, it would be easier to capture electron and positron from vacuum into the corresponding bound states in the collision with an antiproton than to pick up the already existing electron from an atomic hydrogen.

In order to illustrate the very strong dependence of the bound–bound pair production on the charges of the colliding particles in Fig. 7.16, we present the cross section for the (hypothetical) reaction $\overline{U}^{92-} + \mathrm{U}^{92+} \to \overline{U}^{91-}(1s) + \mathrm{U}^{91+}(1s)$.

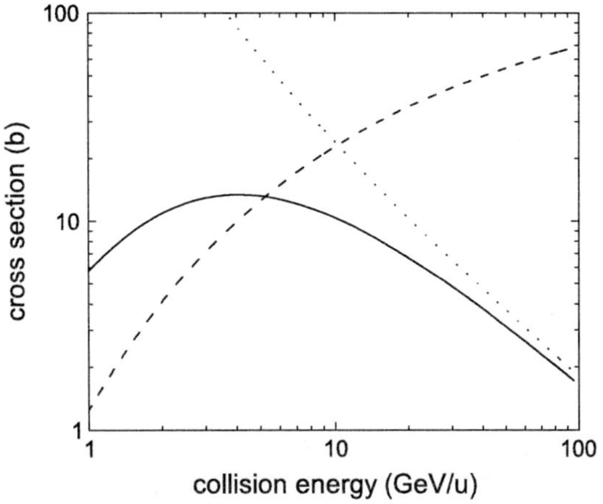

Fig. 7.16 The same as in Fig. 7.15 but for $\overline{U}^{92-}$ + $U^{92+} \to \overline{U}^{91-}(1s)$ + $U^{91+}(1s)$ (*solid curve*), U^{92+} + $U^{92+} \to U^{91+}(1s) + e^{+}$ + U^{92+} (*dash curve*) and $U^{90+}(1s^2) + U^{92+} \to U^{91+}(1s) + U^{91+}(1s)$ (*dot curve*)

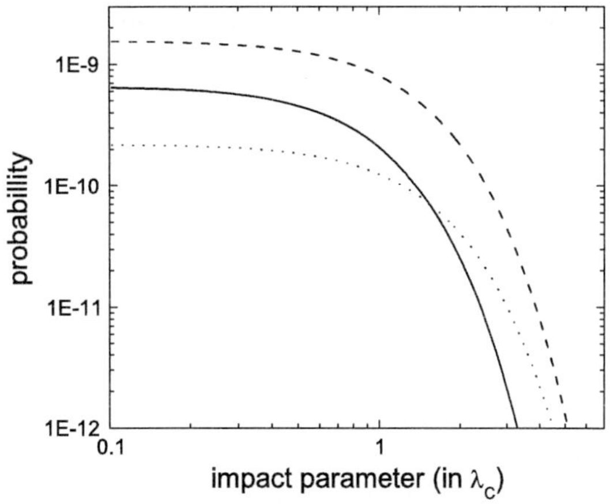

Fig. 7.17 Probability for the reaction $p^- + U^{92+} \to \overline{H}(1s) + U^{91+}(1s)$ given as a function of the impact parameter b. *Solid, dash*, and *dot curves* display the probability for the impact energy of 1, 6, and 100 GeV/u, respectively

It is seen that compared to the case shown in Fig. 7.15, the magnitude of the cross section has increased roughly by 10 orders with the dependence on the collision energy remaining basically the same. In Fig. 7.16, the bound–bound cross section can also be compared with those for the bound–free pair production in $U^{92+} + U^{92+}$ collisions and for the electron capture in collisions between $U^{90+}(1s^2)$ and U^{92+}.

Additional insight into the physics of bound–bound pair production can be obtained by considering the probability $P(b)$ for this process as a function of the impact parameter b ($b = |\mathbf{b}|$). The probability, which is related to the cross section via $\sigma = 2\pi \int_0^\infty db\, b\, P(b)$, is shown in Fig. 7.17 for the reaction $p^- + U^{92+} \to \overline{H}(1s) + U^{91+}(1s)$ for three different impact energies, 1, 6, and 100 GeV/u. In all these cases, the impact parameters characteristic for the process are of the order of λ_C, but the range of b significantly contributing to the cross section slightly broadens with the increase in the impact energy. Both these points can be understood by considering the momentum transfers in the collision. They are given by [60]

$$\mathbf{q} = \left(\mathbf{q}_{\perp}, \frac{mc^2 - I_{\mathrm{e}} + (mc^2 - I_{\mathrm{p}})/\gamma}{v} \right)$$
$$\mathbf{q}' = \left(\mathbf{q}_{\perp}, \frac{mc^2 - I_{\mathrm{p}} + (mc^2 - I_{\mathrm{e}})/\gamma}{v} \right). \tag{7.4}$$

In these equations, I_{e} and I_{p} denote the binding energies of the electron and positron, respectively, in their final states. The quantities $\mathbf{q}$ and $\mathbf{q}'$ have the meaning of the momentum transfers as viewed in the rest frames of the charges Z_1 and Z_2, respectively. The inspection of (7.4) shows that at $v \simeq c$, the longitudinal components of the momentum transfers are of the order of mc and decrease when the impact energy increases. Besides, one should take into account that the transverse and longitudinal components of the momentum transfer in this process are (typically) of the same order of magnitude.

The cross section for bound–bound pair production is very small. Therefore, the detection of this process in laboratory will be feasible; only provided high-luminosity beams of heavy nuclei and antiprotons (or muons) are available.

7.6 Conclusions

We have briefly discussed projectile-electron loss and excitation occurring in relativistic collisions between highly charged hydrogen-, helium-, and lithium-like ions and neutral atoms. A very broad range of impact energies was considered: from $\sim$100 MeV/u, where the typical electron velocities in the initial state of the ion are of the order or even exceed the collision velocity, to 160 GeV/u, where the collision velocity only fractionally differs from the speed of light.

At impact energies below 1 GeV/u, the screening effect of atomic electrons is relatively weak and can be neglected. At the same time, at these energies, the interaction between the nucleus of the atom and the electron(s) of the projectile may be too strong to be considered within the first-order approximation and results, obtained within the latter, can overestimate experimental cross sections by order of magnitude. In such a case, as we have seen, distorted-wave models, which regard loss (or excitation) process as a three-body problem (the projectile electron and the atomic and projectile nuclei), represent a much better (although not perfect) tool for calculating cross sections for projectile-electron loss and excitation and enable one to get reasonable agreement with experimental data.

With increase of impact energy, the role of the higher-order effects in the projectile-target interaction diminishes but that of the screening effect of atomic electrons grows. At impact energies above 1 GeV/u, the application of the first-order perturbation theory already does not overestimate very strongly cross sections for single-electron transitions. Still, such an application may lead to substantially larger values of cross sections for transitions involving two and more electrons of the projectile.

In such a case, the distorted-wave model, discussed in [31], may represent a good alternative. This model takes into account the screening effect of atomic electrons. The distortion of the initial and final states of the projectile electron by the field of the neutral atom is described via introducing corresponding eikonal factors. It is known [4,31] that at asymptotically high impact energies, this model yields the same results as the light-cone approach [62] and, thus, offers an "exact" solution for the transition amplitude. In the case of strong interaction between the projectile-electron and the atom, this solution differs from that given by the first Born approximation. Consequently, even at infinitely high impact energy, this model and the first Born approximation in general do not coincide.

We have also discussed projectile-electron loss from 33 TeV $Pb^{81+}(1s)$ ions penetrating gaseous- and solid-state targets. In case of gas targets results of theoretical calculations for ion–atom collisions are in a good agreement with experiment. In case of solid targets, experimental results exceed the theoretical ones by a factor of 2. This difference can be understood and explained by considering multiple collisions suffered by the projectile-ions in solids which cause excitation of the projectile. Such a profound role of the excitations in the case of very heavy ions is in contrast to the previous experience gained when exploring collisions in the low- and intermediate-relativistic domains of impact energies. In the case under consideration, the excitations become so effective because of the relativistic time dilation which decreases very strongly the spontaneous decay rates of excited states in the ions moving with velocities very closely approaching the speed of light.

In addition to collision-induced transitions between states of electrons with positive total energy (i.e., transitions of electrons, which already exist before the collision occurs), we also discussed collision-induced transitions involving negative and positive energy states. In particular, we have considered bound–free electron–positron pair production in which the electron is created in a bound state of the projectile-ion $Pb^{81+}(1s)$ colliding with a neutral atom at an impact energy of 160 GeV/u.

Besides, we have also discussed bound–bound e^+e^- pair production in which both these particles are created in bound states. Compared to free and bound–free pair productions, which were extensively studied in the past, it represents a qualitatively new subprocess whose cross section has different dependences on the impact energy and charges of the colliding nuclei. In addition, its consideration also enables one to establish an interesting correspondence between the pair production and the more usual atomic collision process of electron capture in which an already existing electron undergoes a transition between colliding centers, but no new particles are created.

In nonrelativistic quantum theory, only the positive energy states exist, and within the nonrelativistic consideration of ion–atom (ion–ion) collisions, only three basic atomic processes appear: excitation, ionization (electron loss), and electron capture. The relativistic theory adds up the negative energy states into consideration. This results in the existence of pair production and the corresponding extension of the group of the basic atomic collision processes to the six. Thus, the bound–bound pair production not only fills in the "vacancy" in the set of the (single-) pair production processes but can also be viewed as completing the whole picture of the basic (single-lepton) atomic processes possible in ion–atom (ion–ion) collisions.

References

1. J. Eichler, W.E. Meyerhof, *Relativistic Atomic Collisions* (Academic, San Diego, 1995)
2. D.S.F. Crothers, *Relativistic Heavy-Particle Collision Theory* (Kluwer Academic/Plenum Publishers, London, 2000)
3. C.A. Bertulani, G. Baur, Phys. Rep. **163**, 299 (1988)
4. A. Voitkiv, J. Ullrich, *Relativistic Collisions of Structured Atomic Particles* (Springer, Berlin, 2008)
5. R. Anholt, W.E. Meyerhof, X.-Y. Xu, H. Gould, B. Feinberg, R.J. McDonald, H.E. Wigner, P. Thieberger, Phys. Rev. **A 36**, 1586 (1987)
6. W.E. Meyerhof, R. Anholt, X.-Y. Xu, H. Gould, B. Feinberg, R.J. McDonald, H.E. Wegner, P. Thieberger, NIM **A 262**, 10 (1987)
7. H.-P. Hülskötter, W.E. Meyerhof, E. Dillard, N. Guardala, Phys. Rev. Lett. **63**, 1938 (1989)
8. H.-P. Hülskötter, B. Feinberg, W.E. Meyerhof, A. Belkacem, J.R. Alonso, L. Blumenfeld, E.A. Dillard, H. Gould, N. Guardala, G.F. Krebs, M.A. McMahan, M.E. Rhoades-Brown, B.S. Rude, J. Schweppe, D.W. Spooner, K. Street, P. Thieberger, H. E. Wegner, Phys. Rev. **A 44**, 1712 (1991)
9. Th. Stöhlker, C.D. Ionesku, P. Rymuza, T. Ludziejewski, P.H. Mokler, C. Scheidenberger, F. Bosch, B. Franzke, H. Geissel, O. Klepper, C. Kozhuharov, R. Moshammer, F. Nickel, H. Reich, Z. Stachura, A. Warczak, Nucl. Instrum. Methods **B 124**, 160 (1997)
10. C. Scheidenberger, H. Geissel, Nucl. Instrum. Methods **B 135** 25 (1998)
11. C. Scheidenberger, Th. Stölker, W.E. Meyerhof, H. Geissel, P.H. Mokler, B. Blank, Nucl. Instrum. Methods **B 142**, 441 (1998)
12. T. Ludziejewsky, T. Stöhlker, C.D. Ionesku, P. Rymuza, H. Beyer, F. Bosch, C. Kozhuharov, A. Krämer, D. Liesen, P.H. Mokler, Z. Stachura, P. Swiat, A. Warczak, R.W. Dunford, Phys. Rev. **A 61**, 052706 (2000)
13. H. Bräuning et al., Phys. Scr. **T92**, 43 (2001)
14. A. Westphal, Y.D. He, Phys. Rev. Lett. **A 71**, 1160 (1993)
15. N. Claytor, A. Belkacem, T. Dinneen, B. Feinberg, H. Gould, Phys. Rev. **A 55**, R842 (1997)
16. H.F. Krause, C.R. Vane, S. Datz, P. Grafström, H. Knudsen, S. Scheidenberger, R.H. Schuch, Phys. Rev. Lett. **80**, 1190 (1998)
17. H.F. Krause, C.R. Vane, S. Datz, P. Grafström, H. Knudsen, U. Mikkelsen, S. Scheidenberger, R.H. Schuch, Z. Vilakazi, Phys. Rev. **A 63**, 032711 (2001)
18. C.R. Vane, H.F. Krause, Nucl. Instrum. Meth. **B 261**, 244 (2007)
19. A.B. Voitkiv, B. Najjari, A. Surzhykov, J. Phys. **B 41**, 111001 (2008)
20. A.B. Voitkiv, N. Grün, W. Scheid, Phys. Rev. **A 61**, 052704 (2000)
21. A.B. Voitkiv, M. Gail, N. Grün, J. Phys. **B 33**, 1299 (2000)
22. A.B. Voitkiv, B. Najjari, J. Ullrich, Phys. Rev. **A 75**, 062716 (2007)
23. A.B. Voitkiv, B. Najjari, J. Ullrich, Phys. Rev. **A 76**, 022709 (2007)
24. A.B. Voitkiv, B. Najjari, J. Phys. **B 40** 3295 (2007)
25. D.S.F. Crothers, J.F. McCann, J. Phys. **B 16**, 3229 (1983)
26. P.D. Fainstein, V.H. Ponce, R.D. Rivarola, J. Phys. **B 24**, 3091 (1991)
27. C. Müller, A.B. Voitkiv, N. Grün, Phys. Rev. **A 66**, 012716 (2002)
28. B. Najjari, A.B. Voitkiv, J. Phys. **B 41**, 115202 (2008)

29. R. Anholt, U. Becker, Phys. Rev. **A 36**, 4628 (1987)
30. A.H. Sørensen, Phys. Rev. A **58**, 2895 (1998)
31. A.B. Voitkiv, B. Najjari, V.P. Shevelko, Phys. Rev. **A 82**, 022707 (2010)
32. FAIR Baseline Technical Report, **A 58** 2895. http://www.gsi.de/fair/reports/btr.html. (1998)
33. G. Baur, I.L. Beigman, I.Yu. Tolstikhina, V.P. Shevelko, Th. Sthlker, Phys. Rev. A **B 80**, 012713 (2009)
34. V.P. Shevelko, I.L. Beigman, M.S. Litsarev, H. Tawara, I.Yu. Tolstikhina, G. Weber, Nucl. Instrum. Meth. B **269**, 1455 (2011)
35. B. Franzke, IEEE Trans. Nucl. Sci. **28**, 2116 (1981)
36. R.E. Olson, R.L. Watson, V. Horvat et al., J. Phys. B **37**, 4539 (2004)
37. G. Weber, C. Omet, R.D. DuBois et al., Phys. Rev. ST Accel. Beams **12**, 099901 (2009)
38. W. Erb, GSI Report GSI-P-78, Darmstadt (1978)
39. A.N. Perumal, V. Horvat, R.L. Watson et al., Nucl. Instrum. Meth. B **227**, 251 (2005)
40. F. Sauter, Z. Phys. **69**, 742 (1931)
41. H.A. Bethe, W. Heitler, Proc. Roy. Soc. A **146**, 83 (1934)
42. L.D. Landau, E.M. Lifshitz, Phys. Z. Sowjet. **6**, 244 (1934)
43. U. Becker, N. Grün, W. Scheid, J. Phys. **B 20**, 2075 (1987)
44. U. Becker, J. Phys. **B 20**, 6563 (1987)
45. G. Baur, K. Hencken, D. Trautmann, S. Sadovsky, Y. Kharlov, Phys. Rep. **364**, 359 (2002)
46. G. Baur, K. Hencken, D. Trautman, Phys. Rep. **453**, 1 (2007)
47. W. Heisenberg, H. Euler, Z. Phys. **98**, 714 (1936)
48. W. Greiner, B. Müller, J. Rafelski, *Quantum Electrodynamics of Strong Fields* (Springer, Berlin, 1985)
49. E. Brézin, C. Itzykson, Phys. Rev. D **2**, 1191 (1970)
50. V.N. Baier, V.M. Katkov, Phys. Rep. **409**, 261 (2005)
51. N.D. Birrell, P.C.W. Davies, *Quantum Fields in Curved Space* (Cambridge University Press, Cambridge, 1984)
52. A. Belkacem, H. Gould, B. Feinberg, R. Bossingham, W.E. Meyerhof, Phys. Rev. Lett. **71**, 1514 (1993)
53. A. Belkacem, H. Gould, B. Feinberg, R. Bossingham, W.E. Meyerhof, Phys. Rev. Lett. **73**, 2432 (1994)
54. A. Belkacem, H. Gould, B. Feinberg, R. Bossingham, W.E. Meyerhof, Phys. Rev. **A 56**, 2806 (1997)
55. A. Belkacem, N. Claytor, T. Dinneen, B. Feinberg, H. Gould, Phys.Rev. **A 58**, 1253 (1998)
56. D.L. Burke et al., Phys. Rev. Lett. **79**, 1626 (1997)
57. H. Chen et al., Phys. Rev. Lett. **102**, 105001 (2009)
58. See the European Light Infrastructure at http://www.extreme-light-infrastructure.eu and the High Power laser Energy Research at http://www.hiperlaser.org
59. R. Anholt, H. Gould, Adv. Atom. Mol. Phys. **22** 315 (1986)
60. A.B. Voitkiv, B. Najjari, A. DiPiazza, New J. Phys. **12**, 063011 (2010)
61. R. J. Hughes, Hyperfine Interact. **76**, 3 (1993)
62. A.J. Baltz, Phys. Rev. Lett. **78**, 1231 (1997)

Chapter 8
Target-Scaling Properties for Electron Loss by Fast Heavy Ions

R.D. DuBois and A.C.F. Santos

Abstract Electron loss by fast heavy ions resulting from interactions with dilute gaseous targets is discussed. Of particular interest is how the cross sections scale as the target nuclear charge increases. Various theoretical models that have been proposed are discussed and compared with available experimental data. It is shown that none of these models yield good agreement with data but that by combining ideas and concepts contained within different models agreement can be obtained.

8.1 Introduction

As fast ions traverse any media, whether the medium is a dilute gas, condensed matter, biological material, or a plasma, inelastic interactions with the various atoms and molecules comprising the media deposit energy, which alters the media and degrades the ion's kinetic energy. In addition, sometimes the projectile is ionized. Projectile ionization, the subject of the present work, is also referred to as projectile stripping or electron loss. Here, we shall use the terms interchangeably. For atomic targets, electron loss interactions have the form:

$$P^{q+} + T \rightarrow P^{(q+n)+} + T^{i+} + (n+i)\mathrm{e}^{-}, \tag{8.1}$$

where P represents the incoming ion (the projectile) with charge q, T is the target atom, and n and i are the number of projectile and target electrons that are liberated. For projectile ionization, $n \geq 1$ whereas the target can be simultaneously ionized or it can remain in its ground or an excited state. Thus, $i \geq 0$. In the case of molecular targets, fragmentation into charged and uncharged components is also possible. These initial, plus any subsequent interactions involving the reaction products, are of extreme importance in understanding radiation damage to materials and/or to biological tissue.

V. Shevelko and H. Tawara (eds.), *Atomic Processes in Basic and Applied Physics*, Springer Series on Atomic, Optical, and Plasma Physics 68,
DOI 10.1007/978-3-642-25569-4_8,

Electron loss processes have been studied for many decades, in part to investigate atomic interactions where the number of active particles can be controlled and systematically varied and in part for practical purposes associated with the creation, acceleration, transport, and storage of ions for research purposes. Electron loss from very heavy, few-electron ions is also used to study processes occurring in the presence of very high electric fields such as those found for inner-shell electrons.

Equation (8.1) shows that electron loss studies can be tailored to study single- or multiple-electron transitions in few- or many-electron systems simply by altering the projectile nuclear, Z_P, and total, q, charges. In addition, passive and active electron effects (as will be described later) can be investigated by comparing electron loss data obtained using different targets or by comparing data obtained for the same target but where simultaneous and independent ionization of the target and projectile are compared.

With regard to the acceleration, transport, and storage of ion beams, electron loss occurring when high-energy beams interact with background gases within beamlines or storage rings has serious detrimental effects. These include loss of beam luminosity (due to scattering), decreased storage times (due to losses at bending magnets or focusing elements because of altered charge states or degraded energies), and possible erosion, heating, and vacuum loading in the accelerator and storage rings (due to interactions with the vacuum walls by the lost beam components). These are crucial technical problems at all high-energy accelerator laboratories, e.g., at GSI, Brookhaven, CERN, Dubna, etc. where considerable effort and expense are being devoted in order to predict and circumvent such problems. See, for example, [1–3].

As an example, the new Facility for Antiproton and Ion Research (FAIR) at the GSI Helmholzzentrum für Schwerionenforschung in Darmstadt, Germany, is being designed to produce very intense beams of high-energy, heavy ions [4]. This requires using low-charge-state ions in order to reduce the space charge. But, as will be shown later, the probability for electron loss is inversely proportional to the energy required to remove bound electrons. Therefore, and for reasons associated with obtaining high velocities plus limitations pertaining to the maximum magnetic fields that can be produced by bending magnets along the beamline and in storage rings, " medium" charge-state ions such as U^{28+} are expected to be used. For such ions traveling with MeV/u to GeV/u velocities, electron loss is the dominant loss mechanism. To achieve the scientific goals, beam intensities will be roughly a factor of 50 larger than what is currently possible. This means that radiation levels, heating, and vacuum loads associated with lost beam components will increase considerably.

Another example where electron loss plays a crucial role is in high-energy-density research. Approximately a decade ago, a US program proposed accelerating intense beams of low-charged heavy ions to GeV energies to indirectly heat small D-T pellets and induce laboratory fusion [5]. To achieve the energy densities that are required, intense, tightly-focused beams have to hit the pellets. Original plans were to employ singly charged heavy ions in order to minimize space charge defocusing of the beams. But because of complications and lack of knowledge about transport and beam losses, particularly in the reaction chamber where a high partial pressure of FLiBe (a molten salt which will coat the inner walls in order to collect energetic neutrons released in the fusion process), the emphasis changed to the use of light ions. However, heavy ions can deliver far more energy to the target. Thus, the original idea may be revived sometime in the future, especially if sufficient experimental data and theoretical understanding of electron loss processes become available.

In both of these examples, intense beams of heavy ions having MeV/u to GeV/u velocities must be transported fairly long distances through beamlines and perhaps stored in storage rings. Even under extremely high-vacuum conditions, interactions with background gases result in portions of the beam suffering energy loss and/or changing charge. These portions are subsequently lost. Localized heating of surfaces occurs; this desorbs gases or atoms from the surfaces; this further degrades the vacuum, which increases the loss processes, etc. In the worst case, the processes avalanche and a rapid loss of beam quality and intensity occurs [1]. To avoid or to minimize these problems requires information relating to (1) total loss probabilities for a broad range of systems and (2) information about the charge states produced and their relative production probabilities. Such information is needed for electron loss resulting from interactions with few-electron targets such as H_2 and He, i.e., the primary components found in high-vacuum environments, as well as with many-electron targets such as N_2, CO, CO_2, H_2O, etc. which occur as lesser components in the vacuum or which are desorbed from heated surfaces. Interactions with heavier gases are especially important because they have larger interaction cross sections. Hence, even though they may be far less abundant, they can still contribute significantly to the total beam losses. Plus, interactions with heavier atoms and molecules are primarily responsible for multiple electron loss from the ion.

This work will concentrate on electron loss by fast ions, with particular emphasis on many-electron systems and how the electron loss cross sections scale. Following a brief description of the various theoretical models available, methods used to obtain experimental data at MeV/u velocities will be outlined. Here, we restrict ourselves to methods and findings from the Federal University in Rio de Janeiro and the GSI Helmholzzentrum für Schwerionenforschung, two laboratories still actively engaged in electron loss measurements. Using selected examples of available data, we will illustrate what scaling properties can be extracted and provide the reader with an overall picture of this field of research. With regard to theory, our intent is only to outline the general methods that have been applied to electron loss and to briefly discuss the results. For detailed information, the reader is referred to the cited references.

8.2 Theoretical Models and Predictions

The basic procedures for calculating electron loss, i.e., projectile ionization, were outlined many years ago by Bates and Griffing [6]. They showed that two interaction channels must be considered. As illustrated in Fig. 8.1, one involves the interaction

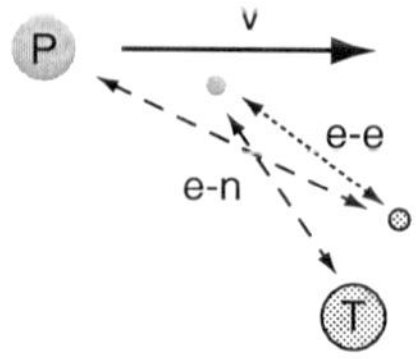

Fig. 8.1 Schematic showing the interaction between the nuclear charge of one of the collision partners and the bound electron of the other (e–n interactions, shown by the *dashed lines*) and the direct interaction between bound projectile and target electrons (e–e interactions, shown by the *dotted line*)

between a bound projectile electron and the partially screened nuclear charge of the target; the other involves a direct interaction between a target and a projectile electron. The first has been referred to as the e–n or the screening or the elastic channel, while the second is called the e–e or the antiscreening or the inelastic channel.

Using the Born approximation, Bates and Griffing showed that the differential electron emission arising from these two channels is given by

$$\mathrm{e-n} \qquad \mathrm{d}^2\sigma(\varepsilon) = \int_{\mathrm{K\,min}} \mathrm{A}(\varepsilon,\mathrm{K}) \left[|\mathrm{Z_T} - \mathrm{N_T F(K)}|^2 \right] \mathrm{dK}, \tag{8.2}$$

$$\mathrm{e-e} \qquad \mathrm{d}^2\sigma(\varepsilon) = \int_{\mathrm{K'\,min}} \mathrm{A}(\varepsilon,\mathrm{K}) \left[\mathrm{N_T} - \mathrm{N_T}\, |\mathrm{F(K)}|^2 \right] \mathrm{dK}. \tag{8.3}$$

For the purposes of the present work, the important quantities to consider are the effective target charges found within the brackets. For the e–n interactions, the target nuclear charge, Z_T, is partially screened by its bound electrons, N_T. But the screening depends on the momentum transfer, *K*. (In an impact parameter representation, the screening would be a function of the impact parameter.) For neutral targets, N_T is equal to Z_T. However, we will continue using N_T as a designation in order to talk about the "effective number of target electrons." Because the electron form factor *F(K)* ranges in value between 0 and 1, the e–n interactions are seen to scale with the square of an effective nuclear charge, Z^2_{Teff}, which has a value somewhere between Z_T^2 and $(Z_\mathrm{T}–N_\mathrm{T})^2$, which for a neutral target, ranges in value from Z_T^2 to 0. In contrast, the e–e interactions scale with the effective number of electrons, N_{Teff}. N_{Teff} ranges in value between N_T and 0, which for a neutral target is also from Z_T to 0. For additional details, the reader is referred to [7].

The other important item to note in (8.2) and (8.3) are the different integration limits for the two channels. Because of their different masses, to transfer the momentum required to remove a projectile electron requires a considerably higher energy for electrons, i.e., for e–e interactions, than for the massive nucleus, i.e., for e–n interactions. These different thresholds are shown in the next figure. To summarize, for few electron systems and energies well above threshold, the Born model predicts that the cross sections will scale as $Z^2_{\mathrm{Teff}} + N_{\mathrm{Teff}}$. With respect to impact velocity, the Born model predicts a v^{-2} dependence.

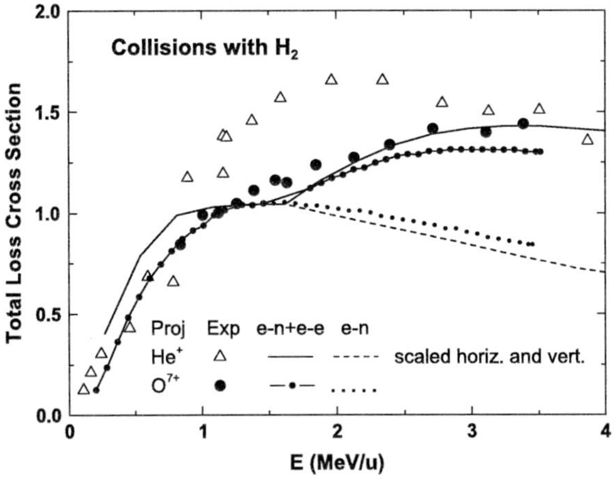

Fig. 8.2 Electron loss by He^{+} and O^{7+} resulting from collisions with molecular hydrogen. Experimental data, *solid dots* and *open triangles*, and PWBA calculations for e–n (*dashed curves*) and e–n + e–e interactions (*solid curves*) are from [8]. Both experiment and theory for He^{+} have been scaled vertically and horizontally for comparison purposes

Meyerhof et al. [8] used a PWBA (plane wave Born approximation) to calculate electron loss from few electron ions colliding with molecular hydrogen. Figure 8.2 shows their results compared with experimental data. For He^{+}, experiment and theory have been scaled horizontally and vertically such that the onset of the e–e channel is roughly the same as for O^{7+}. For ionization of O^{7+}, the onset is clearly seen, whereas for ionization of He^{+} the threshold is less clear.

These scaled data show that even though the binding energies of the ionized electrons are quite different for the two projectiles, the e–e and e–n processes have the same relative magnitudes and energy dependencies. In particular, note that for energies well above threshold, the relative e–e and e–n cross sections are roughly the same. This is because for ionization by H_2 the two target nuclei act independently, i.e., the e–n process scales as two independent ionizations by $Z_T = 1$, e.g., as $1^2 + 1^2$, while the e–e process scales with the total number of target electrons, e.g., as $N_T = 2$. Thus, one would expect equal contributions from each.

In several studies performed at the Federal University in Rio de Janeiro, UFRJ, a semiclassical approximation was used to calculate the e–e (antiscreening) channel [9–14]. In combination with PWBA calculations for the screening channel, cross sections for electron loss from few-electron ions such as He^{+}, C^{3+}, and O^{5+} induced by various targets were calculated. This method predicts the same scaling features as for the Born approximation.

A different approach was used by Voitkiv et al. [15] to calculate electron loss by He^{+} projectiles in collisions with heavy targets. Within the framework of the sudden approximation, which is a nonperturbative, fully unitarized theory, the elastic screening contribution to the electron loss was determined. This method also included the possibility of multielectron transitions in the target atom concomitant with the loss process.

They showed that the elastic part of the total loss cross section can be written as

$$\sigma_{12}^{\text{elas}} = \left[1 - P_{\text{ion}}^{a}(0)\right] S_{\text{loss}}^{\text{sc}}, \tag{8.4}$$

where $S_{\text{loss}}^{\text{sc}}$ is the screening contribution to the total loss cross section and $P_{\text{ion}}^{a}(0)$ is the target ionization probability at zero impact parameter. The magnitude of the screening contribution was found to be much larger than the geometric cross section of $He^{+} \approx 0.8$ a.u, indicating that large impact parameters contribute most to the screening cross section.

Using the independent particle model scenarium, and ignoring postcollision rearrangements, it was shown that [16]

$$\sigma_{\text{inelastic}} \approx \sum_{s} n_s \sigma_{\text{anti}}^{s} = \sigma_{\text{anti}} \approx \langle n \rangle \, \sigma_{\text{e}}. \tag{8.5}$$

Thus, in the projectile frame, the inelastic mode can be viewed as ionization of the projectile by a beam of ⟨n⟩ *target* electrons, each having an ionization cross section σ_{e}. Here, $\langle n \rangle$ is the average number of the active target electrons associated with the inelastic mode. If the projectile velocity is high enough, the ratio $\sigma_{\text{inelasic}}/\sigma_{\text{e}}$ can be used to determine ⟨n⟩, as will be illustrated in a later section.

It should also be mentioned that Voitkiv et al. [17, 18] have extended their calculations to extremely high energies where relativistic effects become important.

To go beyond few-electron ions significantly increases the complexity of the theoretical models. Multiple- as well as single-electron removal must be considered. An independent particle model is often used to model multiple electron removal. In addition, one must account for different binding energies for removing additional electrons. As shown by Santos and DuBois [19], the electron loss cross sections decrease with increasing binding energy of the electron being removed. Different binding energies imply different thresholds for the e–e process. Thus, at any specific impact energy, different numbers of projectile electrons can be removed via the e–e process. Removal of additional, i.e., more tightly bound, electrons requires a larger momentum transfer which implies smaller impact parameters. Thus, at each impact energy, the effective number of target electrons which actively participate in the e–e process or which passively screen the target nuclear charge can change. Such effects must be modeled and incorporated into the screening and antiscreening formulae.

Using a large database of cross sections, Santos and DuBois [19] obtained empirical scaling formulae for single- and multiple-electron loss resulting from collisions with argon and molecular nitrogen. They showed that the cross sections scaled as $N_{\text{Peff}}^{0.4}$, where N_{Peff} is the number of electrons in the outermost shell or shells, depending upon the impact energy.

In 1948, Bohr predicted [20] that in collisions of heavy systems, the cross section for single electron loss should scale as $(Z_{\text{P}}^{1/3} + Z_{\text{T}}^{1/3})^2/v$. This scaling was based upon the Thomas–Fermi model which he used to estimate the screened Coulomb potential between atoms. Firsov [21, 22] numerically derived the interatomic potentials of two colliding Thomas–Fermi atoms and fitted these potentials using the

Thomas–Fermi screening length. He obtained a slightly different scaling, namely, $(Z_P^{1/3} + Z_T^{1/3})^{2/3}$. Several decades later, Alton et al. [23] modified the Bohr formula slightly by multiplying by the number of projectile electrons in the outermost subshell and dividing by their ionization potential. With this modification, they obtained good agreement with their experimental electron loss measurements as a function of Z_T.

Shevelko and coworkers developed a series of codes (LOSS, LOSS-R, and RICODE [24–26], respectively) which use a plane wave Born approximation to calculate single-electron removal cross sections over a wide range of energies including relativistic energies. Cross sections for projectile ionization are calculated using the Schrödinger radial wave functions and with account for atomic structure of both colliding particles. For electron loss from Pb-like ions by various targets [24], the cross sections were found to scale as $(Z_T/I_P)^{1.4}$ and the energy to scale with I_P. Recently, the DEPOSIT code was developed [26] based upon the classical model. The code is intended to calculate single- and multiple-electron removal cross sections at low and intermediate collision energies. Here, the cross sections scale as $\frac{Z_T^{0.85}}{E^{\alpha(Z_T)} I_P^{1.2}}$ where $\alpha(Z_T) \approx 0.8/Z_T^{0.3}$ [26].

Using a nonperturbative theory, Matveev et al. [27] calculated multiple excitation and ionization of structured heavy ions interacting with complex neutral atoms. However, this theoretical approach is rather complex and requires many adjustable parameters. Hence, extracting simple scaling relationships is not possible.

Kaneko used a unitarized impact-parameter method to calculate electron loss and excitation from He^+ colliding with various atoms [28]. With respect to the energy dependence, in the energy range between 4 and 10 MeV, he found that the single loss cross sections scale as E^{-1} for stripping by a helium target, $E^{-0.64}$ for stripping by N_2, $E^{-0.44}$ for stripping by Ar, and $E^{-0.36}$ for stripping by Kr. With regard to the target dependence, he found that the cross sections scale roughly as $Z_T{}^{1.3}$ for $Z_T < 10$. For $Z_T > 10$, the scaling is slower, e.g. with powers decreasing from 0.39 to 0.17 as the impact velocity increases from 0.1 to 0.9 MeV/u.

Using an n-body Classical Trajectory Monte Carlo (nCTMC) method, Olson has calculated single- and multiple-electron loss for a variety of systems [29–33]. This theory includes multiple-electron removal processes and has been applied to a large variety of atomic interactions and many-electron systems. For stripping induced by lighter targets such as H_2, the nCTMC predicts that at high energies, the cross sections decrease as E^{-1}, just as is predicted by first-order perturbation theories. However, for stripping induced by heavier targets, it predicts a much slower dependence, roughly as v^{-1}. This is consistent with Bohr's predictions for heavy systems, although it should be noted that the assumptions used by Bohr are not applicable at high velocities.

As stated in the introduction, accurately modeling and predicting electron loss has significant ramifications with respect to beam transport and storage as well as with accelerator design and performance and radiation safety. However, extrapolation of the various models to high energies leads to differing results, sometimes results that are significantly different. For example, as illustrated in

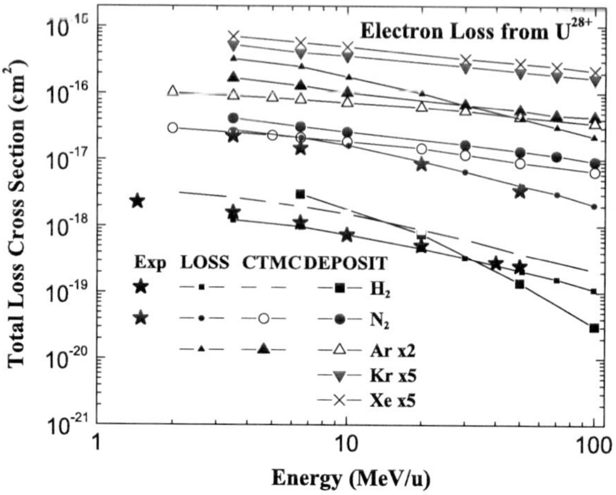

Fig. 8.3 Experimental and theoretical cross sections for electron loss from U^{28+} resulting from collisions with various targets. The experimental data are from [32–36](W. Erb, B. Franzke, 1978, GSI Report No. GSI-J-1-78, unpublished), the LOSS, CTMC, and DEPOSIT theories are from [24, 32, 33], and [26], respectively. The curves for Ar, Kr, and Xe have been shifted upward for display purposes

Fig. 8.3, the LOSS, DEPOSIT, and nCTMC models tend to agree within factors of 2 at low energies, e.g., a few MeV/u, but they predict different velocity dependences. This makes extrapolation to high energies less and less certain. In addition, the figure shows that the predicted velocity dependences depend upon the target. For a low-Z, few-electron, target such as H_2, the velocity dependence is very close to E^{-1}, as expected from first-order perturbation theory and in agreement with the LOSS code of Shevelko et al. but in disagreement with the DEPOSIT code. For a medium-size target such as N_2, the dependence is approximately v^{-1}. This is close to the dependencies predicted by the DEPOSIT code and the higher energy portion of the nCTMC model but in disagreement with the LOSS-R code. For higher-Z, many-electron targets, the predicted dependencies are slower than v^{-1}.

Included in the figure are experimental data obtained at GSI [34–36](W. Erb, B. Franzke, 1978, GSI Report No. GSI-J-1-78, unpublished) and the Texas A&M University cyclotron (TAMU) [32, 33]. As seen, good agreement with the LOSS code and nCTMC model for H_2 is found. But, based upon the highest energy points for N_2, experiment seems to have a faster falloff with impact energy than is predicted by the supposedly accurate DEPOSIT and nCTMC models. The discrepancy with the nCTMC calculations is especially troubling as this method is known to provide excellent results for many applications where many-electron systems are involved. It is not likely that relativistic effects lead to the observed differences since these were simulated by modifying the velocity at which the high-energy calculations were performed. Additional data are needed to determine how, or if, relativistic affects influence the cross sections in the energy region of interest. Plus, data are needed to test the models for heavier targets.

8.3 Experimental Methods and Results

In this section, we describe three experimental methods which have been used to supply information about electron loss at MeV/u velocities. The first describes methods used at the Federal University of Rio de Janeiro (UFRJ) which are very similar to those used at the Texas A&M University (TAMU) cyclotron and for the early studies performed at the GSI UNILAC. The second is the method used for the most recent studies at the UNILAC. The last is the method currently employed at the GSI Experimental Storage Ring (ESR).

8.3.1 UFRJ Method

Aiming to obtain a detailed understanding of electron loss and multiple ionization of atoms, Brazilian studies, formerly at the Pontifical Catholic University (PUC) and now at the Federal University of Rio de Janeiro (UFRJ), have measured absolute cross sections for total electron loss as well as for the projectile loss-target ionization channels. These studies have concentrated on projectiles with one active electron (e.g., He^{+}, C^{3+}, O^{5+}, and B^{2+}) interacting with noble gases (He to Xe) and simple molecules. Data were obtained in the 1.0–4.0 MeV energy region.

8.3.2 UFRJ Experimental Setup

For these studies, collimated, partially dressed ion beams were produced by the 4-MV Van de Graff at PUC or by the 1.7 MV tandem accelerator at UFRJ. Figure 8.4 shows a schematic of the experimental apparatus. Just before entering a windowless target cell, the beams are vertically charge-analyzed (cleaned) by a magnet (PUC) or an electric field (UFRJ) in order to separate the main beam from spurious ones. After the target cell, the emergent beam is horizontally charge-analyzed by a second magnet (PUC) or electric field (UFRJ). The charge states are recorded some meters downstream by surface barrier detectors or by a position-sensitive microchannel plate detector.

Singly and multiply ionized recoil ions are accelerated out of the interaction region by a two-stage electric field and detected by a microchannel plate detector. The recoil ions provide stop signals for a time-to-amplitude converter which are started by the projectile electron loss signal. To prevent signal deterioration, the counting rates of the incident projectiles were kept below 1.5 kHz.

Figure 8.5 shows the postcollision charge states detected by a 2D position-sensitive microchannel plate detector for 200 keV O^{+} colliding with CH_4. These charge states originate from charge-changing collisions of the main beam with CH_4 within the target cell plus with residual gases in the beamline before the gas cell. Because of the vertical field applied before the collision cell, any spurious beams

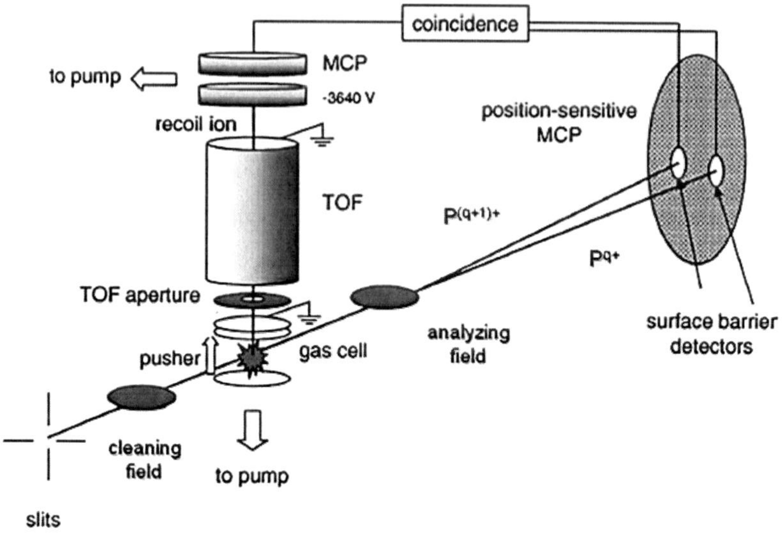

Fig. 8.4 Schematic view of the experimental setup at PUC/UFRJ. See text for details

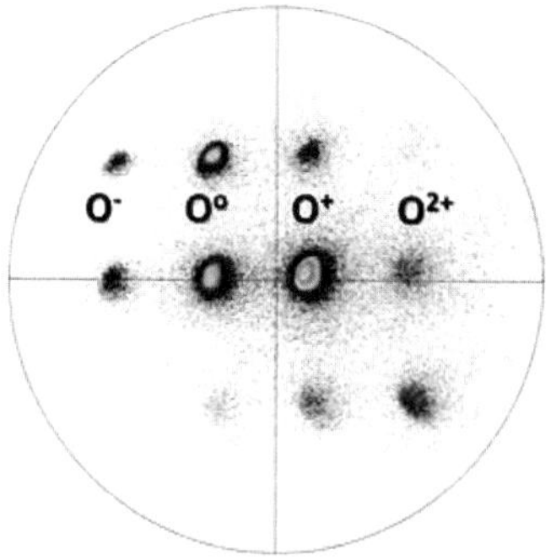

Fig. 8.5 2D MCP spectrum of the various charge states observed for 200 keV O^+ on CH_4 molecule. See text for details

which entered the cell and contributed to the emerging charge states can be identified by their vertical displacements. Hence, in Fig. 8.5 where an O^+ beam was used, the uppermost row shows capture and loss from neutral oxygen which was formed prior to the collision cell via interactions with background gases. The lowermost row shows single and double capture from O^{2+} which was also formed in the beamline. Only the middle row shows interactions involving O^+ ions. However, all of the data can be used to extract capture and loss cross sections for O, O^+, and O^{2+} impact. Electron loss cross sections were obtained using integrated intensities for each charge state as a function of target pressure, i.e., the growth curve method which will be described in the following section. Pressures inside the gas cell were always low enough to ensure single-collision conditions.

Absolute detection efficiencies for the recoil ions, ϵ^{q+}, were determined by simultaneous single and coincidence measurements of single-electron-capture cross sections of C^{3+} on noble gases, using various projectile energies. In the coincidence measurements, the coincidence signal for target ionization associated with single capture by C^{3+}, I_{32}^{q+}, where q is the final charge state of the target, is proportional to the beam intensity I_o, the number of target scattering centers per volume *n(P)*, where P is the target pressure, the effective target length l_{eff}, the cross section, σ_{32}^{q+}, and the recoil ion detection efficiency ϵ^{q+}. Here, a surface barrier detector which is assumed to have unit efficiency is used to detect the postcollision C^{2+} projectile. For the recoil ions, to ensure high efficiencies and efficiencies virtually independent of the impacting charge state or mass, the MCP front plate was biased at a high negative voltage. Thus, we have

$$I_{32}^{q+} = \varepsilon^{q+} l_{\text{eff}} I_o n(P) \sigma_{32}^{q+}. \tag{8.6}$$

In the singles measurement, the absolute single capture cross section, σ_{32}, is determined by measuring the fraction of the main beam that performs single-electron capture. These cross sections are related by

$$\sigma_{32} = \sum_q \sigma_{32}^{q+}. \tag{8.7}$$

Ratios of multiple-to-single ionization cross sections $(\sigma_{ij}^{q+}/\sigma_{ij}^{+})$ were measured for different projectiles, different channels, different projectile energies, and various targets. These were compared with values obtained using cross sections found in the literature. It was found that the detection efficiencies were essentially independent of the recoil-ion charge state. Thus, $\epsilon^{q+} \cong \epsilon$, and can be removed from the summation. Hence,

$$\varepsilon = \frac{1}{l_{\text{eff}} n(P) I_o \sigma_{32}} \sum_q I_{32}^{q+}. \tag{8.8}$$

Using this equation for electron capture from various targets, it was found that ϵ did not depend of the target atomic number. Therefore, absolute cross sections were obtained by modifying (8.8) to apply for electron loss rather than capture and also substituting ϵ for ϵ^{q+} in the equation.

8.3.3 UFRJ Results

As discussed earlier, first-order theories such as the Born theory predict a $Z_{\text{T}}^2 + Z_{\text{T}}$ target dependence with the first term designated as the screening mode and the second term as the antiscreening mode. Thus, as Z_{T} increases, the expected overall

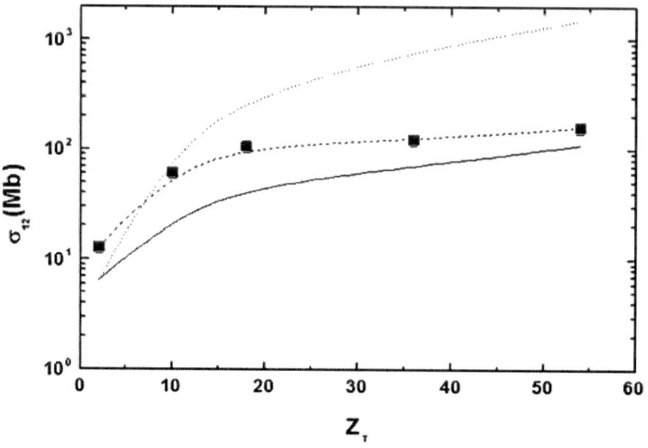

Fig. 8.6 Cross sections for total projectile electron loss as a function of the target atomic number for 2.5 MeV He^+ on noble gases. Experiment: *squares* (*dashed line* drawn for guiding the eyes); *solid curve*, antiscreening; *dotted curve*, screening [11]

dependence is expected to be dominated by the screening mode. However, Fig. 8.6 shows the measurements of Sant'Anna et al. [11] for 2.5 MeV He^+ impinging upon noble gases. One can see a near saturation of the cross sections in the region $18 < Z_2 < 54$. This points toward a much smaller contribution from the screening mode than expected which can possibly be explained by a saturation effect manifested only in the screening channel. A similar saturation was also observed for 1 MeV He^+ impact [12].

Also shown are first-order calculations for the screening and antiscreening channels using the extended-sum-rule method of Montenegro and Meyerhof (W. Erb, GSI Report No. GSI-P-7-78, 1978, unpublished). First-order calculations for the contribution due to the antiscreening channel show a qualitative agreement with the experimental Z_T dependence. In contrast, the calculated dependence for the screening mode is much steeper and severely overestimates the cross sections for the heavier targets. This supports the idea that a saturation effect must exist in the screening channel. To observe such a saturation, a proper theoretical approach using nonperturbative calculations is required.

As a side note, measured cross sections for electron loss to the continuum (ELC) [12] were shown to be in extremely good agreement with the theoretical excitation values calculated by Kaneko [28]. This implies that it might be possible to use electron loss to the continuum measurements to obtain information or test theoretical predictions for excitation processes since, as noted by Sigaud et al. [12], for the ELC process, the ionized electron has near zero velocity in the projectile frame. This means that ELC ionization is quite similar to electron excitation to high n levels.

The target saturation effect was investigated further using C^{3+} and O^{5+} projectiles with velocities ranging from less than 1 to a few atomic units. Figure 8.7 shows results for single electron loss by 3.0 MeV C^{3+} colliding with hydrogen [37]

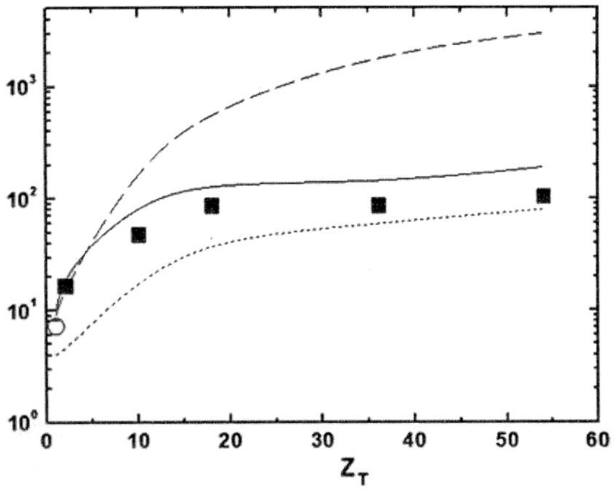

Fig. 8.7 Single-electron-loss cross section of 3.0 MeV C^{3+} on hydrogen and noble gases as a function of target atomic number. Experiment: *squares* [13] and *circles* [37]. Theory: *dashed line*, sum of screening (PWBA) and antiscreening [13]; *solid line*, sum of screening [13] and antiscreening [39].The antiscreening calculations are shown by the *dotted line*

and the noble gases [13]. Again, the experimental data show a saturation effect. These are compared to two different theoretical approaches [13,38]. The behavior of the screening contribution with increasing target atomic number when the extended free-collision model is used is the same as those of the experimental data and of the first-order calculations for the antiscreening (inelastic mode), presenting the same saturation with increasing target atomic number. But, there still exist quite large quantitative discrepancies, which increase as the projectile charge increases.

Although outside the scope of the present work, it should be mentioned that to probe electron loss processes even deeper, measurements of projectile single-electron loss accompanied by multiple target ionization have also been performed at UFRJ [14,16,39,40]. Following the procedure outlined in (8.7), Santos et al. [14] used the ratio $\sigma_{\mathrm{anti}}/\sigma_{\mathrm{e}}$ to provide an estimate of the average number of active target electrons for the antiscreening mode, $\langle n_o \rangle$. Under the assumption that $\langle n_o \rangle$ depends on the collision energy, it can be written as

$$\langle n_o \rangle_E = f(E)\,\langle n_o \rangle_a \,. \tag{8.9}$$

Here, $\langle n_o \rangle_a$ is the value of the average number of electrons at a very high impinging energy (100 MeV was used), and *f(E)* is a function of the collision energy, *E*. They obtained asymptotic values for $\langle n_o \rangle_a$, of 0.63 for He, 2.9 for Ne, 5.6 for Ar, 9.6 for Kr, and 14.1 for Xe. Then they calculated values of *f(E)* for the different targets as a function of the collision energy and observed that all targets except He demonstrated the same behavior: a fast increase just above threshold, followed by a sharp maximum—which lies between 4 and 5 MeV for all targets—and a slow decrease with increasing energy. The He target demonstrated no maximum, but simply decreased monotonically with increasing collision energy. The most notable feature is that the values of *f(E)* converge to the same curve above the maximum region. This is true for all targets, including He. The authors fitted a simple curve which could act as a "universal" function to estimate the average number of active target electrons for the inelastic mode, which we referred to as N_{Teff} in an earlier section, as a function of the collision energy for the C^{3+} projectile ion. The fitted function was

$$f(E) = 4.514 - \frac{3.831E(\mathrm{MeV})}{9.907 + E(\mathrm{MeV})}. \tag{8.10}$$

8.4 UNILAC Measurements

In the 1970s and 1980s there was much activity with respect to measuring electron loss from a variety of ions and energies. A fairly extensive listing of references can be found in [19]. However, the only laboratory where electron loss from MeV/u, low-charge-state, heavy ions could be measured was GSI-Darmstadt (now the GSI Helmholzzentrum für Schwerionenforschung), Germany. But, even there, the maximum velocity was 1.4 MeV/u. Beginning approximately a decade ago, two laboratories have worked on extending the experimental information to higher energies. At the Texas A&M University cyclotron (TAMU) in the USA, the group of Watson and various collaborators [29, 31–33, 41, 47] have measured cross sections up to 6.5 MeV/u for very heavy, low-charge-state ions such as U^{28+} and up to a few tens of MeV/u for lighter ions. These studies used techniques similar to those described above for studies at the Federal University of Rio de Janeiro. The other laboratory currently active in electron-loss studies at MeV/u energies is the GSI Helmholzzentrum für Schwerionenforschung. There, in collaboration with the Missouri University of Science and Technology, detailed cross sections for very low-charge-state ions, for example, $Ar^{1,2+}$, Xe^{3+}, and $U^{4,6,10+}$, were measured at 0.7 and 1.4 MeV/u [30, 41] while cross section and lifetime information for ions such as U^{28+} and Xe^{18+} have been obtained for energies up to many tens of MeV/u [34, 35]. The GSI studies will be described below, with particular emphasis on the experimental methods that were required.

8.4.1 *UNILAC Methods*

Unlike for the early GSI studies performed at the UNILAC, this region is no longer used for experimental studies. For very low-charge-state ions, downstream studies are not possible because of beam transport system magnet limitations. Therefore, the existing facilities were slightly modified in order to perform new measurements at the UNILAC. As illustrated in Fig. 8.8, a pseudostatic gas target was created at the end of the UNILAC accelerator section by inserting slits before and after the gas stripper chamber and valving off the pumps normally used to pump this chamber. These slits collimated and reduced the beam intensity plus provided differential pumping between the target and accelerator/detector regions. Target gases were

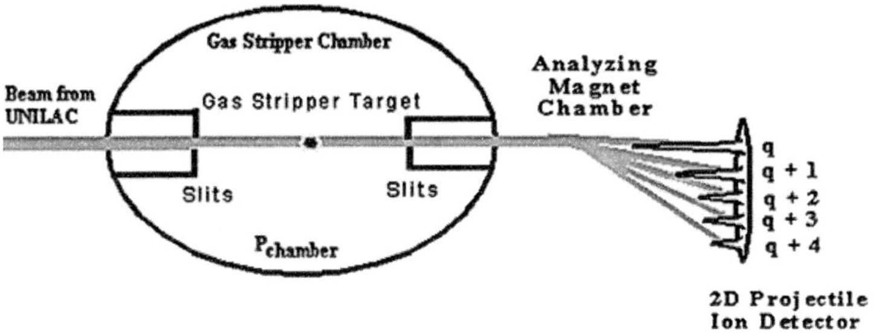

Fig. 8.8 Schematic of apparatus used at the GSI UNILAC

injected through a leak valve, and their pressures were measured at the periphery of the target chamber. Absolute target line densities along the beam path were obtained from normalization measurements using He^+ ions. Immediately following the gas stripper chamber, an analyzing magnet was used to separate the postcollision charge states and direct them onto a high-rate position-sensitive detector that could be inserted or withdrawn.

In contrast to the studies described above, at GSI, the ion sources are designed to provide intense beams that have a time structure resulting in high counting rates for brief periods. Therefore, the beam intensities had to be reduced many orders of magnitude, and a high-rate 2D detector and fast histogramming time-to-digital converter capable of handling rates of 100 kHz or more were used. Additional experimental details can be found in [30] and [41].

The experimental procedure consisted of measuring the postcollision charge state intensities as a function of target pressure, i.e., the growth curve method. Charge state spectra, as shown in Fig. 8.9, were background subtracted and integrated. From these, charge state fractions were calculated and plotted versus the target gas line density as shown in Fig 8.10.

For such, the growth curve data, coupled equations of the type

$$F_q(\pi) = \sum_{q' = q_{in} - \mathrm{maxcap}}^{q_{in} + \mathrm{maxloss}} \pi \; F_{q'}(\pi) \; \sigma_{q' \, q}, \tag{8.11}$$

must be solved. Here, π is the target line density, q' is the final charge state, and the charge state fractions and cross sections are designated by $F_{q'}$ and $\sigma_{q'q}$, with the sum being over all observed charge states, e.g., from the maximum capture channel to the maximum loss channel. Isolating single from multiple collision processes that lead to the same final charge state requires, in principle, solving a matrix of coupled equations, each at a different value of π. However, in practice, if the line density is small enough, multiple collision processes can be ignored. This uncouples the equations, and the cross sections can be determined using the linear portion of the growth curves shown in Fig. 8.10.

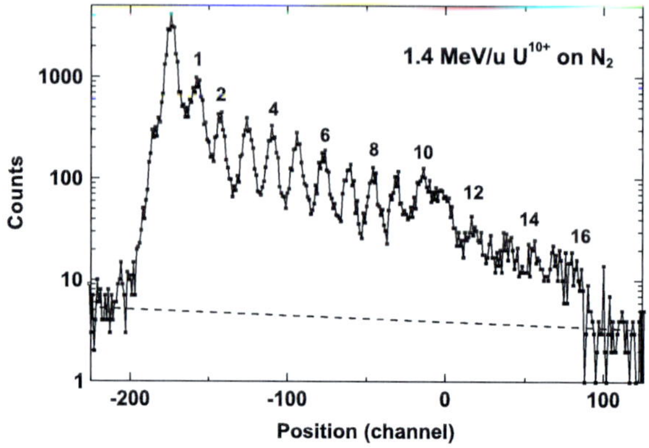

Fig. 8.9 Charge state spectrum for stripping of 1.4 MeV/u U^{10+} ions by molecular nitrogen. The numbers indicate how many electrons are removed from uranium. The *dashed line* indicates the background that was subtracted in order to integrate the peak intensities

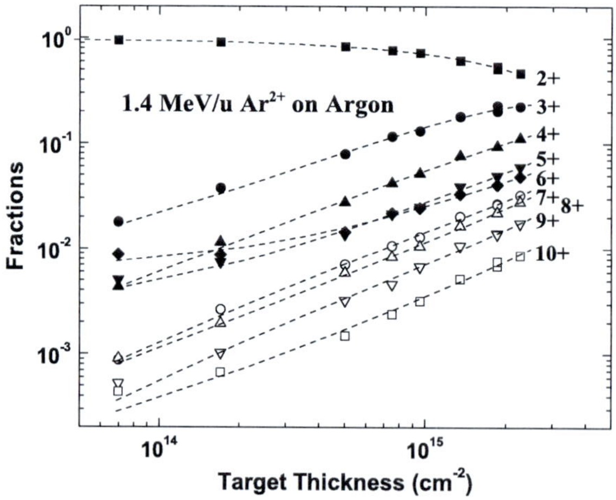

Fig. 8.10 Charge state fractions versus target density measured for electron loss by 1.4 MeV/u Ar^{2+} colliding with Ar

8.4.2 UNILAC Findings

Using the methods outlined above, absolute cross sections were measured for total, single, and multiple electron loss between 0.74 and 1.4 MeV/u for several very low-charge-state ions, e.g., Ar^{+}, Ar^{2+}, Xe^{3+}, and $U^{4,6,10+}$, colliding with neon, molecular nitrogen, and argon targets [28, 47]. For these data, the energy and target Z_T ranges were inadequate for testing the various scalings discussed above. However, the projectile charge dependence on the electron-loss cross sections could be tested. As shown in Fig. 8.11 where the cross sections have been scaled by the effective number of outer-shell electrons, for low-charge-state ions, the scaled cross sections decrease very slowly with increasing charge state and do not depend on the nuclear charge. In contrast, for "highly" charged ions, the cross sections decrease very rapidly with the slope and onset depending on the nuclear charge. The changeover between these two regions appears to be slightly above charge state 10.

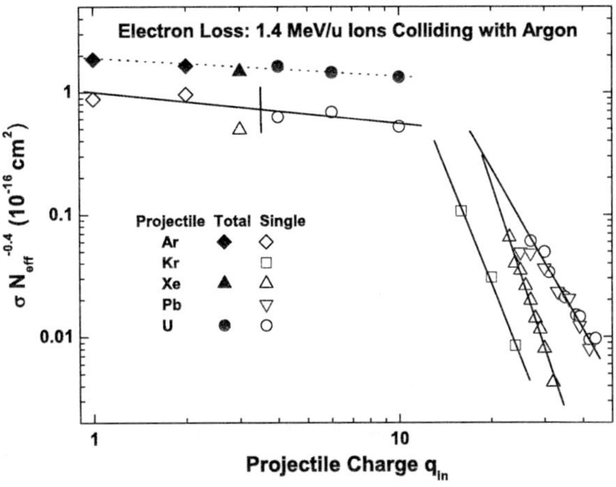

Fig. 8.11 Scaled cross sections for total and single-electron loss by 1.4 MeV/u ions colliding with Ar. Data are from [30, 36, 41] (W. Erb, B. Franzke, 1978, GSI Report No. GSI-J-1-78, unpublished) . The *lines* are to guide the eye. The *vertical line* shows experimental uncertainties for the low-charge-state data

8.5 GSI Storage-Ring Experiments

As shown in a previous section, one of the important discrepancies between the various theories, as well as between theory and available experimental data, is with regard to the energy dependencies at high energies. To extend the electron-loss experimental information to higher velocities, experiments using the GSI storage rings were performed. For these studies, beam lifetimes were measured within the GSI heavy-ion synchrotron (SIS) storage ring and the experimental storage ring (ESR). These provided information about electron loss resulting from collisions with the residual gases. In addition, using the internal gas target at the ESR, lifetimes and total cross sections were measured for H_2 and N_2 targets. Here, only total cross sections could be measured because existing beam upcharge (beam ionization) detector positions plus magnetic field strengths prohibited measuring the individual charge states that were produced.

8.5.1 *Storage-Ring Methods*

For the ESR studies, Xe^{18+} and U^{28+}ion beams were accelerated to approximately 10 MeV/u and injected into the SIS which was used either as a transport device or to provide additional acceleration to higher energies before injecting the ions into the ESR. Bending magnets limited the maximum energies for U^{28+} and Xe^{18+} ions in the ESR to 50 and 70 MeV/u respectively. A minimum energy of 20 MeV/u was determined by the decreasing beam storage lifetimes in the ESR, i.e., the increasing loss cross sections with decreasing energy.

At high energies and for low-charge-state ions, the lifetime τ and the related decay rate λ of the stored ion beams are related to the electron-loss (stripping) cross section by

$$\tau(\mathrm{v}) = \frac{1}{\lambda} = \frac{1}{\rho \mathrm{v} \sigma(\mathrm{v})} \mathrm{f}. \tag{8.12}$$

Here, $\sigma(v)$ corresponds to the weighted total cross sections for electron loss induced by collisions with whatever gases are present, v is the projectile velocity, ρ is the average density of the gas through which the beam travels, and f is the fractional time spent traveling through the gas, e.g., for interactions with background gases in the ring, $f=1$; for interactions with a target gas injected over a certain length, f is the target length, a few mm, divided by the circumference of the storage ring, 108 m for the ESR. Lifetimes were measured for interactions with the residual gases comprising the base vacuum ring as well as with H_2 and N_2 targets injected using supersonic gas jets. The average base pressure of the ESR was 2×10^{-11} mbar with the major components being H_2 (83%), H_2O (10%), CH_4 (5%), CO_2 (1%), and Ar (1%). The H_2 and N_2 targets had densities on the order of 10^{12} particles/cm^3 and jet diameters of approximately 5 mm. Beam lifetimes were measured as a function of time using current transformers. Additional details can be found in [34, 35].

An example of the beam lifetime data is shown in Fig. 8.12. The flat region at times earlier than 9 s corresponds to the slow decay time associated with interactions with the background gases in the ESR. For this dataset, at approximately 9 s, a fast valve is opened admitting the supersonic gas jet. At this time, a rapid decay begins. Note that for an H_2 target, the beam decays away within a few seconds, whereas for an N_2 target, the time is only a couple hundred milliseconds. Also note that these data have been background subtracted as this is essential for establishing the correct decay rates.

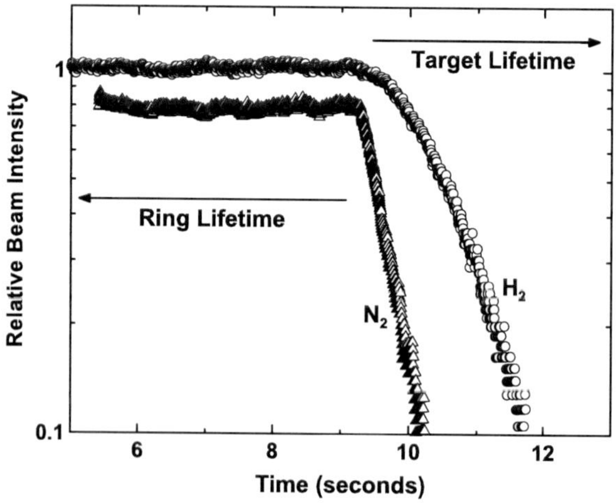

Fig. 8.12 Stored beam currents measured in the ESR. See text for details

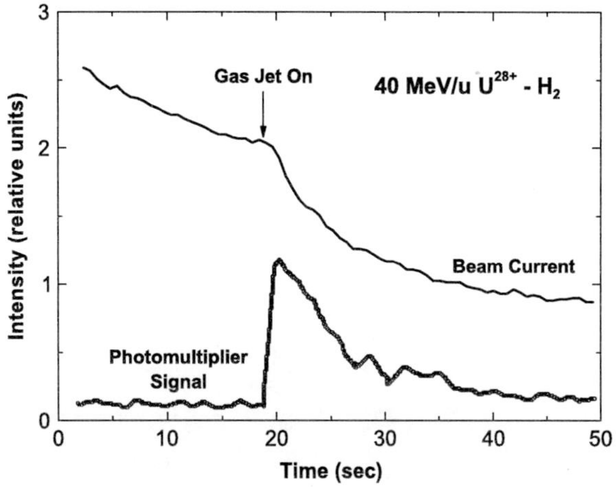

Fig. 8.13 *Upper curve*; stored beam current measured using current transformers in the ESR. *Lower curve*; photon intensity measured by a photomultiplier viewing the beam-gas jet overlap. From [42]

To illustrate the importance of subtracting the correct background plus to show the limitations associated with measuring the small beam currents using the current transformers, the upper curve in Fig. 8.13 shows the raw (not background subtracted) beam current as a function of time. An optional method for measuring the decay times for the target that was tested, but not used, is shown by the lower curve. Here, a photomultiplier recorded light produced in the region of beam-target overlap. As seen, when the gas is injected, the number of emitted photons increases sharply and then exponentially decays away. The advantage of this method is that, in principle, the photomultiplier can be made to have zero background, and therefore the decay process can be followed much longer and the lifetimes determined more accurately.

In the SIS, only lifetime measurements could be performed, and the vacuum is poorer, e.g., the base pressure is approximately 7.7×10^{-11} mbar; the residual gas constituents are H_2(65%), H_2O(17%), CO/N_2(8%), CO_2(1%), Ar, and Cl (each 4%). Also, the ring circumference is larger, 216 m.

8.5.2 *Storage-Ring Findings*

Cross sections for electron loss by U^{28+}, measured at the GSI UNILAC [34, 35], TAMU [32, 33], and those obtained using the measured ESR lifetimes and target densities determined via calibrations of the supersonic gas jet targets [34, 35] are shown in Fig. 8.14. These are compared with Born and nCTMC [32] calculations. As seen, for stripping by a light target such as H_2, both the Born and the nCTMC models predict an E^{-1} dependence at high energies, which is in agreement with the data. For stripping by a many-electron target, the data imply an energy dependence somewhere between the Born E^{-1} dependence and the nCTMC v^{-1} dependence. But, the experimental slope is highly dependent upon the highest energy measurement. Clearly, more data are needed.

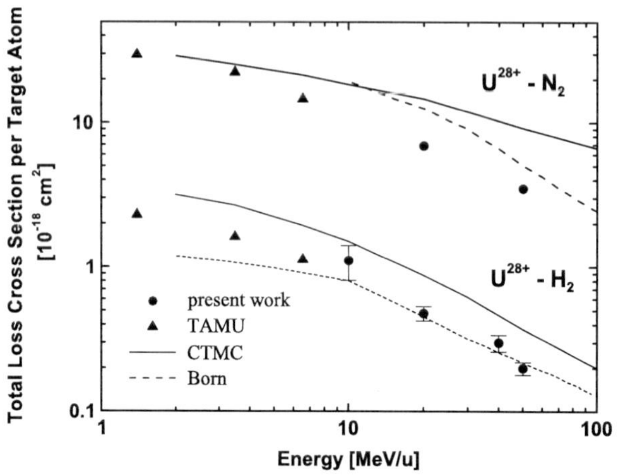

Fig. 8.14 Total loss cross sections for U^{28+}–H_2 and U^{28+}–N_2 collisions. Experimental data from [32–36](W. Erb, B. Franzke, 1978, GSI Report No. GSI-J-1-78, unpublished), CTMC theory, *solid lines*, from [32], Born theory, *dashed*, and *dotted lines*, from Shevelko (private communication)

8.6 Measured Target Scalings

8.6.1 Atomic Targets

To study target-scaling properties, various groups have measured electron-loss cross sections for atomic targets ranging from helium to xenon. These studies have shown that for lighter targets, typically meaning for $Z_T < 10$, the cross sections increase rapidly, e.g., approximately as Z_T^n, where $n \sim 1.4$–1.7. But, for heavier targets, the cross sections increase more slowly and appear to saturate in some cases. In a recent publication, DuBois et al. [43] compared the target dependencies that have been reported. As shown in Fig. 8.15, these data cover an extremely broad range of velocities, e.g., over six orders of magnitude with the highest velocities being approximately a quarter of the speed of light. They also cover a wide range of ion species, e.g., singly charged ions ranging from He^+ to Kr^+, one-electron ions such as He^+ and N^{6+}, low, and medium-charge-state, heavy ions such as $Ar^{6,8+}$, Fe^{4+}, Xe^{18+}, $U^{4,6,10+}$, and Pb^{54+}. In Fig. 8.15, for comparison purposes, the target dependences have been normalized to unity at $Z_T=10$.

In spite of the vast differences in velocities and ion species, these normalized data have very similar target dependences. In particular, note that although there is a slight increase in the overall slope with increasing velocity, data for velocities less than 19 MeV/u all agree in shape within approximately a factor of two. This is in contrast to the much steeper dependencies observed for the two highest velocities where, again, an increase in slope is seen with increasing velocity.

Because the Z_T dependencies for ions with energies less than 19 MeV/u were found to be very similar, average values were calculated for each target. In Fig. 8.16,

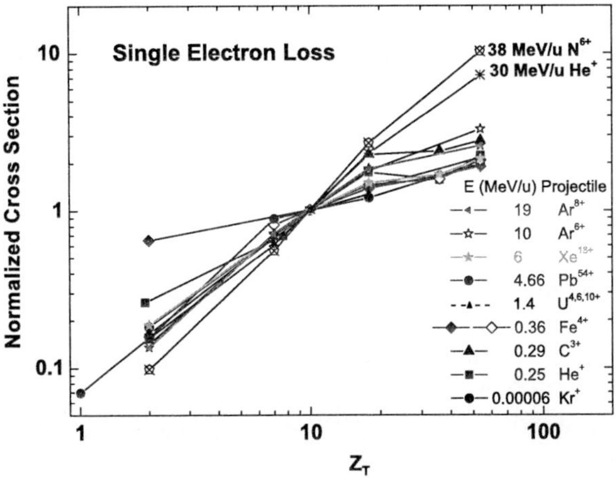

Fig. 8.15 Normalized single electron-loss cross sections measured for various ions and ion energies. The experimental data are from the following sources: Kr^+, [45]; 0.25 MeV/u He^+, [12]; C^{3+}, [14]; Fe^{4+}, [23,46]; $U^{4,6,10+}$, [42]; Pb^{54+}, [48]; Xe^{18+}, [32]; $Ar^{6,8+}$, 30 MeV/u He^+, and 38 MeV/u N^{6+}, [41,47]

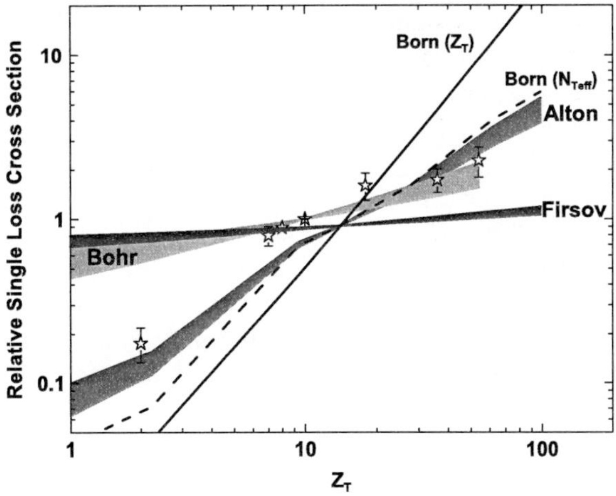

Fig. 8.16 Normalized single electron loss cross sections measured for "low- to medium-energy" ions. The experimental data (*stars*) are average values obtained for the ions listed in the lower corner of Fig. 8.15. The Born(Z_T) and the Born(N_{Teff}) curves are as described in the text. The bands shown are obtained using the Firsov, Bohr, and Alton scaling rules described in an earlier section where the limiting values of 2 and 92 have been used for Z_P

these average values are compared to various theoretical models which have been proposed for describing electron loss in heavy systems. Again, all curves are normalized to unity at $Z_T = 10$. As seen, a pure Born dependence which accounts for stripping by both the full nuclear charge and all of the bound electrons, e.g., $Z_T^2 + Z_T$, predicts a much steeper dependence than is observed. Plus, this dependence shows no change in slope between the lighter and heavier targets. Such an inflection is achieved if, instead of using the full nuclear and bound electron charges, an effective number of electrons is assumed to participate plus to partially screen the nuclear charge. In this case, the Born scaling is given by $N_{\mathrm{Teff}}^2+N_{\mathrm{Teff}}$. Here, for the effective number of electrons, N_{Teff}, the values at an extremely high impact energy suggested by Santos et al. [14] was used. These values (1.4 for He, 5 for N, 5.8 for Ne, 7.8 for Ar, 12.4 for Kr, and 15.1 for Xe) are similar to rough estimates which are obtained by counting all but the most tightly bound inner-shell electrons for each target, e.g., 2 for He, 5 for N, 6 for Ne, 10 for Ar, and 18 for Kr and Xe. As seen, accounting for the effective number of electrons improves the Born model, but still the predicted dependence is steeper than observed.

The Bohr model [20], namely, $(Z_{\mathrm{P}}^{1/3} + Z_{\mathrm{T}}^{1/3})^2$, agrees reasonably well with the data for heavy targets but drastically overestimates the relative cross sections for light targets. Note that a band of values is shown for the Bohr, Alton, and Firsov scalings. These bands show how the projectile charge changes the results and have limiting values for Z_{P} of 2 and 92. The comparison with the scaling suggested by Alton et al. [23] shows much better agreement for lighter targets, but unfortunately poorer agreement for heavy targets. Although they have much different forms, the Alton scaling results are very similar to those obtained using the Born model when the effective number of electrons is taken into account. The final comparison is with the scaling suggested by Firsov [21, 22], e.g., $(Z_{\mathrm{P}}^{1/3} + Z_{\mathrm{T}}^{1/3})^{2/3}$. This scaling has a very flat target dependence and is in poor agreement with experiment. To summarize, none of these scalings is compatible with experiment throughout the entire range of targets studied.

Figure 8.17 shows that various "empirical" scalings could be used to provide better agreement. Here, the Born model, using the full nuclear and electronic charges, Born(Z_{T}), and effective charges, Born (N_{Teff}), are shown by the thin black and thick blue (online) solid lines. The Mod Bohr (Z_{T}), thin red (online) line, uses the Bohr target scaling of $Z_{\mathrm{T}}^{2/3}$ but incorporates the Born concept of ionization by

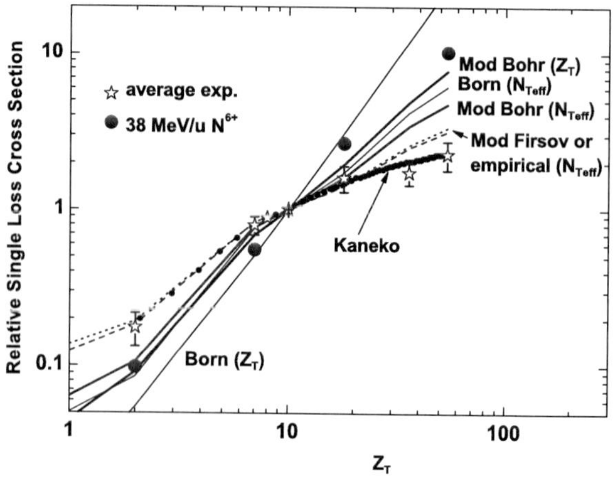

Fig. 8.17 Normalized single electron-loss cross sections compared to various empirical scalings. See text for details. Experimental data are from same references as in Fig. 8.15

both the target nuclear charge and the target electronic charge, i.e., Z_T^2 is replaced with $(Z_T^2 + Z_T)^{1/3}$. Plus, it incorporates the suggestion of Alton et al. by multiplying by N_{Teff}. This yields a scaling of $N_{\mathrm{Teff}}\,(Z_T^2 + Z_T)^{1/3}$. The Mod Bohr ($N_{\mathrm{Teff}}$), thick red (online) line, is similar but replaces the full nuclear and electron charges by their screened and active counterparts, N_{Teff}. This yields $N_{\mathrm{Teff}}(N_{\mathrm{Teff}}^2 + N_{\mathrm{Teff}})^{1/3}$. The Mod Firsov, dashed line, applies the same ideas to the normal Firsov scaling and yields either $N_{\mathrm{Teff}}\,(Z_T^2 + Z_T)^{1/9}$ or $N_{\mathrm{Teff}}(N_{\mathrm{Teff}}^2 + N_{\mathrm{Teff}})^{1/9}$. These were found to produce nearly identical results and to be nearly the same as an empirical scaling, empirical (N_{Teff}), dotted line, which had the form $(N_{\mathrm{Teff}}^2 + N_{\mathrm{Teff}})^{2/3}$. The final theoretical predictions shown in the figure are the quantitative values of Kaneko [28], small black dots, for stripping of 0.9 MeV/u He^+.

These models are compared to the average data for energies less than 19 MeV/u which represent electron loss from "slow- to medium-velocity multi-charged heavy ions," open stars. Also shown are electron loss data for 38 MeV/u N^{6+} which are very similar to data for 30 MeV/u He^+, filled circles. These represent electron loss by " fast" singly charged ions. As shown previously where data for each individual energy and specie were plotted, the target dependence becomes steeper with increasing velocity. Here, we see that the target dependence for electron loss from "fast" ions is predicted quite well by the Mod Bohr (Z_T) scaling, $N_{\mathrm{Teff}}\,(Z_T^2 + Z_T)^{1/3}$, whereas the target dependence for stripping of "slow- to intermediate-velocity" ions can be predicted quite well using either the Mod Firsov scaling of $N_{\mathrm{Teff}}\,(Z_T^2 + Z_T)^{1/9}$ or an empirical scaling of the form $(N_{\mathrm{Teff}}^2 + N_{\mathrm{Teff}})^{2/3}$. An extremely interesting feature is that the target dependence predicted by Kaneko for He^+ seems to apply for a very broad range of ions and velocities. However, his theory predicts that the target dependence becomes flatter (for heavy targets) with increasing velocity whereas the available data indicates just the opposite.

8.7 Concluding Remarks

We have attempted to show what is known to date about electron loss, in particular how the cross sections have been predicted to scale with respect to the target nuclear and electronic charges. As stated in the introduction, the examples provided were selected in order to illustrate specific features. In doing so, we may not have cited, or may have overlooked, important references. However, for those interested in the subject, the present examples and references can be used as a starting point for further investigations.

Acknowledgements R.D.D. gratefully acknowledges the support received from the National Science Foundation, the Fulbright Commission, and the ExtreMe Matter Institute EMMI during the preparation of this manuscript. A.C.F.S. is grateful for the support from CNPq and FAPERJ for performing experimental studies described in this work plus during the preparation of the manuscript.

References

1. C. Omet, P. Spiller, J. Stadlmann, D.H.H. Hoffmann, New J. Phys. **8**, 284 (2006)
2. V. Coco, J. Chamings, A.M. Lombardi, E. Sargsyan, R. Scrivens, Proceedings of LINAC 2004 Conference, Lübeck, Germany, TUP27, p 351, available online: accelconf.web.cern.ch/accelconf/l04/PAPERS/TUP27.PDF
3. A. G. Ruggiero, Accelerator Physics Technical Note No. 42, Brookhaven National Laboratory, 1992, available online: www.rhichome.bnl.gov/RHIC/RAP/rhic_notes/AD-AP-1.../AD-AP-42.pdf
4. FAIR Baseline Technical Report (2006), available online: http://www.gsi.de/fair/reports/btr.html.
5. R.O. Bangerter, Nucl. Instrum. Meth. A **464**, 17 (2001)
6. D.R. Bates, G. Griffing, Proc. Phys. Soc. **A68**, 90 (1955)
7. R.D. DuBois, S.T. Manson, Phys. Rev. A **42**, 1222 (1990)
8. W.E. Meyerhof, H.-P. Hüllskötter, Q. Dai, J.H. McGuire, Y.D. Wang, Phys. Rev. A **43**, 5907 (1991)
9. E.C. Montenegro, W.E. Meyerhof, Phys. Rev. A **44**, 7229 (1991)
10. E.C. Montenegro, W.E. Meyerhof, Phys. Rev. A **46**, 5506 (1992)
11. M.M. Sant'Anna, W.S. Melo, A.C.F. Santos, G.M. Sigaud, E.C. Montenegro, Nucl. Instrum. Meth. B **99**, 46–49 (1995)
12. G.M. Sigaud, F.S. Jorás, A.C.F. Santos, E.C. Montenegro, M.M. Sant'Anna, W.S. Melo, Nucl. Instrum. Meth. B **132**, 312–315 (1997)
13. W.S. Melo, M.M. Sant'Anna, A.C.F. Santos, G.M. Sigaud, E.C. Montenegro, Phys. Rev. A **60**, 1124 (1999)
14. A.C.F. Santos, G.M. Sigaud, W.S. Melo, M.M. Sant'Anna, E.C. Montenegro, Phys. Rev. A **82**, 012704 (2010)
15. A.B. Voitkiv, G.M. Sigaud, E.C. Montenegro, Phys. Rev. A **59**, 2794 (1999)
16. E.C. Montenegro, A.C.F. Santos, W.S. Melo, M.M. Sant'Anna, G.M. Sigaud, Phys. Rev. Lett. **88**, 013201-1 (2002)
17. A. Voitkiv, J. Ullrich, *Collsions of Structured Atomic Particles* (Springer, Berlin, 2008)
18. A.B. Voitkiv, Phys. Rep. **392**, 191 (2004)
19. A.C.F. Santos, R.D. DuBois, Phys. Rev. A **69**, 042709 (2004)
20. N. Bohr, K. Dan. Vidensk. Selsk. Mat. Fys. Medd. **18**(8), (1948)
21. O.B. Firsov, Zh. Eksp. Teor. Fiz. **34**, 447 (1958)
22. O.B. Firsov, JETP **7**, 308 (1958)
23. G.D. Alton, L.B. Bridwell, M. Lucas, C.D. Moak, P.D. Miller, C.M. Jones, Q.C. Kessel, A.A. Antar, M.D. Brown, Phys. Rev. A **23**, 1073 (1981)
24. V.P. Shevelko, Th. Stöhlker, I.Yu. Tolstikhina, Nucl. Instrum. Meth. B **184**, 295 (2001)
25. I.L. Beigman, I.Yu. Tolstikhina, V.P. Shevelko, Tech. Phys. **53**, 546 (2008)
26. V.P. Shevelko, I.L. Beigman, M.S. Litsarev, H. Tawara, I. Yu. Tolstikhina, G. Weber, Nucl. Instrum. Meth. B **269**, 1455 (2011)
27. V.P. Shevelko, D. Kato, M.S. Litsarev, H. Tawara, J. Phys. B **43**, 215202 (2010)
28. V.I. Matveev, D.U. Matrasulov, S.V. Ryabchenko, J. Exp. Theor. Phys. **102**, 1–8 (2006), orig in Zhur. Eksp. noi i Teor. Fiz. **129**, 5–13 (2006)
29. T. Kaneko, Phys. Rev. A **32**, 2175–85 (1985)
30. R.E. Olson, R.L. Watson, V. Horvat, K.E. Zaharakis, J. Phys. B: At. Mol. Opt. Phys. **35**, 1893 (2002)
31. R.D. DuBois, A.C.F. Santos, R.E. Olson, Th. Stöhlker, F. Bosch, A. Bräuning-Demian, A. Gumberidze, S. Hagmann, C. Kozhuharov, R. Mann, A. Oršiæ Muthig, U. Spillmann, S. Tachenov, W. Barth, L. Dahl, B. Franzke, J. Glatz, L. Gröning, S. Richter, D. Wilms, A. Krämer, K. Ullmann, O. Jagutzki, Phys. Rev. A **68**, 042701 (2003)
32. R.L. Watson, Y. Peng, V. Horvat, G.J. Kim, R.E. Olson, Phys. Rev. A **67**, 022706 (2003)

33. R.E. Olson, R.L. Watson, V. Horvat, A.N. Perumal, Y. Peng, Th. Stöhlker, J. Phys. B At. Mol. Opt. Phys. **37**, 4539 (2004)
34. R.E. Olson, R.L. Watson, V. Horvat, K.E. Zaharakis, R.D. DuBois, Th. Stöhlker, Nucl. Instrum. Meth. A **544**, 333–36 (2005)
35. R.D. DuBois, O. de Lucio, M. Thomason, G. Weber, Th. Stöhlker, K. Beckert, P. Beller, F. Bosch, C. Brandau, A. Gumberidze, S. Hagmann, C. Kozhuharov, F. Nolden, R. Reuschl, J. Rzadkjewicz, P. Spiller, U. Spillmann, M. Steck, S. Trotsenko, Nucl. Instrum. Meth. B **261**, 230–233(2007)
36. G. Weber, C. Omet, R.D. DuBois, O. de Lucio, Th. Stöhlker, C. Brandau, A. Gumberidze, S. Hagmann, S. Hess, C. Kozhuharov, R. Reuschl, P. Spiller, U. Spillmann, M. Steck, M. Thomason, S. Trotsenko, Phys. Rev. ST Accel. Beams **12**, 084201 (2009)
37. E.C. Montenegro, W.E. Meyerhof, Phys. Rev. A **43**, 2289 (1991)
38. M.M. Sant' Anna, W.S. Melo, A.C.F. Santos, G.M. Sigaud, E.C. Montenegro, M.B. Shah, W.E. Meyerhof, Phys. Rev. A **58**, 1204 (1998)
39. E.C. Montenegro, W.E. Meyerhof, Phys. Rev. A **43**, 2289 (1991)
40. T. Kirchner, A.C.F. Santos, H. Luna, M.M. Sant'Anna, W.S. Melo, G.M. Sigaud, E.C. Montenegro, Phys. Rev. A, **72**, 012707 (2005)
41. D. Müller, L. Grisham, I. Kaganovich, R.L. Watson, V. Horvat, K.E. Zaharakis, Y. Peng, Princeton Plasma Physics Laboratory Document PPPL-3713 (2002)
42. R.D. DuBois, A.C.F. Santos, Th. Stöhlker, F. Bosch, A. Bräuning-Demian, A. Gumberidze, S. Hagmann, C. Kozhuharov, R. Mann, A. Oršiæ Muthig, U. Spillmann, S. Tachenov, W. Bart, L. Dahl, B. Franzke, J. Glatz, L. Gröning, S. Richter, D. Wilms, K. Ullmann, O. Jagutzki, Phys. Rev. A **70**, 032712 (2004)
43. G. Weber, Diplomarbeit, Ruprecht-Karls Universität Heidelberg (2006)
44. R.D. DuBois, A.C.F. Santos, G. Sigaud, E.C. Montenegro, Phys. Rev. A **84**, 022702 (2011)
45. H. Martínez, A. Amaya-Tapia and J.M. Hernández, J. Phys. B At. Mol. Opt. Phys. **33**, 1935–1942 (2000)
46. H. Knudsen, C.D. Moak, C.M. Jones, P.D. Miller, R.O. Sayer, G.D. Alton, L.B. Bridwell, Phys. Rev. A **19**, 1029 (1979)
47. D. Müller, L. Grisham, I. Kaganovich, R.L. Watson, V. Horvat, K.E. Zaharakis, M.S. Armel, Phys. Plasmas **8**, 1753–1756 (2001)
48. W.G. Graham, K.H. Berkner, R.V. Pyle, A.S. Schlachter, J.W. Stearns, J.A. Tanis, Phys. Rev. A **30**, 722 (1984)

Chapter 8
Target-Scaling Properties for Electron Loss by Fast Heavy Ions

R.D. DuBois and A.C.F. Santos

Abstract Electron loss by fast heavy ions resulting from interactions with dilute gaseous targets is discussed. Of particular interest is how the cross sections scale as the target nuclear charge increases. Various theoretical models that have been proposed are discussed and compared with available experimental data. It is shown that none of these models yield good agreement with data but that by combining ideas and concepts contained within different models agreement can be obtained.

8.1 Introduction

As fast ions traverse any media, whether the medium is a dilute gas, condensed matter, biological material, or a plasma, inelastic interactions with the various atoms and molecules comprising the media deposit energy, which alters the media and degrades the ion's kinetic energy. In addition, sometimes the projectile is ionized. Projectile ionization, the subject of the present work, is also referred to as projectile stripping or electron loss. Here, we shall use the terms interchangeably. For atomic targets, electron loss interactions have the form:

$$P^{q+} + T \rightarrow P^{(q+n)+} + T^{i+} + (n+i)e^{-}, \tag{8.1}$$

where P represents the incoming ion (the projectile) with charge q, T is the target atom, and n and i are the number of projectile and target electrons that are liberated. For projectile ionization, $n \geq 1$ whereas the target can be simultaneously ionized or it can remain in its ground or an excited state. Thus, $i \geq 0$. In the case of molecular targets, fragmentation into charged and uncharged components is also possible. These initial, plus any subsequent interactions involving the reaction products, are of extreme importance in understanding radiation damage to materials and/or to biological tissue.

V. Shevelko and H. Tawara (eds.), *Atomic Processes in Basic and Applied Physics*, Springer Series on Atomic, Optical, and Plasma Physics 68,
DOI 10.1007/978-3-642-25569-4_8,

Electron loss processes have been studied for many decades, in part to investigate atomic interactions where the number of active particles can be controlled and systematically varied and in part for practical purposes associated with the creation, acceleration, transport, and storage of ions for research purposes. Electron loss from very heavy, few-electron ions is also used to study processes occurring in the presence of very high electric fields such as those found for inner-shell electrons.

Equation (8.1) shows that electron loss studies can be tailored to study single- or multiple-electron transitions in few- or many-electron systems simply by altering the projectile nuclear, Z_P, and total, q, charges. In addition, passive and active electron effects (as will be described later) can be investigated by comparing electron loss data obtained using different targets or by comparing data obtained for the same target but where simultaneous and independent ionization of the target and projectile are compared.

With regard to the acceleration, transport, and storage of ion beams, electron loss occurring when high-energy beams interact with background gases within beamlines or storage rings has serious detrimental effects. These include loss of beam luminosity (due to scattering), decreased storage times (due to losses at bending magnets or focusing elements because of altered charge states or degraded energies), and possible erosion, heating, and vacuum loading in the accelerator and storage rings (due to interactions with the vacuum walls by the lost beam components). These are crucial technical problems at all high-energy accelerator laboratories, e.g., at GSI, Brookhaven, CERN, Dubna, etc. where considerable effort and expense are being devoted in order to predict and circumvent such problems. See, for example, [1–3].

As an example, the new Facility for Antiproton and Ion Research (FAIR) at the GSI Helmholzzentrum für Schwerionenforschung in Darmstadt, Germany, is being designed to produce very intense beams of high-energy, heavy ions [4]. This requires using low-charge-state ions in order to reduce the space charge. But, as will be shown later, the probability for electron loss is inversely proportional to the energy required to remove bound electrons. Therefore, and for reasons associated with obtaining high velocities plus limitations pertaining to the maximum magnetic fields that can be produced by bending magnets along the beamline and in storage rings, " medium" charge-state ions such as U^{28+} are expected to be used. For such ions traveling with MeV/u to GeV/u velocities, electron loss is the dominant loss mechanism. To achieve the scientific goals, beam intensities will be roughly a factor of 50 larger than what is currently possible. This means that radiation levels, heating, and vacuum loads associated with lost beam components will increase considerably.

Another example where electron loss plays a crucial role is in high-energy-density research. Approximately a decade ago, a US program proposed accelerating intense beams of low-charged heavy ions to GeV energies to indirectly heat small D-T pellets and induce laboratory fusion [5]. To achieve the energy densities that are required, intense, tightly-focused beams have to hit the pellets. Original plans were to employ singly charged heavy ions in order to minimize space charge defocusing of the beams. But because of complications and lack of knowledge about transport and beam losses, particularly in the reaction chamber where a high partial pressure of FLiBe (a molten salt which will coat the inner walls in order to collect energetic neutrons released in the fusion process), the emphasis changed to the use of light ions. However, heavy ions can deliver far more energy to the target. Thus, the original idea may be revived sometime in the future, especially if sufficient experimental data and theoretical understanding of electron loss processes become available.

In both of these examples, intense beams of heavy ions having MeV/u to GeV/u velocities must be transported fairly long distances through beamlines and perhaps stored in storage rings. Even under extremely high-vacuum conditions, interactions with background gases result in portions of the beam suffering energy loss and/or changing charge. These portions are subsequently lost. Localized heating of surfaces occurs; this desorbs gases or atoms from the surfaces; this further degrades the vacuum, which increases the loss processes, etc. In the worst case, the processes avalanche and a rapid loss of beam quality and intensity occurs [1]. To avoid or to minimize these problems requires information relating to (1) total loss probabilities for a broad range of systems and (2) information about the charge states produced and their relative production probabilities. Such information is needed for electron loss resulting from interactions with few-electron targets such as H_2 and He, i.e., the primary components found in high-vacuum environments, as well as with many-electron targets such as N_2, CO, CO_2, H_2O, etc. which occur as lesser components in the vacuum or which are desorbed from heated surfaces. Interactions with heavier gases are especially important because they have larger interaction cross sections. Hence, even though they may be far less abundant, they can still contribute significantly to the total beam losses. Plus, interactions with heavier atoms and molecules are primarily responsible for multiple electron loss from the ion.

This work will concentrate on electron loss by fast ions, with particular emphasis on many-electron systems and how the electron loss cross sections scale. Following a brief description of the various theoretical models available, methods used to obtain experimental data at MeV/u velocities will be outlined. Here, we restrict ourselves to methods and findings from the Federal University in Rio de Janeiro and the GSI Helmholzzentrum für Schwerionenforschung, two laboratories still actively engaged in electron loss measurements. Using selected examples of available data, we will illustrate what scaling properties can be extracted and provide the reader with an overall picture of this field of research. With regard to theory, our intent is only to outline the general methods that have been applied to electron loss and to briefly discuss the results. For detailed information, the reader is referred to the cited references.

8.2 Theoretical Models and Predictions

The basic procedures for calculating electron loss, i.e., projectile ionization, were outlined many years ago by Bates and Griffing [6]. They showed that two interaction channels must be considered. As illustrated in Fig. 8.1, one involves the interaction

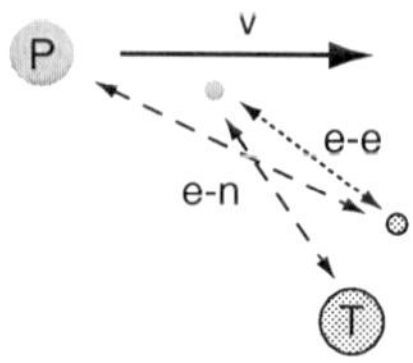

Fig. 8.1 Schematic showing the interaction between the nuclear charge of one of the collision partners and the bound electron of the other (e–n interactions, shown by the *dashed lines*) and the direct interaction between bound projectile and target electrons (e–e interactions, shown by the *dotted line*)

between a bound projectile electron and the partially screened nuclear charge of the target; the other involves a direct interaction between a target and a projectile electron. The first has been referred to as the e–n or the screening or the elastic channel, while the second is called the e–e or the antiscreening or the inelastic channel.

Using the Born approximation, Bates and Griffing showed that the differential electron emission arising from these two channels is given by

$$\mathrm{e-n} \qquad d^2\sigma(\varepsilon) = \int_{K\,min} A(\varepsilon, K)\left[|Z_T - N_T F(K)|^2\right] dK, \tag{8.2}$$

$$\mathrm{e-e} \qquad d^2\sigma(\varepsilon) = \int_{K'\,min} A(\varepsilon, K)\left[N_T - N_T\,|F(K)|^2\right] dK. \tag{8.3}$$

For the purposes of the present work, the important quantities to consider are the effective target charges found within the brackets. For the e–n interactions, the target nuclear charge, Z_T, is partially screened by its bound electrons, N_T. But the screening depends on the momentum transfer, *K*. (In an impact parameter representation, the screening would be a function of the impact parameter.) For neutral targets, N_T is equal to Z_T. However, we will continue using N_T as a designation in order to talk about the "effective number of target electrons." Because the electron form factor *F(K)* ranges in value between 0 and 1, the e–n interactions are seen to scale with the square of an effective nuclear charge, Z^2_{Teff}, which has a value somewhere between Z^2_T and $(Z_T{-}N_T)^2$, which for a neutral target, ranges in value from Z^2_T to 0. In contrast, the e–e interactions scale with the effective number of electrons, N_{Teff}. N_{Teff} ranges in value between N_T and 0, which for a neutral target is also from Z_T to 0. For additional details, the reader is referred to [7].

The other important item to note in (8.2) and (8.3) are the different integration limits for the two channels. Because of their different masses, to transfer the momentum required to remove a projectile electron requires a considerably higher energy for electrons, i.e., for e–e interactions, than for the massive nucleus, i.e., for e–n interactions. These different thresholds are shown in the next figure. To summarize, for few electron systems and energies well above threshold, the Born model predicts that the cross sections will scale as $Z^2_{Teff} + N_{Teff}$. With respect to impact velocity, the Born model predicts a v^{-2} dependence.

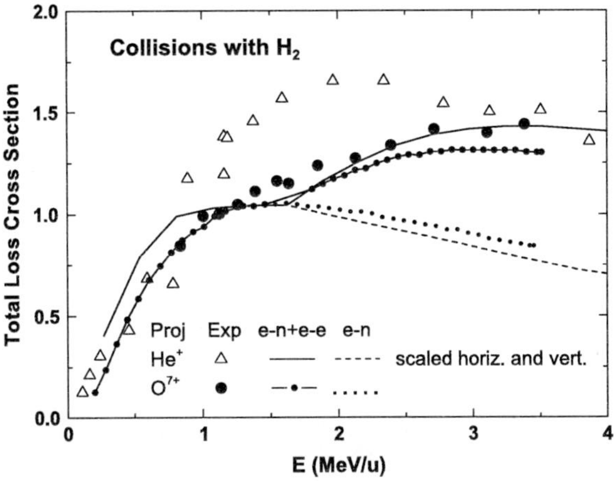

Fig. 8.2 Electron loss by He^+ and O^{7+} resulting from collisions with molecular hydrogen. Experimental data, *solid dots* and *open triangles*, and PWBA calculations for e–n (*dashed curves*) and e–n + e–e interactions (*solid curves*) are from [8]. Both experiment and theory for He^+ have been scaled vertically and horizontally for comparison purposes

Meyerhof et al. [8] used a PWBA (plane wave Born approximation) to calculate electron loss from few electron ions colliding with molecular hydrogen. Figure 8.2 shows their results compared with experimental data. For He^+, experiment and theory have been scaled horizontally and vertically such that the onset of the e–e channel is roughly the same as for O^{7+}. For ionization of O^{7+}, the onset is clearly seen, whereas for ionization of He^+ the threshold is less clear.

These scaled data show that even though the binding energies of the ionized electrons are quite different for the two projectiles, the e–e and e–n processes have the same relative magnitudes and energy dependencies. In particular, note that for energies well above threshold, the relative e–e and e–n cross sections are roughly the same. This is because for ionization by H_2 the two target nuclei act independently, i.e., the e–n process scales as two independent ionizations by $Z_T = 1$, e.g., as $1^2 + 1^2$, while the e–e process scales with the total number of target electrons, e.g., as $N_T = 2$. Thus, one would expect equal contributions from each.

In several studies performed at the Federal University in Rio de Janeiro, UFRJ, a semiclassical approximation was used to calculate the e–e (antiscreening) channel [9–14]. In combination with PWBA calculations for the screening channel, cross sections for electron loss from few-electron ions such as He^+, C^{3+}, and O^{5+} induced by various targets were calculated. This method predicts the same scaling features as for the Born approximation.

A different approach was used by Voitkiv et al. [15] to calculate electron loss by He^+ projectiles in collisions with heavy targets. Within the framework of the sudden approximation, which is a nonperturbative, fully unitarized theory, the elastic screening contribution to the electron loss was determined. This method also included the possibility of multielectron transitions in the target atom concomitant with the loss process.

They showed that the elastic part of the total loss cross section can be written as

$$\sigma_{12}^{\text{elas}} = \left[1 - P_{\text{ion}}^{a}(0)\right] S_{\text{loss}}^{\text{sc}}, \tag{8.4}$$

where $S_{\text{loss}}^{\text{sc}}$ is the screening contribution to the total loss cross section and $P_{\text{ion}}^{a}(0)$ is the target ionization probability at zero impact parameter. The magnitude of the screening contribution was found to be much larger than the geometric cross section of $\text{He}^{+} \approx 0.8$ a.u, indicating that large impact parameters contribute most to the screening cross section.

Using the independent particle model scenarium, and ignoring postcollision rearrangements, it was shown that [16]

$$\sigma_{\text{inelastic}} \approx \sum_{s} n_s \sigma_{\text{anti}}^{s} = \sigma_{\text{anti}} \approx \langle n \rangle \, \sigma_{\text{e}}. \tag{8.5}$$

Thus, in the projectile frame, the inelastic mode can be viewed as ionization of the projectile by a beam of ⟨n⟩ *target* electrons, each having an ionization cross section σ_{e}. Here, $\langle n \rangle$ is the average number of the active target electrons associated with the inelastic mode. If the projectile velocity is high enough, the ratio $\sigma_{\text{inelasic}}/\sigma_{\text{e}}$ can be used to determine ⟨n⟩, as will be illustrated in a later section.

It should also be mentioned that Voitkiv et al. [17, 18] have extended their calculations to extremely high energies where relativistic effects become important.

To go beyond few-electron ions significantly increases the complexity of the theoretical models. Multiple- as well as single-electron removal must be considered. An independent particle model is often used to model multiple electron removal. In addition, one must account for different binding energies for removing additional electrons. As shown by Santos and DuBois [19], the electron loss cross sections decrease with increasing binding energy of the electron being removed. Different binding energies imply different thresholds for the e–e process. Thus, at any specific impact energy, different numbers of projectile electrons can be removed via the e–e process. Removal of additional, i.e., more tightly bound, electrons requires a larger momentum transfer which implies smaller impact parameters. Thus, at each impact energy, the effective number of target electrons which actively participate in the e–e process or which passively screen the target nuclear charge can change. Such effects must be modeled and incorporated into the screening and antiscreening formulae.

Using a large database of cross sections, Santos and DuBois [19] obtained empirical scaling formulae for single- and multiple-electron loss resulting from collisions with argon and molecular nitrogen. They showed that the cross sections scaled as $N_{\text{Peff}}^{0.4}$, where N_{Peff} is the number of electrons in the outermost shell or shells, depending upon the impact energy.

In 1948, Bohr predicted [20] that in collisions of heavy systems, the cross section for single electron loss should scale as $(Z_{\text{P}}^{1/3} + Z_{\text{T}}^{1/3})^2/v$. This scaling was based upon the Thomas–Fermi model which he used to estimate the screened Coulomb potential between atoms. Firsov [21, 22] numerically derived the interatomic potentials of two colliding Thomas–Fermi atoms and fitted these potentials using the

Thomas–Fermi screening length. He obtained a slightly different scaling, namely, $(Z_P^{1/3} + Z_T^{1/3})^{2/3}$. Several decades later, Alton et al. [23] modified the Bohr formula slightly by multiplying by the number of projectile electrons in the outermost subshell and dividing by their ionization potential. With this modification, they obtained good agreement with their experimental electron loss measurements as a function of Z_T.

Shevelko and coworkers developed a series of codes (LOSS, LOSS-R, and RICODE [24–26], respectively) which use a plane wave Born approximation to calculate single-electron removal cross sections over a wide range of energies including relativistic energies. Cross sections for projectile ionization are calculated using the Schrödinger radial wave functions and with account for atomic structure of both colliding particles. For electron loss from Pb-like ions by various targets [24], the cross sections were found to scale as $(Z_T/I_P)^{1.4}$ and the energy to scale with I_P. Recently, the DEPOSIT code was developed [26] based upon the classical model. The code is intended to calculate single- and multiple-electron removal cross sections at low and intermediate collision energies. Here, the cross sections scale as $\frac{Z_T^{0.85}}{E^{\alpha(Z_T)} I_P^{1.2}}$ where $\alpha(Z_T) \approx 0.8/Z_T^{0.3}$ [26].

Using a nonperturbative theory, Matveev et al. [27] calculated multiple excitation and ionization of structured heavy ions interacting with complex neutral atoms. However, this theoretical approach is rather complex and requires many adjustable parameters. Hence, extracting simple scaling relationships is not possible.

Kaneko used a unitarized impact-parameter method to calculate electron loss and excitation from He^+ colliding with various atoms [28]. With respect to the energy dependence, in the energy range between 4 and 10 MeV, he found that the single loss cross sections scale as E^{-1} for stripping by a helium target, $E^{-0.64}$ for stripping by N_2, $E^{-0.44}$ for stripping by Ar, and $E^{-0.36}$ for stripping by Kr. With regard to the target dependence, he found that the cross sections scale roughly as ${Z_T}^{1.3}$ for $Z_T < 10$. For $Z_T > 10$, the scaling is slower, e.g. with powers decreasing from 0.39 to 0.17 as the impact velocity increases from 0.1 to 0.9 MeV/u.

Using an n-body Classical Trajectory Monte Carlo (nCTMC) method, Olson has calculated single- and multiple-electron loss for a variety of systems [29–33]. This theory includes multiple-electron removal processes and has been applied to a large variety of atomic interactions and many-electron systems. For stripping induced by lighter targets such as H_2, the nCTMC predicts that at high energies, the cross sections decrease as E^{-1}, just as is predicted by first-order perturbation theories. However, for stripping induced by heavier targets, it predicts a much slower dependence, roughly as v^{-1}. This is consistent with Bohr's predictions for heavy systems, although it should be noted that the assumptions used by Bohr are not applicable at high velocities.

As stated in the introduction, accurately modeling and predicting electron loss has significant ramifications with respect to beam transport and storage as well as with accelerator design and performance and radiation safety. However, extrapolation of the various models to high energies leads to differing results, sometimes results that are significantly different. For example, as illustrated in

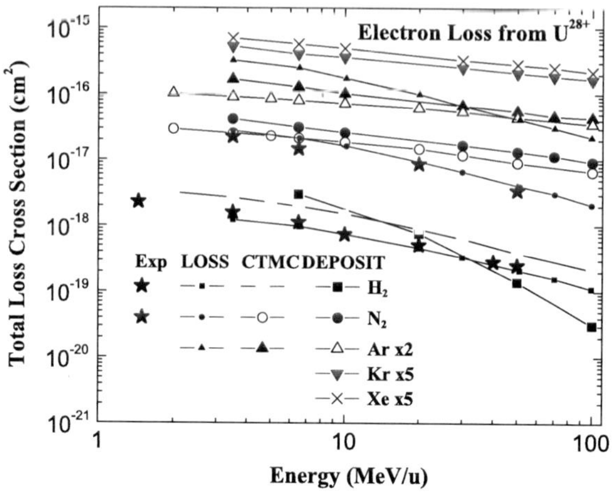

Fig. 8.3 Experimental and theoretical cross sections for electron loss from U^{28+} resulting from collisions with various targets. The experimental data are from [32–36](W. Erb, B. Franzke, 1978, GSI Report No. GSI-J-1-78, unpublished), the LOSS, CTMC, and DEPOSIT theories are from [24, 32, 33], and [26], respectively. The curves for Ar, Kr, and Xe have been shifted upward for display purposes

Fig. 8.3, the LOSS, DEPOSIT, and nCTMC models tend to agree within factors of 2 at low energies, e.g., a few MeV/u, but they predict different velocity dependences. This makes extrapolation to high energies less and less certain. In addition, the figure shows that the predicted velocity dependences depend upon the target. For a low-Z, few-electron, target such as H_2, the velocity dependence is very close to E^{-1}, as expected from first-order perturbation theory and in agreement with the LOSS code of Shevelko et al. but in disagreement with the DEPOSIT code. For a medium-size target such as N_2, the dependence is approximately v^{-1}. This is close to the dependencies predicted by the DEPOSIT code and the higher energy portion of the nCTMC model but in disagreement with the LOSS-R code. For higher-Z, many-electron targets, the predicted dependencies are slower than v^{-1}.

Included in the figure are experimental data obtained at GSI [34–36](W. Erb, B. Franzke, 1978, GSI Report No. GSI-J-1-78, unpublished) and the Texas A&M University cyclotron (TAMU) [32, 33]. As seen, good agreement with the LOSS code and nCTMC model for H_2 is found. But, based upon the highest energy points for N_2, experiment seems to have a faster falloff with impact energy than is predicted by the supposedly accurate DEPOSIT and nCTMC models. The discrepancy with the nCTMC calculations is especially troubling as this method is known to provide excellent results for many applications where many-electron systems are involved. It is not likely that relativistic effects lead to the observed differences since these were simulated by modifying the velocity at which the high-energy calculations were performed. Additional data are needed to determine how, or if, relativistic affects influence the cross sections in the energy region of interest. Plus, data are needed to test the models for heavier targets.

8.3 Experimental Methods and Results

In this section, we describe three experimental methods which have been used to supply information about electron loss at MeV/u velocities. The first describes methods used at the Federal University of Rio de Janeiro (UFRJ) which are very similar to those used at the Texas A&M University (TAMU) cyclotron and for the early studies performed at the GSI UNILAC. The second is the method used for the most recent studies at the UNILAC. The last is the method currently employed at the GSI Experimental Storage Ring (ESR).

8.3.1 UFRJ Method

Aiming to obtain a detailed understanding of electron loss and multiple ionization of atoms, Brazilian studies, formerly at the Pontifical Catholic University (PUC) and now at the Federal University of Rio de Janeiro (UFRJ), have measured absolute cross sections for total electron loss as well as for the projectile loss-target ionization channels. These studies have concentrated on projectiles with one active electron (e.g., He^{+}, C^{3+}, O^{5+}, and B^{2+}) interacting with noble gases (He to Xe) and simple molecules. Data were obtained in the 1.0–4.0 MeV energy region.

8.3.2 UFRJ Experimental Setup

For these studies, collimated, partially dressed ion beams were produced by the 4-MV Van de Graff at PUC or by the 1.7 MV tandem accelerator at UFRJ. Figure 8.4 shows a schematic of the experimental apparatus. Just before entering a windowless target cell, the beams are vertically charge-analyzed (cleaned) by a magnet (PUC) or an electric field (UFRJ) in order to separate the main beam from spurious ones. After the target cell, the emergent beam is horizontally charge-analyzed by a second magnet (PUC) or electric field (UFRJ). The charge states are recorded some meters downstream by surface barrier detectors or by a position-sensitive microchannel plate detector.

Singly and multiply ionized recoil ions are accelerated out of the interaction region by a two-stage electric field and detected by a microchannel plate detector. The recoil ions provide stop signals for a time-to-amplitude converter which are started by the projectile electron loss signal. To prevent signal deterioration, the counting rates of the incident projectiles were kept below 1.5 kHz.

Figure 8.5 shows the postcollision charge states detected by a 2D position-sensitive microchannel plate detector for 200 keV O^{+} colliding with CH_4. These charge states originate from charge-changing collisions of the main beam with CH_4 within the target cell plus with residual gases in the beamline before the gas cell. Because of the vertical field applied before the collision cell, any spurious beams

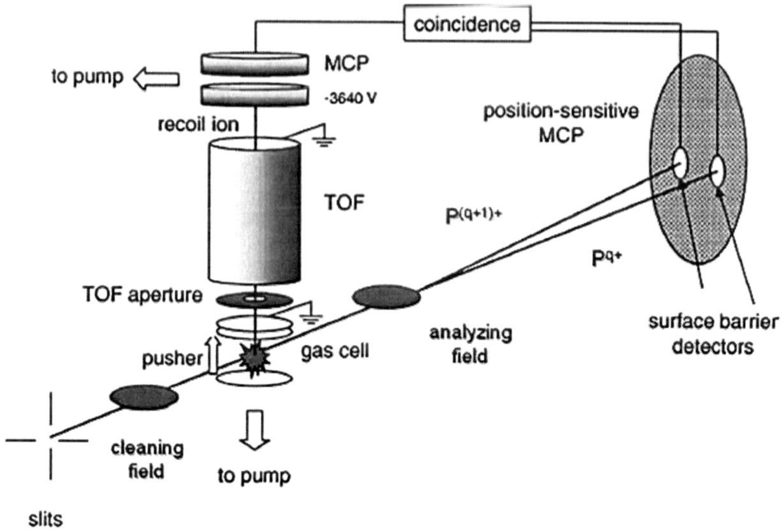

Fig. 8.4 Schematic view of the experimental setup at PUC/UFRJ. See text for details

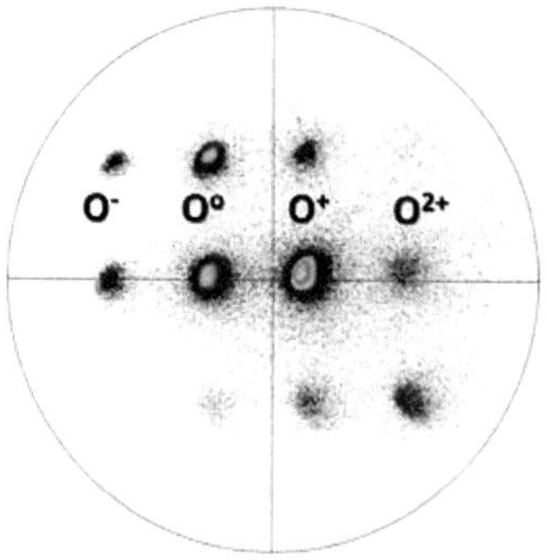

Fig. 8.5 2D MCP spectrum of the various charge states observed for 200 keV O^+ on CH_4 molecule. See text for details

which entered the cell and contributed to the emerging charge states can be identified by their vertical displacements. Hence, in Fig. 8.5 where an O^+ beam was used, the uppermost row shows capture and loss from neutral oxygen which was formed prior to the collision cell via interactions with background gases. The lowermost row shows single and double capture from O^{2+} which was also formed in the beamline. Only the middle row shows interactions involving O^+ ions. However, all of the data can be used to extract capture and loss cross sections for O, O^+, and O^{2+} impact. Electron loss cross sections were obtained using integrated intensities for each charge state as a function of target pressure, i.e., the growth curve method which will be described in the following section. Pressures inside the gas cell were always low enough to ensure single-collision conditions.

Absolute detection efficiencies for the recoil ions, ϵ^{q+}, were determined by simultaneous single and coincidence measurements of single-electron-capture cross sections of C^{3+} on noble gases, using various projectile energies. In the coincidence measurements, the coincidence signal for target ionization associated with single capture by C^{3+}, I_{32}^{q+}, where q is the final charge state of the target, is proportional to the beam intensity I_o, the number of target scattering centers per volume $n(P)$, where P is the target pressure, the effective target length l_{eff}, the cross section, σ_{32}^{q+}, and the recoil ion detection efficiency ϵ^{q+}. Here, a surface barrier detector which is assumed to have unit efficiency is used to detect the postcollision C^{2+} projectile. For the recoil ions, to ensure high efficiencies and efficiencies virtually independent of the impacting charge state or mass, the MCP front plate was biased at a high negative voltage. Thus, we have

$$I_{32}^{q+} = \varepsilon^{q+} l_{\text{eff}} I_o n(P) \sigma_{32}^{q+}. \tag{8.6}$$

In the singles measurement, the absolute single capture cross section, σ_{32}, is determined by measuring the fraction of the main beam that performs single-electron capture. These cross sections are related by

$$\sigma_{32} = \sum_q \sigma_{32}^{q+}. \tag{8.7}$$

Ratios of multiple-to-single ionization cross sections ($\sigma_{ij}^{q+}/\sigma_{ij}^{+}$) were measured for different projectiles, different channels, different projectile energies, and various targets. These were compared with values obtained using cross sections found in the literature. It was found that the detection efficiencies were essentially independent of the recoil-ion charge state. Thus, $\epsilon^{q+} \cong \epsilon$, and can be removed from the summation. Hence,

$$\varepsilon = \frac{1}{l_{\text{eff}} n(P) I_o \sigma_{32}} \sum_q I_{32}^{q+}. \tag{8.8}$$

Using this equation for electron capture from various targets, it was found that ϵ did not depend of the target atomic number. Therefore, absolute cross sections were obtained by modifying (8.8) to apply for electron loss rather than capture and also substituting ϵ for ϵ^{q+} in the equation.

8.3.3 UFRJ Results

As discussed earlier, first-order theories such as the Born theory predict a $Z_T^2 + Z_T$ target dependence with the first term designated as the screening mode and the second term as the antiscreening mode. Thus, as Z_T increases, the expected overall

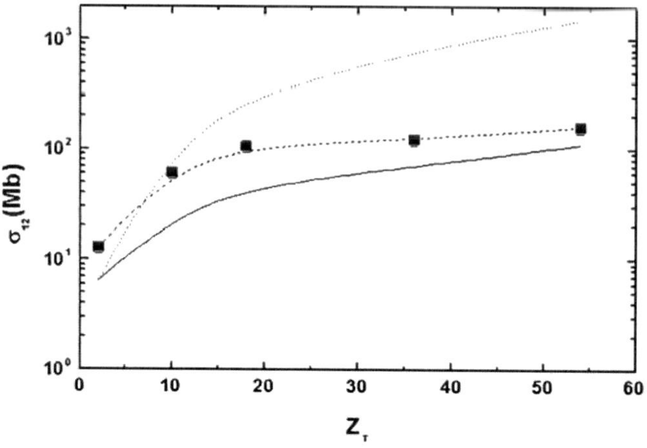

Fig. 8.6 Cross sections for total projectile electron loss as a function of the target atomic number for 2.5 MeV He^+ on noble gases. Experiment: *squares* (*dashed line* drawn for guiding the eyes); *solid curve*, antiscreening; *dotted curve*, screening [11]

dependence is expected to be dominated by the screening mode. However, Fig. 8.6 shows the measurements of Sant'Anna et al. [11] for 2.5 MeV He^+ impinging upon noble gases. One can see a near saturation of the cross sections in the region $18 < Z_2 < 54$. This points toward a much smaller contribution from the screening mode than expected which can possibly be explained by a saturation effect manifested only in the screening channel. A similar saturation was also observed for 1 MeV He^+ impact [12].

Also shown are first-order calculations for the screening and antiscreening channels using the extended-sum-rule method of Montenegro and Meyerhof (W. Erb, GSI Report No. GSI-P-7-78, 1978, unpublished). First-order calculations for the contribution due to the antiscreening channel show a qualitative agreement with the experimental Z_T dependence. In contrast, the calculated dependence for the screening mode is much steeper and severely overestimates the cross sections for the heavier targets. This supports the idea that a saturation effect must exist in the screening channel. To observe such a saturation, a proper theoretical approach using nonperturbative calculations is required.

As a side note, measured cross sections for electron loss to the continuum (ELC) [12] were shown to be in extremely good agreement with the theoretical excitation values calculated by Kaneko [28]. This implies that it might be possible to use electron loss to the continuum measurements to obtain information or test theoretical predictions for excitation processes since, as noted by Sigaud et al. [12], for the ELC process, the ionized electron has near zero velocity in the projectile frame. This means that ELC ionization is quite similar to electron excitation to high n levels.

The target saturation effect was investigated further using C^{3+} and O^{5+} projectiles with velocities ranging from less than 1 to a few atomic units. Figure 8.7 shows results for single electron loss by 3.0 MeV C^{3+} colliding with hydrogen [37]

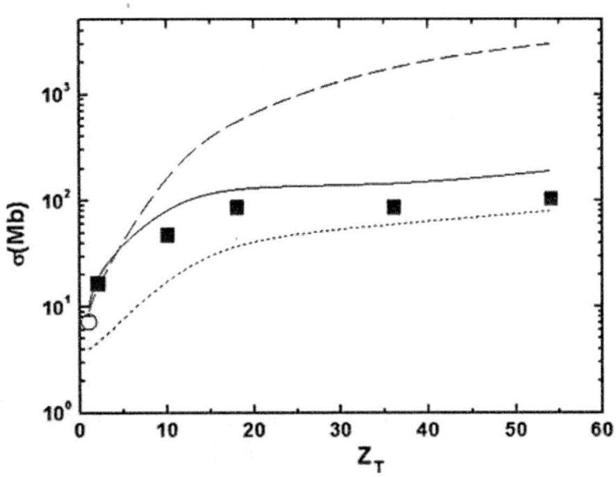

Fig. 8.7 Single-electron-loss cross section of 3.0 MeV C^{3+} on hydrogen and noble gases as a function of target atomic number. Experiment: *squares* [13] and *circles* [37]. Theory: *dashed line*, sum of screening (PWBA) and antiscreening [13]; *solid line*, sum of screening [13] and antiscreening [39].The antiscreening calculations are shown by the *dotted line*

and the noble gases [13]. Again, the experimental data show a saturation effect. These are compared to two different theoretical approaches [13,38]. The behavior of the screening contribution with increasing target atomic number when the extended free-collision model is used is the same as those of the experimental data and of the first-order calculations for the antiscreening (inelastic mode), presenting the same saturation with increasing target atomic number. But, there still exist quite large quantitative discrepancies, which increase as the projectile charge increases.

Although outside the scope of the present work, it should be mentioned that to probe electron loss processes even deeper, measurements of projectile single-electron loss accompanied by multiple target ionization have also been performed at UFRJ [14,16,39,40]. Following the procedure outlined in (8.7), Santos et al. [14] used the ratio $\sigma_{\mathrm{anti}}/\sigma_{\mathrm{e}}$ to provide an estimate of the average number of active target electrons for the antiscreening mode, $\langle \mathrm{n}_o \rangle$. Under the assumption that $\langle \mathrm{n}_o \rangle$ depends on the collision energy, it can be written as

$$\langle n_o \rangle_E = f(E) \langle n_o \rangle_a \,. \tag{8.9}$$

Here, $\langle n_o \rangle_a$ is the value of the average number of electrons at a very high impinging energy (100 MeV was used), and *f(E)* is a function of the collision energy, *E*. They obtained asymptotic values for $\langle n_o \rangle_a$, of 0.63 for He, 2.9 for Ne, 5.6 for Ar, 9.6 for Kr, and 14.1 for Xe. Then they calculated values of *f(E)* for the different targets as a function of the collision energy and observed that all targets except He demonstrated the same behavior: a fast increase just above threshold, followed by a sharp maximum—which lies between 4 and 5 MeV for all targets—and a slow decrease with increasing energy. The He target demonstrated no maximum, but simply decreased monotonically with increasing collision energy. The most notable feature is that the values of *f(E)* converge to the same curve above the maximum region. This is true for all targets, including He. The authors fitted a simple curve which could act as a "universal" function to estimate the average number of active target electrons for the inelastic mode, which we referred to as N_{Teff} in an earlier section, as a function of the collision energy for the C^{3+} projectile ion. The fitted function was

$$f(E) = 4.514 - \frac{3.831E(\mathrm{MeV})}{9.907 + E(\mathrm{MeV})}. \tag{8.10}$$

8.4 UNILAC Measurements

In the 1970s and 1980s there was much activity with respect to measuring electron loss from a variety of ions and energies. A fairly extensive listing of references can be found in [19]. However, the only laboratory where electron loss from MeV/u, low-charge-state, heavy ions could be measured was GSI-Darmstadt (now the GSI Helmholzzentrum für Schwerionenforschung), Germany. But, even there, the maximum velocity was 1.4 MeV/u. Beginning approximately a decade ago, two laboratories have worked on extending the experimental information to higher energies. At the Texas A&M University cyclotron (TAMU) in the USA, the group of Watson and various collaborators [29, 31–33, 41, 47] have measured cross sections up to 6.5 MeV/u for very heavy, low-charge-state ions such as U^{28+} and up to a few tens of MeV/u for lighter ions. These studies used techniques similar to those described above for studies at the Federal University of Rio de Janeiro. The other laboratory currently active in electron-loss studies at MeV/u energies is the GSI Helmholzzentrum für Schwerionenforschung. There, in collaboration with the Missouri University of Science and Technology, detailed cross sections for very low-charge-state ions, for example, $Ar^{1,2+}$, Xe^{3+}, and $U^{4,6,10+}$, were measured at 0.7 and 1.4 MeV/u [30, 41] while cross section and lifetime information for ions such as U^{28+} and Xe^{18+} have been obtained for energies up to many tens of MeV/u [34, 35]. The GSI studies will be described below, with particular emphasis on the experimental methods that were required.

8.4.1 UNILAC Methods

Unlike for the early GSI studies performed at the UNILAC, this region is no longer used for experimental studies. For very low-charge-state ions, downstream studies are not possible because of beam transport system magnet limitations. Therefore, the existing facilities were slightly modified in order to perform new measurements at the UNILAC. As illustrated in Fig. 8.8, a pseudostatic gas target was created at the end of the UNILAC accelerator section by inserting slits before and after the gas stripper chamber and valving off the pumps normally used to pump this chamber. These slits collimated and reduced the beam intensity plus provided differential pumping between the target and accelerator/detector regions. Target gases were

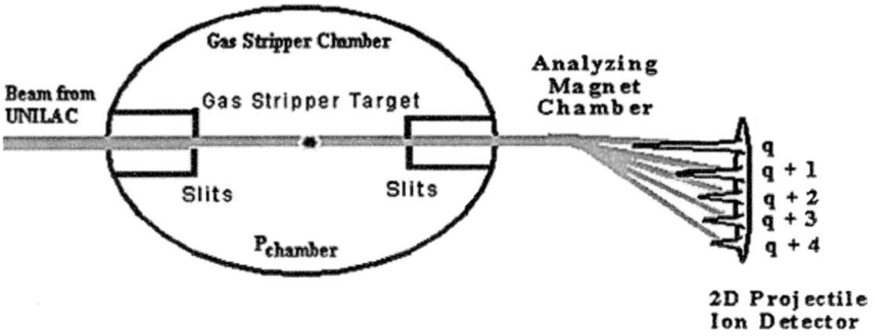

Fig. 8.8 Schematic of apparatus used at the GSI UNILAC

injected through a leak valve, and their pressures were measured at the periphery of the target chamber. Absolute target line densities along the beam path were obtained from normalization measurements using He^+ ions. Immediately following the gas stripper chamber, an analyzing magnet was used to separate the postcollision charge states and direct them onto a high-rate position-sensitive detector that could be inserted or withdrawn.

In contrast to the studies described above, at GSI, the ion sources are designed to provide intense beams that have a time structure resulting in high counting rates for brief periods. Therefore, the beam intensities had to be reduced many orders of magnitude, and a high-rate 2D detector and fast histogramming time-to-digital converter capable of handling rates of 100 kHz or more were used. Additional experimental details can be found in [30] and [41].

The experimental procedure consisted of measuring the postcollision charge state intensities as a function of target pressure, i.e., the growth curve method. Charge state spectra, as shown in Fig. 8.9, were background subtracted and integrated. From these, charge state fractions were calculated and plotted versus the target gas line density as shown in Fig 8.10.

For such, the growth curve data, coupled equations of the type

$$F_q(\pi) = \sum_{q' = q_{in} - \text{maxcap}}^{q_{in} + \text{maxloss}} \pi \; F_{q'}(\pi) \; \sigma_{q'\,q}, \tag{8.11}$$

must be solved. Here, π is the target line density, q' is the final charge state, and the charge state fractions and cross sections are designated by $F_{q'}$ and $\sigma_{q'q}$, with the sum being over all observed charge states, e.g., from the maximum capture channel to the maximum loss channel. Isolating single from multiple collision processes that lead to the same final charge state requires, in principle, solving a matrix of coupled equations, each at a different value of π. However, in practice, if the line density is small enough, multiple collision processes can be ignored. This uncouples the equations, and the cross sections can be determined using the linear portion of the growth curves shown in Fig. 8.10.

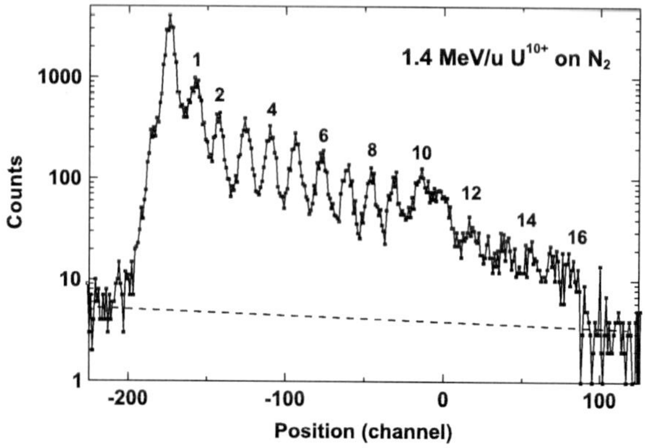

Fig. 8.9 Charge state spectrum for stripping of 1.4 MeV/u U^{10+} ions by molecular nitrogen. The numbers indicate how many electrons are removed from uranium. The *dashed line* indicates the background that was subtracted in order to integrate the peak intensities

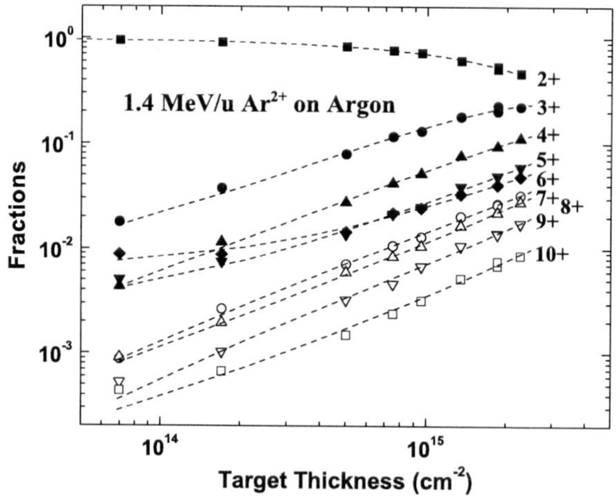

Fig. 8.10 Charge state fractions versus target density measured for electron loss by 1.4 MeV/u Ar^{2+} colliding with Ar

8.4.2 UNILAC Findings

Using the methods outlined above, absolute cross sections were measured for total, single, and multiple electron loss between 0.74 and 1.4 MeV/u for several very low-charge-state ions, e.g., Ar^{+}, Ar^{2+}, Xe^{3+}, and $U^{4,6,10+}$, colliding with neon, molecular nitrogen, and argon targets [28, 47]. For these data, the energy and target Z_T ranges were inadequate for testing the various scalings discussed above. However, the projectile charge dependence on the electron-loss cross sections could be tested. As shown in Fig. 8.11 where the cross sections have been scaled by the effective number of outer-shell electrons, for low-charge-state ions, the scaled cross sections decrease very slowly with increasing charge state and do not depend on the nuclear charge. In contrast, for "highly" charged ions, the cross sections decrease very rapidly with the slope and onset depending on the nuclear charge. The changeover between these two regions appears to be slightly above charge state 10.

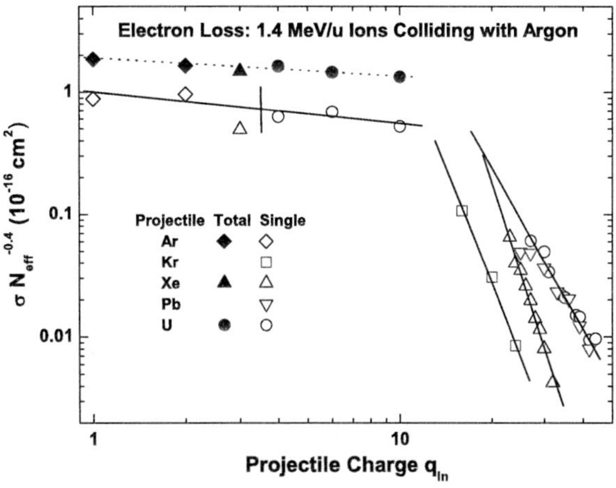

Fig. 8.11 Scaled cross sections for total and single-electron loss by 1.4 MeV/u ions colliding with Ar. Data are from [30, 36, 41] (W. Erb, B. Franzke, 1978, GSI Report No. GSI-J-1-78, unpublished) . The *lines* are to guide the eye. The *vertical line* shows experimental uncertainties for the low-charge-state data

8.5 GSI Storage-Ring Experiments

As shown in a previous section, one of the important discrepancies between the various theories, as well as between theory and available experimental data, is with regard to the energy dependencies at high energies. To extend the electron-loss experimental information to higher velocities, experiments using the GSI storage rings were performed. For these studies, beam lifetimes were measured within the GSI heavy-ion synchrotron (SIS) storage ring and the experimental storage ring (ESR). These provided information about electron loss resulting from collisions with the residual gases. In addition, using the internal gas target at the ESR, lifetimes and total cross sections were measured for H_2 and N_2 targets. Here, only total cross sections could be measured because existing beam upcharge (beam ionization) detector positions plus magnetic field strengths prohibited measuring the individual charge states that were produced.

8.5.1 *Storage-Ring Methods*

For the ESR studies, Xe^{18+} and U^{28+}ion beams were accelerated to approximately 10 MeV/u and injected into the SIS which was used either as a transport device or to provide additional acceleration to higher energies before injecting the ions into the ESR. Bending magnets limited the maximum energies for U^{28+} and Xe^{18+} ions in the ESR to 50 and 70 MeV/u respectively. A minimum energy of 20 MeV/u was determined by the decreasing beam storage lifetimes in the ESR, i.e., the increasing loss cross sections with decreasing energy.

At high energies and for low-charge-state ions, the lifetime τ and the related decay rate λ of the stored ion beams are related to the electron-loss (stripping) cross section by

$$\tau(\mathrm{v}) = \frac{1}{\lambda} = \frac{1}{\rho \mathrm{v} \sigma(\mathrm{v})}\mathrm{f}. \tag{8.12}$$

Here, $\sigma(v)$ corresponds to the weighted total cross sections for electron loss induced by collisions with whatever gases are present, v is the projectile velocity, ρ is the average density of the gas through which the beam travels, and f is the fractional time spent traveling through the gas, e.g., for interactions with background gases in the ring, $f=1$; for interactions with a target gas injected over a certain length, f is the target length, a few mm, divided by the circumference of the storage ring, 108 m for the ESR. Lifetimes were measured for interactions with the residual gases comprising the base vacuum ring as well as with H_2 and N_2 targets injected using supersonic gas jets. The average base pressure of the ESR was 2×10^{-11} mbar with the major components being H_2 (83%), H_2O (10%), CH_4 (5%), CO_2 (1%), and Ar (1%). The H_2 and N_2 targets had densities on the order of 10^{12} particles/cm^3 and jet diameters of approximately 5 mm. Beam lifetimes were measured as a function of time using current transformers. Additional details can be found in [34, 35].

An example of the beam lifetime data is shown in Fig. 8.12. The flat region at times earlier than 9 s corresponds to the slow decay time associated with interactions with the background gases in the ESR. For this dataset, at approximately 9 s, a fast valve is opened admitting the supersonic gas jet. At this time, a rapid decay begins. Note that for an H_2 target, the beam decays away within a few seconds, whereas for an N_2 target, the time is only a couple hundred milliseconds. Also note that these data have been background subtracted as this is essential for establishing the correct decay rates.

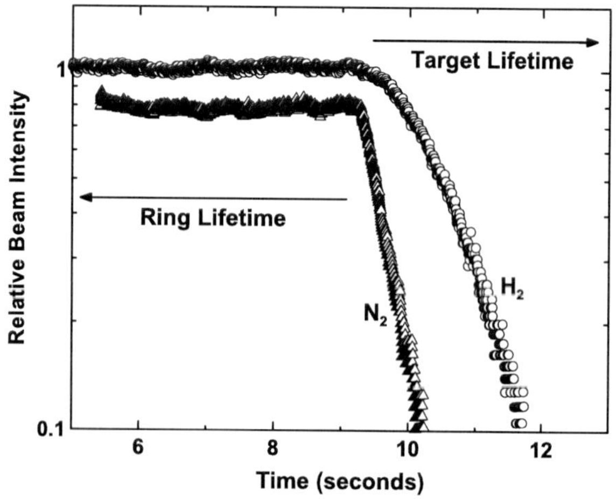

Fig. 8.12 Stored beam currents measured in the ESR. See text for details

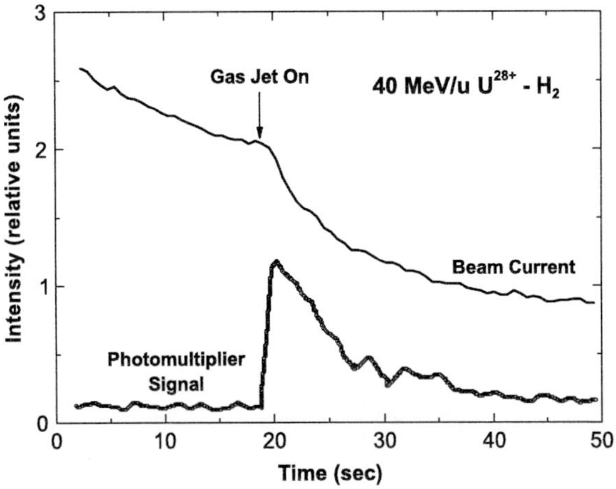

Fig. 8.13 *Upper curve*; stored beam current measured using current transformers in the ESR. *Lower curve*; photon intensity measured by a photomultiplier viewing the beam-gas jet overlap. From [42]

To illustrate the importance of subtracting the correct background plus to show the limitations associated with measuring the small beam currents using the current transformers, the upper curve in Fig. 8.13 shows the raw (not background subtracted) beam current as a function of time. An optional method for measuring the decay times for the target that was tested, but not used, is shown by the lower curve. Here, a photomultiplier recorded light produced in the region of beam-target overlap. As seen, when the gas is injected, the number of emitted photons increases sharply and then exponentially decays away. The advantage of this method is that, in principle, the photomultiplier can be made to have zero background, and therefore the decay process can be followed much longer and the lifetimes determined more accurately.

In the SIS, only lifetime measurements could be performed, and the vacuum is poorer, e.g., the base pressure is approximately 7.7×10^{-11} mbar; the residual gas constituents are H_2(65%), H_2O(17%), CO/N_2(8%), CO_2(1%), Ar, and Cl (each 4%). Also, the ring circumference is larger, 216 m.

8.5.2 *Storage-Ring Findings*

Cross sections for electron loss by U^{28+}, measured at the GSI UNILAC [34, 35], TAMU [32, 33], and those obtained using the measured ESR lifetimes and target densities determined via calibrations of the supersonic gas jet targets [34, 35] are shown in Fig. 8.14. These are compared with Born and nCTMC [32] calculations. As seen, for stripping by a light target such as H_2, both the Born and the nCTMC models predict an E^{-1} dependence at high energies, which is in agreement with the data. For stripping by a many-electron target, the data imply an energy dependence somewhere between the Born E^{-1} dependence and the nCTMC v^{-1} dependence. But, the experimental slope is highly dependent upon the highest energy measurement. Clearly, more data are needed.

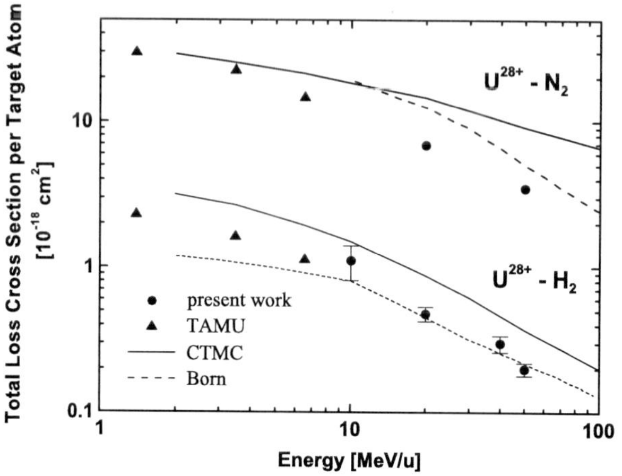

Fig. 8.14 Total loss cross sections for U^{28+}–H_2 and U^{28+}–N_2 collisions. Experimental data from [32–36](W. Erb, B. Franzke, 1978, GSI Report No. GSI-J-1-78, unpublished), CTMC theory, *solid lines*, from [32], Born theory, *dashed*, and *dotted lines*, from Shevelko (private communication)

8.6 Measured Target Scalings

8.6.1 Atomic Targets

To study target-scaling properties, various groups have measured electron-loss cross sections for atomic targets ranging from helium to xenon. These studies have shown that for lighter targets, typically meaning for $Z_T < 10$, the cross sections increase rapidly, e.g., approximately as Z_T^n, where $n \sim 1.4$–1.7. But, for heavier targets, the cross sections increase more slowly and appear to saturate in some cases. In a recent publication, DuBois et al. [43] compared the target dependencies that have been reported. As shown in Fig. 8.15, these data cover an extremely broad range of velocities, e.g., over six orders of magnitude with the highest velocities being approximately a quarter of the speed of light. They also cover a wide range of ion species, e.g., singly charged ions ranging from He^+ to Kr^+, one-electron ions such as He^+ and N^{6+}, low, and medium-charge-state, heavy ions such as $Ar^{6,8+}$, Fe^{4+}, Xe^{18+}, $U^{4,6,10+}$, and Pb^{54+}. In Fig. 8.15, for comparison purposes, the target dependences have been normalized to unity at $Z_T=10$.

In spite of the vast differences in velocities and ion species, these normalized data have very similar target dependences. In particular, note that although there is a slight increase in the overall slope with increasing velocity, data for velocities less than 19 MeV/u all agree in shape within approximately a factor of two. This is in contrast to the much steeper dependencies observed for the two highest velocities where, again, an increase in slope is seen with increasing velocity.

Because the Z_T dependencies for ions with energies less than 19 MeV/u were found to be very similar, average values were calculated for each target. In Fig. 8.16,

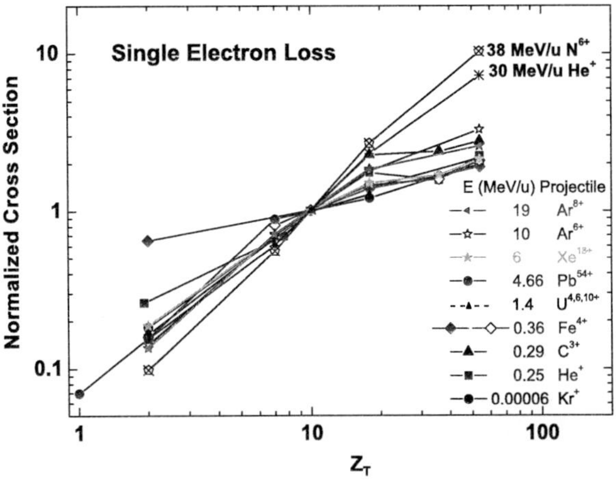

Fig. 8.15 Normalized single electron-loss cross sections measured for various ions and ion energies. The experimental data are from the following sources: Kr^{+}, [45]; 0.25 MeV/u He^{+}, [12]; C^{3+}, [14]; Fe^{4+}, [23,46]; $U^{4,6,10+}$, [42]; Pb^{54+}, [48]; Xe^{18+}, [32]; $Ar^{6,8+}$, 30 MeV/u He^{+}, and 38 MeV/u N^{6+}, [41,47]

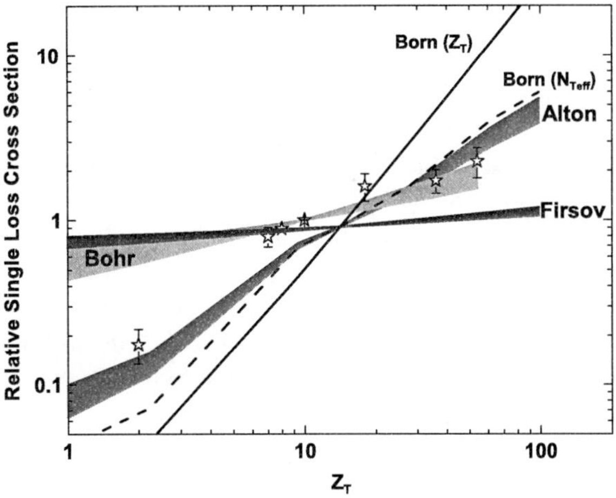

Fig. 8.16 Normalized single electron loss cross sections measured for "low- to medium-energy" ions. The experimental data (*stars*) are average values obtained for the ions listed in the lower corner of Fig. 8.15. The Born(Z_T) and the Born(N_{Teff}) curves are as described in the text. The bands shown are obtained using the Firsov, Bohr, and Alton scaling rules described in an earlier section where the limiting values of 2 and 92 have been used for Z_P

these average values are compared to various theoretical models which have been proposed for describing electron loss in heavy systems. Again, all curves are normalized to unity at $Z_T = 10$. As seen, a pure Born dependence which accounts for stripping by both the full nuclear charge and all of the bound electrons, e.g., $Z_T^2 + Z_T$, predicts a much steeper dependence than is observed. Plus, this dependence shows no change in slope between the lighter and heavier targets. Such an inflection is achieved if, instead of using the full nuclear and bound electron charges, an effective number of electrons is assumed to participate plus to partially screen the nuclear charge. In this case, the Born scaling is given by $N_{\mathrm{Teff}}^2 + N_{\mathrm{Teff}}$. Here, for the effective number of electrons, N_{Teff}, the values at an extremely high impact energy suggested by Santos et al. [14] was used. These values (1.4 for He, 5 for N, 5.8 for Ne, 7.8 for Ar, 12.4 for Kr, and 15.1 for Xe) are similar to rough estimates which are obtained by counting all but the most tightly bound inner-shell electrons for each target, e.g., 2 for He, 5 for N, 6 for Ne, 10 for Ar, and 18 for Kr and Xe. As seen, accounting for the effective number of electrons improves the Born model, but still the predicted dependence is steeper than observed.

The Bohr model [20], namely, $(Z_{\mathrm{P}}^{1/3} + Z_{\mathrm{T}}^{1/3})^2$, agrees reasonably well with the data for heavy targets but drastically overestimates the relative cross sections for light targets. Note that a band of values is shown for the Bohr, Alton, and Firsov scalings. These bands show how the projectile charge changes the results and have limiting values for Z_{P} of 2 and 92. The comparison with the scaling suggested by Alton et al. [23] shows much better agreement for lighter targets, but unfortunately poorer agreement for heavy targets. Although they have much different forms, the Alton scaling results are very similar to those obtained using the Born model when the effective number of electrons is taken into account. The final comparison is with the scaling suggested by Firsov [21, 22], e.g., $(Z_{\mathrm{P}}^{1/3} + Z_{\mathrm{T}}^{1/3})^{2/3}$. This scaling has a very flat target dependence and is in poor agreement with experiment. To summarize, none of these scalings is compatible with experiment throughout the entire range of targets studied.

Figure 8.17 shows that various "empirical" scalings could be used to provide better agreement. Here, the Born model, using the full nuclear and electronic charges, Born(Z_{T}), and effective charges, Born (N_{Teff}), are shown by the thin black and thick blue (online) solid lines. The Mod Bohr (Z_{T}), thin red (online) line, uses the Bohr target scaling of $Z_{\mathrm{T}}^{2/3}$ but incorporates the Born concept of ionization by

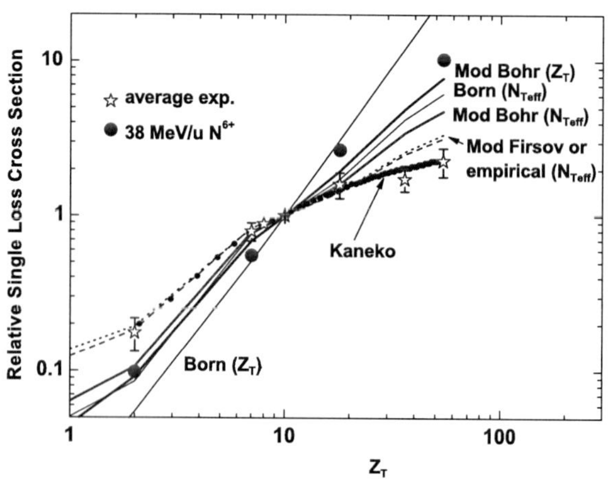

Fig. 8.17 Normalized single electron-loss cross sections compared to various empirical scalings. See text for details. Experimental data are from same references as in Fig. 8.15

both the target nuclear charge and the target electronic charge, i.e., Z_T^2 is replaced with $(Z_T^2+Z_T)^{1/3}$. Plus, it incorporates the suggestion of Alton et al. by multiplying by N_{Teff}. This yields a scaling of $N_{\mathrm{Teff}}\ (Z_T^2 + Z_T)^{1/3}$. The Mod Bohr ($N_{\mathrm{Teff}}$), thick red (online) line, is similar but replaces the full nuclear and electron charges by their screened and active counterparts, N_{Teff}. This yields $N_{\mathrm{Teff}}(N_{\mathrm{Teff}}^2+N_{\mathrm{Teff}})^{1/3}$. The Mod Firsov, dashed line, applies the same ideas to the normal Firsov scaling and yields either $N_{\mathrm{Teff}}\ (Z_T^2 + Z_T)^{1/9}$ or $N_{\mathrm{Teff}}(N_{\mathrm{Teff}}^2 + N_{\mathrm{Teff}})^{1/9}$. These were found to produce nearly identical results and to be nearly the same as an empirical scaling, empirical (N_{Teff}), dotted line, which had the form $(N_{\mathrm{Teff}}^2 + N_{\mathrm{Teff}})^{2/3}$. The final theoretical predictions shown in the figure are the quantitative values of Kaneko [28], small black dots, for stripping of 0.9 MeV/u He^+.

These models are compared to the average data for energies less than 19 MeV/u which represent electron loss from "slow- to medium-velocity multi-charged heavy ions," open stars. Also shown are electron loss data for 38 MeV/u N^{6+} which are very similar to data for 30 MeV/u He^+, filled circles. These represent electron loss by " fast" singly charged ions. As shown previously where data for each individual energy and specie were plotted, the target dependence becomes steeper with increasing velocity. Here, we see that the target dependence for electron loss from "fast" ions is predicted quite well by the Mod Bohr (Z_T) scaling, $N_{\mathrm{Teff}}\ (Z_T^2+Z_T)^{1/3}$, whereas the target dependence for stripping of "slow- to intermediate-velocity" ions can be predicted quite well using either the Mod Firsov scaling of $N_{\mathrm{Teff}}\ (Z_T^2 + Z_T)^{1/9}$ or an empirical scaling of the form $(N_{\mathrm{Teff}}^2 + N_{\mathrm{Teff}})^{2/3}$. An extremely interesting feature is that the target dependence predicted by Kaneko for He^+ seems to apply for a very broad range of ions and velocities. However, his theory predicts that the target dependence becomes flatter (for heavy targets) with increasing velocity whereas the available data indicates just the opposite.

8.7 Concluding Remarks

We have attempted to show what is known to date about electron loss, in particular how the cross sections have been predicted to scale with respect to the target nuclear and electronic charges. As stated in the introduction, the examples provided were selected in order to illustrate specific features. In doing so, we may not have cited, or may have overlooked, important references. However, for those interested in the subject, the present examples and references can be used as a starting point for further investigations.

Acknowledgements R.D.D. gratefully acknowledges the support received from the National Science Foundation, the Fulbright Commission, and the ExtreMe Matter Institute EMMI during the preparation of this manuscript. A.C.F.S. is grateful for the support from CNPq and FAPERJ for performing experimental studies described in this work plus during the preparation of the manuscript.

References

1. C. Omet, P. Spiller, J. Stadlmann, D.H.H. Hoffmann, New J. Phys. **8**, 284 (2006)
2. V. Coco, J. Chamings, A.M. Lombardi, E. Sargsyan, R. Scrivens, Proceedings of LINAC 2004 Conference, Lübeck, Germany, TUP27, p 351, available online: accelconf.web.cern.ch/accelconf/l04/PAPERS/TUP27.PDF
3. A. G. Ruggiero, Accelerator Physics Technical Note No. 42, Brookhaven National Laboratory, 1992, available online: www.rhichome.bnl.gov/RHIC/RAP/rhic_notes/AD-AP-1.../AD-AP-42.pdf
4. FAIR Baseline Technical Report (2006), available online: http://www.gsi.de/fair/reports/btr.html.
5. R.O. Bangerter, Nucl. Instrum. Meth. A **464**, 17 (2001)
6. D.R. Bates, G. Griffing, Proc. Phys. Soc. **A68**, 90 (1955)
7. R.D. DuBois, S.T. Manson, Phys. Rev. A **42**, 1222 (1990)
8. W.E. Meyerhof, H.-P. Hüllskötter, Q. Dai, J.H. McGuire, Y.D. Wang, Phys. Rev. A **43**, 5907 (1991)
9. E.C. Montenegro, W.E. Meyerhof, Phys. Rev. A **44**, 7229 (1991)
10. E.C. Montenegro, W.E. Meyerhof, Phys. Rev. A **46**, 5506 (1992)
11. M.M. Sant'Anna, W.S. Melo, A.C.F. Santos, G.M. Sigaud, E.C. Montenegro, Nucl. Instrum. Meth. B **99**, 46–49 (1995)
12. G.M. Sigaud, F.S. Jorás, A.C.F. Santos, E.C. Montenegro, M.M. Sant'Anna, W.S. Melo, Nucl. Instrum. Meth. B **132**, 312–315 (1997)
13. W.S. Melo, M.M. Sant'Anna, A.C.F. Santos, G.M. Sigaud, E.C. Montenegro, Phys. Rev. A **60**, 1124 (1999)
14. A.C.F. Santos, G.M. Sigaud, W.S. Melo, M.M. Sant'Anna, E.C. Montenegro, Phys. Rev. A **82**, 012704 (2010)
15. A.B. Voitkiv, G.M. Sigaud, E.C. Montenegro, Phys. Rev. A **59**, 2794 (1999)
16. E.C. Montenegro, A.C.F. Santos, W.S. Melo, M.M. Sant'Anna, G.M. Sigaud, Phys. Rev. Lett. **88**, 013201-1 (2002)
17. A. Voitkiv, J. Ullrich, *Collsions of Structured Atomic Particles* (Springer, Berlin, 2008)
18. A.B. Voitkiv, Phys. Rep. **392**, 191 (2004)
19. A.C.F. Santos, R.D. DuBois, Phys. Rev. A **69**, 042709 (2004)
20. N. Bohr, K. Dan. Vidensk. Selsk. Mat. Fys. Medd. **18**(8), (1948)
21. O.B. Firsov, Zh. Eksp. Teor. Fiz. **34**, 447 (1958)
22. O.B. Firsov, JETP **7**, 308 (1958)
23. G.D. Alton, L.B. Bridwell, M. Lucas, C.D. Moak, P.D. Miller, C.M. Jones, Q.C. Kessel, A.A. Antar, M.D. Brown, Phys. Rev. A **23**, 1073 (1981)
24. V.P. Shevelko, Th. Stöhlker, I.Yu. Tolstikhina, Nucl. Instrum. Meth. B **184**, 295 (2001)
25. I.L. Beigman, I.Yu. Tolstikhina, V.P. Shevelko, Tech. Phys. **53**, 546 (2008)
26. V.P. Shevelko, I.L. Beigman, M.S. Litsarev, H. Tawara, I. Yu. Tolstikhina, G. Weber, Nucl. Instrum. Meth. B **269**, 1455 (2011)
27. V.P. Shevelko, D. Kato, M.S. Litsarev, H. Tawara, J. Phys. B **43**, 215202 (2010)
28. V.I. Matveev, D.U. Matrasulov, S.V. Ryabchenko, J. Exp. Theor. Phys. **102**, 1–8 (2006), orig in Zhur. Eksp. noi i Teor. Fiz. **129**, 5–13 (2006)
29. T. Kaneko, Phys. Rev. A **32**, 2175–85 (1985)
30. R.E. Olson, R.L. Watson, V. Horvat, K.E. Zaharakis, J. Phys. B: At. Mol. Opt. Phys. **35**, 1893 (2002)
31. R.D. DuBois, A.C.F. Santos, R.E. Olson, Th. Stöhlker, F. Bosch, A. Bräuning-Demian, A. Gumberidze, S. Hagmann, C. Kozhuharov, R. Mann, A. Oršiæ Muthig, U. Spillmann, S. Tachenov, W. Barth, L. Dahl, B. Franzke, J. Glatz, L. Gröning, S. Richter, D. Wilms, A. Krämer, K. Ullmann, O. Jagutzki, Phys. Rev. A **68**, 042701 (2003)
32. R.L. Watson, Y. Peng, V. Horvat, G.J. Kim, R.E. Olson, Phys. Rev. A **67**, 022706 (2003)

33. R.E. Olson, R.L. Watson, V. Horvat, A.N. Perumal, Y. Peng, Th. Stöhlker, J. Phys. B At. Mol. Opt. Phys. **37**, 4539 (2004)
34. R.E. Olson, R.L. Watson, V. Horvat, K.E. Zaharakis, R.D. DuBois, Th. Stöhlker, Nucl. Instrum. Meth. A **544**, 333–36 (2005)
35. R.D. DuBois, O. de Lucio, M. Thomason, G. Weber, Th. Stöhlker, K. Beckert, P. Beller, F. Bosch, C. Brandau, A. Gumberidze, S. Hagmann, C. Kozhuharov, F. Nolden, R. Reuschl, J. Rzadkjewicz, P. Spiller, U. Spillmann, M. Steck, S. Trotsenko, Nucl. Instrum. Meth. B **261**, 230–233(2007)
36. G. Weber, C. Omet, R.D. DuBois, O. de Lucio, Th. Stöhlker, C. Brandau, A. Gumberidze, S. Hagmann, S. Hess, C. Kozhuharov, R. Reuschl, P. Spiller, U. Spillmann, M. Steck, M. Thomason, S. Trotsenko, Phys. Rev. ST Accel. Beams **12**, 084201 (2009)
37. E.C. Montenegro, W.E. Meyerhof, Phys. Rev. A **43**, 2289 (1991)
38. M.M. Sant' Anna, W.S. Melo, A.C.F. Santos, G.M. Sigaud, E.C. Montenegro, M.B. Shah, W.E. Meyerhof, Phys. Rev. A **58**, 1204 (1998)
39. E.C. Montenegro, W.E. Meyerhof, Phys. Rev. A **43**, 2289 (1991)
40. T. Kirchner, A.C.F. Santos, H. Luna, M.M. Sant'Anna, W.S. Melo, G.M. Sigaud, E.C. Montenegro, Phys. Rev. A, **72**, 012707 (2005)
41. D. Müller, L. Grisham, I. Kaganovich, R.L. Watson, V. Horvat, K.E. Zaharakis, Y. Peng, Princeton Plasma Physics Laboratory Document PPPL-3713 (2002)
42. R.D. DuBois, A.C.F. Santos, Th. Stöhlker, F. Bosch, A. Bräuning-Demian, A. Gumberidze, S. Hagmann, C. Kozhuharov, R. Mann, A. Oršiæ Muthig, U. Spillmann, S. Tachenov, W. Bart, L. Dahl, B. Franzke, J. Glatz, L. Gröning, S. Richter, D. Wilms, K. Ullmann, O. Jagutzki, Phys. Rev. A **70**, 032712 (2004)
43. G. Weber, Diplomarbeit, Ruprecht-Karls Universität Heidelberg (2006)
44. R.D. DuBois, A.C.F. Santos, G. Sigaud, E.C. Montenegro, Phys. Rev. A **84**, 022702 (2011)
45. H. Martínez, A. Amaya-Tapia and J.M. Hernández, J. Phys. B At. Mol. Opt. Phys. **33**, 1935–1942 (2000)
46. H. Knudsen, C.D. Moak, C.M. Jones, P.D. Miller, R.O. Sayer, G.D. Alton, L.B. Bridwell, Phys. Rev. A **19**, 1029 (1979)
47. D. Müller, L. Grisham, I. Kaganovich, R.L. Watson, V. Horvat, K.E. Zaharakis, M.S. Armel, Phys. Plasmas **8**, 1753–1756 (2001)
48. W.G. Graham, K.H. Berkner, R.V. Pyle, A.S. Schlachter, J.W. Stearns, J.A. Tanis, Phys. Rev. A **30**, 722 (1984)

Part III
Atomic X-Ray Physics for Laboratory and Astrophysical Plasmas

Chapter 10
On Spectroscopic Diagnostics of Hot Optically Thin Plasmas

A.M. Urnov, F. Goryaev, and S. Oparin

Abstract X-ray and extreme ultraviolet (XUV) emission spectra of highly charged ions in hot plasmas contain diverse information on both elementary processes and the ambient medium. The theoretical analysis of spectra and spectral images of laboratory and astrophysical sources of short-wave radiation, based on the modern methods of atomic data calculations of spectral and collisional ion characteristics, allows one to determine various physical parameters of the emitting plasma. Here, we consider and discuss some basic principles on which the spectroscopic diagnostics of hot optically thin plasmas emitting XUV spectra is based. In order to obtain information about the internal structure of a physical system under study, one generally needs to solve inverse problems for determining the physical conditions in the plasma. Using concepts from the probability theory, we formulate the spectral inverse problem in the framework of the probabilistic approach to be used for the temperature diagnostics of hot plasma structures. We then demonstrate applications of our diagnostics methods to hot plasmas in laboratory (tokamak plasma) and astrophysical (solar corona) conditions.

10.1 Introduction

Spectroscopic methods of investigation are of great importance for numerous problems related to hot astrophysical and laboratory plasmas. Spectroscopic diagnostics of hot plasmas is a very effective and in many cases unique (for instance, for astrophysical plasma) way for deriving information on structure and dynamics of plasma sources. The analysis of X-ray and extreme ultraviolet (XUV) emission spectra based on calculations of spectral and collisional characteristics of highly charged ions allows us to determine various plasma parameters: electron temperature and density, element abundances, ionization state, temperature structure of hot plasma sources, etc. Plasma characteristics derived from XUV emission are needed to constrain the classes of relevant plasma models and to enable

V. Shevelko and H. Tawara (eds.), *Atomic Processes in Basic and Applied Physics*,
Springer Series on Atomic, Optical, and Plasma Physics 68,
DOI 10.1007/978-3-642-25569-4_10,

quantitative simulations of plasma processes—spatial and temporal dynamics of plasma parameters in the emitting regions.

Among astrophysical objects to be actually widely exploited, one can mark X-ray binary sources, interplanetary hot gases, coronae of the Sun, and other stars. The solar corona due to low electron densities and large range of temperatures is an important source of information on spectra and excitation processes of highly charged ions. The solar atmosphere is also of great interest due to the complex structure (active regions, coronal holes, bright points, coronal condensations) and its activity behavior (flares, coronal mass ejections, explosive protuberances, jets, and others). The active phenomena are prominent manifestations of nonstationary processes leading to the transformation of magnetic energy to its other forms; however, till now, their nature is not completely understood. Understanding mechanisms of these local processes is also important for solving fundamental problems of the physics of the solar atmosphere such as the coronal-heating problem and acceleration of the solar wind.

A number of conditions characterizing the astrophysical plasma can be reproduced in laboratory devices. This "laboratory astrophysics" can be used for the purpose of studying properties of short-wave emission of highly charged ions, in particular the verification (estimation of accuracy) of atomic data and methods of spectroscopic diagnostics. The precision of spectroscopic methods of plasma diagnostics and even possibility of their use depend on both the accuracy of atomic data and adopted models of emitting plasma based on the equations of atomic kinetics and plasma dynamics. The spectra of low-density (coronal) plasma from electron beam plasma devices and tokamaks are important sources of information about both binary atomic and hydrodynamic processes. The topicality of problems of the XUV spectroscopy is also defined by numerous applications in the atomic spectroscopy, requirements for diagnostics of emitting plasma objects, and necessity of developing short-wave emission sources for applying to the biology, medicine, materials technology, and in other domains of the modern science and technology.

In order to solve the main problem of the spectroscopy, that is, to identify and to interpret line spectra of the emitting plasma, a lot of atomic characteristics are needed, as well as information on the plasma sources under study. On the other hand, when having reliably identified spectra, one should build up models of the emitting plasmas and, using them, determine plasma macroparameters—spatial distributions of temperature, density, ionic composition, and other characteristics as well as their temporal dynamics, that is, to solve another problem of the spectroscopy—plasma diagnostics. The latter task is closely dealt with the necessity to formulate and to solve inverse problems, which frequently arise in practical applications to interpret the results of experiments and observations and to obtain information about the internal structure of a physical system under study. It is however worth to note that all spectroscopic methods based on the solution of the spectral inverse problem require definite model assumptions. These assumptions made explicitly or implicitly lead to a formulation of the appropriate model for the emitting plasma. Thus, the results of diagnostics of plasma parameters depend on the adopted model. In order to avoid the ambiguity, the chosen fitting parameters should be consistent with the physical model and all available information about the radiation source.

This review is devoted to some spectroscopic problems related to diagnostics of hot plasma sources using their XUV emission spectra. Here, we only consider the case of optically thin sources and low density (in particular coronal) plasmas. The latter means that the emission fluxes are formed by the damping due to the binary atomic processes and, as consequence, depend linearly on both electron and ionic densities. Low-density conditions are realized in many astrophysical objects; these also are often applicable to the interpretation of line spectra observed in laboratory devices, for example, tokamaks, laser plasma, pinches, etc. Firstly, we discuss basic principles on which the spectroscopic diagnostics of hot plasmas is based and how these lead to the formulation of the inverse problems for the distribution of radiating plasma material. Then we briefly describe two mathematical formalizations of the spectral inverse problem and formulate our inversion method developed in the frame of the probabilistic approach. Finally, we demonstrate some applications of developed diagnostics methods to the interpretation of spectral data from tokamak and solar plasmas.

10.2 Basic Principles of Spectroscopic Diagnostics of Hot Optically Thin Plasmas

The spectroscopic diagnostics is based on the sensitivity of the distribution of the emission spectrum over the photon energies to the physical conditions in plasma. In order to extract the information about plasma macroparameters from the line or/and continuum spectra, one should generally solve the spectral inverse problem. This problem, however, could not be solved or even formulated for an arbitrary case without an additional knowledge concerning the state of the plasma source. The shape of the spectra depends on the properties of the emitting plasma, which should be stipulated with the help of complementary experimental and theoretical analysis of the source properties. Resulting assumptions made explicitly or implicitly make it possible to express the spectroscopic characteristics of XUV emission in close analytical form providing mathematical formulation of the inverse problem. These assumptions make up a basis for a physical model of the emitting plasma. The results of spectroscopic diagnostics depend on the accepted model and therefore the same parameters of plasma could be different for various models.

The basic model assumptions used for diagnostics purposes usually include some opposite plasma conditions, for example, steady state or transient plasma, thermal or nonthermal conditions, or optically thin- or thick-emitting sources. In order to provide a comparative analysis of diverse models, one should in fact consider and analyze the dependence of spectroscopic characteristics of emission on these assumptions.

The steady-state conditions in plasma imply that characteristic times of relaxation τ_e, τ_i, and τ_z (for electrons, ions, and ionization equilibrium, respectively) are much less than the observational time of spectra τ. Furthermore, it is also assumed that the distribution functions of electrons and ions, as well as the distribution of ion species N_z do not depend on time. The opposite case of non-steady-state conditions ($\tau \ll \tau_e, \tau_i, \tau_z$) is a subject of a special study and is out of the scope of our overview. At the intermediate situation, when the plasma is in a transient state, the conditions $\tau_e \ll \tau < \tau_i, \tau_z$ are assumed to be fulfilled.

Thermal plasma condition strictly speaking implies the presence of Maxwellian velocity distribution for all sorts of plasma particles with the same temperature T. However, in applications, one often considers the quasi-steady state plasma characterized by Maxwellian distribution functions with different electron and ion temperatures, $T_e \neq T_i$. The ionization equilibrium for these plasmas is usually described by means of the parameter T_z defined as the temperature corresponding to the observed ion densities. Then the condition $T_z = T_e$ indicates the plasma in steady state, $T_z < T_e$ corresponds to the ionizing plasma, and $T_z > T_e$ to the recombining one.

Further in this paragraph, we will define some important spectral characteristics and notions, which are widely used in applications to diagnostics of hot optically thin plasmas.

10.2.1 Intensities of Spectral Lines

The total line intensity (or flux) emitted by an optically thin plasma source of the volume V in the transition $i \rightarrow f$ and observed at a distance R can be expressed as

$$I_{if} = \frac{1}{4\pi R^2} \int_V \varepsilon_{if}(\mathbf{r})\, \mathrm{d}V . \quad (\mathrm{phot\,cm^{-2}\,s^{-1}}). \tag{10.1}$$

Here, the total volume emissivity function $\varepsilon_{if}(\mathbf{r})$ (phot cm^{-3} s^{-1}) is given by

$$\varepsilon_{if} = N_i A_{if}, \tag{10.2}$$

where N_i (cm^{-3}) is the population density for the upper level (i) of the emitting ion, and A_{if} (s^{-1}) is the radiative spontaneous probability for the transition $i \rightarrow f$. The number density N_i is generally determined by solving a system of kinetic equations of balance:

$$\frac{\mathrm{d}N_i}{\mathrm{d}t} = \sum_{m \neq i} N_m W_{mn} - N_i \sum_{n \neq i} W_{in}, \tag{10.3}$$

where each of integer indices $n, m, i, \ldots$ is used for the enumeration of a state of the ion with charge z and a set of quantum numbers $\{\alpha\}$ characterizing this state. The matrix element W_{mn} denotes the total probability rate coefficient for the transition $m \rightarrow n$ and is a sum of all radiative and collisional rate coefficients contributing to this transition.

Numerical solution of the system of (10.3) can be obtained in the framework of an accepted atomic collisional–radiative model, which implies the specification of elementary processes relevant to particular plasma source under consideration. Through the rate coefficients W_{mn}, level populations N_i are the functions of electron temperature T_e and density N_e, and possibly of parameters of distribution velocity functions of plasma particles. At steady state (or quasi-steady state) plasma conditions, one assumes that $dN_i/dt = 0$ in (10.3). For the general case, it is necessary to add members connected with plasma movements leading to space-temporal nonequilibrium of ionic populations in the left side of (10.3).

In the real plasmas, the spectral lines are not monochromatic, and their fluxes are distributed over a wavelength range. Lines can be broadened by various mechanisms, for example, natural and Doppler broadening and the influence of the instrumentation ("apparatus function"). In order to take into account the line profile, one has to introduce the spectral density of the volume emissivity, $\varepsilon_l(\lambda)$ (phot cm^{-3} s^{-1} Å^{-1}), in the spectral line labeled l for the transition $i \rightarrow f$ as follows:

$$\varepsilon_l(\lambda) = N_{\mathrm{i}}\, A_{if}\, \varphi(\lambda - \lambda_l), \tag{10.4}$$

where the function $\varphi(\lambda - \lambda_l)$ is the line profile normalized to unity when integrated over λ, and λ_l is the central wavelength of the photon flux distribution $\varphi(\lambda - \lambda_l)$.

Introducing also the spectral density $I(\lambda - \lambda_l)$ for the line intensity

$$I(\lambda - \lambda_l) = \frac{1}{4\pi R^2} \int\limits_V \varepsilon_l(\mathbf{r})\, \varphi(\lambda - \lambda_l)\, \mathrm{d}V, \tag{10.5}$$

one has for the whole spectrum $I(\lambda) = \sum_l I(\lambda - \lambda_l)$

$$I(\lambda) = \frac{1}{4\pi R^2} \int\limits_V \widetilde{F}(\lambda; \mathbf{r})\, \mathrm{d}V, \tag{10.6}$$

where the total spectral density $\widetilde{F}(\lambda; \mathbf{r})$ for the emissivity functions $\varepsilon_l(\mathbf{r})$ is given by

$$\widetilde{F}(\lambda; \mathbf{r}) = \sum_l \varepsilon_l(\mathbf{r})\, \varphi(\lambda - \lambda_l)\ . \tag{10.7}$$

In the sequel, we will be generally interested in the total flux and emissivity, given by (10.1) and (10.2) respectively, without going into the details of the line shapes.

10.2.2 Concept of Differential Emission Measure

It is useful to define another important diagnostic tool widely used in applications for determining physical conditions in hot plasma sources, namely, differential emission measure (DEM) function apparently introduced for the first time by Pottasch [1]. In order to define this quantity, it is convenient to rewrite the flux in (10.1), radiated in a particular spectral line labeled l for the transition $i \to f$, as follows:

$$I_l = \frac{1}{4\pi R^2} \int_V G_l\,(T_e(\mathbf{r}), N_e(\mathbf{r}))\, N_e^2(\mathbf{r})\,\mathrm{d}V, \qquad (\mathrm{phot\,cm^{-2}\,s^{-1}}), \tag{10.8}$$

where $T_e(\mathbf{r})$ and $N_e(\mathbf{r})$ are the electron temperature and density spatial distributions in the plasma volume V, and the function G_l, usually called in astrophysical literature *contribution function* (see, e.g., [2] for more details concerning the definition of this function), can be expressed as

$$G_l\,(T_e(\mathbf{r}), N_e(\mathbf{r})) = j_l(\mathbf{r})\,\beta(\mathbf{r}), \qquad (\mathrm{phot\,cm^3\,s^{-1}}), \tag{10.9}$$

where the factor $\beta = N(\mathrm{X})/N_e$ is the relative abundance of atoms for the considered element X to the electron density, and j_l ($\mathrm{cm^{-3}\,s^{-1}}$) is the luminosity function per atom and per free electron:

$$j_l(\mathbf{r}) = \frac{1}{N(\mathrm{X})\,N_e}\,\varepsilon_l(\mathbf{r}). \tag{10.10}$$

The contribution functions G_l as functions of temperature and density are calculated by solving the adopted systems of balance equations (10.3) and include all atomic parameters contributing to the line formation, as well as ion and element abundances. It also exist atomic databases for these quantities, for example, the well-known CHIANTI database from the Solar Soft library [3,4].

The factor β in (10.9) is usually a slow varying function of T_e within the interval of temperatures where the luminosity function j_l gives the main contribution to (10.9) due to the very sharp character of its temperature dependence. For hot astrophysical plasmas, this factor is believed to be not dependent on temperature, because in astrophysical conditions the electron density is mainly caused by the amounts of hydrogen $N(\mathrm{H})$ and helium $N(\mathrm{He})$, which are almost completely ionized. For this case, the value β is usually expressed as

$$\beta = \frac{N(\mathrm{X})}{N(\mathrm{H})}\,\frac{N(\mathrm{H})}{N_e}, \tag{10.11}$$

where $N(\mathrm{X})/N(\mathrm{H})$ is the abundance of the element X relative to hydrogen, and $N(\mathrm{H})/N_e$ is the density of hydrogen atoms to the electron density estimated to be ≈ 0.83 for hot regions of the solar atmosphere.

The quantity $\mathrm{d}Y = N_e^2\,\mathrm{d}V$ ($\mathrm{cm^{-3}}$) in (10.8) is the *emission measure* (EM) of the plasma volume element $\mathrm{d}V$. This differential form is proportional to the number of free electrons and to the electron density in the volume $\mathrm{d}V$ and hence is related to the physical conditions in the plasma. The total EM in the plasma volume V is defined as $Y = \int_V N_e^2 \mathrm{d}V$.

At the steady-state conditions, the emitting plasma source may be described by a temperature distribution $T(\mathbf{r})$ in space within the volume V. Since the density of emission power for the stationary plasma is dependent on temperature, it appears to be convenient to use the inverse function characterizing the plasma volume $V(T)$ with a definite temperature T. The electron density N_e in this volume is then characterized by the same temperature. In this case, we can define the *emission measure differential in temperature*, $y(T)$, by the expression:

$$y(T)\,\mathrm{d}T = \overline{N_e^2}(T)\,\mathrm{d}V, \tag{10.12}$$

where $\overline{N_e^2}(T)$ is the mean square of the electron density over all the plasma volume elements $\mathrm{d}V_i$ at temperature T ($\mathrm{d}V = \sum \mathrm{d}V_i$) inside the total emitting volume V. The quantity $y(T)$ (cm^{-3} K^{-1}) is also called DEM. The defined DEM function describes the distribution of emitting material as a function of temperature and allows one to study the temperature content of plasma structures. In low density, plasma conditions, there are many lines, for which the sensitivity of the contribution functions G_l to plasma density is absent or small, so that the dependence on N_e can be ignored. For these lines, using the definition (10.12), the expression (10.8) can be written in the form:

$$I_l = \frac{1}{4\pi R^2} \int_T G(l,T)\, y(T)\,\mathrm{d}T, \tag{10.13}$$

where we have used the designation $G(l,T)$ for the contribution function independent on N_e. Note that more strict definitions of the DEM distribution, as well as the general case of the emission measure differential in both temperature and density can be found in [5–7]. Correct mathematical definitions of the EM and DEM quantities were given in [8,9] using the mathematical notion of the Stieltjes integral.

Thus, the radiative model for the emitting plasma source at steady-state conditions for hot optically thin low-density plasma can be formulated in terms of the convolution of the contribution function $G(l,T)$ and the DEM distribution $y(T)$. The first one is determined through luminosity functions $j_l(T)$, which should be calculated in the framework of some adopted model assumptions. The latter one, $y(T)$, has to be derived by formulating and solving the inverse problem with a given spectrum $I(\lambda)$ (see (10.6)) acquired from observations or experiments.

10.2.3 XUV Emission in Wide Wavelength Bands from the Sun

There are two main instrumentation actually used in observations of the solar atmosphere in the XUV part of the spectrum. These are (1) spectrometers recording individual spectral lines and (2) imaging instruments ("imagers") operating over wide wavelength ranges (channels), which may contain many spectral lines and the continuum. The imagers usually have good spatial and temporal resolution, but are not well fitted for multitemperature analysis of plasma structures because of bad spectral resolution. In contrast, the spectrometers can give relatively accurate determination of coronal temperature and density distributions, but have larger pixels and slower time cadence. In this context, the availability of reliable diagnostic methods using broadband-imaging data is actually important.

Consider the emission plasma model in the case of broadband spectral channels. For the sake of convenience, we will also consider the plasma radiation spectra in terms of the total power of emission $F_l = 4\pi R^2 I_l$ in a particular spectral channel l. Then the total power of emission F_l (erg s^{-1}) in the wavelength band $\Delta\lambda(l)$ of the spectral channel l produced by the temperature interval ΔT from an optically thin plasma region with the volume V may be written similarly to (10.13):

$$F_l = \int\limits_{\Delta T} G(l, T)\ y(T)\,\mathrm{d}T, \tag{10.14}$$

where $G(l, T)$ is the *temperature response function* defined as the spectral emissivity function $\widetilde{F}(\lambda; T, N_\mathrm{e})$ (see (10.6) and (10.7)) calculated using definite model assumptions about emitting plasma (including the coronal approximation and the optically thin condition) and integrated over $\Delta\lambda(l)$ range with the known filter ("apparatus") function $f(l, \lambda)$:

$$G(l, T) = \frac{1}{N_\mathrm{e}^2} \int\limits_{\Delta\lambda(l)} f(l, \lambda)\, \widetilde{F}(\lambda; T, N_\mathrm{e})\,\mathrm{d}\lambda. \tag{10.15}$$

For the line spectrum, the function $f(l, \lambda) = 1$ and $G(l, T)$ coincides with the contribution function of the line l.

10.3 Spectral Inverse Problem: Two Mathematical Formalizations

There exist a variety of methods to solve inverse problems using spectroscopic data (see, e.g., [2, 10]). In principle, all these regularization and optimization techniques could be classified using diverse ways. We consider the classification including two mathematical formalizations of the spectral inverse problem, namely, the standard (or "algebraic") and probabilistic approaches. In the standard approach traditionally used in applications, the inverse problem is formulated in terms of the Fredholm integral equation of the first kind. In numerical techniques based on the standard approach, regularization constraints are usually used to derive a physically meaningful stable solution of this integral equation. The reason of the regularization procedure comes from the inherent ill-posedness of the inverse problem (see, e.g., [10] for details). In the probabilistic approach based on another mathematical formalization (see, e.g., [11]), the spectral inverse problem is formulated using relative fluxes and normalized functions treated as probability distributions for some random variables. One of the main features of the probabilistic approach is that the language of the probability theory and mathematical statistics is used and the spectral inverse problem is consequently formulated in terms of distribution functions, hypotheses, confidence level, etc.

Below, we briefly consider these two approaches and formulate the Bayesian iterative method (BIM) developed in the frame of the probabilistic approach to the spectral inverse problem. In the following paragraph, we will consider a few examples of the BIM applications to laboratory and solar coronal plasmas for the DEM diagnostic analysis of spectroscopic data.

10.3.1 Standard Approach to the Spectral Inverse Problem

As mentioned above, the formulation of the spectral inverse problem in the standard approach gives rise to the general Fredholm integral equation of the first kind given by the formula (10.14), where F_l is the measured emission flux in the channel l, $G(l, T)$ is the kernel of the integral transformation (10.14), and $y(T)$ is the unknown function to be derived. In order to calculate the temperature distribution $y(T)$, we need to solve a system of integral equations of the form (10.14) for an available set of spectral channels $\{l\}$.

It is well known that the inverse problem is "ill-posed" and usually poses questions of existence, uniqueness, and stability of its solution. Mathematical formalization is realized by means of the theory of mappings and functional analysis. A variety of numerical techniques based on the standard approach have been developed to solve inverse problems. Methods usually used in this approach require regularization constraints to derive a physically meaningful approximate stable solution $y(T)$ (see, e.g., [10]).

10.3.2 Probabilistic Approach and Formulation of the Bayesian Iterative Method

In this paragraph, we are going to use the Bayesian formalism described in Appendix to formulate the spectral inverse problem in the frame of the probabilistic approach. Mathematical formalization of this approach can be performed using the formalism and notions of the mathematical logic and probability theory [11]. Methods for the solution of the corresponding equations include those of the mathematical statistics (in particular Bayesian analysis) and information theory (see, e.g., [12, 13]).

In the probabilistic approach, (10.14), is transformed to the formula of the total probability (see [9] for details):

$$P(l) = \int_{\Delta T} P(l\,|T)\,P(T)\,\mathrm{d}T, \tag{10.16}$$

for probability distributions $P(l)$, $P(l\,|T)$, and $P(T)$, which are positively determined and satisfy the normalization conditions

$$\sum_l P(l) = 1, \quad \sum_l P(l\,|T) = 1, \quad \int_{\Delta T} P(T)\,\mathrm{d}T = 1, \tag{10.17}$$

where the summation is extended over all spectral lines or channels under consideration. These functions are interpreted as probability distributions for some random variables defined on the field of events $\{l\}$ and $\{T\}$: $P(l)$ is the probability of a photon being emitted in the channel l, $P(T)$ the probability density of a photon being emitted by plasma at temperature T, and $P(l\,|T)$ the conditional probability of the event l at the condition T.

If we want to use the probability distributions $P(l)$, $P(l\,|T)$, and $P(T)$ from (10.16) to interpret experimental data, we need to establish their relationships with measured and calculated physical values. This can be done by means of the following normalized relations:

$$P(l) = \frac{F_l}{\sum_l F_l}, \quad P(l\,|T) = \frac{G(l,T)}{\sum_l G(l,T)}, \tag{10.18}$$
$$P(T) = \frac{y(T)\sum_l G(l,T)}{\sum_l F_l}.$$

In order to derive the unknown function $P(T)$ as the solution of the spectral inverse problem, we use the BIM algorithm (see Appendix). Using the known (measured) emission fluxes $F_l^{(\mathrm{exp})}$ for $P(l)$ values in (10.16), i.e., for $P^{(\mathrm{exp})}(l)$, and applying Bayes' theorem, one can obtain a recurrence relation for the $P(T)$ distribution:

$$P^{(n+1)}(T) = P^{(n)}(T)\sum_l \frac{P^{(\mathrm{exp})}(l)}{P^{(n)}(l)}\,P(l\,|T), \tag{10.19}$$

where $P^{(n)}(l)$ is calculated from (10.16) using the nth approximation $P^{(n)}(T)$. The correction values $P^{(\mathrm{exp})}(l)/P^{(n)}(l)$ and the χ^2 minimization criterion for emission fluxes are used as a measure of the accuracy for the distribution $P(l)$ over a set of spectral channels under study. When the solution for the $P(T)$ distribution is evaluated, the corresponding DEM profile is deduced using (10.18) as follows:

$$y(T) = \frac{P(T) \sum_l F_l^{(\mathrm{exp})}}{\sum_l G(l,T)}. \tag{10.20}$$

In real applications, it is necessary to have information about the accuracy (confidence level) of the solution derived because the experimental fluxes $F_l^{(\mathrm{exp})}$ are known with some accuracies. In order to obtain the solution of the inverse problem, one can use the procedure of the Monte Carlo simulations. The reconstruction procedure is repeated several times for M different generations of randomly distributed experimental data $F_l^{(\mathrm{exp})}$ using a given noise distribution (gaussian, poisson, etc.). This finally provides the solution in the form of the mean value:

$$\overline{y(T)} = \frac{1}{M} \sum_{i=1}^{M} y^{[i]}(T), \tag{10.21}$$

and the dispersion

$$\sigma\left[y(T)\right] = \sqrt{\frac{1}{M-1} \sum_{i=1}^{M} \left[y^{[i]}(T) - \overline{y(T)}\right]^2}, \tag{10.22}$$

which are evaluated over different sets of reconstructed distributions $y^{[i]}(T)(i = 1, 2, \ldots, M)$.

Finally, it is worth to enumerate some important features of the BIM. This method has been deduced in a regular way on the basis of Bayes' theorem, and, in contrast to the standard optimization techniques, the regularization constraints are not needed. The BIM algorithm, being assumed to be a sequential estimate based on a current hypothesis for the DEM distribution, corresponds to the maximum likelihood criterion (i.e., it brings the most likely solution of the inverse problem) and provides the ultimate resolution enhancement (super-resolution) as compared with linear and other nonparametric methods (see [13], and references therein).

10.4 Applications of the BIM to Laboratory and Solar Corona Plasmas

The BIM was applied effectively in solving a series of problems, such as the image restoration [14], the signal recovery of noisy experimental data [12, 13], and the deconvolution of initial X-ray spectra of the Sun recorded by Bragg spectrometers [15]. For the last decade, we have used this diagnostic tool for analyzing and interpreting XUV spectral data from hot tokamak and solar corona plasmas. In the first case, we have analyzed X-ray emission spectra of highly charged helium- and lithium-like ions at the tokamak plasma to develop a self-consistent approach (SCA) for deriving information on plasma parameters and the verification (i.e., the estimation of the accuracy) of atomic data and of methods of their calculation (see [16, 17]). In the second case, we have used the BIM for determining the temperature contents of the solar corona plasma structures, namely, the DEM temperature analysis of XUV spectra and imaging data observed in the space experiments onboard satellites (see, e.g., [8,9,18]). Further, we will give an overview of the methods developed for applying the BIM to the aforementioned problems and a discussion of some important results derived by means of this diagnostic technique. More detailed consideration of these results can be found in [8, 9, 16–18].

10.4.1 X-Ray Spectroscopy of Highly Charged Ions at the TEXTOR Tokamak

Studies of the highly charged ions spectra by means of the X-ray spectroscopy allow important information about elementary processes and plasma parameters to be derived. For the last decades, K-spectra of highly charged ions, associated with the transitions $nl - 1s$ of the optical electron, are effectively used for measuring parameters of hot plasmas. At present, the X-ray spectroscopy is one of the main methods for diagnostics of astrophysical objects and fusion plasma at modern tokamaks. However, the accuracy of these methods, as well as the possibility of the unambiguous interpretation of spectra, is critically dependent on the accuracy of the atomic data used for modeling X-ray emission sources. This problem, being fundamental for the theoretical spectroscopy, was not analyzed widely. For this reason, possible errors in collisional and radiative characteristics of highly charged ions remain uncertain. Thus, the verification of the accuracy of atomic data is crucially needed for correctly determining plasma parameters and for unambiguous description of the mechanisms of spectra formation. Besides the fundamental importance for the atomic physics, it is also necessary for future investigations of fast and non-Maxwellian phenomena in low-density plasmas in tokamak and astrophysical sources, in particular in solar flares.

Since crossed-beam experiments, offering direct measurements of spectroscopic and collisional characteristics of highly charged ions, are now practically absent, their accuracy is usually estimated by comparing them either with the most accurate (theoretical) calculations or with the data of beam-plasma experiments at EBIT-like setups [19, 20]. Due to narrow spectral lines, the EBIT sources are traditionally used to measure wavelengths, lifetimes of metastable states and electron–ion cross sections. However, because of the low photon statistics, they are not always valid for the verification of collisional and radiative data with a precision sufficient for diagnostic purposes. At the same time, the problem of the quantitative verification of atomic data employing the emission spectra from tokamak plasma was not practically discussed. In this connection, the problem of the atomic data verification is rather actual and important both for atomic physics, stimulating improvements of their calculation methods, and for solving the plasma physics problems based on the results of the X-ray spectral diagnostics.

Studies of the high-resolution K-spectra of Ar^{16+} and Ar^{15+} ions at the TEXTOR (Torus Experiment for Technology Oriented Research) tokamak plasma [16, 17, 21, 22] showed that the tokamak spectra can be efficiently used not only for the diagnostic purposes but also both for the verification of the accuracy of atomic data—wavelengths and rates (cross sections averaged over the Maxwellian velocity distribution) of elementary processes in hot plasmas—and the diagnostic methods used. As it was shown, the analysis of the accuracy and the correction of atomic and plasma characteristics can be made with the help of a new SCA based on the solution of the inverse problem for these spectra in the framework of the semiempirical "spectroscopic" model. Self-consistency of the model parameters is provided by a number of functional relationships, which are determined by requirements of the coincidence (within the limits of experimental errors) of the theoretical and measured spectra under essentially various conditions in the plasma. The necessary conditions for applying this approach are ensured by the following features of the experiment at the TEXTOR tokamak: (1) the high accuracy for measuring the fluxes and high photon statistics, (2) the detection of the spectra under various plasma conditions, and (3) the existence of additional diagnostic techniques for measuring the radial profiles of the electron temperature and density simultaneously with the spectra.

10.4.1.1 Formulation of the Problem

The spectrum of Ar^{16+} (helium-like) and Ar^{15+} (lithium-like) ions under study covers the wavelength region 3.94–4.02 Å and consists of a set of well-resolved intensity peaks $\{\mathbf{L}\}$, which are associated with corresponding spectral lines l giving a dominant contribution to their intensities. The most prominent and intense features of the spectrum result mainly from the lines caused by transitions in Ar^{16+} ions: resonance ($1s2p(^1P_1)$–$1s^2(^1S_0)$), magnetic quadrupole ($1s2p(^3P_2)$–$1s^2(^1S_0)$), intercombination ($1s2p(^3P_1)$–$1s^2(^1S_0)$), and forbidden ($1s2s(^3S_1)$–$1s^2(^1S_0)$), designated as **w**, **x**, **y**, and **z** lines, respectively (these notations are used following [23]). These lines are produced primarily due to direct electron impact excitation including cascades from higher levels, as well as have contributions from radiative recombination of Ar^{17+} (hydrogen-like) ions, from inner-shell ionization of Ar^{15+} ions (contributing to the **z** line), and from resonance scattering via doubly excited autoionizing states $1snln'l'$ $(n, n' > 2)$.

In addition to the helium-like lines, there are numerous lines, called dielectronic satellites, to the lines of helium-like argon ions due to transitions $1s^2nl$–$1s2pnl$ in lithium-like ions and, to a lesser extent, $1s^22snl$–$1s2s2pnl$ in beryllium-like ions. The most prominent lines emitted by the Ar^{15+} ion $(n = 2)$ and resolved in the spectrum are the **q** and **r** satellites excited predominantly by collisions with electrons, and the **k** and **a** ones excited by means of the dielectronic mechanism from Ar^{16+} ion. The most intense **j** satellite is blended with the **z** line. There are also two groups of satellites in the long wavelength wing of the **w** resonance line corresponding to dielectronically excited $n = 3$ and $n = 3, 4$ satellites and denoted here as $\mathbf{N}_3$ and $\mathbf{N}_4$ peaks, respectively. The remaining less-intense satellites densely fill the spectral range, forming series converging to helium- and lithium-like lines. For the purpose of quantitative analysis of spectral fluxes, the whole spectral range was divided into the intervals $\Delta\lambda_{\mathbf{L}} = [\mathbf{L}]$, and the following set of peaks was identified: $\{\mathbf{L}\} = \{\mathbf{W}, \mathbf{N}_4, \mathbf{N}_3, \mathbf{X}, \mathbf{Y}, \mathbf{Q}, \mathbf{R}, \mathbf{A}, \mathbf{K}, \mathbf{Z}\}$.

The theoretical (synthetic) spectrum $I(\lambda)$ and the emission flux $F_{[\mathbf{L}]} = \int_{[\mathbf{L}]} I(\lambda)\mathrm{d}\lambda$ in a spectral interval $[\mathbf{L}]$ are functions (or functionals) of two sets of physical characteristics: (1) the atomic data (AD) including the sets of atomic constants (wavelengths, transition probabilities, branching ratios, etc.) and the effective rates (collisional characteristics) of elementary processes, $C_1^z(T_e)$, and (2) the plasma parameters (PP), namely, radial profiles of the electron T_e and ion T_i temperatures, electron density N_e, argon ion densities N_z for the charges z, and neutral atom density of the working gas (hydrogen, deuterium, or helium) N_a.

The synthetic emission fluxes $F_{[\mathbf{L}]}$ depend on a particular model based on equations of atomic and plasma kinetics. For the adopted model, the main, basic or "key", parameters $\mathbf{D} = \{\mathbf{D}_i\}$ and $\mathbf{P} = \{\mathbf{P}_i\}$ were defined from the corresponding sets AD and PP (e.g., ratios of line excitation rates or central temperature $T_e^{(0)}$ in the plasma core), which identically (with a given accuracy) simulate the fluxes of the synthetic spectrum $F_{[\mathbf{L}]}^{(\mathrm{syn})} = F_{[\mathbf{L}]}(\mathbf{D}, \mathbf{P})$. In the frame of semiempirical models, key plasma parameters $\mathbf{P}$ have to be independently predetermined (calculated or measured) for a direct "ab initio" calculations of the synthetic spectra, while for diagnostic purposes they can be derived from the measured spectra by solving the inverse problem using a set of equalities $F_{[\mathbf{L}]}^{(\exp)} = F_{[\mathbf{L}]}(\mathbf{D}, \mathbf{P})$, where $F_{[\mathbf{L}]}^{(\exp)}$ are the fluxes measured in a given spectrum. These equalities impose restrictions on the possible values of the model parameters, relating them through implicit functional relations.

The sets **D** and **P** must also satisfy to additional physical constraints. In particular, the quantities from the set **D** derived by the inversion of different spectra have to coincide (within the experimental errors), i.e., have to be independent on plasma conditions. The quantities from the set **P** for a given spectrum (e.g., the central temperature) do not have to depend on **L**, while the radial profiles of relative ionic abundances $n_z = N_z / \sum_z N_z$, derived from any measured spectra, have to obey the continuity condition

$$n(r) = \sum_z n_z(r) = 1, \tag{10.23}$$

which may not be satisfied for arbitrary atomic data from **D**. As a result, the presented conditions and relations ensure the consistency of atomic and plasma parameters, which means the existence of the range of their values satisfying to these conditions for a given accuracy of measured fluxes $F_{[\mathbf{L}]}^{(\exp)}$, and can be used for the atomic data verification problem.

By definition, the term "verify" means to prove to be true or to check for accuracy. Both sets of characteristics, **D** and **P**, being the variable parameters of the spectroscopic model (SM) developed, may be optimized by using (10.23), which was shown to be the necessary and sufficient conditions for these parameters to be correct [17].

10.4.1.2 X-Ray Spectroscopy at the TEXTOR Tokamak

TEXTOR is a medium-sized tokamak experiment with a major radius of 1.75 m and a minor radius of 0.46 m. It operates with toroidal magnetic fields of up to 2.7 T and plasma currents up to 580 kA. In addition to ohmic heating of about 0.5 MW, which is obtained from the plasma current, auxiliary heating is provided by the injection of neutral hydrogen beams with the total power of up to 2 MW. Further information on TEXTOR and its diagnostic equipment can be found in [24–26].

The K-spectra from the Ar^{16+} and Ar^{15+} impurity ions were investigated on the TEXTOR tokamak equipped with a high-resolution X-ray spectrometer/polarimeter [24, 25] consisting of two (horizontal and vertical) Bragg spectrometers in the Johann scheme, designed for the polarization measurements of the radiation from the same central region of the tokamak plasma. Figure 10.1 shows a schematic diagram of the experimental setup. The horizontal instrument was used to measure the K_α-spectra (formed by transitions $n = 2 \rightarrow n = 1$ of the optical electron), while the vertical (perpendicularly arranged) spectrometer was used in the experiment for recording the K_β-emission lines (emitted due to transitions $n = 3 \rightarrow n = 1$).

The electron temperatures in the plasma core region were derived from the K_α-spectra and also compared with the results of the Thomson scattering diagnostics [27] on the TEXTOR tokamak. The results of both measurements agree to within 5–10%. In our plasma diagnostic analysis, we also used the radial profiles of the electron temperature $T_e(r)$ and density $N_e(r)$ measured for each spectrum by

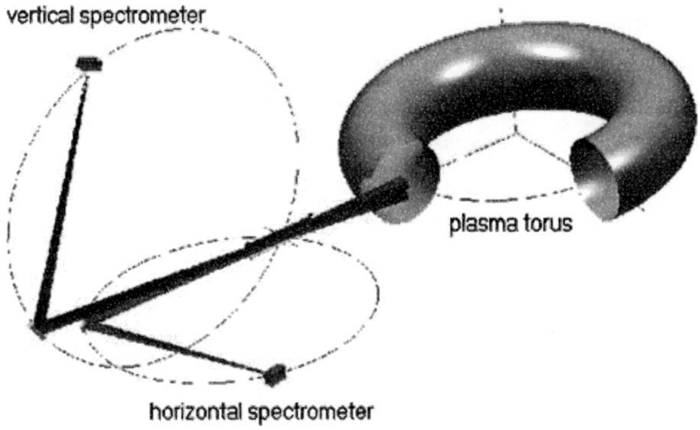

Fig. 10.1 Schematic diagram of the X-ray spectrometer installed at the TEXTOR tokamak (by courtesy of G. Bertschinger)

means of an electron cyclotron emission (ECE) polychromator and a far-infrared interferometer/polarimeter (FIIP), respectively [25].

10.4.1.3 Self-consistent Approach and Atomic Data Calculations

The general concept of the SCA used for the verification problem includes several aspects or levels of consistency. The main idea is dealt with the aforementioned condition of "intrinsic" consistency of key parameters **D** and **P** in the SM. Another aspect is connected with a coordination of the atomic data evaluated in the frame of a uniform method to avoid their compilation and/or extrapolation. Thus, the verification of atomic characteristics means at the same time the verification of the corresponding method of atomic data calculations. The third aspect concerns a consistency of plasma parameters of the SM with those measured by means of other diagnostic techniques. From the experimental point of view, the accuracy of the verification procedure depends on the following: (1) the accuracy of the experimental data, (2) the number of spectral features in the selected spectral range, and (3) the number of spectra measured under significantly different conditions. In our particular case of the argon ion K_α-spectrum, there were 10 prominent well-resolved peaks consisting of numerous lines in selected experimental spectra, which were measured for a wide range of temperatures $T_e = 0.8$–$2.5\,\text{keV}$ and densities $N_e = 10^{13}$–$10^{14}\,\text{cm}^{-3}$.

The atomic data needed to model the synthetic spectra of argon ions were calculated by means of the atomic codes ATOM and MZ [28]. These data include wavelengths, radiative, and autoionization (for autoionizing states) decay probabilities, as well as collisional characteristics of elementary processes: cross sections and rates for direct (potential or background) processes of electron–ion impact excitation, ionization, and radiative recombination. The contribution of resonance scattering was accounted for as cascade processes to autoionizing levels caused by dielectronic capture with the following autoionization (resonance excitation) or radiation (dielectronic recombination). The effective rate coefficients for excitation and recombination processes including cascades via radiative and autoionizing levels were obtained by solving equations of atomic kinetics in the frame of a radiative–collisional model. The calculations were carried out under the assumptions of the quasisteady plasma ($\mathrm{d}N_z^{(i)}/\mathrm{d}t = 0$ for ion populations $N_z^{(i)}$) and Maxwellian velocity distribution for plasma particles. More detailed description of the atomic data calculations and radiative–collisional model used can be found in [16, 17].

10.4.1.4 Spectroscopic Model

The aforementioned semiempirical spectroscopic model (SM) is based on the following assumptions:

(1) The conditions of coronal approximation are satisfied for the populations $N_z^{(k)}$ of the excited ion states k $(k > 1)$.
(2) The profiles of the relative ion abundances $n_z(r)$ are the parameters of the model and satisfy to the continuity relation (10.23), while the profiles of the electron temperature $T_e(r)$ and density $N_e(r)$ are known (i.e., measured).

Under the coronal equilibrium conditions, implying the quasi-steady state of the plasma $(\mathrm{d}N_z^{(k)}/\mathrm{d}t = 0)$, the effects of the ionic and radiative transfer for the excited ion states are negligible, and their populations are low, $\left(N_z^{(k)} \ll N_z^{(1)}\right)$. Therefore, the ion abundances are practically equal to the populations of the ground states, $N_z \cong N_z^{(1)}$. The calculations in the collisional–radiative model for argon ions [16, 17] showed that deviations from the coronal approximation are negligible ($< 1\%$) for densities $\leq 10^{14}\,\mathrm{cm}^{-3}$; thus, for these densities, the condition (1) is justified. It is also worth noting that the ion abundances N_z under conditions typical for the tokamak plasma do not coincide with coronal equilibrium concentrations $N_z^{(c)}$ used for solar corona plasma (coronal ionization equilibrium), but obey kinetic equations of balance taking account for the effects of the transfer and charge exchange of the impurity ions on neutral atoms. The role of these effects was investigated in [22] using an impurity transport model.

In the coronal approximation, the emission flux in the spectral range **[L]** from the plasma column along the radius $r = \rho a$ (a is the small radius of the plasma torus, ρ—dimensionless radius) can be written as the sum $F_{[\mathrm{L}]} = \sum_z F_{[\mathrm{L}]}^z$ over the partial fluxes $F_{[\mathrm{L}]}^z$ given by [17]

$$F_{[\mathrm{L}]}^z = C \int_0^1 J_{[\mathrm{L}]}^z(P(\rho))\, n_z(\rho)\ [N_e(\rho)]^2\, \mathrm{d}\rho, \qquad (10.24)$$

where $J_{[\mathrm{L}]}^z$ are the partial excitation rates

$$J_{[\mathrm{L}]}^z(T_e, T_i, n_a) = \sum_l C_l^z(T_e, n_a)\, \varphi_{[\mathrm{L}]}^l(T_i). \qquad (10.25)$$

Here, P stands for the set of plasma parameters, $P = \{T_e, T_i, n_a\}$; C_l^z are the effective rates corresponding to the excitation processes of the lines from the ions with the charge z (in our case for argon ions with $z = 15$–17) and also including radiative cascades from the upper levels and branching coefficients k_l for the lines l; C is the conversion coefficient defined by the condition of equality between the measured and calculated fluxes; $\varphi_{[\mathrm{L}]}^l(T_i) = \int_{[\mathrm{L}]} \varphi_l(T_i; \lambda - \lambda_l)\mathrm{d}\lambda$ is the correction factor associated with the line profile $\varphi_l(T_i; \lambda)$ due to the natural width, Doppler broadening, and the instrumental function; $n_a = N_a/N_e$ is the relative density of hydrogen atoms.

The partial excitation rate $J_{[\mathrm{L}]}^z$ for a feature **L** describes the total effective rate of the formation for all the lines, contributing to its intensity and excited from the ions with the charge z. Thus, for $z = 16$ (denoted also as he), the partial emissivity $J_{[\mathrm{L}]}^{\mathrm{he}}$ contains the effective collisional excitation rates for the lines emitted by the helium-like ions, and the dielectronic capture rates for lithium-like dielectronic satellites resulted from the helium-like ions; for the lines excited from lithium-like ions (z = 15 or li), $J^{\text{li}}_{[\mathbf{L}]}$ includes the effective rate for the inner-shell excitation of lithium-like satellites, as well as the inner-shell ionization of the 1s electron (contributing to the excitation of the **z** line); for the hydrogen-like ions (**z** = 17 or h), $J^{h}_{[\mathbf{L}]}$ contains the total rate of recombination (both radiative and dielectronic) from the hydrogen-like ions to the excited states of helium-like lines, as well as the charged exchange rate on the neutral hydrogen atoms. Thus, the partial excitation rates $J^{z}_{[\mathbf{L}]}$ depend on both atomic characteristics and the radial distributions of plasma parameters.

To formulate the SM and to derive its parameters by solving the inverse problem, the dependence on the radial profiles of the plasma parameters $n_z(\rho)$, $N_e(\rho)$, and $T_e(\rho)$ was factorized by changing the radial variable ρ to the dimensionless temperature variable $\beta(\rho) = \left[T_e(\rho)/T_e^{(0)}\right]^{-1} - 1$ (the electron temperature in the plasma core $T_e^{(0)}$ is one of the main "key" parameters of the SM) in (10.24) for the partial emission flux:

$$F^{z}_{[\mathbf{L}]} = C \int_0^b J^{z}_{[\mathbf{L}]}\left(T_e^{(0)}, \beta\right) n_z(\beta)\, y(\beta)\, \mathrm{d}\beta, \tag{10.26}$$

where $y(\beta) = [N(\rho(\beta))]^2\, |\mathrm{d}\rho/\mathrm{d}\beta|$ is the DEM distribution and $b = \beta(1) \gg 1$. Note that, in contrast to the ρ-representation given in (10.24), in the β-representation, the values $J^{z}_{[\mathbf{L}]}$ are sensitive to the atomic characteristics and are independent on the radial profiles of the electron temperature $T_e(\rho)$.

In order to use the BIM for the inversion procedure, it is also necessary to pass from the absolute total, $F_{[\mathbf{L}]}$, and partial, $F^{z}_{[\mathbf{L}]}$, fluxes in the peaks to the relative fluxes $\Gamma_{[\mathbf{L}]}(\sigma_z) = F_{[\mathbf{L}]}/F_{[\sigma_z]}$ and $P^{z}_{[\mathbf{L}]} = F^{z}_{[\mathbf{L}]}/F^{z}_{[\sigma_z]}$, normalized in three spectral regions $[\sigma_z]$ chosen for each z. To choose these regions, we consider three sets of peaks denoted through $\mathbf{Z} = \{\mathbf{L}_i\}$ ($\mathbf{Z} = Li, He, H$) and corresponding to three (generally speaking overlapping) wavelength regions $[\sigma_z] = [\mathbf{Z}]$, which include lines excited from the ions with the charge z (z = li, he, h). The choice of the sets $[\sigma_z]$ was stipulated by two demands: to provide (1) the maximum contribution of partial fluxes $F^{z}_{[\mathbf{L}]}$ ($\mathbf{L} \subset \mathbf{Z}$) in each region $[\mathbf{Z}]$ and (2) the minimum contribution of partial fluxes $F^{k}_{[\mathbf{L}]}$ from ions with other charges $k \neq z$. The following peak groups were used: **He**={**W**, $\mathbf{N}_4$, $\mathbf{N}_3$, **K**}; **Li**={**Q**, **R**, **Z**}; and **H**={**W**, **X**, **Y**, **Z**}.

The total flux $F_{[\mathbf{Z}]}$ in the spectral range $[\mathbf{Z}]$ can be written as a sum of partial fluxes $F^{z}_{[\mathbf{Z}]}$ similar to that in the peak $[\mathbf{L}]$ (see (10.24)), namely, $F_{[\mathbf{Z}]} = \sum_z F^{z}_{[\mathbf{Z}]}$, and the partial excitation rate $J^{z}_{[\mathbf{Z}]}$ as a sum over all $\mathbf{L} \subset \mathbf{Z}$

$$J^{z}_{[\mathbf{Z}]}\left(T_e^{(0)}, \beta\right) = \sum_{\mathbf{L} \subset \mathbf{Z}} J^{z}_{[\mathbf{L}]}\left(T_e^{(0)}, \beta\right). \tag{10.27}$$

Define also the relative partial excitation rates $p^{k}_{[\mathbf{L}]}$ of the peaks $\mathbf{L} \subset \mathbf{Z}$ for each range $[\mathbf{Z}]$ and the functions Φ_z (associated with the plasma parameters) by the following equalities:

$$p^{k}_{[\mathbf{L}]}\left(T^{(0)}_{\mathrm{e}},\beta\right)=\frac{J^{k}_{[\mathbf{L}]}\left(T^{(0)}_{\mathrm{e}},\beta\right)}{J^{k}_{[\mathbf{K}]}\left(T^{(0)}_{\mathrm{e}},\beta\right)}, \tag{10.28}$$

$$\Phi_{z}\left(T^{(0)}_{\mathrm{e}},\beta\right)=C\,J^{z}_{[\mathbf{Z}]}\left(T^{(0)}_{\mathrm{e}},\beta\right)\,n_{z}\left(T^{(0)}_{\mathrm{e}},\beta\right)\,y(\beta)/\,F^{z}_{[\mathbf{Z}]}, \tag{10.29}$$

which satisfy to the normality conditions

$$\int_{0}^{b}\Phi_{z}\left(T^{(0)}_{\mathrm{e}},\beta\right)\,d\beta=1,\quad \sum_{\mathbf{L}\subset\mathbf{Z}}p^{z}_{[\mathbf{L}]}\left(T^{(0)}_{\mathrm{e}}\right)=1. \tag{10.30}$$

Further, the relative flux $\Gamma_{[\mathbf{L}]}=\sum_{k}F^{k}_{[\mathbf{L}]}/F_{[\mathbf{Z}]}$ for the peak $\mathbf{L}\subset\mathbf{Z}$ in the range $[\mathbf{Z}]$ can be presented in the form:

$$\Gamma_{[\mathbf{L}]}=\sum_{k=\mathbf{K}}\frac{F_{[\mathbf{K}]}}{F_{[\mathbf{Z}]}}\,\frac{F^{k}_{[\mathbf{K}]}}{F_{[\mathbf{K}]}}\,\frac{F^{k}_{[\mathbf{L}]}}{F^{k}_{[\mathbf{K}]}}=\sum_{k=\mathbf{K}}R^{\mathbf{K}}_{\mathbf{Z}}\,\Gamma^{k}_{[\mathbf{K}]}\,P^{k}_{[\mathbf{L}]}, \tag{10.31}$$

where $R^{\mathbf{K}}_{\mathbf{Z}}$ is the ratio of the fluxes in two peak groups $[\mathbf{K}]$ and $[\mathbf{Z}]$, and $\Gamma^{z}_{[\mathbf{Z}]}$ is the ratio of the partial flux $F^{k}_{[\mathbf{K}]}$ to the corresponding total flux in the range $[\mathbf{K}]$. The partial fluxes $P^{k}_{[\mathbf{L}]}\left(T^{(0)}_{\mathrm{e}}\right)$ normalized in the ranges $[\mathbf{Z}]$ are expressed as

$$\begin{aligned}P^{k}_{[\mathbf{L}]}\left(T^{(0)}_{\mathrm{e}}\right)&=\int_{0}^{b}p^{k}_{[\mathbf{L}]}\left(T^{(0)}_{\mathrm{e}},\beta\right)\,\Phi_{k}\left(T^{(0)}_{\mathrm{e}},\beta\right)\,\mathrm{d}\beta,\\ P^{z}_{[\mathbf{Z}]}\left(T^{(0)}_{\mathrm{e}}\right)&=\sum_{\mathbf{L}\subset\mathbf{Z}}P^{z}_{[\mathbf{L}]}\left(T^{(0)}_{\mathrm{e}}\right)=1.\end{aligned} \tag{10.32}$$

Note that the relative partial fluxes $\Gamma^{z}_{[\mathbf{Z}]}$ in three spectral ranges $[\mathbf{Z}]$ can be expressed through the flux ratios $R^{\mathbf{K}}_{\mathbf{Z}}$ and the normalized integrals $P^{k}_{[\mathbf{Z}]}\left(T^{(0)}_{\mathrm{e}}\right)=\sum_{\mathbf{L}\subset\mathbf{Z}}P^{k}_{[\mathbf{L}]}\left(T^{(0)}_{\mathrm{e}}\right)$ (see (10.32)) by solving the system of equations (10.31) for all $\mathbf{Z}$ and using the normality relation $\Gamma_{[\mathbf{Z}]}=\sum_{\mathbf{L}}\Gamma_{[\mathbf{L}]}=1$:

$$1=\sum_{k=\mathbf{K}}\Gamma^{k}_{[\mathbf{K}]}\left(T^{(0)}_{\mathrm{e}}\right)\,R^{\mathbf{K}}_{\mathbf{Z}}\,P^{k}_{[\mathbf{Z}]}\left(T^{(0)}_{\mathrm{e}}\right),\quad \mathbf{Z}=\mathbf{Li},\mathbf{He},\mathbf{H}. \tag{10.33}$$

Thus, the relative fluxes $\Gamma_{[\mathbf{L}]}$ in (10.31) are determined (according to (10.32)) by the three key plasma parameters (β-profiles) $\Phi_{z}(\beta)$ and by the kernels $p^{z}_{[\mathbf{L}]}(\beta)$ of the integral operator. These kernels depend on the set of atomic data, the central temperatures T_{e} and T_{i} of the plasma core, and (for $z=h$) the profile $n_{a}(\beta)$. If the relative fluxes $\Gamma_{[\mathbf{L}]}$ in (10.31) are fixed by the condition $\Gamma_{[\mathbf{L}]}=\Gamma^{(\mathrm{exp})}_{[\mathbf{L}]}$, the normalized profiles $\Phi_{z}(\beta)$ depend only on the arguments of the quantities $p^{z}_{[\mathbf{L}]}$.

If all relative fluxes in the peaks $\Gamma_{[\mathbf{L}]}$ and AD are known (i.e., fixed), the $\Phi_{z}\left(T^{(0)}_{\mathrm{e}},\beta\right)$ functions can be found by solving the inverse problem for the set of relations (10.27)–(10.33) for given $T^{(0)}_{\mathrm{e}}$, T_{i}, and $n_{a}(\beta)$ values. In order to determine these parameters, it is necessary to use the additional model condition given by (10.23), which limits the class of possible formal solutions of these equations, $\{\Phi_z^*(\beta)\}$. Since this criterion is expressed in terms of the relative ion abundances $n_z(\beta)$, rather than in terms of the functions $\Phi_z(\beta)$ related to $n_z(\beta)$ by (10.29), one has to adopt additional (empirical for the SM) information about plasma parameters, namely, about the DEM profile $y(\beta)$ depending on the temperature $T_e(\beta)$ and density $N_e(\beta)$ distributions. The abundances $n_z(\beta)$ and their sum $n(\beta)$ for $z = 15, 16$, and 17 (li, he, and h, respectively) can be expressed through the values $\Phi_z(\beta)$ and $n_{he}(\beta)$ using (10.29) as follows:

$$
\begin{aligned}
n_z\left(T_e^{(0)}, \beta\right) &= \widetilde{n}_z\left(T_e^{(0)}, \beta\right) n_{he}\left(T_e^{(0)}, \beta\right), \\
n\left(T_e^{(0)}, \beta\right) &= n_{he}\left(T_e^{(0)}, \beta\right) \sum_z \widetilde{n}_z\left(T_e^{(0)}, \beta\right), \\
\widetilde{n}_z\left(T_e^{(0)}, \beta\right) &= \frac{\Phi_z}{\Phi_{he}} \frac{J_{[\mathbf{He}]}^{he}}{J_{[\mathbf{Z}]}^{z}} \frac{\Gamma_{[\mathbf{Z}]}^{z}}{\Gamma_{[\mathbf{He}]}^{he}} R_{\mathbf{He}}^{\mathbf{Z}}.
\end{aligned}
\tag{10.34}
$$

The continuity relation (10.23) is also used in the SM to optimize model parameters $T_e^{(0)}$ and AD, if the DEM profile $y(\beta)$ is known.

10.4.1.5 BIM Inversion

To solve the inverse problem for determining the $\Phi_z(\beta)$ profiles in the framework of the SM, the BIM is adopted as follows. Defining the quantities $P_{[\mathbf{L}]}^z(\exp)$ and $\Gamma_{[\mathbf{L}]}^z(\exp)$ by means of the expressions

$$
\begin{aligned}
P_{[\mathbf{L}]}^z(\exp) &= \Gamma_{[\mathbf{L}]}^z(\exp)/\Gamma_{[\mathbf{Z}]}^z, \\
\Gamma_{[\mathbf{L}]}^z(\exp) &= \Gamma_{[\mathbf{L}]}^{(\exp)} - \sum_{k \neq z} \Gamma_{[\mathbf{L}]}^k \quad \text{for } \mathbf{L} \subset [\mathbf{Z}],
\end{aligned}
\tag{10.35}
$$

where $\Gamma_{[\mathbf{L}]}^{(\exp)}$ stands for the measured flux ratios of the peaks, and using the relation (10.31) and the definition of the $P_{[\mathbf{L}]}^z$ in (10.32) under the condition $P_{[\mathbf{L}]}^z(\exp) = P_{[\mathbf{L}]}^z\left(T_e^{(0)}\right)$, one arrives at the following system of self-consistent equations for $z =$ li, he, h, and $\mathbf{L} \subset [\mathbf{Z}]$ for $\mathbf{Z} = z$:

$$
P_{[\mathbf{L}]}^z(\exp) = \int_0^h p_{[\mathbf{L}]}^z\left(T_e^{(0)}, \beta\right) \Phi_z\left(T_e^{(0)}, \beta\right) \mathrm{d}\beta, \tag{10.36}
$$

which, due to normalization conditions (10.30) and according to the BIM formulation, can be considered as the system of Bayesian relations between the probability $P_{[\mathbf{L}]}^z\left(T_e^{(0)}\right)$ for a photon to be emitted in the peak $\mathbf{L}$ and the product of the conditional probability $p^{z}_{[\mathbf{L}]}\left(T^{(0)}_{\mathrm{e}}, \beta\right)$ for a certain β value (local temperature) on the probability density $\Phi_{\mathrm{z}}\left(T^{(0)}_{\mathrm{e}}, \beta\right)$ for this value.

For each set of (10.36) corresponding to $z = \mathbf{Z}$, the BIM procedure takes the form (see Appendix):

$$\Phi^{(n+1)}_{\mathrm{z}}\left(T^{(0)}_{\mathrm{e}}, \beta\right) = \Phi^{(n)}_{\mathrm{z}}\left(T^{(0)}_{\mathrm{e}}, \beta\right) \sum_{\mathrm{L} \subset \mathrm{Z}} \frac{P^{z}_{[\mathbf{L}]}(\exp)\, p^{z}_{[\mathbf{L}]}\left(T^{(0)}_{\mathrm{e}}, \beta\right)}{\int\limits_{0}^{b} p^{z}_{[\mathbf{L}]}\left(T^{(0)}_{\mathrm{e}}, \beta\right) \Phi^{(n)}_{\mathrm{z}}\left(T^{(0)}_{\mathrm{e}}, \beta\right) \mathrm{d}\beta}. \tag{10.37}$$

The χ^2 criterion is applied to estimate the convergence of the iterative procedure. The ratios of the partial fluxes $\Gamma^{k}_{[\mathbf{L}]}(n)$ and $\Gamma^{z}_{[\mathbf{Z}]}(n)$ for the nth iteration are calculated using the formulas from the system of (10.27)–(10.33), where the ratios for the measured fluxes $\Gamma^{(\exp)}_{[\mathbf{L}]}$ and $R^{\mathbf{K}}_{\mathbf{Z}}(\exp) = F^{(\exp)}_{[\mathbf{K}]} / F^{(\exp)}_{[\mathbf{Z}]}$ have to be used. For the zeroth approximation $\Phi^{(0)}_{\mathrm{z}}(\beta)$, we used the coronal abundances $n^{(c)}_{\mathrm{z}}(\beta)$, and the $n_a(\beta)$ profiles were used from [22] for the recombination rates $J^{h}_{[\mathbf{L}]}$. The parameter $T^{(0)}_{\mathrm{e}}$ and correction factors for the atomic data are determined by minimizing χ^2 and $\delta = \mid n(\beta) - 1 \mid$ quantities.

10.4.1.6 Results of Plasma Diagnostics and Verification of Atomic Data

On the basis of the developed SCA and using K_{α}- and K_{β}-spectra of impurity argon ions, we estimated the accuracy of atomic data necessary for modeling K-spectra and verified the corresponding methods of their calculation, as well as performed diagnostics of plasma parameters (central temperature and relative ion abundances) at the TEXTOR tokamak. Figures 10.2 and 10.3 demonstrate examples of determining some key parameters of the SM, in particular the core temperature $T^{(0)}_{\mathrm{e}}$ and effective excitation rates for the satellite group $\mathbf{N}_3$. These results showed that the χ^2 and δ optimization conditions for the BIM solutions give rise to rather strong constraints for possible $\Phi^{*}_{\mathrm{z}}(\beta)$ profiles under 5% variations of the SM parameters. Due to these constraints, the values of the SM parameters beyond their optimization region of definition cannot be simultaneously consistent with the SM system of (10.27)–(10.33) and minimization conditions χ^2 and δ. The analysis of the $\Phi^{*}_{\mathrm{z}}(\beta)$ distributions allows to derive the key parameters and to find the region of their consistency with the measured spectra: the set $\mathbf{D} = \{\alpha_l\}$ includes the ratios of the effective excitation rates for the **x**, **y**, **z** lines, **k** satellite, and satellite group $\mathbf{N}_3$, while the plasma set is $\mathbf{P} = \left\{T^{(0)}_{\mathrm{e}}, \Phi_{\mathrm{z}}(\beta)\right\}$.

It was shown that the developed approach for interpreting the experimental results from the TEXTOR tokamak made it possible to verify the methods for calculating the atomic data with an accuracy of ~5–10%. The calculations per-

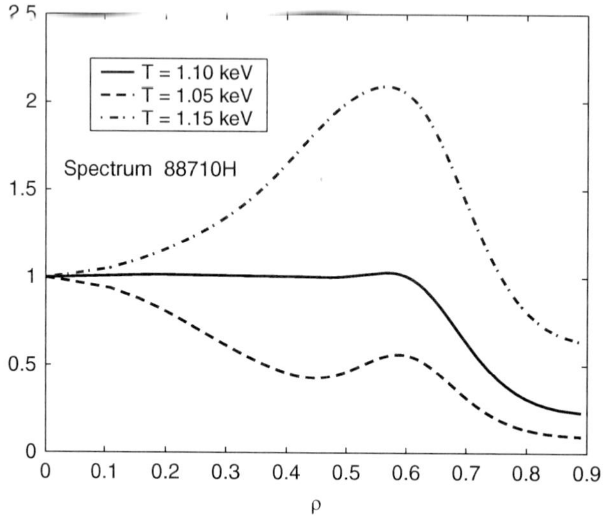

Fig. 10.2 Sum of the relative ion abundances $n(\rho)$ given by (10.23), for various values of the central temperature $T_e^{(0)}$

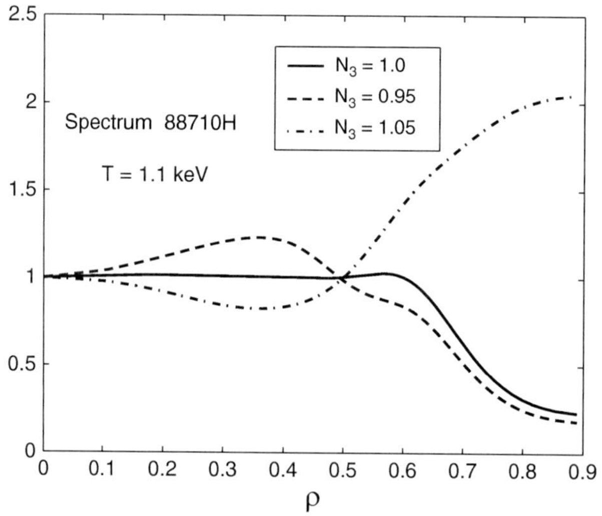

Fig. 10.3 Sum of the relative ion abundances $n(\rho)$ given by (10.23), for various relative effective excitation rates $\alpha_l\left(T_e^{(0)}\right)$ corresponding to the satellite group $\mathbf{N_3}$

formed by means of the atomic codes ATOM and MZ [28] require ~10% correction of the effective excitation rate ratios for the inter-combination line **y** and the dielectronic satellite group $\mathbf{N}_3$ to the resonance line **w**. It was established that the reason of considerable disagreement between the theory and the experiment for $\mathbf{N}_3$ and $\mathbf{N}_4$ satellite groups consists in calculations of autoionization decay probabilities obtained by means of the MZ code. The MZ code based on the Z-expansion method (Z is the nuclear charge) was modified to account for the first-order corrections (in comparison with the previous zeroth order) in powers of $1/Z$, corresponding to the screening effects, in calculations of autoionization rates. Atomic data calculations by Z-expansion method taking into account these effects for doubly excited states 2lnl′ of helium-like ions and $1s2lnl'$ of lithium-like ions with $Z = 6$–36 can be found in [29]. The spectra calculated with corrected atomic data are in agreement with the spectra measured in the wide range of plasma conditions within the experimental accuracy of $\leq 10\%$. Figure 10.4 demonstrates an example of the spectrum calculated with the central temperature $T_e^{(0)} = 1.21$ keV before and after the correction of the atomic data in comparison with a measured spectrum. In the frame of the SM, the parameter $T_e^{(0)}$ can be derived with the accuracy of ~5%, whereas the relative ion abundances n_z differ considerably (in 2–5 times) from ionization equilibrium values in coronal conditions because of the effects of the transfer and charge exchange of the impurity argon ions on neutral atoms.

Data obtained on the TEXTOR tokamak for the K_β-emission spectrum of Ar^{16+} ions at temperatures $T_e \sim 1$ keV made it possible to study the temperature dependence of the ratio $G_3 = I[K_{\beta_2}]/I[K_{\beta_1}]$ for the intercombination and resonance lines corresponding respectively to the transitions $1s3p(^3P_1) \to 1s^2(^1S_0)(K_{\beta_2}$ line) and $1s3p(^1P_1) \to 1s^2(^1S_0)(K_{\beta_1}$ line) [16]. Figure 10.5 shows an example of a measured spectrum containing the K_{β_1} and K_{β_2} lines. The emissivity functions for these lines were calculated as functions of electron temperature and density for the equilibrium plasma with the Maxwellian electron velocity distribution. Using these quantities, the relative K_β intensities were derived and compared with experimental data (see [16] for details). The calculations carried out within the framework of the radiative–collisional model using the ATOM and MZ atomic codes are in agreement, to the experimental accuracy, with all the experimental data obtained on the PLT (Princeton, USA) [19], ALCATOR-C (Cambridge, USA) [30], and TEXTOR tokamaks. It was shown that the previously observed discrepancies by factor of 1.3–2 between the measured and calculated G_3 ratios and exceeding appreciably the experimental errors are caused by the use of inaccurate atomic data and simplified atomic models in those works [19, 30]. These results are evidence for a high accuracy of atomic data used and for the possibility of effectively using, on their basis, the K_β lines for diagnostics of the electron temperature and density in the laboratory and astrophysical coronal plasma sources.

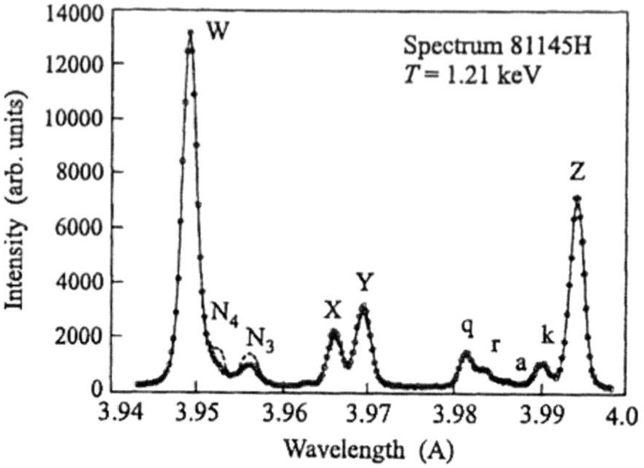

Fig. 10.4 Argon K_α spectrum in the range 3.94–4.0 Å measured at the TEXTOR tokamak (*points*) and calculated (*solid line*) for the central temperature $T_e^{(0)} = 1.21$ keV taking into account correction of the atomic data. The *dashed line* shows the spectral zones calculated without correction of the atomic data, for which the experimental data noticeably differ from the calculations (**N**$_4$, **N**$_3$, **X**, **Y**, **Z**)

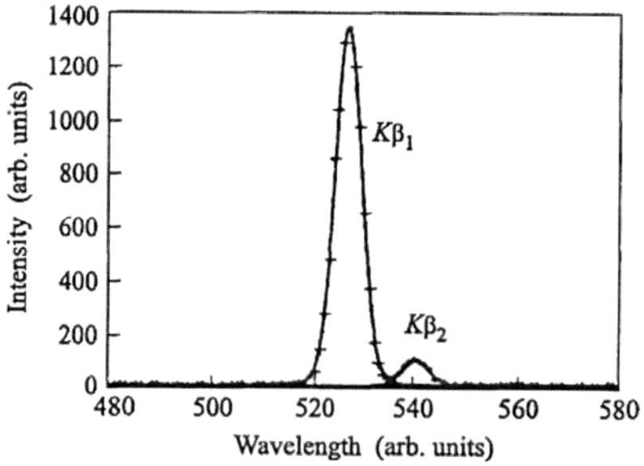

Fig. 10.5 Measured (*crosses*) K_β-emission spectrum of Ar^{16+} ions and its approximation by two Voigt profiles (*solid line*) taking into account the instrumental function and the radial distribution of ion temperature

10.4.2 XUV Spectra and Imaging Data from the Solar Corona

10.4.2.1 XUV Imaging Spectroscopy for the Sun

High-resolution imaging spectroscopy, giving rise to the "golden age" of the XUV solar astronomy, made it possible to visualize high inhomogeneity and temporal variability of coronal plasma structures. Observations with full-Sun broadband soft X-ray telescope (SXT) on the Yohkoh satellite discovered a complex structure and plasma dynamics with high temperatures $T > 2\,\text{MK}$. These structures, associated with magnetic configurations of various topology, were observed later with unprecedented spatial resolution by X-ray telescope (XRT) onboard the Hinode satellite. Implementation of imaging devices with narrow EUV spectral bands on SOHO (Solar and Heliospheric Observatory), CORONAS-F, TRACE (Transition Region and Coronal Explorer), STEREO, and other missions helped to study spatial and kinematic properties of $\sim$1–3 MK plasma of quiescent as well as "active" coronal loops, manifesting in explosive and eruptive phenomena (flares, coronal mass ejections, EIT-waves, etc.). Hard X-ray telescope (HXT) with high spatial and temporal resolution on Yohkoh and RHESSI (Reuven Ramaty High-Energy Solar Spectroscopic Imager) experiments allowed to localize superhot thermal ($T > 10\,\text{MK}$) and high-energy nonthermal plasma sources and to study their temporal dynamics during flare events. Spatial and temporal characteristics of plasma derived from aforementioned experiments resulted in plasma modeling and testing theoretical scenarios for flare and eruptive phenomena. Nevertheless, despite significant progress achieved in solar corona physics in the last decades owing to intensive ground-based and space studies, a number of basic questions related to specific mechanisms of energy release during solar flare events and its transformation remain debatable.

The main reason that restricts further progress in theoretically describing active events on the Sun is the shortage of information either about the spectral composition of XUV emission in broadband filter images or about the spatial localization of the monochromatic emission in the full-Sun line spectra. It leads to significant uncertainties in diagnosing the basic plasma characteristics—spatial distributions of emission measure, temperature, and density, as well as their temporal dynamics. This fact significantly hampers plasma modeling and requires further experimental and theoretical studies.

The advent of XUV full-Sun monochromatic imaging spectroscopy in the SPIRIT (Spectroheliographic X-ray Imaging Telescope) experiment onboard the CORONAS-F satellite (functioning during 2001–2005) allowed to disclose a new class of 4–20 MK plasma strictures characterized by specific morphologic and temporal features. Inverse spectroscopic methods developed and verified by employing temporal series of SPIRIT images showed a principle possibility to infer spatial and temporal properties of hot plasma (temperature, density, and EM distributions) needed for quantitative description of transient phenomena revealed in monochromatic XUV images [8, 18].

10.4.2.2 SPIRIT Experiment Onboard the CORONAS-F Satellite

The multichannel RES spectroheliograph of the SPIRIT instrumentation was designed to acquire monochromatic images of the full solar disc and the adjacent corona with relatively high spatial (up to 5″), temporal (up to 7 s), and spectral resolutions in the following spectral bands: (1) 8.41–8.43 Å (two MgXII X-ray channels including the MgXII resonance line 8.42 Å) and (2) two EUV spectral channels, 176–207 Å and 280–330 Å [18, 31]. Various programs of observations in the MgXII and EUV channels, including simultaneous observations, were realized in the SPIRIT experiment.

Note that the RES spectroheliograph in the MgXII channel register the emission of hot plasma at temperatures $T > 4$ MK because the emissivity function of this line is sensitive at high temperatures and has the maximum at $\approx$10 MK. This implies that the monochromatic full-Sun MgXII 8.42 Å images provide direct confirmation of the presence of hot plasmas. The Sun monochromatic X-ray images revealed regions in the solar corona with high temperatures $T > 4$ MK and, thus, allowed a new class of highly dynamic plasma structures with various characteristic sizes and lifetimes from several minutes to several days to be discovered (see [32–34]). In particular, long-lived (up to several days) plasma structures located high in the corona (up to 3×10^5 km) and resembling spiders in shape were observed. Examples of such "spiders" observed on 2001 November 12 (left image) and on 2001 December 29 (right image) in the MgXII channel are shown in Fig. 10.6.

Figure 10.7 demonstrates a comparison of images acquired on 2001 November 12 in the broadband SXT/Yohkoh channel sensitive to temperatures in a wide range of about $\sim$2.5–25 MK (left panel) and in the monochromatic MgXII channel (right panel) whose sensitivity range is narrower ($\sim$5–15 MK). This comparison shows a substantial difference between the spatial scales of plasma structures with different temperature contents: $T > 2$ MK plasma occupying a large area in the SXT image and well-localized plasma features at temperatures $T > 4$ MK, fuzzy in the SXT broadband channel, but clearly developed in the monochromatic MgXII image.

10.4.2.3 Results of the EUV Spectra Analysis

During the SPIRIT experiment carrying out onboard the CORONAS-F satellite, several thousands of spectroheliograms in two spectral bands 176–207 Å and

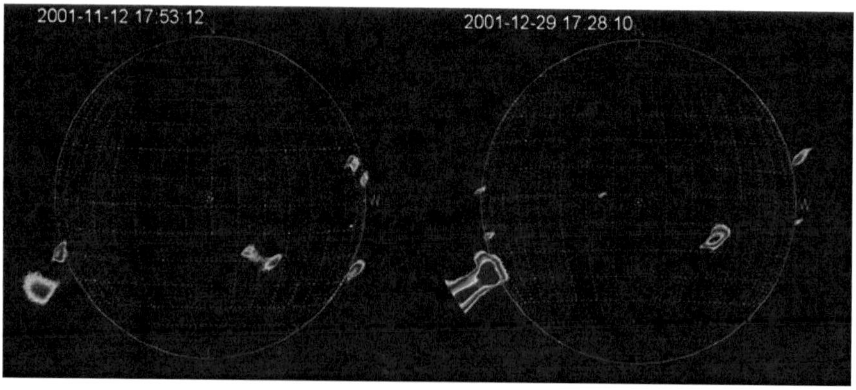

Fig. 10.6 Examples of the hot coronal structures ("spiders") in the monochromatic MgXII 8.42 Å line on 2001 November 12 and on 2001 December 29

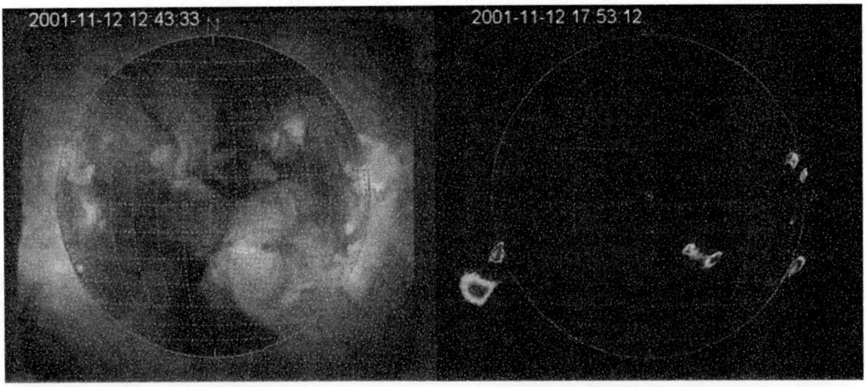

Fig. 10.7 Solar corona images recorded with the broadband SXT/Yohkoh filters (*left panel*) and in the monochromatic MgXII/CORONAS-F channel (*right panel*) on 2001 November 12

280–330 Å were acquired by the XUV channels of the spectroheliograph RES. The spectra of various solar regions (active regions, quiet sun areas on the disc, flares, etc.) were analyzed and a catalogue of the spectral lines was composed [35].

Using the lines with the relative intensities sensitive to the temperature in the range ~0.5–20 MK but independent on the electron density, the DEM temperature distributions were calculated by means of the BIM. The contribution functions $G(l,T)$ for the lines were derived from the CHIANTI atomic database [3, 4]. Results of these calculations for a series of active events are presented in [8, 18]. An important result is the similarity between the temperature behaviors of the DEM profiles for a number of active regions in a wide temperature range up to ~8 MK. It is also worth to note that a few active regions had a pronounced peak of plasma material toward temperatures of ~10–12 MK.

Another important result of diagnosing the hot plasma in active coronal structures was the determination of the soft X-ray emission mechanism of the "spider" plasma. This was made by comparing data measured in the EUV and soft X-ray spectral ranges [8]. The calculations were performed for the event on 2001 December 29 (see right image in Fig. 10.6). In this figure, one may see a radially (along the solar radius) elongated "spider" structure in the X-ray (hot) MgXII RES image. Due to the favorable angular orientation of the RES instrument in the observing period (in the XUV channel corresponding to the spectral band 280–330 Å, the "spider" was directed along the axis perpendicular to the direction of dispersion), it was possible to analyze the radial dependence of the DEM temperature distributions along the solar radius. Figure 10.8 shows these distributions for three altitudes above the solar limb and for an active region on the limb. From this figure, one can clearly see a nonuniform radial distribution of plasma material (DEM) in the temperature range 4–10 MK. For the sake of comparison, a DEM distribution is presented for a flare event of the class X3.4 on 2001 December 28, which is extremely peaked at temperatures ~10–12 MK.

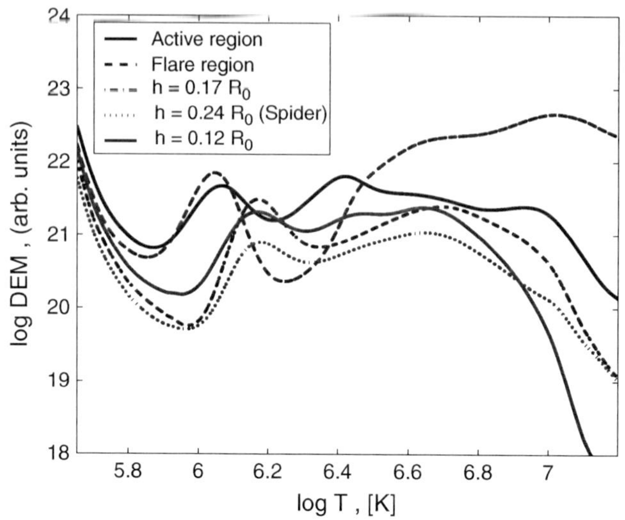

Fig. 10.8 DEM temperature distributions (**a**) for the "spider" on December 28–29, 2001 (radial dependence for three altitudes above the limb along the solar radius), (**b**) for the active region on the limb NOAA 9765, and (**c**) for a flare of the class X3.4 on December 28, 2001

Figure 10.9 shows the radial distributions of the intensities (normalized to the values on the solar limb) in the individual EUV lines and in the X-ray MgXII line recorded when there were no flare events. A fundamentally important result of this comparison is that the observed intensities of the EUV CaXVIII line and the X-ray MgXII line with the close formation temperatures behave quite differently: the intensity of the former line decreases, while the intensity of the latter one increases. At the same time, the relative intensities in the MgXII line calculated using the distributions in Fig. 10.8 for the active region on the limb and three regions of the spider (marked by the crosses in Fig. 10.9) behave similarly to the radial intensity distribution observed in the CaXVIII line.

The following explanation of this fact was given [8]. Since the excitation threshold of the EUV lines is low enough, their intensities are proportional to the emission measure formed by thermal (Maxwellian) electrons. The X-ray line of the MgXII ion having a considerably higher excitation threshold is formed by high-energy nonthermal electrons ($\sim$2 keV and more) whose densities can reach several percent of the total electron density. As a consequence, the EUV line intensities do not depend on the presence of a small admixture of nonthermal electrons and are proportional to the "thermal" emission measure. In contrast, the X-ray line intensities are determined by the relative contribution functions, which can exceed the thermal part by many times due to nonthermal electrons. Thus, the observational data from the RES EUV channels clearly point to the nonthermal (nonstationary) character of the emission mechanism of the hot plasma for the event under study. This was in close agreement with the conclusions drawn from an independent analysis of the time profiles of the emission in the MgXII channel (see [8, 18] for details).

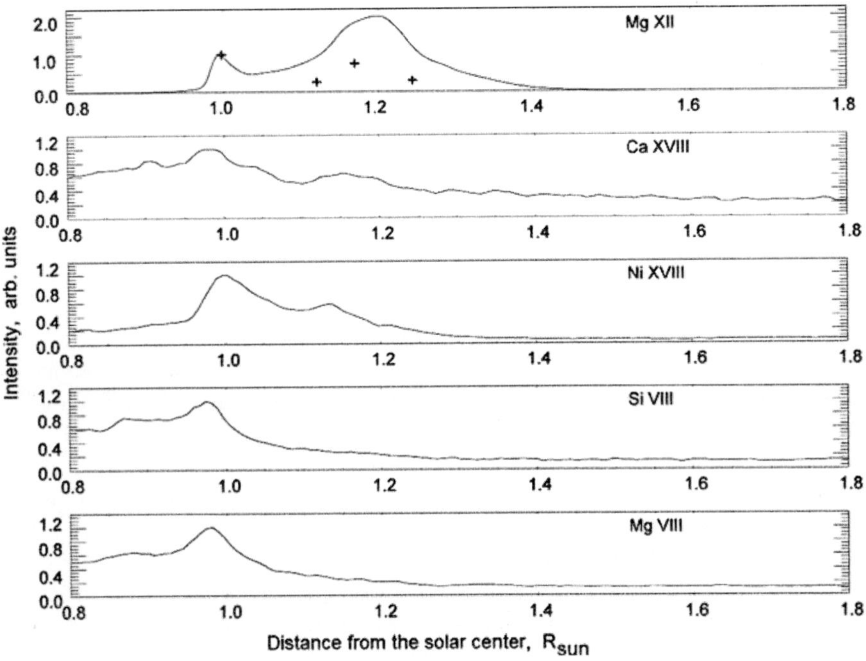

Fig. 10.9 Radial (along the solar radius) distribution of intensities in the EUV lines and in the MgXII channel in the region of the "spider" on December 29, 2001

A deep investigation of the BIM abilities as a temperature diagnostic technique for analyzing XUV solar data was carried out in [9]. A series of numerical tests and model simulations were realized in the case of both line spectra and broadband imaging data. Applications to high-resolution line spectra (from SUMER/SOHO and SPIRIT/CORONAS-F data) and broadband-imaging data (provided by XRT/Hinode instrument) were also considered. These studies confirmed the robustness and effectiveness of the BIM as a tool for the temperature analysis of hot plasma structures.

10.5 Conclusions

Studies of hot plasmas by means of spectroscopic diagnostics methods have wide development related to the necessity to solve important applied problems, such as the controlled fusion, X-ray lasers, processes in atmospheres of the Sun, and other stars. On the other hand, spectroscopic methods force to formulate and to solve inverse problems, which are crucially dependent on input data. In particular, the results of spectroscopic diagnostics based on the XUV emission spectra depend considerably on the accuracy of atomic data, as well as on adopted models of the emitting sources. In this context, we have given a review to show our developments of spectroscopic methods for diagnostics of hot optically thin plasmas and have demonstrated their applications to studying plasma structures in astrophysical and laboratory conditions.

We formulated and developed a SCA allowing to estimate (or to verify) the accuracy of atomic data needed for modeling and diagnosing hot coronal plasmas. This approach is based on solving the spectral inverse problem by means of the BIM algorithm in the framework of the adopted "spectroscopic model." High-resolution K-spectra of highly charged ions measured in the TEXTOR tokamak plasma were used for the verification of atomic data (spectral and collisional characteristics) and methods of their calculation. In particular, it was shown that the calculations carried out by means of the ATOM and MZ codes require essential corrections of atomic data for dielectronic satellites (~30–50% for autoionization probabilities). Corrected atomic data made it possible to perform an accurate diagnostics of plasma parameters, namely, to determine plasma temperatures and argon ion densities in the tokamak plasma. Atomic data verified employing the high resolution laboratory spectra may essentially improve the accuracy of plasma parameters derived from inverse diagnostics techniques in the case of astrophysical sources.

We also used our diagnostics techniques for studying the solar coronal plasma. Investigations and detailed analysis of the BIM abilities for the DEM temperature analysis through numerical tests and simulations were performed by modeling various plasma structures in the solar atmosphere. The diagnostics of active regions in the solar corona plasma was carried out using EUV spectral data acquired onboard the CORONAS-F satellite. The most reliable lines were used to reconstruct DEM temperature distributions by means of the BIM. These results allowed us to infer the presence of hot plasma at temperature range ~4–10 MK in a number of active regions on the Sun, as well as to confirm the nonthermal nature of the soft X-ray emission mechanism in active coronal structures.

Appendix: Bayesian Iterative Scheme

Let us consider two related complete systems of events $\{X_i\}$ and $\{Y_k\}(i = 1,\ldots,n; k = 1,\ldots,m)$, and corresponding sets of probability distributions $\{P(X_i)\}$ and $\{P(Y_k)\}$ for them. In applications, these distributions may also be ones for some random variables $\mathbf{X}$ and $\mathbf{Y}$. The probability distributions $\{P(X_i)\}$ and $\{P(Y_k)\}$ are related by the formulas of the total probability:

$$P(Y_k) = \sum_i P(Y_k|X_i)\, P(X_i), \tag{10.38}$$

where $P(Y_k|X_i)$ is the conditional probability of the event Y_k at the condition X_i. If the distribution $\{P(Y_k)\}$ is known, one can formulate the problem for deriving the $\{P(X_i)\}$ one from the relations (10.38). Below, we will state an iterative procedure called the BIM to resolve this task.

The BIM is based on Bayes' theorem for the a posteriori conditional probability connecting two random variables defined on the fields of events $\{X_i\}$ and $\{Y_k\}$ as follows:

$$P(X_i|Y_k) = \frac{P(Y_k|X_i)\,P(X_i)}{\sum_j P(Y_k|X_j)\,P(X_j)}. \tag{10.39}$$

The formula of the total probability for the distribution $P(X_i)$ is then given by the expression inverse to (10.38):

$$P(X_i) = \sum_k P(X_i|Y_k)\,P(Y_k). \tag{10.40}$$

Substituting (10.39) in (10.40) gives the identity

$$P(X_i) = P(X_i)\sum_k \frac{P(Y_k|X_i)\,P(Y_k)}{\sum_j P(Y_k|X_j)\,P(X_j)}. \tag{10.41}$$

The expression (10.41) can be used for formulating an iterative scheme. For this purpose, the value $P(X_i)$ in the right side of (10.41) is interpreted as step n, and in the left side as step $(n+1)$ of the iterative procedure. Thus, one obtains the following recurrence relation for the distribution $P(X_i)$:

$$P^{(n+1)}(X_i) = P^{(n)}(X_i)\sum_k \frac{P(Y_k|X_i)\,P(Y_k)}{\sum_j P(Y_k|X_j)\,P^{(n)}(X_j)}. \tag{10.42}$$

It is worth to make some comments regarding the formula (10.42). The left side of (10.42) can be considered as the estimate of the nth hypothesis for the probability distribution $P(X_i)$. The initial approximation for a priori distribution $P^{(0)}(X_i)$ may be taken in accordance with any prior information. If such information is absent, according to the Bayes' postulate, one assumes a uniform distribution (corresponding to equal lack of knowledge). It is also possible, using relation (10.42), to show that the normalizing condition for the distribution $\{P(X_i)\}$ is automatically conserved at any step of the iterative procedure (10.42).

Acknowledgements We would like to thank our colleague Vladimir Slemzin for fruitfully collaborating on the subject of the present review. We also gratefully acknowledge the financial support from the European Commission Programme under the grant agreement 218816 (FP-7 SOTERIA project) and the Russian Foundation for Basic Research (project 11-02-01079-a)—Programme of the Presidium of the Russian Academy of Sciences "Plasma Processes in the Solar System."

References

1. S.R. Pottasch, Space Sci. Rev. **3**, 816 (1964)
2. K.J.H. Phillips, U. Feldman, E. Landi, *Ultraviolet and X-ray Spectroscopy of the Solar Atmosphere* (Cambridge University Press, New York, 2008)

3. K.P. Dere, E. Landi, H.E. Mason, B.C. Monsignori Fossi, P.R. Young, Astron. Astrophys. Suppl. Ser. **125**, 149 (1997)
4. E. Landi, G. Del Zanna, P.R. Young, K.P. Dere, H.E. Mason, M. Landini, Astrophys. J. Suppl. Ser. **162**, 261 (2006)
5. J.T. Jefferies, F.Q. Orrall, J.B. Zirker, Sol. Phys. **22**, 307 (1972)
6. I.J.D. Craig, J.C. Brown, Astron. Astrophys. **49**, 239 (1976)
7. J.C. Brown, B.N. Dwivedi, Y.M. Almleaky, P.A. Sweet, Astron. Astrophys. **249**, 277 (1991)
8. A.M. Urnov, S.V. Shestov, S.A. Bogachev, F.F. Goryaev, I.. Zhitnik, S.V. Kuzin, Astron. Lett. **33**, 396 (2007)
9. F.F. Goryaev, S. Parenti, A.M. Urnov, S.N. Oparin, J.-F. Hochedez, F. Reale, Astron. Astrophys. **523**, id.A44 (2010)
10. I.J.D. Craig, J.C. Brown, *Inverse Problems in Astronomy* (Hilger, Bristol, 1986)
11. A. Tarantola, *Inverse Problem Theory and Methods for Model Parameter Estimation* (SIAM, Philadelphia, 2005)
12. E.L. Kosarev, Inverse Probl. **6**, 55 (1990)
13. V.I. Gelfgat, E.L. Kosarev, E.R. Podolyak, Comp. Phys. Commun. **74**, 349 (1993)
14. W.H. Richardson, JOSA **62**, 55 (1972)
15. I.A. Zhitnik, V.V. Korneev, V.V. Krutov, S.N. Oparin, A.M. Urnov, TrSSR (Trudy Akademiia Nauk SSSR Fizicheskii Institut) **179**, 39 (1987)
16. F.F. Goryaev, A.M. Urnov, G. Bertschinger, A.G. Marchuk, H.-J. Kunze, J. Dubau, JETP Lett. **78**, 363 (2003)
17. A.M. Urnov, F.F. Goryaev, G. Bertschinger, H.-J. Kunze, O. Marchuk, JETP Lett. **85**, 374 (2007)
18. I.A. Zhitnik, S.V. Kuzin, A.M. Urnov, S.A. Bogachev, F.F. Goryaev, S.V. Shestov, Sol. Sys. Res. **40**, 272 (2006)
19. P. Beiersdorfer, A.L. Osterheld, T.W. Phillips, M. Bitter, K.W. Hill, S. von Goeler, Phys. Rev. E **52**, 1980 (1995)
20. A.J. Smith, P. Beiersdorfer, V. Decaux, K. Widmann, K.J. Reed, M.H. Chen, Phys. Rev. A **54**, 462 (1996)
21. O. Marchuk, G. Bertschinger, A. Urnov, F. Goryaev, N.R. Badnell, A.D. Whiteford, J. Plasma Fusion Res. Ser. **7**, 274 (2006)
22. O. Marchuk, M. Tokar, G. Bertschinger, A. Urnov, H.-J. Kunze, D. Pilipenko, D. Kalupin, D. Reiter, A. Pospieszczyk, M. Biel, M. Goto, F. Goryaev, Plasma Phys. Contr. Fusion **48**, 1633 (2006)
23. A.H. Gabriel, Mon. Not. R. Astron. Soc. **160**, 99 (1972)
24. G. Bertschinger, W. Biel, the TEXTOR-94 Team, O. Herzog, J. Weinheimer, H.-J. Kunze, M. Bitter, Phys. Scr. T **83**, 132 (1999)
25. J. Weinheimer, I. Ahmad, O. Herzog, H.-J. Kunze, G. Bertschinger, W. Biel, G. Borchert, M. Bitter, Rev. Sci. Instrum. **72**, 2566 (2001)
26. G. Bertschinger, O. Marchuk, in *High-temperature plasmas diagnostics by X-ray spectroscopy in the low density limit*, ed. by R.E.H. Clark, D.H. Reiter. Nuclear Fusion Research (Springer, Berlin, 2005), pp. 183–199
27. M.Yu. Kantor, A.J.H. Donné, R. Jaspers, H.J. van der Meiden, TEXTOR Team, Plasma Phys. Contr. Fusion **51**, 055002 (2009)
28. V.P. Shevelko, L.A. Vainshtein, *Atomic Physics for Hot Plasmas* (IOP Publishing, Bristol, 1993)
29. F.F. Goryaev, A.M. Urnov, and L.A. Vainshtein, ArXiv, physics/0603164 (2006)
30. J.E. Rice, K.B. Fournier, U.I. Safronova, J.A. Goetz, S. Gutmann, A.E. Hubbard, J. Irby, B. LaBombard, E.S. Marmar, J.L. Terry, New J. Phys. **1**, 19 (1999)
31. S.V. Kuzin, I.A. Zhitnik, A.A. Pertsov, V.A. Slemzin, A.V. Mitrofanov, A.P. Ignatiev, V.V. Korneev, V.V. Krutov, I.I. Sobelman, E.N. Ragozin, R.J. Thomas, J. X-ray Sci. Technol. **7**, 233 (1997)
32. I. Zhitnik, S. Kuzin, A. Afanas'ev, O. Bugaenko, A. Ignat'ev, V. Krutov, A. Mitrofanov, S. Oparin, A. Pertsov, V. Slemzin, N. Sukhodrev, A. Urnov, Adv. Space Res. **32**, 473 (2003a)

33. I.A. Zhitnik, O.I. Bugaenko, A.P. Ignat'ev, V.V. Krutov, S.V. Kuzin, A.V. Mitrofanov, S.N. Oparin, A.A. Pertsov, V.A. Slemzin, A.I. Stepanov, A.M. Urnov, Mon. Not. R. Astron. Soc. **338**, 67 (2003b)
34. I. Zhitnik, S. Kuzin, O. Bugaenko, A. Ignat'ev, V. Krutov, D. Lisin, A. Mitrofanov, S. Oparin, A. Pertsov, V. Slemzin, A. Urnov, Adv. Space Res. **32**, 2573 (2003c)
35. S.V. Shestov, S.A. Bozhenkov, I.A. Zhitnik, S.V. Kuzin, A.M. Urnov, I.L. Beigman, F.F. Goryaev, I.Yu. Tolstikhina, Astron. Lett. **34**, 33 (2008)

Chapter 10
On Spectroscopic Diagnostics of Hot Optically Thin Plasmas

A.M. Urnov, F. Goryaev, and S. Oparin

Abstract X-ray and extreme ultraviolet (XUV) emission spectra of highly charged ions in hot plasmas contain diverse information on both elementary processes and the ambient medium. The theoretical analysis of spectra and spectral images of laboratory and astrophysical sources of short-wave radiation, based on the modern methods of atomic data calculations of spectral and collisional ion characteristics, allows one to determine various physical parameters of the emitting plasma. Here, we consider and discuss some basic principles on which the spectroscopic diagnostics of hot optically thin plasmas emitting XUV spectra is based. In order to obtain information about the internal structure of a physical system under study, one generally needs to solve inverse problems for determining the physical conditions in the plasma. Using concepts from the probability theory, we formulate the spectral inverse problem in the framework of the probabilistic approach to be used for the temperature diagnostics of hot plasma structures. We then demonstrate applications of our diagnostics methods to hot plasmas in laboratory (tokamak plasma) and astrophysical (solar corona) conditions.

10.1 Introduction

Spectroscopic methods of investigation are of great importance for numerous problems related to hot astrophysical and laboratory plasmas. Spectroscopic diagnostics of hot plasmas is a very effective and in many cases unique (for instance, for astrophysical plasma) way for deriving information on structure and dynamics of plasma sources. The analysis of X-ray and extreme ultraviolet (XUV) emission spectra based on calculations of spectral and collisional characteristics of highly charged ions allows us to determine various plasma parameters: electron temperature and density, element abundances, ionization state, temperature structure of hot plasma sources, etc. Plasma characteristics derived from XUV emission are needed to constrain the classes of relevant plasma models and to enable

V. Shevelko and H. Tawara (eds.), *Atomic Processes in Basic and Applied Physics*, Springer Series on Atomic, Optical, and Plasma Physics 68,
DOI 10.1007/978-3-642-25569-4_10,

quantitative simulations of plasma processes spatial and temporal dynamics of plasma parameters in the emitting regions.

Among astrophysical objects to be actually widely exploited, one can mark X-ray binary sources, interplanetary hot gases, coronae of the Sun, and other stars. The solar corona due to low electron densities and large range of temperatures is an important source of information on spectra and excitation processes of highly charged ions. The solar atmosphere is also of great interest due to the complex structure (active regions, coronal holes, bright points, coronal condensations) and its activity behavior (flares, coronal mass ejections, explosive protuberances, jets, and others). The active phenomena are prominent manifestations of nonstationary processes leading to the transformation of magnetic energy to its other forms; however, till now, their nature is not completely understood. Understanding mechanisms of these local processes is also important for solving fundamental problems of the physics of the solar atmosphere such as the coronal-heating problem and acceleration of the solar wind.

A number of conditions characterizing the astrophysical plasma can be reproduced in laboratory devices. This "laboratory astrophysics" can be used for the purpose of studying properties of short-wave emission of highly charged ions, in particular the verification (estimation of accuracy) of atomic data and methods of spectroscopic diagnostics. The precision of spectroscopic methods of plasma diagnostics and even possibility of their use depend on both the accuracy of atomic data and adopted models of emitting plasma based on the equations of atomic kinetics and plasma dynamics. The spectra of low-density (coronal) plasma from electron beam plasma devices and tokamaks are important sources of information about both binary atomic and hydrodynamic processes. The topicality of problems of the XUV spectroscopy is also defined by numerous applications in the atomic spectroscopy, requirements for diagnostics of emitting plasma objects, and necessity of developing short-wave emission sources for applying to the biology, medicine, materials technology, and in other domains of the modern science and technology.

In order to solve the main problem of the spectroscopy, that is, to identify and to interpret line spectra of the emitting plasma, a lot of atomic characteristics are needed, as well as information on the plasma sources under study. On the other hand, when having reliably identified spectra, one should build up models of the emitting plasmas and, using them, determine plasma macroparameters—spatial distributions of temperature, density, ionic composition, and other characteristics as well as their temporal dynamics, that is, to solve another problem of the spectroscopy—plasma diagnostics. The latter task is closely dealt with the necessity to formulate and to solve inverse problems, which frequently arise in practical applications to interpret the results of experiments and observations and to obtain information about the internal structure of a physical system under study. It is however worth to note that all spectroscopic methods based on the solution of the spectral inverse problem require definite model assumptions. These assumptions made explicitly or implicitly lead to a formulation of the appropriate model for the emitting plasma. Thus, the results of diagnostics of plasma parameters depend on the adopted model. In order to avoid the ambiguity, the chosen fitting parameters should be consistent with the physical model and all available information about the radiation source.

This review is devoted to some spectroscopic problems related to diagnostics of hot plasma sources using their XUV emission spectra. Here, we only consider the case of optically thin sources and low density (in particular coronal) plasmas. The latter means that the emission fluxes are formed by the damping due to the binary atomic processes and, as consequence, depend linearly on both electron and ionic densities. Low-density conditions are realized in many astrophysical objects; these also are often applicable to the interpretation of line spectra observed in laboratory devices, for example, tokamaks, laser plasma, pinches, etc. Firstly, we discuss basic principles on which the spectroscopic diagnostics of hot plasmas is based and how these lead to the formulation of the inverse problems for the distribution of radiating plasma material. Then we briefly describe two mathematical formalizations of the spectral inverse problem and formulate our inversion method developed in the frame of the probabilistic approach. Finally, we demonstrate some applications of developed diagnostics methods to the interpretation of spectral data from tokamak and solar plasmas.

10.2 Basic Principles of Spectroscopic Diagnostics of Hot Optically Thin Plasmas

The spectroscopic diagnostics is based on the sensitivity of the distribution of the emission spectrum over the photon energies to the physical conditions in plasma. In order to extract the information about plasma macroparameters from the line or/and continuum spectra, one should generally solve the spectral inverse problem. This problem, however, could not be solved or even formulated for an arbitrary case without an additional knowledge concerning the state of the plasma source. The shape of the spectra depends on the properties of the emitting plasma, which should be stipulated with the help of complementary experimental and theoretical analysis of the source properties. Resulting assumptions made explicitly or implicitly make it possible to express the spectroscopic characteristics of XUV emission in close analytical form providing mathematical formulation of the inverse problem. These assumptions make up a basis for a physical model of the emitting plasma. The results of spectroscopic diagnostics depend on the accepted model and therefore the same parameters of plasma could be different for various models.

The basic model assumptions used for diagnostics purposes usually include some opposite plasma conditions, for example, steady state or transient plasma, thermal or nonthermal conditions, or optically thin- or thick-emitting sources. In order to provide a comparative analysis of diverse models, one should in fact consider and analyze the dependence of spectroscopic characteristics of emission on these assumptions.

The steady-state conditions in plasma imply that characteristic times of relaxation τ_e, τ_i, and τ_z (for electrons, ions, and ionization equilibrium, respectively) are much less than the observational time of spectra τ. Furthermore, it is also assumed that the distribution functions of electrons and ions, as well as the distribution of ion species N_z do not depend on time. The opposite case of non-steady-state conditions ($\tau \ll \tau_e, \tau_i, \tau_z$) is a subject of a special study and is out of the scope of our overview. At the intermediate situation, when the plasma is in a transient state, the conditions $\tau_e \ll \tau < \tau_i, \tau_z$ are assumed to be fulfilled.

Thermal plasma condition strictly speaking implies the presence of Maxwellian velocity distribution for all sorts of plasma particles with the same temperature T. However, in applications, one often considers the quasi-steady state plasma characterized by Maxwellian distribution functions with different electron and ion temperatures, $T_e \neq T_i$. The ionization equilibrium for these plasmas is usually described by means of the parameter T_z defined as the temperature corresponding to the observed ion densities. Then the condition $T_z = T_e$ indicates the plasma in steady state, $T_z < T_e$ corresponds to the ionizing plasma, and $T_z > T_e$ to the recombining one.

Further in this paragraph, we will define some important spectral characteristics and notions, which are widely used in applications to diagnostics of hot optically thin plasmas.

10.2.1 Intensities of Spectral Lines

The total line intensity (or flux) emitted by an optically thin plasma source of the volume V in the transition $i \to f$ and observed at a distance R can be expressed as

$$I_{if} = \frac{1}{4\pi R^2} \int_V \varepsilon_{if}(\mathbf{r})\, dV . \quad (\text{phot cm}^{-2}\,\text{s}^{-1}). \tag{10.1}$$

Here, the total volume emissivity function $\varepsilon_{if}(\mathbf{r})$ (phot cm^{-3} s^{-1}) is given by

$$\varepsilon_{if} = N_i\, A_{if}, \tag{10.2}$$

where N_i (cm^{-3}) is the population density for the upper level (i) of the emitting ion, and A_{if} (s^{-1}) is the radiative spontaneous probability for the transition $i \to f$. The number density N_i is generally determined by solving a system of kinetic equations of balance:

$$\frac{dN_i}{dt} = \sum_{m \neq i} N_m W_{mn} - N_i \sum_{n \neq i} W_{in}, \tag{10.3}$$

where each of integer indices n, m, i, ... is used for the enumeration of a state of the ion with charge z and a set of quantum numbers $\{\alpha\}$ characterizing this state. The matrix element W_{mn} denotes the total probability rate coefficient for the transition $m \to n$ and is a sum of all radiative and collisional rate coefficients contributing to this transition.

Numerical solution of the system of (10.3) can be obtained in the framework of an accepted atomic collisional–radiative model, which implies the specification of elementary processes relevant to particular plasma source under consideration. Through the rate coefficients W_{mn}, level populations N_i are the functions of electron temperature T_e and density N_e, and possibly of parameters of distribution velocity functions of plasma particles. At steady state (or quasi-steady state) plasma conditions, one assumes that $dN_i/dt = 0$ in (10.3). For the general case, it is necessary to add members connected with plasma movements leading to space-temporal nonequilibrium of ionic populations in the left side of (10.3).

In the real plasmas, the spectral lines are not monochromatic, and their fluxes are distributed over a wavelength range. Lines can be broadened by various mechanisms, for example, natural and Doppler broadening and the influence of the instrumentation ("apparatus function"). In order to take into account the line profile, one has to introduce the spectral density of the volume emissivity, $\varepsilon_l(\lambda)$ (phot $\mathrm{cm}^{-3}\,\mathrm{s}^{-1}\,\text{Å}^{-1}$), in the spectral line labeled l for the transition $i \rightarrow f$ as follows:

$$\varepsilon_l(\lambda) = N_{\mathrm{i}}\, A_{if}\, \varphi(\lambda - \lambda_l), \tag{10.4}$$

where the function $\varphi(\lambda - \lambda_l)$ is the line profile normalized to unity when integrated over λ, and λ_l is the central wavelength of the photon flux distribution $\varphi(\lambda - \lambda_l)$.

Introducing also the spectral density $I(\lambda - \lambda_l)$ for the line intensity

$$I(\lambda - \lambda_l) = \frac{1}{4\pi R^2} \int\limits_V \varepsilon_l(\mathbf{r})\, \varphi(\lambda - \lambda_l)\, \mathrm{d}V, \tag{10.5}$$

one has for the whole spectrum $I(\lambda) = \sum_l I(\lambda - \lambda_l)$

$$I(\lambda) = \frac{1}{4\pi R^2} \int\limits_V \widetilde{F}(\lambda; \mathbf{r})\, \mathrm{d}V, \tag{10.6}$$

where the total spectral density $\widetilde{F}(\lambda; \mathbf{r})$ for the emissivity functions $\varepsilon_l(\mathbf{r})$ is given by

$$\widetilde{F}(\lambda; \mathbf{r}) = \sum_l \varepsilon_l(\mathbf{r})\, \varphi(\lambda - \lambda_l)\ . \tag{10.7}$$

In the sequel, we will be generally interested in the total flux and emissivity, given by (10.1) and (10.2) respectively, without going into the details of the line shapes.

10.2.2 Concept of Differential Emission Measure

It is useful to define another important diagnostic tool widely used in applications for determining physical conditions in hot plasma sources, namely, differential emission measure (DEM) function apparently introduced for the first time by Pottasch [1]. In order to define this quantity, it is convenient to rewrite the flux in (10.1), radiated in a particular spectral line labeled l for the transition $i \rightarrow f$, as follows:

$$I_l = \frac{1}{4\pi R^2} \int_V G_l\,(T_e(\mathbf{r}), N_e(\mathbf{r}))\, N_e^2(\mathbf{r})\,\mathrm{d}V, \qquad (\mathrm{phot\,cm^{-2}\,s^{-1}}), \tag{10.8}$$

where $T_e(\mathbf{r})$ and $N_e(\mathbf{r})$ are the electron temperature and density spatial distributions in the plasma volume V, and the function G_l, usually called in astrophysical literature *contribution function* (see, e.g., [2] for more details concerning the definition of this function), can be expressed as

$$G_l\,(T_e(\mathbf{r}), N_e(\mathbf{r})) = j_l(\mathbf{r})\,\beta(\mathbf{r}), \qquad (\mathrm{phot\,cm^3\,s^{-1}}), \tag{10.9}$$

where the factor $\beta = N(\mathrm{X})/N_e$ is the relative abundance of atoms for the considered element X to the electron density, and j_l ($\mathrm{cm^{-3}\,s^{-1}}$) is the luminosity function per atom and per free electron:

$$j_l(\mathbf{r}) = \frac{1}{N(\mathrm{X})\,N_e}\,\varepsilon_l(\mathbf{r}). \tag{10.10}$$

The contribution functions G_l as functions of temperature and density are calculated by solving the adopted systems of balance equations (10.3) and include all atomic parameters contributing to the line formation, as well as ion and element abundances. It also exist atomic databases for these quantities, for example, the well-known CHIANTI database from the Solar Soft library [3, 4].

The factor β in (10.9) is usually a slow varying function of T_e within the interval of temperatures where the luminosity function j_l gives the main contribution to (10.9) due to the very sharp character of its temperature dependence. For hot astrophysical plasmas, this factor is believed to be not dependent on temperature, because in astrophysical conditions the electron density is mainly caused by the amounts of hydrogen $N(\mathrm{H})$ and helium $N(\mathrm{He})$, which are almost completely ionized. For this case, the value β is usually expressed as

$$\beta = \frac{N(\mathrm{X})}{N(\mathrm{H})}\,\frac{N(\mathrm{H})}{N_e}, \tag{10.11}$$

where $N(\mathrm{X})/N(\mathrm{H})$ is the abundance of the element X relative to hydrogen, and $N(\mathrm{H})/N_e$ is the density of hydrogen atoms to the electron density estimated to be ≈ 0.83 for hot regions of the solar atmosphere.

The quantity $\mathrm{d}Y = N_e^2\,\mathrm{d}V$ ($\mathrm{cm^{-3}}$) in (10.8) is the *emission measure* (EM) of the plasma volume element $\mathrm{d}V$. This differential form is proportional to the number of free electrons and to the electron density in the volume $\mathrm{d}V$ and hence is related to the physical conditions in the plasma. The total EM in the plasma volume V is defined as $Y = \int_V N_e^2 \mathrm{d}V$.

At the steady-state conditions, the emitting plasma source may be described by a temperature distribution $T(\mathbf{r})$ in space within the volume V. Since the density of emission power for the stationary plasma is dependent on temperature, it appears to be convenient to use the inverse function characterizing the plasma volume $V(T)$ with a definite temperature T. The electron density N_e in this volume is then characterized by the same temperature. In this case, we can define the *emission measure differential in temperature*, $y(T)$, by the expression:

$$y(T)\,\mathrm{d}T = \overline{N_e^2}(T)\,\mathrm{d}V, \tag{10.12}$$

where $\overline{N_e^2}(T)$ is the mean square of the electron density over all the plasma volume elements $\mathrm{d}V_i$ at temperature T ($\mathrm{d}V = \sum \mathrm{d}V_i$) inside the total emitting volume V. The quantity $y(T)$ (cm^{-3} K^{-1}) is also called DEM. The defined DEM function describes the distribution of emitting material as a function of temperature and allows one to study the temperature content of plasma structures. In low density, plasma conditions, there are many lines, for which the sensitivity of the contribution functions G_l to plasma density is absent or small, so that the dependence on N_e can be ignored. For these lines, using the definition (10.12), the expression (10.8) can be written in the form:

$$I_l = \frac{1}{4\pi R^2} \int_T G(l,T)\,y(T)\,\mathrm{d}T, \tag{10.13}$$

where we have used the designation $G(l,T)$ for the contribution function independent on N_e. Note that more strict definitions of the DEM distribution, as well as the general case of the emission measure differential in both temperature and density can be found in [5–7]. Correct mathematical definitions of the EM and DEM quantities were given in [8,9] using the mathematical notion of the Stieltjes integral.

Thus, the radiative model for the emitting plasma source at steady-state conditions for hot optically thin low-density plasma can be formulated in terms of the convolution of the contribution function $G(l,T)$ and the DEM distribution $y(T)$. The first one is determined through luminosity functions $j_l(T)$, which should be calculated in the framework of some adopted model assumptions. The latter one, $y(T)$, has to be derived by formulating and solving the inverse problem with a given spectrum $I(\lambda)$ (see (10.6)) acquired from observations or experiments.

10.2.3 XUV Emission in Wide Wavelength Bands from the Sun

There are two main instrumentation actually used in observations of the solar atmosphere in the XUV part of the spectrum. These are (1) spectrometers recording individual spectral lines and (2) imaging instruments ("imagers") operating over wide wavelength ranges (channels), which may contain many spectral lines and the continuum. The imagers usually have good spatial and temporal resolution, but are not well fitted for multitemperature analysis of plasma structures because of bad spectral resolution. In contrast, the spectrometers can give relatively accurate determination of coronal temperature and density distributions, but have larger pixels and slower time cadence. In this context, the availability of reliable diagnostic methods using broadband-imaging data is actually important.

Consider the emission plasma model in the case of broadband spectral channels. For the sake of convenience, we will also consider the plasma radiation spectra in terms of the total power of emission $F_l = 4\pi R^2 I_l$ in a particular spectral channel l. Then the total power of emission F_l (erg s^{-1}) in the wavelength band $\Delta\lambda(l)$ of the spectral channel l produced by the temperature interval ΔT from an optically thin plasma region with the volume V may be written similarly to (10.13):

$$F_l = \int\limits_{\Delta T} G\,(l,T)\ y(T)\,\mathrm{d}T, \tag{10.14}$$

where $G(l,T)$ is the *temperature response function* defined as the spectral emissivity function $\widetilde{F}(\lambda;T,N_\mathrm{e})$ (see (10.6) and (10.7)) calculated using definite model assumptions about emitting plasma (including the coronal approximation and the optically thin condition) and integrated over $\Delta\lambda(l)$ range with the known filter ("apparatus") function $f(l,\lambda)$:

$$G(l,T) = \frac{1}{N_\mathrm{e}^2} \int\limits_{\Delta\lambda(l)} f(l,\lambda)\,\widetilde{F}(\lambda;T,N_\mathrm{e})\,\mathrm{d}\lambda. \tag{10.15}$$

For the line spectrum, the function $f(l,\lambda) = 1$ and $G(l,T)$ coincides with the contribution function of the line l.

10.3 Spectral Inverse Problem: Two Mathematical Formalizations

There exist a variety of methods to solve inverse problems using spectroscopic data (see, e.g., [2, 10]). In principle, all these regularization and optimization techniques could be classified using diverse ways. We consider the classification including two mathematical formalizations of the spectral inverse problem, namely, the standard (or "algebraic") and probabilistic approaches. In the standard approach traditionally used in applications, the inverse problem is formulated in terms of the Fredholm integral equation of the first kind. In numerical techniques based on the standard approach, regularization constraints are usually used to derive a physically meaningful stable solution of this integral equation. The reason of the regularization procedure comes from the inherent ill-posedness of the inverse problem (see, e.g., [10] for details). In the probabilistic approach based on another mathematical formalization (see, e.g., [11]), the spectral inverse problem is formulated using relative fluxes and normalized functions treated as probability distributions for some random variables. One of the main features of the probabilistic approach is that the language of the probability theory and mathematical statistics is used and the spectral inverse problem is consequently formulated in terms of distribution functions, hypotheses, confidence level, etc.

Below, we briefly consider these two approaches and formulate the Bayesian iterative method (BIM) developed in the frame of the probabilistic approach to the spectral inverse problem. In the following paragraph, we will consider a few examples of the BIM applications to laboratory and solar coronal plasmas for the DEM diagnostic analysis of spectroscopic data.

10.3.1 Standard Approach to the Spectral Inverse Problem

As mentioned above, the formulation of the spectral inverse problem in the standard approach gives rise to the general Fredholm integral equation of the first kind given by the formula (10.14), where F_l is the measured emission flux in the channel l, $G(l, T)$ is the kernel of the integral transformation (10.14), and $y(T)$ is the unknown function to be derived. In order to calculate the temperature distribution $y(T)$, we need to solve a system of integral equations of the form (10.14) for an available set of spectral channels $\{l\}$.

It is well known that the inverse problem is "ill-posed" and usually poses questions of existence, uniqueness, and stability of its solution. Mathematical formalization is realized by means of the theory of mappings and functional analysis. A variety of numerical techniques based on the standard approach have been developed to solve inverse problems. Methods usually used in this approach require regularization constraints to derive a physically meaningful approximate stable solution $y(T)$ (see, e.g., [10]).

10.3.2 Probabilistic Approach and Formulation of the Bayesian Iterative Method

In this paragraph, we are going to use the Bayesian formalism described in Appendix to formulate the spectral inverse problem in the frame of the probabilistic approach. Mathematical formalization of this approach can be performed using the formalism and notions of the mathematical logic and probability theory [11]. Methods for the solution of the corresponding equations include those of the mathematical statistics (in particular Bayesian analysis) and information theory (see, e.g., [12, 13]).

In the probabilistic approach, (10.14), is transformed to the formula of the total probability (see [9] for details):

$$P(l) = \int_{\Delta T} P(l\,|T)\, P(T)\, \mathrm{d}T, \tag{10.16}$$

for probability distributions $P(l)$, $P(l\,|T)$, and $P(T)$, which are positively determined and satisfy the normalization conditions

$$\sum_l P(l) = 1, \quad \sum_l P(l\,|T) = 1, \quad \int_{\Delta T} P(T)\, \mathrm{d}T = 1, \tag{10.17}$$

where the summation is extended over all spectral lines or channels under consideration. These functions are interpreted as probability distributions for some random variables defined on the field of events $\{l\}$ and $\{T\}$: $P(l)$ is the probability of a photon being emitted in the channel l, $P(T)$ the probability density of a photon being emitted by plasma at temperature T, and $P(l\,|T)$ the conditional probability of the event l at the condition T.

If we want to use the probability distributions $P(l)$, $P(l\,|T)$, and $P(T)$ from (10.16) to interpret experimental data, we need to establish their relationships with measured and calculated physical values. This can be done by means of the following normalized relations:

$$P(l) = \frac{F_l}{\sum_l F_l}, \quad P(l\,|T) = \frac{G(l,T)}{\sum_l G(l,T)}, \tag{10.18}$$
$$P(T) = \frac{y(T)\sum_l G(l,T)}{\sum_l F_l}.$$

In order to derive the unknown function $P(T)$ as the solution of the spectral inverse problem, we use the BIM algorithm (see Appendix). Using the known (measured) emission fluxes $F_l^{(\mathrm{exp})}$ for $P(l)$ values in (10.16), i.e., for $P^{(\mathrm{exp})}(l)$, and applying Bayes' theorem, one can obtain a recurrence relation for the $P(T)$ distribution:

$$P^{(n+1)}(T) = P^{(n)}(T) \sum_l \frac{P^{(\mathrm{exp})}(l)}{P^{(n)}(l)}\, P(l\,|T), \tag{10.19}$$

where $P^{(n)}(l)$ is calculated from (10.16) using the nth approximation $P^{(n)}(T)$. The correction values $P^{(\mathrm{exp})}(l)/P^{(n)}(l)$ and the χ^2 minimization criterion for emission fluxes are used as a measure of the accuracy for the distribution $P(l)$ over a set of spectral channels under study. When the solution for the $P(T)$ distribution is evaluated, the corresponding DEM profile is deduced using (10.18) as follows:

$$y(T) = \frac{P(T) \sum_l F_l^{(\exp)}}{\sum_l G(l,T)}. \tag{10.20}$$

In real applications, it is necessary to have information about the accuracy (confidence level) of the solution derived because the experimental fluxes $F_l^{(\exp)}$ are known with some accuracies. In order to obtain the solution of the inverse problem, one can use the procedure of the Monte Carlo simulations. The reconstruction procedure is repeated several times for M different generations of randomly distributed experimental data $F_l^{(\exp)}$ using a given noise distribution (gaussian, poisson, etc.). This finally provides the solution in the form of the mean value:

$$\overline{y(T)} = \frac{1}{M} \sum_{i=1}^{M} y^{[i]}(T), \tag{10.21}$$

and the dispersion

$$\sigma\left[y(T)\right] = \sqrt{\frac{1}{M-1} \sum_{i=1}^{M} \left[y^{[i]}(T) - \overline{y(T)}\right]^2}, \tag{10.22}$$

which are evaluated over different sets of reconstructed distributions $y^{[i]}(T)(i = 1, 2, \ldots, M)$.

Finally, it is worth to enumerate some important features of the BIM. This method has been deduced in a regular way on the basis of Bayes' theorem, and, in contrast to the standard optimization techniques, the regularization constraints are not needed. The BIM algorithm, being assumed to be a sequential estimate based on a current hypothesis for the DEM distribution, corresponds to the maximum likelihood criterion (i.e., it brings the most likely solution of the inverse problem) and provides the ultimate resolution enhancement (super-resolution) as compared with linear and other nonparametric methods (see [13], and references therein).

10.4 Applications of the BIM to Laboratory and Solar Corona Plasmas

The BIM was applied effectively in solving a series of problems, such as the image restoration [14], the signal recovery of noisy experimental data [12, 13], and the deconvolution of initial X-ray spectra of the Sun recorded by Bragg spectrometers [15]. For the last decade, we have used this diagnostic tool for analyzing and interpreting XUV spectral data from hot tokamak and solar corona plasmas. In the first case, we have analyzed X-ray emission spectra of highly charged helium- and lithium-like ions at the tokamak plasma to develop a self-consistent approach (SCA) for deriving information on plasma parameters and the verification (i.e., the estimation of the accuracy) of atomic data and of methods of their calculation (see [16, 17]). In the second case, we have used the BIM for determining the temperature contents of the solar corona plasma structures, namely, the DEM temperature analysis of XUV spectra and imaging data observed in the space experiments onboard satellites (see, e.g., [8,9,18]). Further, we will give an overview of the methods developed for applying the BIM to the aforementioned problems and a discussion of some important results derived by means of this diagnostic technique. More detailed consideration of these results can be found in [8, 9, 16–18].

10.4.1 X-Ray Spectroscopy of Highly Charged Ions at the TEXTOR Tokamak

Studies of the highly charged ions spectra by means of the X-ray spectroscopy allow important information about elementary processes and plasma parameters to be derived. For the last decades, K-spectra of highly charged ions, associated with the transitions $nl - 1s$ of the optical electron, are effectively used for measuring parameters of hot plasmas. At present, the X-ray spectroscopy is one of the main methods for diagnostics of astrophysical objects and fusion plasma at modern tokamaks. However, the accuracy of these methods, as well as the possibility of the unambiguous interpretation of spectra, is critically dependent on the accuracy of the atomic data used for modeling X-ray emission sources. This problem, being fundamental for the theoretical spectroscopy, was not analyzed widely. For this reason, possible errors in collisional and radiative characteristics of highly charged ions remain uncertain. Thus, the verification of the accuracy of atomic data is crucially needed for correctly determining plasma parameters and for unambiguous description of the mechanisms of spectra formation. Besides the fundamental importance for the atomic physics, it is also necessary for future investigations of fast and non-Maxwellian phenomena in low-density plasmas in tokamak and astrophysical sources, in particular in solar flares.

Since crossed-beam experiments, offering direct measurements of spectroscopic and collisional characteristics of highly charged ions, are now practically absent, their accuracy is usually estimated by comparing them either with the most accurate (theoretical) calculations or with the data of beam-plasma experiments at EBIT-like setups [19, 20]. Due to narrow spectral lines, the EBIT sources are traditionally used to measure wavelengths, lifetimes of metastable states and electron–ion cross sections. However, because of the low photon statistics, they are not always valid for the verification of collisional and radiative data with a precision sufficient for diagnostic purposes. At the same time, the problem of the quantitative verification of atomic data employing the emission spectra from tokamak plasma was not practically discussed. In this connection, the problem of the atomic data verification is rather actual and important both for atomic physics, stimulating improvements of their calculation methods, and for solving the plasma physics problems based on the results of the X-ray spectral diagnostics.

Studies of the high-resolution K-spectra of Ar^{16+} and Ar^{15+} ions at the TEXTOR (Torus Experiment for Technology Oriented Research) tokamak plasma [16, 17, 21, 22] showed that the tokamak spectra can be efficiently used not only for the diagnostic purposes but also both for the verification of the accuracy of atomic data—wavelengths and rates (cross sections averaged over the Maxwellian velocity distribution) of elementary processes in hot plasmas—and the diagnostic methods used. As it was shown, the analysis of the accuracy and the correction of atomic and plasma characteristics can be made with the help of a new SCA based on the solution of the inverse problem for these spectra in the framework of the semiempirical "spectroscopic" model. Self-consistency of the model parameters is provided by a number of functional relationships, which are determined by requirements of the coincidence (within the limits of experimental errors) of the theoretical and measured spectra under essentially various conditions in the plasma. The necessary conditions for applying this approach are ensured by the following features of the experiment at the TEXTOR tokamak: (1) the high accuracy for measuring the fluxes and high photon statistics, (2) the detection of the spectra under various plasma conditions, and (3) the existence of additional diagnostic techniques for measuring the radial profiles of the electron temperature and density simultaneously with the spectra.

10.4.1.1 Formulation of the Problem

The spectrum of Ar^{16+} (helium-like) and Ar^{15+} (lithium-like) ions under study covers the wavelength region 3.94–4.02 Å and consists of a set of well-resolved intensity peaks $\{\mathbf{L}\}$, which are associated with corresponding spectral lines l giving a dominant contribution to their intensities. The most prominent and intense features of the spectrum result mainly from the lines caused by transitions in Ar^{16+} ions: resonance ($1s2p(^1P_1)$–$1s^2(^1S_0)$), magnetic quadrupole ($1s2p(^3P_2)$–$1s^2(^1S_0)$), intercombination ($1s2p(^3P_1)$–$1s^2(^1S_0)$), and forbidden ($1s2s(^3S_1)$–$1s^2(^1S_0)$), designated as **w**, **x**, **y**, and **z** lines, respectively (these notations are used following [23]). These lines are produced primarily due to direct electron impact excitation including cascades from higher levels, as well as have contributions from radiative recombination of Ar^{17+} (hydrogen-like) ions, from inner-shell ionization of Ar^{15+} ions (contributing to the **z** line), and from resonance scattering via doubly excited autoionizing states $1snln'l'$ $(n, n' > 2)$.

In addition to the helium-like lines, there are numerous lines, called dielectronic satellites, to the lines of helium-like argon ions due to transitions $1s^2nl$–$1s2pnl$ in lithium-like ions and, to a lesser extent, $1s^22snl$–$1s2s2pnl$ in beryllium-like ions. The most prominent lines emitted by the Ar^{15+} ion ($n = 2$) and resolved in the spectrum are the **q** and **r** satellites excited predominantly by collisions with electrons, and the **k** and **a** ones excited by means of the dielectronic mechanism from Ar^{16+} ion. The most intense **j** satellite is blended with the **z** line. There are also two groups of satellites in the long-wavelength wing of the **w** resonance line corresponding to dielectronically excited $n = 3$ and $n = 3, 4$ satellites and denoted here as $\mathbf{N}_3$ and $\mathbf{N}_4$ peaks, respectively. The remaining less-intense satellites densely fill the spectral range, forming series converging to helium- and lithium-like lines. For the purpose of quantitative analysis of spectral fluxes, the whole spectral range was divided into the intervals $\Delta\lambda_{\mathrm{L}} = [\mathbf{L}]$, and the following set of peaks was identified: $\{\mathbf{L}\} = \{\mathbf{W}, \mathbf{N}_4, \mathbf{N}_3, \mathbf{X}, \mathbf{Y}, \mathbf{Q}, \mathbf{R}, \mathbf{A}, \mathbf{K}, \mathbf{Z}\}$.

The theoretical (synthetic) spectrum $I(\lambda)$ and the emission flux $F_{[\mathbf{L}]} = \int_{[\mathbf{L}]} I(\lambda)\mathrm{d}\lambda$ in a spectral interval $[\mathbf{L}]$ are functions (or functionals) of two sets of physical characteristics: (1) the atomic data (AD) including the sets of atomic constants (wavelengths, transition probabilities, branching ratios, etc.) and the effective rates (collisional characteristics) of elementary processes, $C_1^z(T_e)$, and (2) the plasma parameters (PP), namely, radial profiles of the electron T_e and ion T_i temperatures, electron density N_e, argon ion densities N_z for the charges z, and neutral atom density of the working gas (hydrogen, deuterium, or helium) N_a.

The synthetic emission fluxes $F_{[\mathbf{L}]}$ depend on a particular model based on equations of atomic and plasma kinetics. For the adopted model, the main, basic or "key", parameters $\mathbf{D} = \{\mathbf{D}_i\}$ and $\mathbf{P} = \{\mathbf{P}_i\}$ were defined from the corresponding sets AD and PP (e.g., ratios of line excitation rates or central temperature $T_e^{(0)}$ in the plasma core), which identically (with a given accuracy) simulate the fluxes of the synthetic spectrum $F_{[\mathbf{L}]}^{(\mathrm{syn})} = F_{[\mathbf{L}]}(\mathbf{D}, \mathbf{P})$. In the frame of semiempirical models, key plasma parameters $\mathbf{P}$ have to be independently predetermined (calculated or measured) for a direct "ab initio" calculations of the synthetic spectra, while for diagnostic purposes they can be derived from the measured spectra by solving the inverse problem using a set of equalities $F_{[\mathbf{L}]}^{(\mathrm{exp})} = F_{[\mathbf{L}]}(\mathbf{D}, \mathbf{P})$, where $F_{[\mathbf{L}]}^{(\mathrm{exp})}$ are the fluxes measured in a given spectrum. These equalities impose restrictions on the possible values of the model parameters, relating them through implicit functional relations.

The sets **D** and **P** must also satisfy to additional physical constraints. In particular, the quantities from the set **D** derived by the inversion of different spectra have to coincide (within the experimental errors), i.e., have to be independent on plasma conditions. The quantities from the set **P** for a given spectrum (e.g., the central temperature) do not have to depend on **L**, while the radial profiles of relative ionic abundances $n_z = N_z / \sum_z N_z$, derived from any measured spectra, have to obey the continuity condition

$$n(r) = \sum_{z} n_z(r) = 1, \tag{10.23}$$

which may not be satisfied for arbitrary atomic data from **D**. As a result, the presented conditions and relations ensure the consistency of atomic and plasma parameters, which means the existence of the range of their values satisfying to these conditions for a given accuracy of measured fluxes $F_{[\mathbf{L}]}^{(\mathrm{exp})}$, and can be used for the atomic data verification problem.

By definition, the term "verify" means to prove to be true or to check for accuracy. Both sets of characteristics, **D** and **P**, being the variable parameters of the spectroscopic model (SM) developed, may be optimized by using (10.23), which was shown to be the necessary and sufficient conditions for these parameters to be correct [17].

10.4.1.2 X-Ray Spectroscopy at the TEXTOR Tokamak

TEXTOR is a medium-sized tokamak experiment with a major radius of 1.75 m and a minor radius of 0.46 m. It operates with toroidal magnetic fields of up to 2.7 T and plasma currents up to 580 kA. In addition to ohmic heating of about 0.5 MW, which is obtained from the plasma current, auxiliary heating is provided by the injection of neutral hydrogen beams with the total power of up to 2 MW. Further information on TEXTOR and its diagnostic equipment can be found in [24–26].

The K-spectra from the Ar^{16+} and Ar^{15+} impurity ions were investigated on the TEXTOR tokamak equipped with a high-resolution X-ray spectrometer/polarimeter [24, 25] consisting of two (horizontal and vertical) Bragg spectrometers in the Johann scheme, designed for the polarization measurements of the radiation from the same central region of the tokamak plasma. Figure 10.1 shows a schematic diagram of the experimental setup. The horizontal instrument was used to measure the K_α-spectra (formed by transitions $n = 2 \to n = 1$ of the optical electron), while the vertical (perpendicularly arranged) spectrometer was used in the experiment for recording the K_β-emission lines (emitted due to transitions $n = 3 \to n = 1$).

The electron temperatures in the plasma core region were derived from the K_α-spectra and also compared with the results of the Thomson scattering diagnostics [27] on the TEXTOR tokamak. The results of both measurements agree to within 5–10%. In our plasma diagnostic analysis, we also used the radial profiles of the electron temperature $T_e(r)$ and density $N_e(r)$ measured for each spectrum by

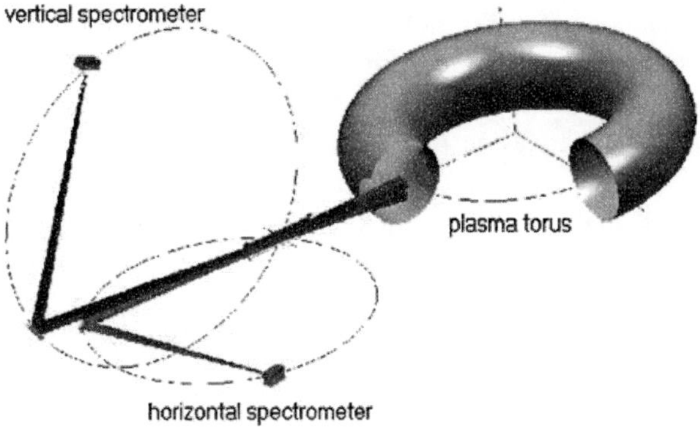

Fig. 10.1 Schematic diagram of the X-ray spectrometer installed at the TEXTOR tokamak (by courtesy of G. Bertschinger)

means of an electron cyclotron emission (ECE) polychromator and a far-infrared interferometer/polarimeter (FIIP), respectively [25].

10.4.1.3 Self-consistent Approach and Atomic Data Calculations

The general concept of the SCA used for the verification problem includes several aspects or levels of consistency. The main idea is dealt with the aforementioned condition of "intrinsic" consistency of key parameters **D** and **P** in the SM. Another aspect is connected with a coordination of the atomic data evaluated in the frame of a uniform method to avoid their compilation and/or extrapolation. Thus, the verification of atomic characteristics means at the same time the verification of the corresponding method of atomic data calculations. The third aspect concerns a consistency of plasma parameters of the SM with those measured by means of other diagnostic techniques. From the experimental point of view, the accuracy of the verification procedure depends on the following: (1) the accuracy of the experimental data, (2) the number of spectral features in the selected spectral range, and (3) the number of spectra measured under significantly different conditions. In our particular case of the argon ion K_α-spectrum, there were 10 prominent well-resolved peaks consisting of numerous lines in selected experimental spectra, which were measured for a wide range of temperatures $T_e = 0.8$–$2.5\,\mathrm{keV}$ and densities $N_e = 10^{13}$–$10^{14}\,\mathrm{cm}^{-3}$.

The atomic data needed to model the synthetic spectra of argon ions were calculated by means of the atomic codes ATOM and MZ [28]. These data include wavelengths, radiative, and autoionization (for autoionizing states) decay probabilities, as well as collisional characteristics of elementary processes: cross sections and rates for direct (potential or background) processes of electron–ion impact excitation, ionization, and radiative recombination. The contribution of resonance scattering was accounted for as cascade processes to autoionizing levels caused by dielectronic capture with the following autoionization (resonance excitation) or radiation (dielectronic recombination). The effective rate coefficients for excitation and recombination processes including cascades via radiative and autoionizing levels were obtained by solving equations of atomic kinetics in the frame of a radiative–collisional model. The calculations were carried out under the assumptions of the quasisteady plasma ($\mathrm{d}N_z^{(i)}/\mathrm{d}t = 0$ for ion populations $N_z^{(i)}$) and Maxwellian velocity distribution for plasma particles. More detailed description of the atomic data calculations and radiative–collisional model used can be found in [16, 17].

10.4.1.4 Spectroscopic Model

The aforementioned semiempirical spectroscopic model (SM) is based on the following assumptions:

(1) The conditions of coronal approximation are satisfied for the populations $N_z^{(k)}$ of the excited ion states k $(k > 1)$.
(2) The profiles of the relative ion abundances $n_z(r)$ are the parameters of the model and satisfy to the continuity relation (10.23), while the profiles of the electron temperature $T_e(r)$ and density $N_e(r)$ are known (i.e., measured).

Under the coronal equilibrium conditions, implying the quasi-steady state of the plasma $(\mathrm{d}N_z^{(k)}/\mathrm{d}t = 0)$, the effects of the ionic and radiative transfer for the excited ion states are negligible, and their populations are low, $\left(N_z^{(k)} \ll N_z^{(1)}\right)$. Therefore, the ion abundances are practically equal to the populations of the ground states, $N_z \cong N_z^{(1)}$. The calculations in the collisional–radiative model for argon ions [16, 17] showed that deviations from the coronal approximation are negligible ($< 1\%$) for densities $\leq 10^{14}\,\mathrm{cm}^{-3}$; thus, for these densities, the condition (1) is justified. It is also worth noting that the ion abundances N_z under conditions typical for the tokamak plasma do not coincide with coronal equilibrium concentrations $N_z^{(c)}$ used for solar corona plasma (coronal ionization equilibrium), but obey kinetic equations of balance taking account for the effects of the transfer and charge exchange of the impurity ions on neutral atoms. The role of these effects was investigated in [22] using an impurity transport model.

In the coronal approximation, the emission flux in the spectral range [**L**] from the plasma column along the radius $r = \rho a$ (a is the small radius of the plasma torus, ρ—dimensionless radius) can be written as the sum $F_{[\mathrm{L}]} = \sum_z F_{[\mathrm{L}]}^z$ over the partial fluxes $F_{[\mathrm{L}]}^z$ given by [17]

$$F_{[\mathrm{L}]}^z = C \int_0^1 J_{[\mathrm{L}]}^z(P(\rho))\, n_z(\rho)\, [N_e(\rho)]^2\, \mathrm{d}\rho, \tag{10.24}$$

where $J_{[\mathrm{L}]}^z$ are the partial excitation rates

$$J_{[\mathrm{L}]}^z(T_e, T_i, n_a) = \sum_l C_l^z(T_e, n_a)\, \varphi_{[\mathrm{L}]}^l(T_i). \tag{10.25}$$

Here, P stands for the set of plasma parameters, $P = \{T_e, T_i, n_a\}$; C_l^z are the effective rates corresponding to the excitation processes of the lines from the ions with the charge z (in our case for argon ions with $z = 15$–17) and also including radiative cascades from the upper levels and branching coefficients k_l for the lines l; C is the conversion coefficient defined by the condition of equality between the measured and calculated fluxes; $\varphi_{[\mathrm{L}]}^l(T_i) = \int_{[\mathrm{L}]} \varphi_l(T_i; \lambda - \lambda_l)\mathrm{d}\lambda$ is the correction factor associated with the line profile $\varphi_l(T_i; \lambda)$ due to the natural width, Doppler broadening, and the instrumental function; $n_a = N_a/N_e$ is the relative density of hydrogen atoms.

The partial excitation rate $J_{[\mathrm{L}]}^z$ for a feature **L** describes the total effective rate of the formation for all the lines, contributing to its intensity and excited from the ions with the charge z. Thus, for $z = 16$ (denoted also as he), the partial emissivity $J_{[\mathrm{L}]}^{\mathrm{he}}$ contains the effective collisional excitation rates for the lines emitted by the helium-like ions, and the dielectronic capture rates for lithium-like dielectronic satellites resulted from the helium-like ions; for the lines excited from lithium-like ions ($z = 15$ or li), $J^{\mathrm{li}}_{[\mathbf{L}]}$ includes the effective rate for the inner-shell excitation of lithium-like satellites, as well as the inner-shell ionization of the 1s electron (contributing to the excitation of the **z** line); for the hydrogen-like ions ($\mathbf{z} = 17$ or h), $J^{h}_{[\mathbf{L}]}$ contains the total rate of recombination (both radiative and dielectronic) from the hydrogen-like ions to the excited states of helium-like lines, as well as the charged exchange rate on the neutral hydrogen atoms. Thus, the partial excitation rates $J^{z}_{[\mathbf{L}]}$ depend on both atomic characteristics and the radial distributions of plasma parameters.

To formulate the SM and to derive its parameters by solving the inverse problem, the dependence on the radial profiles of the plasma parameters $n_z(\rho)$, $N_e(\rho)$, and $T_e(\rho)$ was factorized by changing the radial variable ρ to the dimensionless temperature variable $\beta(\rho) = \left[T_e(\rho)/T_e^{(0)}\right]^{-1} - 1$ (the electron temperature in the plasma core $T_e^{(0)}$ is one of the main "key" parameters of the SM) in (10.24) for the partial emission flux:

$$F^{z}_{[\mathbf{L}]} = C \int_0^b J^{z}_{[\mathbf{L}]}\left(T_e^{(0)}, \beta\right) n_z(\beta)\, y(\beta)\, \mathrm{d}\beta, \qquad (10.26)$$

where $y(\beta) = [N(\rho(\beta))]^2 \, |\mathrm{d}\rho/\mathrm{d}\beta|$ is the DEM distribution and $b = \beta(1) \gg 1$. Note that, in contrast to the ρ-representation given in (10.24), in the β-representation, the values $J^{z}_{[\mathbf{L}]}$ are sensitive to the atomic characteristics and are independent on the radial profiles of the electron temperature $T_e(\rho)$.

In order to use the BIM for the inversion procedure, it is also necessary to pass from the absolute total, $F_{[\mathbf{L}]}$, and partial, $F^{z}_{[\mathbf{L}]}$, fluxes in the peaks to the relative fluxes $\Gamma_{[\mathbf{L}]}(\sigma_z) = F_{[\mathbf{L}]}/F_{[\sigma_z]}$ and $P^{z}_{[\mathbf{L}]} = F^{z}_{[\mathbf{L}]}/F^{z}_{[\sigma_z]}$, normalized in three spectral regions $[\sigma_z]$ chosen for each z. To choose these regions, we consider three sets of peaks denoted through $\mathbf{Z} = \{\mathbf{L}_i\}$ ($\mathbf{Z} = Li, He, H$) and corresponding to three (generally speaking overlapping) wavelength regions $[\sigma_z] = [\mathbf{Z}]$, which include lines excited from the ions with the charge z ($z = \mathrm{li, he, h}$). The choice of the sets $[\sigma_z]$ was stipulated by two demands: to provide (1) the maximum contribution of partial fluxes $F^{z}_{[\mathbf{L}]}$ ($\mathbf{L} \subset \mathbf{Z}$) in each region $[\mathbf{Z}]$ and (2) the minimum contribution of partial fluxes $F^{k}_{[\mathbf{L}]}$ from ions with other charges $k \neq z$. The following peak groups were used: **He**=$\{\mathbf{W}, \mathbf{N}_4, \mathbf{N}_3, \mathbf{K}\}$; **Li**=$\{\mathbf{Q}, \mathbf{R}, \mathbf{Z}\}$; and **H**=$\{\mathbf{W}, \mathbf{X}, \mathbf{Y}, \mathbf{Z}\}$.

The total flux $F_{[\mathbf{Z}]}$ in the spectral range $[\mathbf{Z}]$ can be written as a sum of partial fluxes $F^{z}_{[\mathbf{Z}]}$ similar to that in the peak $[\mathbf{L}]$ (see (10.24)), namely, $F_{[\mathbf{Z}]} = \sum_z F^{z}_{[\mathbf{Z}]}$, and the partial excitation rate $J^{z}_{[\mathbf{Z}]}$ as a sum over all $\mathbf{L} \subset \mathbf{Z}$

$$J^{z}_{[\mathbf{Z}]}\left(T_e^{(0)}, \beta\right) = \sum_{\mathbf{L} \subset \mathbf{Z}} J^{z}_{[\mathbf{L}]}\left(T_e^{(0)}, \beta\right). \qquad (10.27)$$

Define also the relative partial excitation rates $p^{k}_{[\mathbf{L}]}$ of the peaks $\mathbf{L} \subset \mathbf{Z}$ for each range $[\mathbf{Z}]$ and the functions Φ_z (associated with the plasma parameters) by the following equalities:

$$p^k_{[\mathbf{L}]}\left(T^{(0)}_\mathrm{e},\beta\right) = \frac{J^k_{[\mathbf{L}]}\left(T^{(0)}_\mathrm{e},\beta\right)}{J^k_{[\mathbf{K}]}\left(T^{(0)}_\mathrm{e},\beta\right)}, \tag{10.28}$$

$$\Phi_\mathrm{z}\left(T^{(0)}_\mathrm{e},\beta\right) = C\, J^z_{[\mathbf{Z}]}\left(T^{(0)}_\mathrm{e},\beta\right)\, n_z\left(T^{(0)}_\mathrm{e},\beta\right)\, y(\beta)/\, F^z_{[\mathbf{Z}]}, \tag{10.29}$$

which satisfy to the normality conditions

$$\int_0^b \Phi_\mathrm{z}\left(T^{(0)}_\mathrm{e},\beta\right)\, d\beta = 1, \quad \sum_{\mathbf{L}\subset\mathbf{Z}} p^z_{[\mathbf{L}]}\left(T^{(0)}_\mathrm{e}\right) = 1. \tag{10.30}$$

Further, the relative flux $\Gamma_{[\mathbf{L}]} = \sum_k F^k_{[\mathbf{L}]}/F_{[\mathbf{Z}]}$ for the peak $\mathbf{L}\subset\mathbf{Z}$ in the range $[\mathbf{Z}]$ can be presented in the form:

$$\Gamma_{[\mathbf{L}]} = \sum_{k=\mathbf{K}} \frac{F_{[\mathbf{K}]}}{F_{[\mathbf{Z}]}}\,\frac{F^k_{[\mathbf{K}]}}{F_{[\mathbf{K}]}}\,\frac{F^k_{[\mathbf{L}]}}{F^k_{[\mathbf{K}]}} = \sum_{k=\mathbf{K}} R^\mathbf{K}_\mathbf{Z}\,\Gamma^k_{[\mathbf{K}]}\,P^k_{[\mathbf{L}]}, \tag{10.31}$$

where $R^\mathbf{K}_\mathbf{Z}$ is the ratio of the fluxes in two peak groups $[\mathbf{K}]$ and $[\mathbf{Z}]$, and $\Gamma^z_{[\mathbf{Z}]}$ is the ratio of the partial flux $F^k_{[\mathbf{K}]}$ to the corresponding total flux in the range $[\mathbf{K}]$. The partial fluxes $P^k_{[\mathbf{L}]}\left(T^{(0)}_\mathrm{e}\right)$ normalized in the ranges $[\mathbf{Z}]$ are expressed as

$$\begin{gathered} P^k_{[\mathbf{L}]}\left(T^{(0)}_\mathrm{e}\right) = \int_0^b p^k_{[\mathbf{L}]}\left(T^{(0)}_\mathrm{e},\beta\right)\,\Phi_k\left(T^{(0)}_\mathrm{e},\beta\right)\,\mathrm{d}\beta, \\ P^z_{[\mathbf{Z}]}\left(T^{(0)}_\mathrm{e}\right) = \sum_{\mathbf{L}\subset\mathbf{Z}} P^z_{[\mathbf{L}]}\left(T^{(0)}_\mathrm{e}\right) = 1. \end{gathered} \tag{10.32}$$

Note that the relative partial fluxes $\Gamma^z_{[\mathbf{Z}]}$ in three spectral ranges $[\mathbf{Z}]$ can be expressed through the flux ratios $R^\mathbf{K}_\mathbf{Z}$ and the normalized integrals $P^k_{[\mathbf{Z}]}\left(T^{(0)}_\mathrm{e}\right) = \sum_{\mathbf{L}\subset\mathbf{Z}} P^k_{[\mathbf{L}]}\left(T^{(0)}_\mathrm{e}\right)$ (see (10.32)) by solving the system of equations (10.31) for all $\mathbf{Z}$ and using the normality relation $\Gamma_{[\mathbf{Z}]} = \sum_\mathbf{L}\Gamma_{[\mathbf{L}]} = 1$:

$$1 = \sum_{k=\mathbf{K}} \Gamma^k_{[\mathbf{K}]}\left(T^{(0)}_\mathrm{e}\right)\, R^\mathbf{K}_\mathbf{Z}\, P^k_{[\mathbf{Z}]}\left(T^{(0)}_\mathrm{e}\right), \quad \mathbf{Z} = \mathbf{Li}, \mathbf{He}, \mathbf{H}. \tag{10.33}$$

Thus, the relative fluxes $\Gamma_{[\mathbf{L}]}$ in (10.31) are determined (according to (10.32)) by the three key plasma parameters (β-profiles) $\Phi_\mathrm{z}(\beta)$ and by the kernels $p^z_{[\mathbf{L}]}(\beta)$ of the integral operator. These kernels depend on the set of atomic data, the central temperatures T_e and T_i of the plasma core, and (for $z = h$) the profile $n_a(\beta)$. If the relative fluxes $\Gamma_{[\mathbf{L}]}$ in (10.31) are fixed by the condition $\Gamma_{[\mathbf{L}]} = \Gamma^{(\mathrm{exp})}_{[\mathbf{L}]}$, the normalized profiles $\Phi_\mathrm{z}(\beta)$ depend only on the arguments of the quantities $p^z_{[\mathbf{L}]}$.

If all relative fluxes in the peaks $\Gamma_{[\mathbf{L}]}$ and AD are known (i.e., fixed), the $\Phi_\mathrm{z}\left(T^{(0)}_\mathrm{e},\beta\right)$ functions can be found by solving the inverse problem for the set of relations (10.27)–(10.33) for given $T^{(0)}_\mathrm{e}$, T_i, and $n_a(\beta)$ values. In order to determine these parameters, it is necessary to use the additional model condition given by (10.23), which limits the class of possible formal solutions of these equations, $\{\Phi_z^*(\beta)\}$. Since this criterion is expressed in terms of the relative ion abundances $n_z(\beta)$, rather than in terms of the functions $\Phi_z(\beta)$ related to $n_z(\beta)$ by (10.29), one has to adopt additional (empirical for the SM) information about plasma parameters, namely, about the DEM profile $y(\beta)$ depending on the temperature $T_e(\beta)$ and density $N_e(\beta)$ distributions. The abundances $n_z(\beta)$ and their sum $n(\beta)$ for $z = 15, 16$, and 17 (li, he, and h, respectively) can be expressed through the values $\Phi_z(\beta)$ and $n_{he}(\beta)$ using (10.29) as follows:

$$n_z\left(T_e^{(0)},\beta\right) = \widetilde{n}_z\left(T_e^{(0)},\beta\right) n_{he}\left(T_e^{(0)},\beta\right),$$
$$n\left(T_e^{(0)},\beta\right) = n_{he}\left(T_e^{(0)},\beta\right) \sum_z \widetilde{n}_z\left(T_e^{(0)},\beta\right), \tag{10.34}$$
$$\widetilde{n}_z\left(T_e^{(0)},\beta\right) = \frac{\Phi_z}{\Phi_{he}} \frac{J_{[\mathbf{He}]}^{he}}{J_{[\mathbf{Z}]}^{z}} \frac{\Gamma_{[\mathbf{Z}]}^{z}}{\Gamma_{[\mathbf{He}]}^{he}} R_{\mathbf{He}}^{\mathbf{Z}}.$$

The continuity relation (10.23) is also used in the SM to optimize model parameters $T_e^{(0)}$ and AD, if the DEM profile $y(\beta)$ is known.

10.4.1.5 BIM Inversion

To solve the inverse problem for determining the $\Phi_z(\beta)$ profiles in the framework of the SM, the BIM is adopted as follows. Defining the quantities $P_{[\mathbf{L}]}^z(\exp)$ and $\Gamma_{[\mathbf{L}]}^z(\exp)$ by means of the expressions

$$P_{[\mathbf{L}]}^z(\exp) = \Gamma_{[\mathbf{L}]}^z(\exp)/\Gamma_{[\mathbf{Z}]}^z,$$
$$\Gamma_{[\mathbf{L}]}^z(\exp) = \Gamma_{[\mathbf{L}]}^{(\exp)} - \sum_{k\neq z} \Gamma_{[\mathbf{L}]}^k \quad \text{for } \mathbf{L} \subset [\mathbf{Z}], \tag{10.35}$$

where $\Gamma_{[\mathbf{L}]}^{(\exp)}$ stands for the measured flux ratios of the peaks, and using the relation (10.31) and the definition of the $P_{[\mathbf{L}]}^z$ in (10.32) under the condition $P_{[\mathbf{L}]}^z(\exp) = P_{[\mathbf{L}]}^z\left(T_e^{(0)}\right)$, one arrives at the following system of self-consistent equations for $z = $ li, he, h, and $\mathbf{L} \subset [\mathbf{Z}]$ for $\mathbf{Z} = z$:

$$P_{[\mathbf{L}]}^z(\exp) = \int_0^b p_{[\mathbf{L}]}^z\left(T_e^{(0)},\beta\right) \Phi_z\left(T_e^{(0)},\beta\right) \mathrm{d}\beta, \tag{10.36}$$

which, due to normalization conditions (10.30) and according to the BIM formulation, can be considered as the system of Bayesian relations between the probability $P_{[\mathbf{L}]}^z\left(T_e^{(0)}\right)$ for a photon to be emitted in the peak $\mathbf{L}$ and the product of the conditional probability $p^{z}_{[L]}\left(T^{(0)}_{e},\beta\right)$ for a certain β value (local temperature) on the probability density $\Phi_{z}\left(T^{(0)}_{e},\beta\right)$ for this value.

For each set of (10.36) corresponding to $z = \mathbf{Z}$, the BIM procedure takes the form (see Appendix):

$$\Phi^{(n+1)}_{z}\left(T^{(0)}_{e},\beta\right) = \Phi^{(n)}_{z}\left(T^{(0)}_{e},\beta\right)\sum_{L\subset Z}\frac{P^{z}_{[L]}(\exp)\, p^{z}_{[L]}\left(T^{(0)}_{e},\beta\right)}{\int\limits_{0}^{b} p^{z}_{[L]}\left(T^{(0)}_{e},\beta\right)\,\Phi^{(n)}_{z}\left(T^{(0)}_{e},\beta\right)\,d\beta}. \tag{10.37}$$

The χ^2 criterion is applied to estimate the convergence of the iterative procedure. The ratios of the partial fluxes $\Gamma^{k}_{[L]}(n)$ and $\Gamma^{z}_{[Z]}(n)$ for the nth iteration are calculated using the formulas from the system of (10.27)–(10.33), where the ratios for the measured fluxes $\Gamma^{(\exp)}_{[L]}$ and $R^{\mathbf{K}}_{\mathbf{Z}}(\exp) = F^{(\exp)}_{[\mathbf{K}]}/F^{(\exp)}_{[\mathbf{Z}]}$ have to be used. For the zeroth approximation $\Phi^{(0)}_{z}(\beta)$, we used the coronal abundances $n^{(c)}_{z}(\beta)$, and the $n_{a}(\beta)$ profiles were used from [22] for the recombination rates $J^{h}_{[L]}$. The parameter $T^{(0)}_{e}$ and correction factors for the atomic data are determined by minimizing χ^2 and $\delta = \mid n(\beta) - 1 \mid$ quantities.

10.4.1.6 Results of Plasma Diagnostics and Verification of Atomic Data

On the basis of the developed SCA and using K_{α}- and K_{β}-spectra of impurity argon ions, we estimated the accuracy of atomic data necessary for modeling K-spectra and verified the corresponding methods of their calculation, as well as performed diagnostics of plasma parameters (central temperature and relative ion abundances) at the TEXTOR tokamak. Figures 10.2 and 10.3 demonstrate examples of determining some key parameters of the SM, in particular the core temperature $T^{(0)}_{e}$ and effective excitation rates for the satellite group $\mathbf{N}_3$. These results showed that the χ^2 and δ optimization conditions for the BIM solutions give rise to rather strong constraints for possible $\Phi^{*}_{z}(\beta)$ profiles under 5% variations of the SM parameters. Due to these constraints, the values of the SM parameters beyond their optimization region of definition cannot be simultaneously consistent with the SM system of (10.27)–(10.33) and minimization conditions χ^2 and δ. The analysis of the $\Phi^{*}_{z}(\beta)$ distributions allows to derive the key parameters and to find the region of their consistency with the measured spectra: the set $\mathbf{D} = \{\alpha_l\}$ includes the ratios of the effective excitation rates for the **x**, **y**, **z** lines, **k** satellite, and satellite group $\mathbf{N}_3$, while the plasma set is $\mathbf{P} = \left\{T^{(0)}_{e}, \Phi_{z}(\beta)\right\}$.

It was shown that the developed approach for interpreting the experimental results from the TEXTOR tokamak made it possible to verify the methods for calculating the atomic data with an accuracy of ~5–10%. The calculations per-

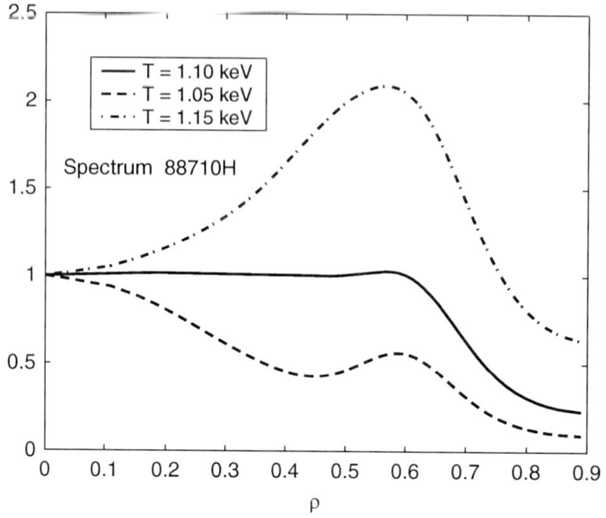

Fig. 10.2 Sum of the relative ion abundances $n(\rho)$ given by (10.23), for various values of the central temperature $T_e^{(0)}$

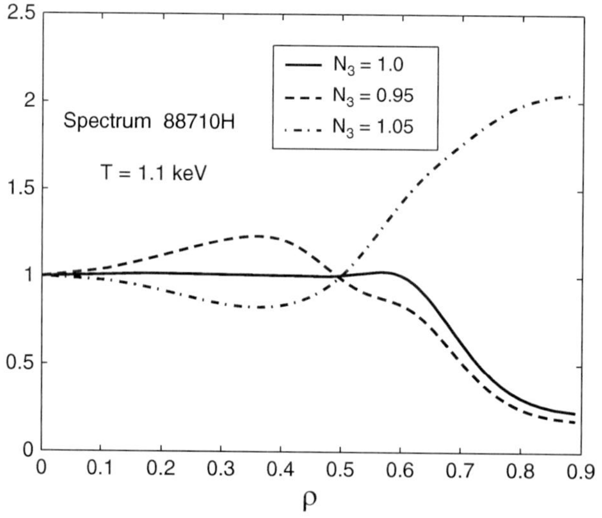

Fig. 10.3 Sum of the relative ion abundances $n(\rho)$ given by (10.23), for various relative effective excitation rates $\alpha_l\left(T_e^{(0)}\right)$ corresponding to the satellite group $\mathbf{N_3}$

formed by means of the atomic codes ATOM and MZ [28] require $\sim$10% correction of the effective excitation rate ratios for the inter-combination line **y** and the dielectronic satellite group $\mathbf{N}_3$ to the resonance line **w**. It was established that the reason of considerable disagreement between the theory and the experiment for $\mathbf{N}_3$ and $\mathbf{N}_4$ satellite groups consists in calculations of autoionization decay probabilities obtained by means of the MZ code. The MZ code based on the Z-expansion method (Z is the nuclear charge) was modified to account for the first-order corrections (in comparison with the previous zeroth order) in powers of $1/Z$, corresponding to the screening effects, in calculations of autoionization rates. Atomic data calculations by Z-expansion method taking into account these effects for doubly excited states 2lnl′ of helium-like ions and $1s2lnl'$ of lithium-like ions with $Z = 6$–36 can be found in [29]. The spectra calculated with corrected atomic data are in agreement with the spectra measured in the wide range of plasma conditions within the experimental accuracy of $\leq$10%. Figure 10.4 demonstrates an example of the spectrum calculated with the central temperature $T_e^{(0)} = 1.21$ keV before and after the correction of the atomic data in comparison with a measured spectrum. In the frame of the SM, the parameter $T_e^{(0)}$ can be derived with the accuracy of $\sim$5%, whereas the relative ion abundances n_z differ considerably (in 2–5 times) from ionization equilibrium values in coronal conditions because of the effects of the transfer and charge exchange of the impurity argon ions on neutral atoms.

Data obtained on the TEXTOR tokamak for the K_β-emission spectrum of Ar^{16+} ions at temperatures $T_e \sim 1$ keV made it possible to study the temperature dependence of the ratio $G_3 = I[K_{\beta_2}]/I[K_{\beta_1}]$ for the intercombination and resonance lines corresponding respectively to the transitions $1s3p(^3P_1) \to 1s^2(^1S_0)(K_{\beta_2}$ line) and $1s3p(^1P_1) \to 1s^2(^1S_0)(K_{\beta_1}$ line) [16]. Figure 10.5 shows an example of a measured spectrum containing the K_{β_1} and K_{β_2} lines. The emissivity functions for these lines were calculated as functions of electron temperature and density for the equilibrium plasma with the Maxwellian electron velocity distribution. Using these quantities, the relative K_β intensities were derived and compared with experimental data (see [16] for details). The calculations carried out within the framework of the radiative–collisional model using the ATOM and MZ atomic codes are in agreement, to the experimental accuracy, with all the experimental data obtained on the PLT (Princeton, USA) [19], ALCATOR-C (Cambridge, USA) [30], and TEXTOR tokamaks. It was shown that the previously observed discrepancies by factor of 1.3–2 between the measured and calculated G_3 ratios and exceeding appreciably the experimental errors are caused by the use of inaccurate atomic data and simplified atomic models in those works [19,30]. These results are evidence for a high accuracy of atomic data used and for the possibility of effectively using, on their basis, the K_β lines for diagnostics of the electron temperature and density in the laboratory and astrophysical coronal plasma sources.

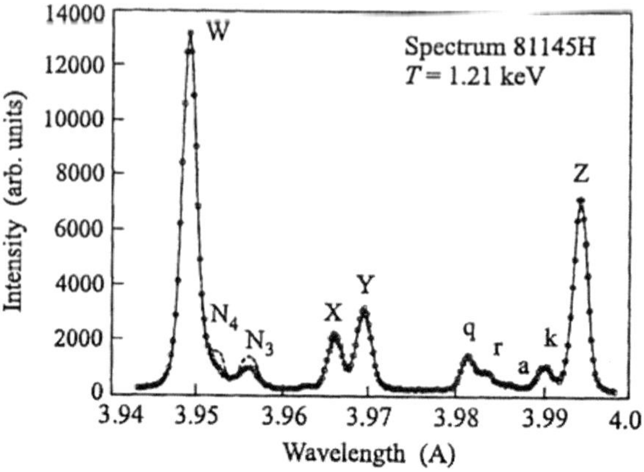

Fig. 10.4 Argon K_α spectrum in the range 3.94–4.0 Å measured at the TEXTOR tokamak (*points*) and calculated (*solid line*) for the central temperature $T_e^{(0)} = 1.21$ keV taking into account correction of the atomic data. The *dashed line* shows the spectral zones calculated without correction of the atomic data, for which the experimental data noticeably differ from the calculations (**N**$_4$, **N**$_3$, **X**, **Y**, **Z**)

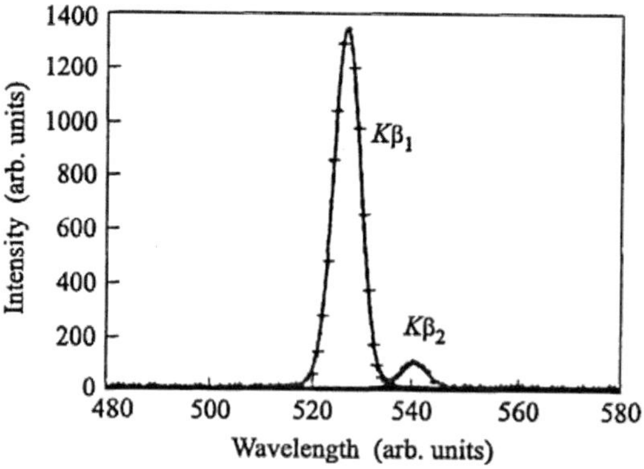

Fig. 10.5 Measured (*crosses*) K_β-emission spectrum of Ar^{16+} ions and its approximation by two Voigt profiles (*solid line*) taking into account the instrumental function and the radial distribution of ion temperature

10.4.2 XUV Spectra and Imaging Data from the Solar Corona

10.4.2.1 XUV Imaging Spectroscopy for the Sun

High-resolution imaging spectroscopy, giving rise to the "golden age" of the XUV solar astronomy, made it possible to visualize high inhomogeneity and temporal variability of coronal plasma structures. Observations with full-Sun broadband soft X-ray telescope (SXT) on the Yohkoh satellite discovered a complex structure and plasma dynamics with high temperatures $T > 2\,\mathrm{MK}$. These structures, associated with magnetic configurations of various topology, were observed later with unprecedented spatial resolution by X-ray telescope (XRT) onboard the Hinode satellite. Implementation of imaging devices with narrow EUV spectral bands on SOHO (Solar and Heliospheric Observatory), CORONAS-F, TRACE (Transition Region and Coronal Explorer), STEREO, and other missions helped to study spatial and kinematic properties of $\sim$1–3 MK plasma of quiescent as well as "active" coronal loops, manifesting in explosive and eruptive phenomena (flares, coronal mass ejections, EIT-waves, etc.). Hard X-ray telescope (HXT) with high spatial and temporal resolution on Yohkoh and RHESSI (Reuven Ramaty High-Energy Solar Spectroscopic Imager) experiments allowed to localize superhot thermal ($T > 10\,\mathrm{MK}$) and high-energy nonthermal plasma sources and to study their temporal dynamics during flare events. Spatial and temporal characteristics of plasma derived from aforementioned experiments resulted in plasma modeling and testing theoretical scenarios for flare and eruptive phenomena. Nevertheless, despite significant progress achieved in solar corona physics in the last decades owing to intensive ground-based and space studies, a number of basic questions related to specific mechanisms of energy release during solar flare events and its transformation remain debatable.

The main reason that restricts further progress in theoretically describing active events on the Sun is the shortage of information either about the spectral composition of XUV emission in broadband filter images or about the spatial localization of the monochromatic emission in the full-Sun line spectra. It leads to significant uncertainties in diagnosing the basic plasma characteristics—spatial distributions of emission measure, temperature, and density, as well as their temporal dynamics. This fact significantly hampers plasma modeling and requires further experimental and theoretical studies.

The advent of XUV full-Sun monochromatic imaging spectroscopy in the SPIRIT (Spectroheliographic X-ray Imaging Telescope) experiment onboard the CORONAS-F satellite (functioning during 2001–2005) allowed to disclose a new class of 4–20 MK plasma strictures characterized by specific morphologic and temporal features. Inverse spectroscopic methods developed and verified by employing temporal series of SPIRIT images showed a principle possibility to infer spatial and temporal properties of hot plasma (temperature, density, and EM distributions) needed for quantitative description of transient phenomena revealed in monochromatic XUV images [8, 18].

10.4.2.2 SPIRIT Experiment Onboard the CORONAS-F Satellite

The multichannel RES spectroheliograph of the SPIRIT instrumentation was designed to acquire monochromatic images of the full solar disc and the adjacent corona with relatively high spatial (up to 5″), temporal (up to 7 s), and spectral resolutions in the following spectral bands: (1) 8.41–8.43 Å (two MgXII X-ray channels including the MgXII resonance line 8.42 Å) and (2) two EUV spectral channels, 176–207 Å and 280–330 Å [18, 31]. Various programs of observations in the MgXII and EUV channels, including simultaneous observations, were realized in the SPIRIT experiment.

Note that the RES spectroheliograph in the MgXII channel register the emission of hot plasma at temperatures $T > 4$ MK because the emissivity function of this line is sensitive at high temperatures and has the maximum at $\approx$10 MK. This implies that the monochromatic full-Sun MgXII 8.42 Å images provide direct confirmation of the presence of hot plasmas. The Sun monochromatic X-ray images revealed regions in the solar corona with high temperatures $T > 4$ MK and, thus, allowed a new class of highly dynamic plasma structures with various characteristic sizes and lifetimes from several minutes to several days to be discovered (see [32–34]). In particular, long-lived (up to several days) plasma structures located high in the corona (up to 3×10^5 km) and resembling spiders in shape were observed. Examples of such "spiders" observed on 2001 November 12 (left image) and on 2001 December 29 (right image) in the MgXII channel are shown in Fig. 10.6.

Figure 10.7 demonstrates a comparison of images acquired on 2001 November 12 in the broadband SXT/Yohkoh channel sensitive to temperatures in a wide range of about ~2.5–25 MK (left panel) and in the monochromatic MgXII channel (right panel) whose sensitivity range is narrower (~5–15 MK). This comparison shows a substantial difference between the spatial scales of plasma structures with different temperature contents: $T > 2$ MK plasma occupying a large area in the SXT image and well-localized plasma features at temperatures $T > 4$ MK, fuzzy in the SXT broadband channel, but clearly developed in the monochromatic MgXII image.

10.4.2.3 Results of the EUV Spectra Analysis

During the SPIRIT experiment carrying out onboard the CORONAS-F satellite, several thousands of spectroheliograms in two spectral bands 176–207 Å and

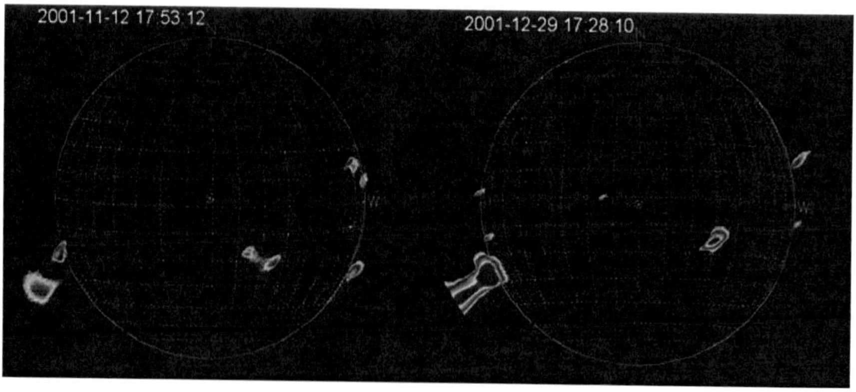

Fig. 10.6 Examples of the hot coronal structures ("spiders") in the monochromatic MgXII 8.42 Å line on 2001 November 12 and on 2001 December 29

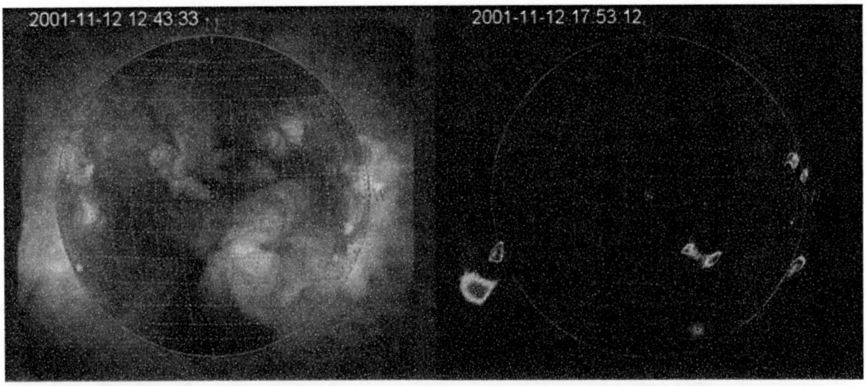

Fig. 10.7 Solar corona images recorded with the broadband SXT/Yohkoh filters (*left panel*) and in the monochromatic MgXII/CORONAS-F channel (*right panel*) on 2001 November 12

280–330 Å were acquired by the XUV channels of the spectroheliograph RES. The spectra of various solar regions (active regions, quiet sun areas on the disc, flares, etc.) were analyzed and a catalogue of the spectral lines was composed [35].

Using the lines with the relative intensities sensitive to the temperature in the range ~0.5–20 MK but independent on the electron density, the DEM temperature distributions were calculated by means of the BIM. The contribution functions $G(l, T)$ for the lines were derived from the CHIANTI atomic database [3, 4]. Results of these calculations for a series of active events are presented in [8, 18]. An important result is the similarity between the temperature behaviors of the DEM profiles for a number of active regions in a wide temperature range up to ~8 MK. It is also worth to note that a few active regions had a pronounced peak of plasma material toward temperatures of ~10–12 MK.

Another important result of diagnosing the hot plasma in active coronal structures was the determination of the soft X-ray emission mechanism of the "spider" plasma. This was made by comparing data measured in the EUV and soft X-ray spectral ranges [8]. The calculations were performed for the event on 2001 December 29 (see right image in Fig. 10.6). In this figure, one may see a radially (along the solar radius) elongated "spider" structure in the X-ray (hot) MgXII RES image. Due to the favorable angular orientation of the RES instrument in the observing period (in the XUV channel corresponding to the spectral band 280–330 Å, the "spider" was directed along the axis perpendicular to the direction of dispersion), it was possible to analyze the radial dependence of the DEM temperature distributions along the solar radius. Figure 10.8 shows these distributions for three altitudes above the solar limb and for an active region on the limb. From this figure, one can clearly see a nonuniform radial distribution of plasma material (DEM) in the temperature range 4–10 MK. For the sake of comparison, a DEM distribution is presented for a flare event of the class X3.4 on 2001 December 28, which is extremely peaked at temperatures ~10–12 MK.

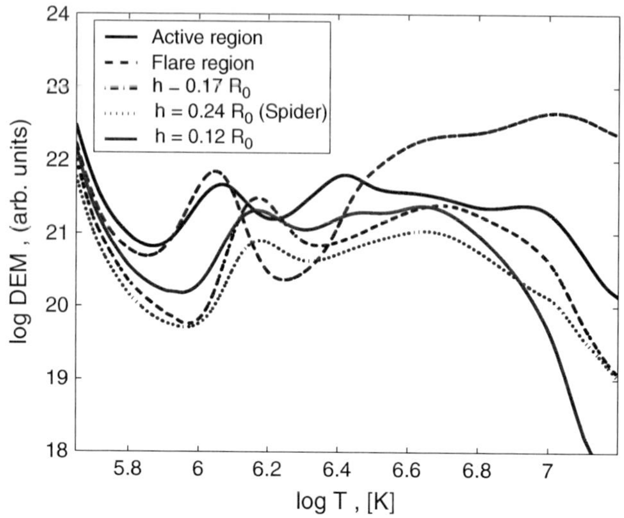

Fig. 10.8 DEM temperature distributions (**a**) for the "spider" on December 28–29, 2001 (radial dependence for three altitudes above the limb along the solar radius), (**b**) for the active region on the limb NOAA 9765, and (**c**) for a flare of the class X3.4 on December 28, 2001

Figure 10.9 shows the radial distributions of the intensities (normalized to the values on the solar limb) in the individual EUV lines and in the X-ray MgXII line recorded when there were no flare events. A fundamentally important result of this comparison is that the observed intensities of the EUV CaXVIII line and the X-ray MgXII line with the close formation temperatures behave quite differently: the intensity of the former line decreases, while the intensity of the latter one increases. At the same time, the relative intensities in the MgXII line calculated using the distributions in Fig. 10.8 for the active region on the limb and three regions of the spider (marked by the crosses in Fig. 10.9) behave similarly to the radial intensity distribution observed in the CaXVIII line.

The following explanation of this fact was given [8]. Since the excitation threshold of the EUV lines is low enough, their intensities are proportional to the emission measure formed by thermal (Maxwellian) electrons. The X-ray line of the MgXII ion having a considerably higher excitation threshold is formed by high-energy nonthermal electrons ($\sim$2 keV and more) whose densities can reach several percent of the total electron density. As a consequence, the EUV line intensities do not depend on the presence of a small admixture of nonthermal electrons and are proportional to the "thermal" emission measure. In contrast, the X-ray line intensities are determined by the relative contribution functions, which can exceed the thermal part by many times due to nonthermal electrons. Thus, the observational data from the RES EUV channels clearly point to the nonthermal (nonstationary) character of the emission mechanism of the hot plasma for the event under study. This was in close agreement with the conclusions drawn from an independent analysis of the time profiles of the emission in the MgXII channel (see [8, 18] for details).

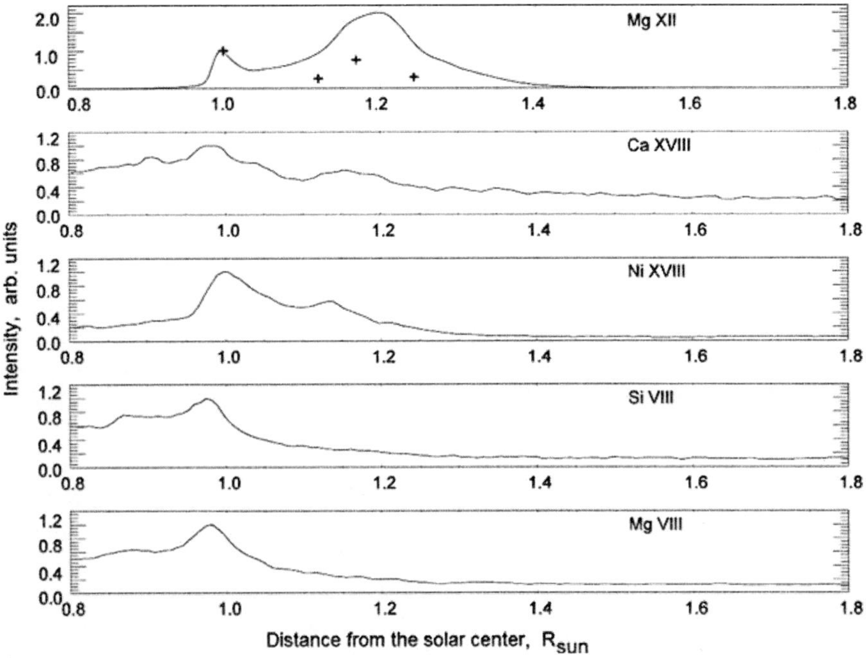

Fig. 10.9 Radial (along the solar radius) distribution of intensities in the EUV lines and in the MgXII channel in the region of the "spider" on December 29, 2001

A deep investigation of the BIM abilities as a temperature diagnostic technique for analyzing XUV solar data was carried out in [9]. A series of numerical tests and model simulations were realized in the case of both line spectra and broadband imaging data. Applications to high-resolution line spectra (from SUMER/SOHO and SPIRIT/CORONAS-F data) and broadband-imaging data (provided by XRT/Hinode instrument) were also considered. These studies confirmed the robustness and effectiveness of the BIM as a tool for the temperature analysis of hot plasma structures.

10.5 Conclusions

Studies of hot plasmas by means of spectroscopic diagnostics methods have wide development related to the necessity to solve important applied problems, such as the controlled fusion, X-ray lasers, processes in atmospheres of the Sun, and other stars. On the other hand, spectroscopic methods force to formulate and to solve inverse problems, which are crucially dependent on input data. In particular, the results of spectroscopic diagnostics based on the XUV emission spectra depend considerably on the accuracy of atomic data, as well as on adopted models of the emitting sources. In this context, we have given a review to show our developments of spectroscopic methods for diagnostics of hot optically thin plasmas and have demonstrated their applications to studying plasma structures in astrophysical and laboratory conditions.

We formulated and developed a SCA allowing to estimate (or to verify) the accuracy of atomic data needed for modeling and diagnosing hot coronal plasmas. This approach is based on solving the spectral inverse problem by means of the BIM algorithm in the framework of the adopted "spectroscopic model." High-resolution K-spectra of highly charged ions measured in the TEXTOR tokamak plasma were used for the verification of atomic data (spectral and collisional characteristics) and methods of their calculation. In particular, it was shown that the calculations carried out by means of the ATOM and MZ codes require essential corrections of atomic data for dielectronic satellites (~30–50% for autoionization probabilities). Corrected atomic data made it possible to perform an accurate diagnostics of plasma parameters, namely, to determine plasma temperatures and argon ion densities in the tokamak plasma. Atomic data verified employing the high resolution laboratory spectra may essentially improve the accuracy of plasma parameters derived from inverse diagnostics techniques in the case of astrophysical sources.

We also used our diagnostics techniques for studying the solar coronal plasma. Investigations and detailed analysis of the BIM abilities for the DEM temperature analysis through numerical tests and simulations were performed by modeling various plasma structures in the solar atmosphere. The diagnostics of active regions in the solar corona plasma was carried out using EUV spectral data acquired onboard the CORONAS-F satellite. The most reliable lines were used to reconstruct DEM temperature distributions by means of the BIM. These results allowed us to infer the presence of hot plasma at temperature range ~4–10 MK in a number of active regions on the Sun, as well as to confirm the nonthermal nature of the soft X-ray emission mechanism in active coronal structures.

Appendix: Bayesian Iterative Scheme

Let us consider two related complete systems of events $\{X_i\}$ and $\{Y_k\}(i = 1,\ldots,n; k = 1,\ldots,m)$, and corresponding sets of probability distributions $\{P(X_i)\}$ and $\{P(Y_k)\}$ for them. In applications, these distributions may also be ones for some random variables $\mathbf{X}$ and $\mathbf{Y}$. The probability distributions $\{P(X_i)\}$ and $\{P(Y_k)\}$ are related by the formulas of the total probability:

$$P(Y_k) = \sum_i P(Y_k|X_i)\,P(X_i), \tag{10.38}$$

where $P(Y_k|X_i)$ is the conditional probability of the event Y_k at the condition X_i. If the distribution $\{P(Y_k)\}$ is known, one can formulate the problem for deriving the $\{P(X_i)\}$ one from the relations (10.38). Below, we will state an iterative procedure called the BIM to resolve this task.

The BIM is based on Bayes' theorem for the a posteriori conditional probability connecting two random variables defined on the fields of events $\{X_i\}$ and $\{Y_k\}$ as follows:

$$P(X_i|Y_k) = \frac{P(Y_k|X_i)\,P(X_i)}{\sum_j P(Y_k|X_j)\,P(X_j)}. \tag{10.39}$$

The formula of the total probability for the distribution $P(X_i)$ is then given by the expression inverse to (10.38):

$$P(X_i) = \sum_k P(X_i|Y_k)\,P(Y_k). \tag{10.40}$$

Substituting (10.39) in (10.40) gives the identity

$$P(X_i) = P(X_i) \sum_k \frac{P(Y_k|X_i)\,P(Y_k)}{\sum_j P(Y_k|X_j)\,P(X_j)}. \tag{10.41}$$

The expression (10.41) can be used for formulating an iterative scheme. For this purpose, the value $P(X_i)$ in the right side of (10.41) is interpreted as step n, and in the left side as step $(n + 1)$ of the iterative procedure. Thus, one obtains the following recurrence relation for the distribution $P(X_i)$:

$$P^{(n+1)}(X_i) = P^{(n)}(X_i) \sum_k \frac{P(Y_k|X_i)\,P(Y_k)}{\sum_j P(Y_k|X_j)\,P^{(n)}(X_j)}. \tag{10.42}$$

It is worth to make some comments regarding the formula (10.42). The left side of (10.42) can be considered as the estimate of the nth hypothesis for the probability distribution $P(X_i)$. The initial approximation for a priori distribution $P^{(0)}(X_i)$ may be taken in accordance with any prior information. If such information is absent, according to the Bayes' postulate, one assumes a uniform distribution (corresponding to equal lack of knowledge). It is also possible, using relation (10.42), to show that the normalizing condition for the distribution $\{P(X_i)\}$ is automatically conserved at any step of the iterative procedure (10.42).

Acknowledgements We would like to thank our colleague Vladimir Slemzin for fruitfully collaborating on the subject of the present review. We also gratefully acknowledge the financial support from the European Commission Programme under the grant agreement 218816 (FP-7 SOTERIA project) and the Russian Foundation for Basic Research (project 11-02-01079-a)—Programme of the Presidium of the Russian Academy of Sciences "Plasma Processes in the Solar System."

References

1. S.R. Pottasch, Space Sci. Rev. **3**, 816 (1964)
2. K.J.H. Phillips, U. Feldman, E. Landi, *Ultraviolet and X-ray Spectroscopy of the Solar Atmosphere* (Cambridge University Press, New York, 2008)

3. K.P. Dere, E. Landi, H.E. Mason, B.C. Monsignori Fossi, P.R. Young, Astron. Astrophys. Suppl. Ser. **125**, 149 (1997)
4. E. Landi, G. Del Zanna, P.R. Young, K.P. Dere, H.E. Mason, M. Landini, Astrophys. J. Suppl. Ser. **162**, 261 (2006)
5. J.T. Jefferies, F.Q. Orrall, J.B. Zirker, Sol. Phys. **22**, 307 (1972)
6. I.J.D. Craig, J.C. Brown, Astron. Astrophys. **49**, 239 (1976)
7. J.C. Brown, B.N. Dwivedi, Y.M. Almleaky, P.A. Sweet, Astron. Astrophys. **249**, 277 (1991)
8. A.M. Urnov, S.V. Shestov, S.A. Bogachev, F.F. Goryaev, I.. Zhitnik, S.V. Kuzin, Astron. Lett. **33**, 396 (2007)
9. F.F. Goryaev, S. Parenti, A.M. Urnov, S.N. Oparin, J.-F. Hochedez, F. Reale, Astron. Astrophys. **523**, id.A44 (2010)
10. I.J.D. Craig, J.C. Brown, *Inverse Problems in Astronomy* (Hilger, Bristol, 1986)
11. A. Tarantola, *Inverse Problem Theory and Methods for Model Parameter Estimation* (SIAM, Philadelphia, 2005)
12. E.L. Kosarev, Inverse Probl. **6**, 55 (1990)
13. V.I. Gelfgat, E.L. Kosarev, E.R. Podolyak, Comp. Phys. Commun. **74**, 349 (1993)
14. W.H. Richardson, JOSA **62**, 55 (1972)
15. I.A. Zhitnik, V.V. Korneev, V.V. Krutov, S.N. Oparin, A.M. Urnov, TrSSR (Trudy Akademiia Nauk SSSR Fizicheskii Institut) **179**, 39 (1987)
16. F.F. Goryaev, A.M. Urnov, G. Bertschinger, A.G. Marchuk, H.-J. Kunze, J. Dubau, JETP Lett. **78**, 363 (2003)
17. A.M. Urnov, F.F. Goryaev, G. Bertschinger, H.-J. Kunze, O. Marchuk, JETP Lett. **85**, 374 (2007)
18. I.A. Zhitnik, S.V. Kuzin, A.M. Urnov, S.A. Bogachev, F.F. Goryaev, S.V. Shestov, Sol. Sys. Res. **40**, 272 (2006)
19. P. Beiersdorfer, A.L. Osterheld, T.W. Phillips, M. Bitter, K.W. Hill, S. von Goeler, Phys. Rev. E **52**, 1980 (1995)
20. A.J. Smith, P. Beiersdorfer, V. Decaux, K. Widmann, K.J. Reed, M.H. Chen, Phys. Rev. A **54**, 462 (1996)
21. O. Marchuk, G. Bertschinger, A. Urnov, F. Goryaev, N.R. Badnell, A.D. Whiteford, J. Plasma Fusion Res. Ser. **7**, 274 (2006)
22. O. Marchuk, M. Tokar, G. Bertschinger, A. Urnov, H.-J. Kunze, D. Pilipenko, D. Kalupin, D. Reiter, A. Pospieszczyk, M. Biel, M. Goto, F. Goryaev, Plasma Phys. Contr. Fusion **48**, 1633 (2006)
23. A.H. Gabriel, Mon. Not. R. Astron. Soc. **160**, 99 (1972)
24. G. Bertschinger, W. Biel, the TEXTOR-94 Team, O. Herzog, J. Weinheimer, H.-J. Kunze, M. Bitter, Phys. Scr. T **83**, 132 (1999)
25. J. Weinheimer, I. Ahmad, O. Herzog, H.-J. Kunze, G. Bertschinger, W. Biel, G. Borchert, M. Bitter, Rev. Sci. Instrum. **72**, 2566 (2001)
26. G. Bertschinger, O. Marchuk, in *High-temperature plasmas diagnostics by X-ray spectroscopy in the low density limit*, ed. by R.E.H. Clark, D.H. Reiter. Nuclear Fusion Research (Springer, Berlin, 2005), pp. 183–199
27. M.Yu. Kantor, A.J.H. Donné, R. Jaspers, H.J. van der Meiden, TEXTOR Team, Plasma Phys. Contr. Fusion **51**, 055002 (2009)
28. V.P. Shevelko, L.A. Vainshtein, *Atomic Physics for Hot Plasmas* (IOP Publishing, Bristol, 1993)
29. F.F. Goryaev, A.M. Urnov, and L.A. Vainshtein, ArXiv, physics/0603164 (2006)
30. J.E. Rice, K.B. Fournier, U.I. Safronova, J.A. Goetz, S. Gutmann, A.E. Hubbard, J. Irby, B. LaBombard, E.S. Marmar, J.L. Terry, New J. Phys. **1**, 19 (1999)
31. S.V. Kuzin, I.A. Zhitnik, A.A. Pertsov, V.A. Slemzin, A.V. Mitrofanov, A.P. Ignatiev, V.V. Korneev, V.V. Krutov, I.I. Sobelman, E.N. Ragozin, R.J. Thomas, J. X-ray Sci. Technol. **7**, 233 (1997)
32. I. Zhitnik, S. Kuzin, A. Afanas'ev, O. Bugaenko, A. Ignat'ev, V. Krutov, A. Mitrofanov, S. Oparin, A. Pertsov, V. Slemzin, N. Sukhodrev, A. Urnov, Adv. Space Res. **32**, 473 (2003a)

33. I.A. Zhitnik, O.I. Bugaenko, A.P. Ignat'ev, V.V. Krutov, S.V. Kuzin, A.V. Mitrofanov, S.N. Oparin, A.A. Pertsov, V.A. Slemzin, A.I. Stepanov, A.M. Urnov, Mon. Not. R. Astron. Soc. **338**, 67 (2003b)
34. I. Zhitnik, S. Kuzin, O. Bugaenko, A. Ignat'ev, V. Krutov, D. Lisin, A. Mitrofanov, S. Oparin, A. Pertsov, V. Slemzin, A. Urnov, Adv. Space Res. **32**, 2573 (2003c)
35. S.V. Shestov, S.A. Bozhenkov, I.A. Zhitnik, S.V. Kuzin, A.M. Urnov, I.L. Beigman, F.F. Goryaev, I.Yu. Tolstikhina, Astron. Lett. **34**, 33 (2008)

Chapter 12
Atomic Physics Using Ultra-Intense X-Ray Pulses

M. Martins, M. Meyer, M. Richter, A.A. Sorokin, and K. Tiedtke

Abstract The combination of short wavelengths and ultrahigh intensities as provided by the new soft and hard X-ray free electron laser sources opens the doorway to totally new experiments on photon-matter interaction. It concerns, in particular, new classes of nonlinear inner-atomic processes. In the present contribution, recent results on sequential and nonsequential multi-photon ionization of gas phase targets are presented and discussed; including processes where also inner shells are affected. Moreover, examples are given how linear and nonlinear photoionization may be used for online photon diagnostics at these new radiation sources.

12.1 Introduction

With the advent of new powerful soft X-ray sources based on free electron laser (FEL) like FLASH in Germany [1, 2], SCSS and SACLA in Japan [3], the LCLS in the USA [4], and high-harmonic generation (HHG) techniques [5, 6], multiphoton processes can be studied for the first time at photon energies above atomic ionization thresholds. In contrast to conventional laser sources for optical radiation, the new soft X-ray lasers can also access inner valence and core level electrons. Thus, all advantages of core level spectroscopy, for example, its site and element specificity, can now be combined with the potentials of multiphoton processes, for example, the study of fast dynamical processes by pump-probe spectroscopy.

An outstanding feature in core level spectroscopy are resonant inner-shell excitations because they can increase the core-level photoionization cross sections quite dramatically. Furthermore, they open a way to obtain a chemical contrast or to excite, in a complex multiatom system, a well-defined site. In this article, we discuss some results of the photoionization of rare gas atoms in these strong X-ray fields, after giving a short overview of the properties of the Free Electron Laser at Hamburg FLASH. After a discussion of the basic nonlinear interaction processes of intense X-rays beams with atoms, we focus on direct and sequential multi-photon processes.

V. Shevelko and H. Tawara (eds.), *Atomic Processes in Basic and Applied Physics*, Springer Series on Atomic, Optical, and Plasma Physics 68,
DOI 10.1007/978-3-642-25569-4_12,

A special emphasis will be given to core level resonances and their unique effect on nonlinear processes.

These new laser-like sources for soft X-rays offer ways to study so far unknown processes in rather simple systems as free atoms or molecules. However, for atoms with their well-known properties in the linear regime, photoionization can serve as a probe to study the properties of the ultra-intense X-ray laser sources. Some of these applications of atomic physics will be discussed in Sect. 12.6.

In this review, we concentrate our attention on experiments carried out at FLASH as the first soft X-ray FEL source. A recent review on further experiments including also results from other FEL sources can be found in [7].

12.2 Experiments at FLASH

In mid-2005 the Free electron LASer in Hamburg (FLASH) started regular user operation [8, 9], providing uniquely intense, short-pulsed radiation that initially could be tuned from 47 to 6.9 nm. With the upgrade in 2009, the maximum electron energy has been increased from 750 MeV to 1.25 GeV, extending the wavelength range down to 4.2 nm.

Peak and average brilliance of this user facility exceed both modern synchrotron facilities and laser plasma sources by many orders of magnitude. The soft X-ray output possesses unprecedented flux of about 10^{13} photons per pulse with pulse durations in the femtosecond range, and hence, combined with appropriate focusing optics, peak irradiance levels of more than $10^{16}\,\mathrm{W\,cm^{-2}}$ can be achieved [10]. Brilliance, coherence, and an ultrashort pulse length down to the femtosecond regime are the outstanding properties opening a new era in the study of soft X-ray radiation–matter interaction.

FLASH is a single-pass FEL lasing in the soft X-ray regime based superconducting linear accelerator described in detail by Ackermann et al. ([2] and references therein). A photoinjector generates very-high-quality electron bunch trains which are accelerated to relativistic energies of up to 1.25 GeV and produce laser-like soft X-ray radiation during a single pass through a 30-m long undulator, a periodic magnetic structure. The generation of soft X-ray laser radiation is based on the so-called self-amplified spontaneous emission (SASE) process.

Briefly, in the undulator, the electron bunches undergo a sinusoidal motion and emit synchrotron radiation. The radiation moves faster than the electron bunch and interacts with electrons further up leading to a charge density modulation within the bunch with a period corresponding to the fundamental in the wavelength spectrum of the undulator. This well-defined periodicity in the emitting bunch enhances the coherence and power of the laser-like radiation field exponentially while the electron and the resulting photon bunch travel once through the long undulator without the need for a resonator.

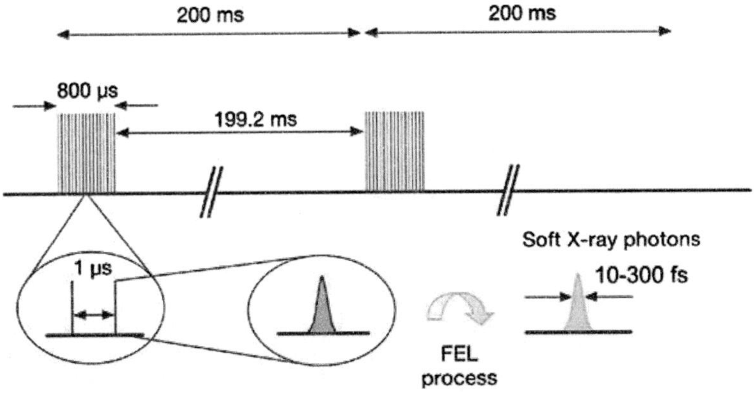

Fig. 12.1 Electron bunch time pattern of FLASH with 5-Hz repetition rate and up to 800 bunches in a 800-μs-long bunch train. The separation of electron bunches within a train can be set to 1, 2, 4, 5, or 10 μs. The pattern is directly transferred to the photon pulses as each electron bunch produces one photon pulse

Table 12.1 Performance of FLASH

Parameter of FLASH	
Wavelength range of fundamental	4.2–60 nm
Higher harmonics	3rd ~1.3 nm
Pulse energy average	10–250 μJ
Peak power	Several GW
Pulse duration (FWHM)	10–300 fs
Spectral width (FWHM)	0.5–1%
Angular divergence (FWHM)	90 ± 10 μrad[a]
Peak brilliance	10^{29}–10^{30} photons/s/mrad^2/mm^2/0.1% bw

[a]SASE in saturation @ 30 nm

Figure 12.1 depicts the electron bunch and the resulting photon pulse pattern of FLASH at a repetition rate of 5 Hz. Table 12.1 summarizes the performance of FLASH in May 2011.

Since the exponential amplification process in a SASE FEL starts from spontaneous emission (shot noise) in the electron bunch, the SASE FEL radiation itself is of stochastic nature, meaning that individual radiation pulses differ in their intensity, temporal structure, and spectral distribution. Therefore, exploitation of the unique properties of the FEL radiation requires suitable pulse-resolved diagnostic tools. Furthermore, online determination of these important photon beam parameters, such as intensity, spectral distribution, and temporal structure, is mandatory for most experiments. This requires diagnostics tools which operate in parallel to the user experiments and in a nondestructive way. To fulfill these demands, new diagnostics concepts, such as an online spectrometer and intensity monitors, have been developed for FLASH as will be discussed in Sect. 12.6.

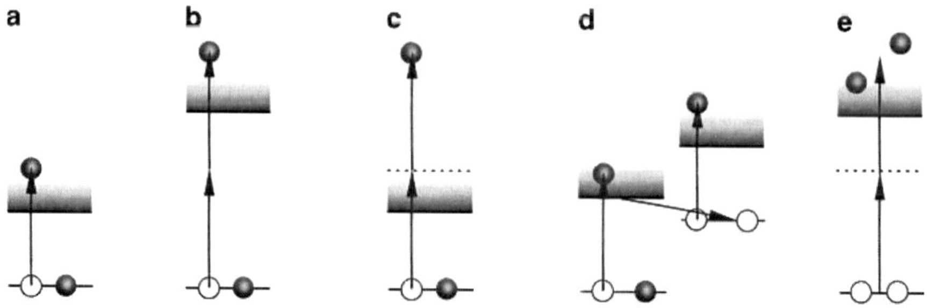

Fig. 12.2 Direct one- and two-photon excitations. (**a**) is the classical photoeffect, and (**b**) is the corresponding direct two-photon single ionization via a virtual state. (**c**) is called above threshold ionization (ATI). (**d**) is sequential two-photon double ionization via an ionic state, and (**e**) is direct two-photon double ionization

12.3 Atoms in Strong X-Ray Fields

In Fig. 12.2, the simplest direct, nonresonant one- and two-photon processes in atoms are depicted. Figure 12.2a represents the classical photoeffect, where a single electron with the binding energy E_{bind} is excited by a single photon with photon energy $\hbar\omega$. The kinetic energy of the electron is given by $E_{\text{kin}} = \hbar\omega - E_{\text{bind}}$ [11]. The corresponding direct two-photon process is described in Fig. 12.2b, where two photons are necessary to eject a single electron. Here, the kinetic energy of the electron is given by $E_{\text{kin}} = n \cdot \hbar\omega - E_{\text{bind}}$, where $n(= 2)$ is the number of photons involved in the ionization. A variation of the process in Fig. 12.2b is the so-called above threshold ionization (ATI) in Fig. 12.2c [12]. Here, already a single photon can ionize the atom. However, if the photon density is huge enough, the emitted electron can absorb a further photon giving rise to a larger kinetic energy. Similar to this process a single photon may ionize an atom; however, the second photon will now further excite the remaining ion giving rise to a doubly ionized atom as shown in Fig. 12.2d. This is the so-called sequential two-photon double ionization of an atom via an ionic state. In Fig. 12.2e, finally, a simultaneous (or direct) two-photon double ionization process is depicted.

The schemes given show clearly the complexity already for two-photon processes. Regarding n-photon processes, all types of combinations of the processes shown in Fig. 12.2 are possible [13–19, and references therein]. In the following section, results on such processes will be presented and, in particular, how these processes can be distinguished.

12.3.1 Saturation Effects

If a considerable percentage of the targets within the interaction volume is ionized by a single photon pulse, a saturation effect may occur which is due to

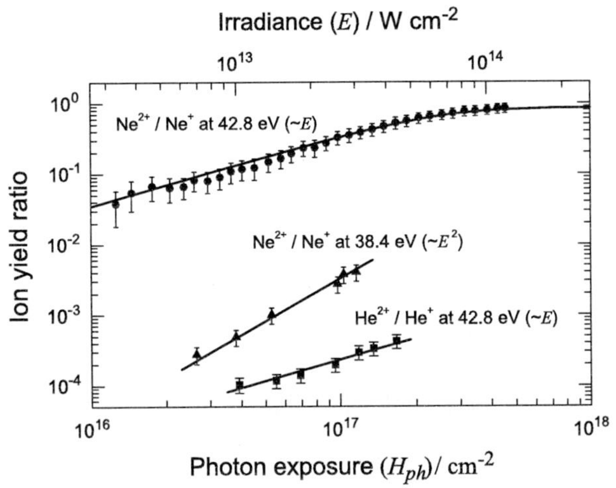

Fig. 12.3 Ratio of different ion yields at two photon energies depending on the photon exposure. Different slopes for the Ne^{2+}/Ne^{+} ratio are found for two different wavelength exciting above and below the ionization threshold of Ne^{+} ions. The initial slope is a signature for the number of absorbed photons [13]

target depletion. While for the first photons within a pulse 100% of the targets is available to be photoionized, the target number may already be strongly reduced for the last photons of a pulse, which explains this type of nonlinearity. In the simplified case of a rectangular beam profile, the number of ions $N_{\rm ion}$ generated per pulse is described by an exponential saturation function of the number of photon per pulse $N_{\rm ph}$ [20]:

$$\frac{N_{\rm ion}}{nzA} = 1 - exp\left(-N_{\rm ph}\frac{\sigma}{A}\right). \tag{12.1}$$

An example for this nonlinearity is found in Fig. 12.3 for the ratio of Ne^{2+}/Ne^{+} at 42.8 eV. Above a photon exposure of $H_{\rm ph} = N_{\rm ph}/A \cong 10^{17}{\rm cm}^{-2}$, the slope of the curve is decreasing due to the saturation of the target density (see also Fig. 12.12b in Sect. 12.4 for the yield of helium 1s electrons).

Saturation due to target depletion does not depend on the initial target gas density n but on the photoionization cross section σ and the photon beam cross section A only. z is the length of the interaction volume along the photon beam. The effect can be used as a method to derive the photon beam cross section A and, by that, the focus size and beam waist of a focused FEL beam [20]. By fitting the saturation curves according to (12.1), using known one-photon single ionization cross sections and measuring the absolute photon number per pulse, for example, with a calibrated gas monitor detector (GMD) [21], A can be obtained.

12.3.2 Sequential Multiphoton Ionization

If a considerable percentage of the target within the interaction volume is ionized by a single photon pulse, the generated ions also represent a new sort of target which can be further ionized within the same pulse as sketched in Fig. 12.2d. In particular, sequences of one-photon ionization processes may occur as long as the photon pulse duration is short enough, so that the atomic and ionic targets may be regarded to be frozen. This type of sequential multiphoton ionization is restricted to photon energies above the respective ionization thresholds and, hence, not possible in the optical regime. The sequential multiphoton multiple ionization

$$X \longrightarrow X^{+} \longrightarrow X^{++} \longrightarrow \cdots, \tag{12.2}$$

was studied at FLASH firstly on molecular nitrogen [22] and later on rare gas atoms in the spectral range from 38 to 43 eV [13] for the case of two-photon double ionization. Solving the rate equations for generation and annihilation of the different atomic and ionic species results in an X^{2+} to X^{1+} ion yield ratio within the regime free from any target depletion [13, 23]:

$$\frac{N_2}{N_1} = \frac{\sigma_{1\rightarrow 2}}{2}\frac{N_{\text{ph}}}{A}, \tag{12.3}$$

that is, a linear increase with the photon exposure $H_{\text{ph}} = N_{\text{ph}}/A$. $\sigma_{1\rightarrow 2}$ denotes the one-photon single ionization cross section of X^{1+} leading to X^{2+}. For neon, Fig. 12.3 shows a corresponding experimental result which was obtained at the photon energy of (42.8 ± 0.2) eV, that is, well above the threshold for the $Ne^{1+} \rightarrow Ne^{2+}$ process at 41.0 eV [24]. From a fit of (12.3) to the data, a cross section $\sigma_{1\rightarrow 2}(42.8\,\text{eV}) = (7.0 \pm 1.0) \times 10^{18}]\text{cm}^2$ is found, which is in fair agreement with results obtained from an Ne^{+} ion-beam photoionization experiment using synchrotron radiation [25]. Nevertheless, the latter comparison is difficult to perform because there is a number of resonances in Ne^{+} around 42.8 eV which are partly covered by the 1% spectral bandpass of the FLASH photon beam [8]. Resonant multi-photon processes will be further discussed in Sects. 12.3.4 and 12.4.

12.3.3 Direct Multiphoton Ionization

A direct two-photon double ionization process as depicted in Fig. 12.2e was observed in the soft X-ray regime firstly on helium using a HHG source at 41.8 eV photon energy [26, 27]. Figure 12.4 shows the results of ion TOF spectroscopy, which has been obtained on He with 42.8 eV photons at FLASH [28]. The He^{2+}/He^{1+} ratio depends strongly on the photon beam cross section A, which has been proven by changing the focus size within the interaction volume, hence,

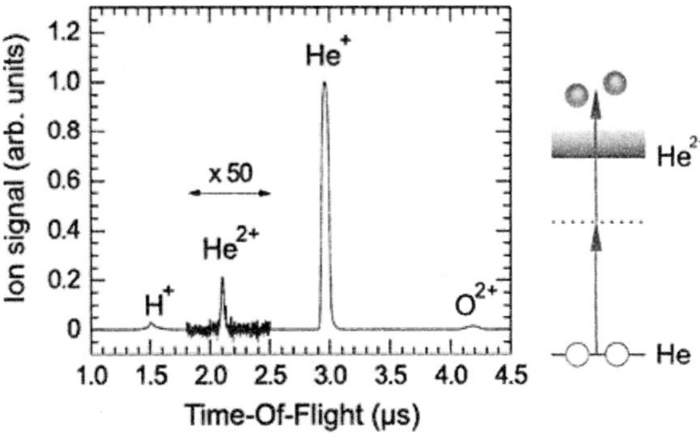

Fig. 12.4 Ion TOF spectra of helium (He) taken at 42.8 eV photon energy. He^{2+} ions are produced by a direct two-photon process

changing the mean irradiance E given by

$$E = \frac{\hbar\omega N_{\text{ph}}}{A\Delta t}. \tag{12.4}$$

This behavior indicates the He^{2+} signal being dominated by nonlinear multiphoton ionization as described by the power law of perturbation theory:

$$\frac{\mathrm{d}N^{(m)}}{\mathrm{d}t} = nzA\sigma^{(m)}\left(\frac{E}{\hbar\omega}\right)^{m}. \tag{12.5}$$

Here, $\sigma^{(m)}$ denotes the generalized m-photon ionization cross section. In particular, the linear increase of the He^{2+} to He^{1+} ion yield ratio with irradiance as demonstrated in Fig. 12.3 identifies the He^{2+} generation as a two-photon process. He^{2+} generation arising from one-photon double ionization by higher FEL harmonics at higher photon energies [29] whose contributions do not vary along the photon beam plays a negligible role. Also, a sequential two-photon process according to Fig. 12.2d via He^{+} can be excluded here, because the photon energy of 42.8 eV is by far lower than the ionization energy of 54.4 eV for the second step from He^{+} to He^{2+}.

Fitting the measured data for the He^{2+} to He^{1+} ratio as shown in Fig. 12.3 by a linear function of irradiance and using the known one-photon single ionization cross section $\sigma^{(1)}(42.8\,\text{eV}) = (2.75 \pm 0.08) \cdot 10^{18}\,\text{cm}^2$, one obtains the experimental result for the direct two-photon double ionization cross section of helium of $\sigma^{(2)}(42.8\,\text{eV}) = (1.6 \pm 0.6) \cdot 10^{52}\,\text{cm}^4\,\text{s}$ [13, 28]. This value agrees fairly well with the estimation of $10^{52}\,\text{cm}^4\,\text{s}$ derived for the total He^{2+} yield from the HHG measurements at 41.8 eV photon energy [27]. Both values lie perfectly in between the two main groups of theoretical predictions for this cross section [30–38] as shown by Fig. 12.5. A further recent theoretical analysis on two-photon double ionization of helium can be found in [39].

Since the strength of direct multiphoton ionization depends according to (12.5) on the irradiance E and, hence, also on the pulse duration t, the latter may be

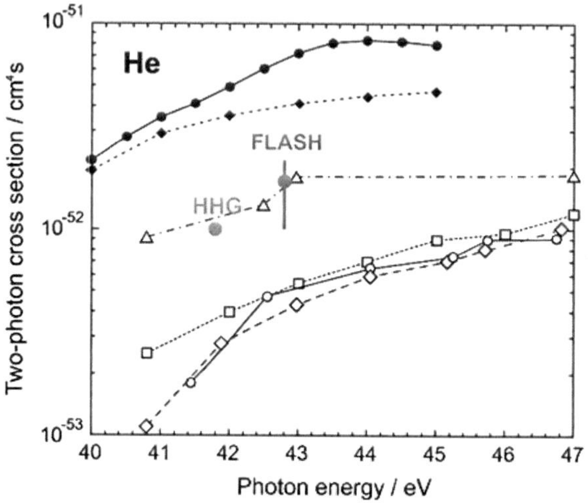

Fig. 12.5 Experimental and theoretical two-photon double ionization cross section of helium [42]. The HHG value is taken from [27]. The theoretical data are taken from *filled circle*, *open circle* [30], *filled diamond* [31], *open triangle* [37], *open square* [34], and *open diamond* [33]

experimentally determined by means of autocorrelation technique [26]. Splitting an FEL pulse with the temporal pulse shape $f(t)$ into two parts with the irradiance amplitudes E_1 and E_2 and a temporal delay of $t'(E(t) = E_1 f(t) + E_2 f(t - t')$, one obtains according to (12.5) for a direct two-photon process:

$$\frac{N^{(2)}(t')}{N^{(2)}(t' \to \infty)} = 1 + \alpha \int f(t) f(t - t') \mathrm{d}t. \tag{12.6}$$

At FLASH, the determination of the pulse shape function $f(t)$ and, by that, the pulse duration has been performed for the first time by Mitzner et al. [40] by investigating the direct two-photon double ionization of helium as a function of the pulse delay t' [41], using a two-pulse correlator for the pulse splitting, and evaluating the convolution integral in (12.6).

Regarding its experimental and theoretical interest, the two-photon double ionization of He certainly represents a prime example for a nonlinear process in the soft X-ray regime. However, a few FEL studies refer also to two-photon single ionization processes of He at 38.5 eV [43] and the two-photon 4d ATI of Xe at 93 eV photon energy [44], both studied by electron spectroscopy. The square dependence of the Ne^{2+} to Ne^{+} ratio at 38.4 eV as depicted in Fig. 12.3 indicates, again, a two-photon single ionization process, from Ne^{+} to Ne^{2+}, here studied by ion spectroscopy [13].

12.3.4 Multiple Ionization and Strong Field Effects

The FLASH results of nonlinear photoionization presented up to here can be well explained in terms of target depletion, sequential, and direct multi-photon schemes

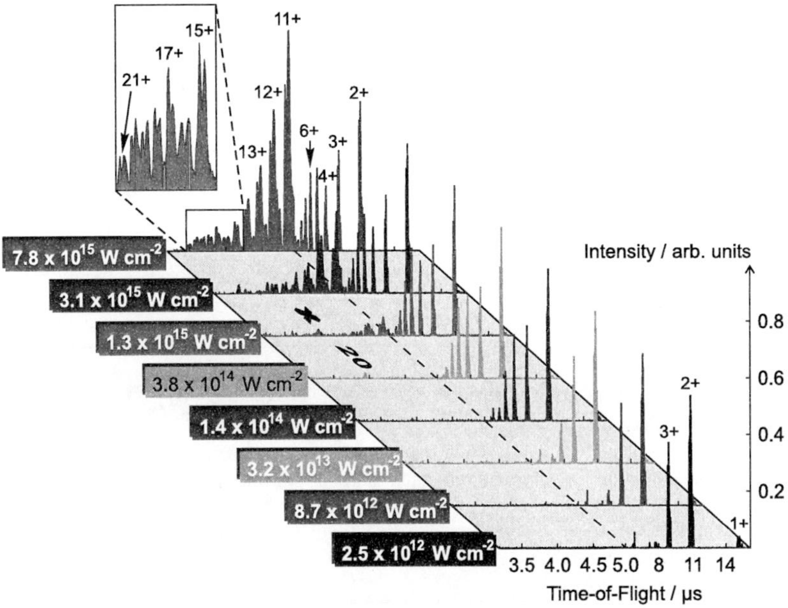

Fig. 12.6 Ion TOF mass/charge spectra of Xe taken at the photon energy of 93 eV and different pulse irradiance levels [46]

as sketched in Fig. 12.2. In the EUV, however, exceptional behavior of light matter interaction was found. With the aid of a high-quality industrial grade spherical Si/Mo multi-layer mirror of the type used in [45], one can realize focal spots of 3–5 μm in diameter at photon energies from 90 to 93 eV and, hence, irradiance levels up to 10^{16} W cm^2 [46, 47]. In the case of Xe, the situation is rather complex. Figure 12.6 shows the nonlinear appearance of higher charge states with increasing EUV irradiance E obtained in a multi-layer focus at 93 eV photon energy [46]. At $E = 7.8 \times 10^{15}$ W cm^2, Xe^{21+} clearly occurs whose generation, starting from neutral Xe, requires a total energy of at least 5 keV. Plasma effects like electron impact from neighboring atoms can be excluded because the spectra do not depend on the target density. This means that almost 60 EUV photons at the energy of 93 eV must have been absorbed by an individual atom within the estimated FLASH pulse duration of 10–20 fs.

In a first theoretical study of the Xe results, the high degree of photoionization was explained by a sequence of one- and multi-photon ionization processes according to (12.2) in which an ion created in a preceding step represents the target for a subsequent step [48]. The corresponding ion yield results are shown in Fig. 12.7 which describes the experimental data up to the charge state Xe^{15+}. However, this approach has to consider at least 19 steps to generate Xe^{21+} from atomic Xe, which is related to 19 coupled differential rate equations with 19 individual multi-photon ionization cross sections. The latter were not calculated

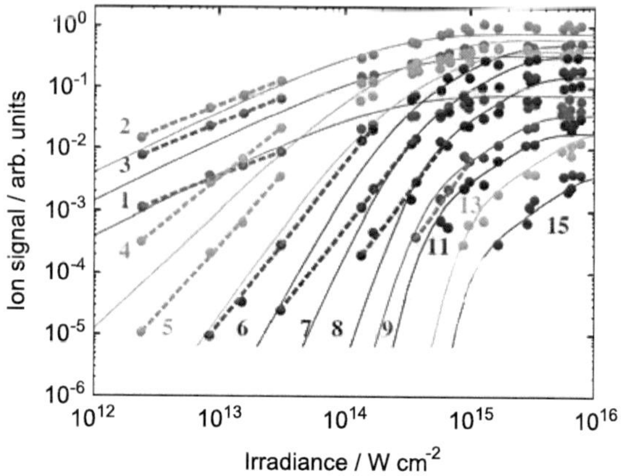

Fig. 12.7 Calculated [48] and measured [46] relative Xe^{q+} ion signal intensities as a function of the peak irradiance for different charge states $1 \leq q \leq 15$

Table 12.2 Highest charge state q_{max} observed at irradiance levels in the range from 1.5 to 1.8 × 10^{15} W cm^2, ionization energy I required to reach this state starting from the atomic ground state, corresponding to the minimum number N_{ph} of EUV photons of 90.5 eV photon energy which must have been absorbed within a single FEL pulse by an individual atom to deliver this amount of energy, and one-photon ionization cross section σ at 90.5 eV photon energy for the rare gases Ne, Ar, Kr, and Xe, respectively [47]

Gas	q_{max}	I/eV	N_{ph}	σ/Mbarn
Ne	7+	715	8	4.4
Ar	7+	434	5	1.35
Kr	7+	383	5	0.55
Xe	14+	1,930	22	24

from first principles but obtained through a technique of scaling, with occasional adjustment to the experimental data.

In order to obtain a deeper insight into the Xe photoionization process in the EUV at irradiance levels beyond 10^{15} W cm^{-2}, Xe was compared in a further experimental study with Ne, Ar, and Kr [47]. The gases were investigated at FLASH under equivalent conditions at 90.5 eV photon energy and irradiance levels in the range from 1.5 to 1.8 × 10^{15} W cm^{-2}. The main result is summarized in Table 12.2: The highest charge state observable in that experiment was 7+ for Ne, Ar, and Kr but 14+ in the case of Xe. The minimum number N_{ph} of EUV photons with 90.5 eV photon energy, which must have been absorbed within a single FEL pulse by an individual atom to deliver the amount of energy for the highest charge state observable, is listed in the fourth row of Table 12.2. It strongly varies from $N_{ph} = 5$ for Kr and Ar, $N_{ph} = 8$ for Ne, and $N_{ph} = 22$ for Xe. This is obviously reflected by the respective one-photon ionization cross section values for 90.5 eV photons which are listed in the last row. In the case of Xe, the photoionization cross section is strongly enhanced by the so-called giant $4d \rightarrow \epsilon f$ continuum resonance and amounts to 24×10^{-18} cm^2.

On the role of the $4d$ giant resonance for the interpretation of the outstanding Xe results, a controversial discussion has started recently [48]. In comparison to the

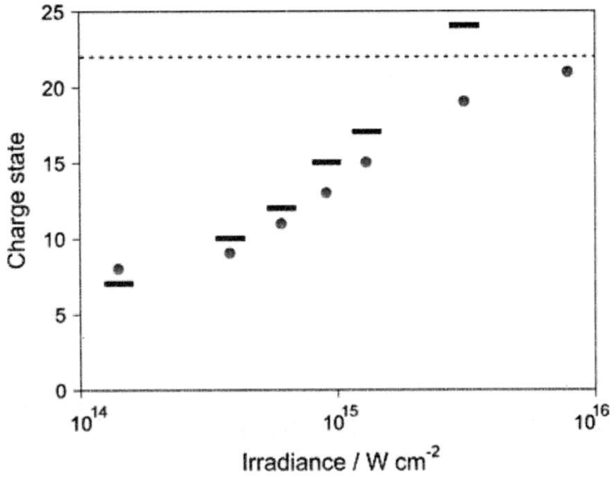

Fig. 12.8 Highest charge states of Xe generated by (15 ± 5) fs EUV pulses as a function of irradiance (*filled circles*: experimental data [33, 35]; *bars*: calculated within the harmonic oscillator model of the Xe *4d* giant resonance [52])

other rare gases, it is a particular feature of Xe and arises in the photon energy range from about 85–115 eV ([49] and references therein). For many years, this resonance has represented a prime example of the impact of electron correlation on inner-shell photoionization at low irradiance [50, and references therein]; and was described in some theoretical works by collective electron oscillations in the $4d$ shell [51]. Recently, this idea was used to express the energy transfer from a high-intensity EUV field to the Xe atom in terms of its giant resonance and by the equations of a damped harmonic oscillator [52]. In this context, fair agreement was obtained with the experimental results of the high charge states in Xe as shown in Fig. 12.8.

12.4 Resonant Processes

The process behind the special behavior of xenon atoms is still under discussion, but it shows that resonant excitations in the soft X-ray regime can have a strong influence on the nonlinear process. Regarding resonant processes some of the principle processes are sketched in Fig. 12.10. Clearly, the complexity of possible excitation processes is increasing as compared to the direct multi-photon processes or one-photon resonant processes depending in which ionic stage the resonance will be excited. The simplest process is shown in Fig. 12.10a describing the direct excitation of a two-photon resonance with the same parity as the ground state, which is not accessible by one-photon excitations. Such an excitation can be regarded as an excitation via a virtual, intermediate state. In Fig. 12.10b the corresponding process for a real resonant excitation is sketched. An important difference between the process Figs. 12.10a, b is the finite lifetime of the intermediate state.

The direct resonant two-photon excitation Fig. 12.10a has been demonstrated at FLASH for the case of the resonant Auger decay of the Kr $3d^{-1}4d$ resonance [53]. This resonance is located at 92 eV excitation energy and can be reached from

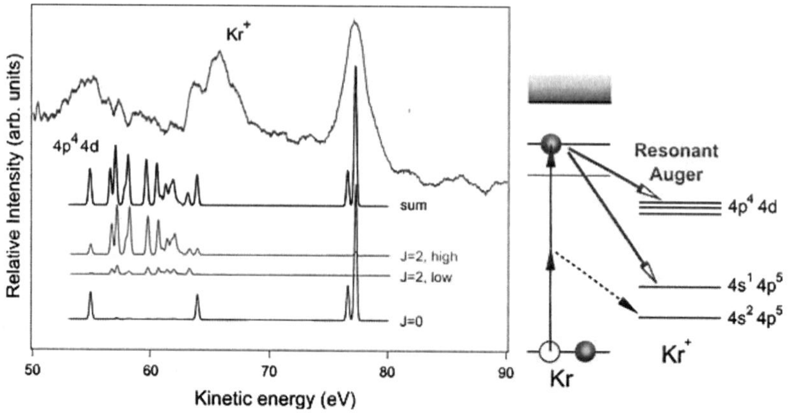

Fig. 12.9 High-resolution electron spectrum of atomic Kr compared to a theoretical spectrum taking into account the relative probabilities for the excitation of different $3d^{-1}4d,\ J = 0,2$ resonances and their decay rates to the $\mathrm{Kr}^+\ 4p^4 4d\ ^{2,4}L_J$ final ionic states (from [53])

the electronic ground state via simultaneous absorption of two FEL photons of 46 eV photon energy. In order to induce the nonlinear process, the FEL beam was focused to diameter of about 5 μm, which enabled us to produce intensities of up to $10^{15}\,\mathrm{W\,cm^{-2}}$ in the interaction volume. As a result of the two-photon process, the Auger decay to the $\mathrm{Kr}^+\ 4p^4 4d$ ionic states was recorded by means of high-resolution electron spectroscopy (Fig. 12.9). By means of two-photon excitation, a completely new class of resonances can be accessed and investigated, which were out of reach by conventional one-photon processes from the ground state. The main structures in the experimental spectrum are well reproduced by theoretical simulations including the two-photon excitation to the Kr $3d^{-1}4d,\ J = 0,2$ core-excited states and their decay to the Kr^+ $4p^4 4d\ ^{2,4}L_J$ multiplet. Most of the intensity of the resonant Auger lines is attributed to the decay of the higher lying $J = 2$ component of the resonance, whereas the excitation of the $J = 0$ component enhances mainly the intensity of the $4p$ and $4s$ autoionization lines as well as of the $4p^4(^1D)4d\ ^2S_{1/2}$, which is strongly mixed with the $4s$ line.

Even at these high intensities, the probability for the observed two-photon process is about 1,000 times smaller than for a one-photon process causing very intense electron lines in the kinetic energy region below 46 eV. As indication for the strong production of singly charged ions via one-photon ionization, the broad structure at about 65 eV kinetic energy shows up, which can be attributed to the $4p$ ATI Above Threshold Ionization, Fig. 12.2c of the Kr^+ ion, that is, to the emission of the $4p$ electron from the ion after absorption of two FEL photons. This manifold of processes taking place during the same FEL pulse, that is, one-photon and multiphoton processes of the neutral as well as of the ionic species, represents the real experimental challenge for studying the nonlinear processes in the XUV regime; the already known one-photon processes produce a huge "background" signal and cause fast depletion of the neutral target.

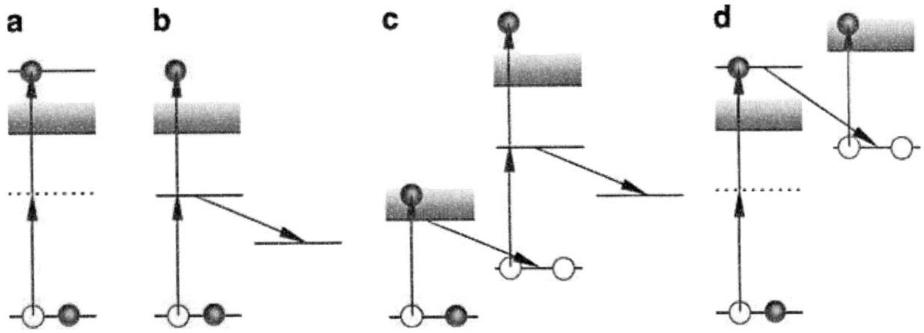

Fig. 12.10 Sketch of some possible resonant multiphoton excitations. Process (**a**) is similar to the direct two-photon ionization; however, the final state of this process is a resonance, for example, a bound state of the atom in the continuum

A resonant excitation can take place not only in the first excitation step, but also in an ionic state. A process of the type Fig. 12.10c has been observed, for example, for the case of neon atoms [54]. For the ratio of the Ne^{2+}/Ne^{1+} ions, an increase is found depending on the wavelength of the exciting FEL radiation above 41.5 eV photon energy (see Fig. 12.11). This enhancement is attributed to a two (three) photon process, where the first photon is ionizing a neon atom to singly ionized Ne^+. A second photon will ionize directly Ne^+ to Ne^{2+} below 41.4 eV. However, around 41.5 eV, several Ne^+ $2s^2 4p^4 6\ell$ resonances have been observed in a merged beam experiment with rather large cross sections above 30 Mbarn [25]. Thus, the second photon can excite these states with a lifetime of $\cong$130 fs, and a third photon will ionize this resonant state further to Ne^{2+} or the resonance will decay via autoionization. Furthermore, a deexcitation process by stimulated emission will become possible.

Also the process Fig. 12.10d, where subsequent to a resonant, direct two-photon excitation, a further photon excitation in an excited states of the atoms occurs, has been observed by photoelectron spectroscopy [43]. In Fig. 12.12a, the photoelectron spectrum of helium atoms excited with 38.5 eV is depicted. The strong *1s* photoline is due to normal one-photon ionization, which can be clearly seen by the almost linear dependence of the electron intensity from the pulse energy. The decreasing slope for higher photon intensities (Fig. 12.12b) can be attributed to saturation of the target (see Sect. 12.3.1). At higher kinetic energies around 25 eV, a very weak signal is observed, where the electron intensity has a square dependence on the irradiance, hence, a two-photon excitation (Fig. 12.12b). The observed signal can be explained by the process sketched in (Fig. 12.10d). Doubly excited helium $n\ell m\ell'(n = 5, 6)$ states of even parity at $\cong$77 eV excitation energy are excited by 2×38.5 eV photons. These states will decay either by Auger decay to He^+ or to a small amount by fluorescence to lower lying doubly excited states with $n = 2, 3$ of neutral helium, which can be ionized by a third photon. Interestingly, this ionization will take place by the next FEL pulse following 1 μs after the first one due to the long lifetime of the metastable states, decaying by fluorescence [43].

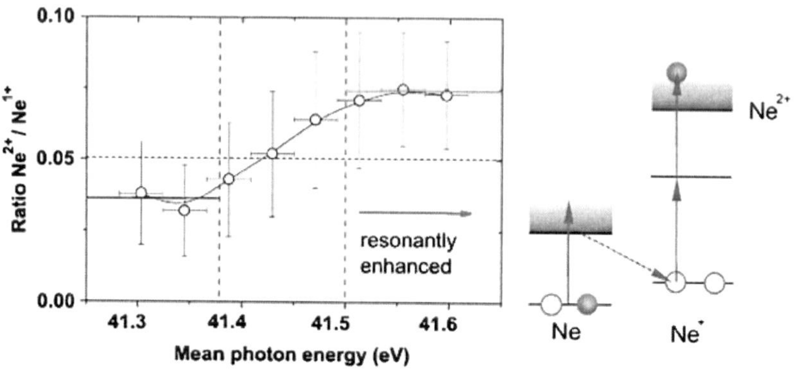

Fig. 12.11 Ratio of doubly to singly ionized neon, depending on the photon energy of the exciting FLASH pulses

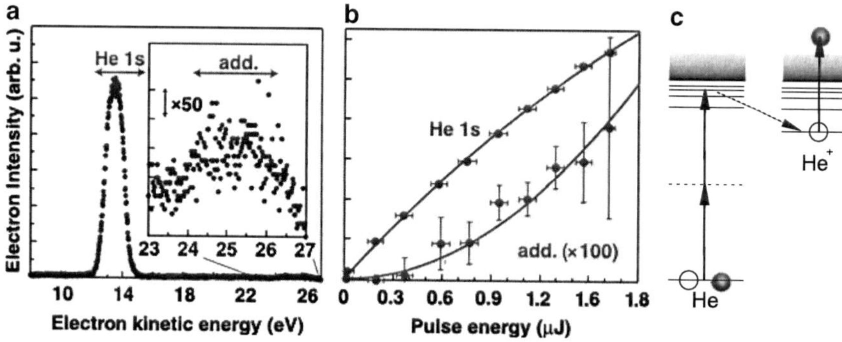

Fig. 12.12 Photoelectron spectrum of helium taken with intense XUV pulses at 38.5 eV. A weak photoelectron signal around 25 eV kinetic energy (**a**) shows a quadratic dependence on the pulse energy (**b**) which is attributed to a resonant two-photon process as sketched in (**c**) [43]. For the He *1s* main line, a saturation according to (12.1) is found

Thus, this experiment shows a very interesting point in multi-photon studies of resonant states. Due to the finite lifetime of the excited states the time dependent dynamic of the decay of the core excited atoms can be studied.

12.5 Time Resolved Studies

12.5.1 Two-Color Above Threshold Ionization

A wealth of new experiments focusing on the time-evolution of the photoionization processes can be explored by combing the intense of the FEL to an external optical femtosecond pulses laser. Some first experiments have been performed

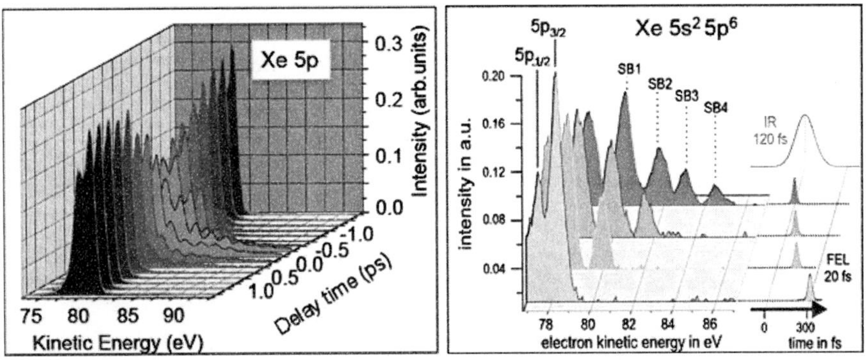

Fig. 12.13 Two-color above threshold ionization of atomic Xe. (*left*) Electron spectra recorded by averaging over 100 FEL pulses as a function of the temporal delay between the FEL (91.8 eV) and the IR laser (4×10^{13} W cm^{-2}). (*right*) Series of single-shot spectra of atomic Xe recorded with FLASH lasing at 90 eV and for nominal maximal temporal overlap with the IR pulses (from [56])

demonstrating the huge potential of this type of investigation in the future. Some general description can be found in recent reviews [55, 56]. The synchronization of both sources on the femtosecond time scale remains still the main challenge and difficulty for these experiments since it affects directly the achievable temporal resolution. A suitable process to measure and monitor the temporal stability and the relative temporal delay between two femtosecond pulses is given by the above threshold ionization [57, 58]. If during the ionization with the XUV pulse an additional optical or infrared field is present in the interaction region, the photoelectron will gain or loose energy in this dressing field. As a consequence, additional lines show up in the photoelectron spectra on both sides of the main photoline separated by the photon energy of the dressing laser. This approach developed using femtosecond high-order harmonic generation sources has been successfully transferred to experiments at FEL sources [59, 60].

A typical series of spectra is shown in Fig. 12.13. Here, the photoionization of the *5p* electron in Xe was investigated. When scanning the delay between the FLASH and IR pulses, a decrease of intensity of the photoline and an increase of additional structures are observe in the region of overlap being the most intense for perfect overlap of both pulses. This is directly used to find the position of zero time delay, which is the principal parameter for many pump–probe experiments. In addition, by using single pulse detection mode, also a more precise timing can be given in the region of overlap between both pulses. Due to the temporal jitter of the FEL with respect to the IR pulse, the photoionization takes place at different position underneath the IR pulse, that is, at different intensity of the dressing laser. This leads to a different number of sidebands, no sidebands for small (or no) overlap and to maximal number of sidebands for perfect overlap. In this way, the spectroscopy of the two-color above threshold ionization was used in some studies to characterize the pulse duration and the temporal stability of FLASH [61, 62].

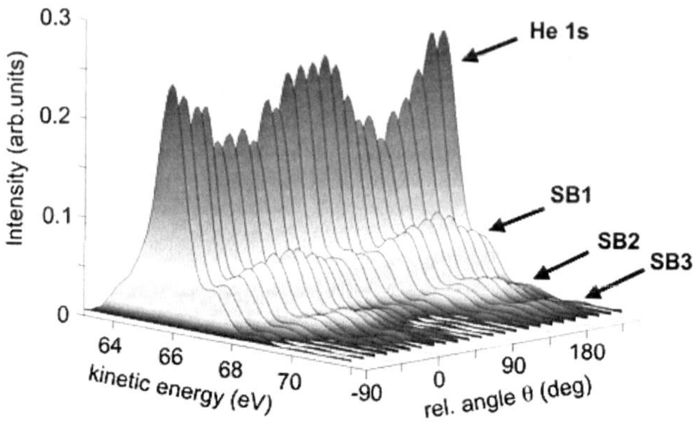

Fig. 12.14 Photoelectron spectra of atomic He recorded for different relative orientations between the polarization vectors of FLASH lasing at 90.5 eV and for a dressing field of about 6×10^{11} W cm^{-2} (from [63])

In a first application, of this technique it was possible to determine the partial photoionization cross section in the two-photon ionization by making use of the well-defined linear polarization of the FEL and the IR laser pulses [63]. In the region of perfect temporal overlap, the relative orientation between the polarization vectors of both sources is varied. This leads to a characteristic variation of the relative intensity between photoline and sidebands, maximum sideband intensity for parallel orientation of both vectors, and minimum for perpendicular orientation (Fig. 12.14). In the case of two-photon ionization of the *1s* electron in atomic He, the relative probability of emission of an electron with *s* or *d* symmetry could be determined, with the surprising result that for photoionization at 90 eV photon energy, the *s* emission is favored with respect to *d* emission, in contrast to the general propensity rules, predicting a higher probability for a change to higher angular momenta.

12.5.2 Two-Color Resonant Excitations

In future studies, the possibility to tune the wavelength of the FEL and of the optical laser will become more and more important. In particular, it will enable the excitation of highly excited autoionization states. By combining both the intense XUV and optical radiation, and adjusting precisely their photon energies, resonances can be excited which are inaccessible by a one-photon process from the electronic ground state due to parity conservation rules. The relaxation pathways and dynamics of these resonances are still almost completely unexplored. The possibility to vary widely the wavelength, the temporal delay, the polarization, and the intensity of both pulses provides a unique basis for a detailed characterization of these resonances and their relaxation dynamics. Beside the

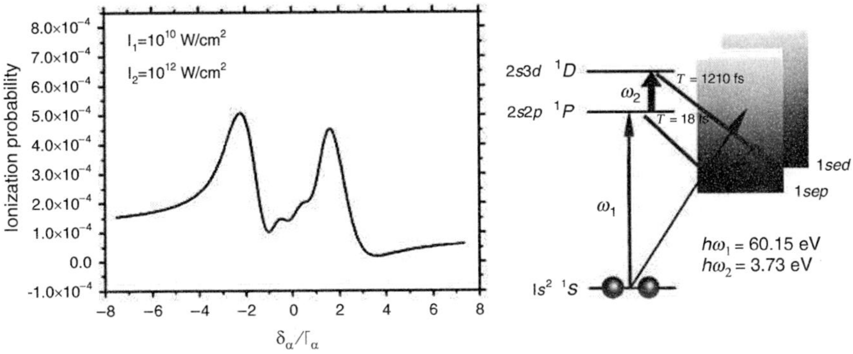

Fig. 12.15 Photoionization probability of atomic He in the region of the $2s2p\ ^1P$ resonance in presence of a resonant optical laser with an intensity of 10^{12} W/cm^2 (from [65]). The X-axis corresponds to XUV photon energy detuning δ_a from resonance in units of the Fano resonance width Γ_a. The intensities of the XUV (I_1) and the optical (I_2) lasers is given in the figure while the duration of the pulses is 200 fs and 150 fs, respectively

investigation of the electronic relaxation of these two-photon resonances, the strong optical field can also be used to shift the energy position of the resonances, that is, to control the transparency of the medium at a particular wavelength. Moreover, the wavelength of both photons can be chosen to coincide with a doubly resonant excitation. The underlying physics for such doubly optically resonant processes has already been investigated already in many theoretical papers (e.g., [64–66]), but only one experiment has been performed up to now, and only at very low excitation energies [67]. An illustration of the expected effect of optical coupling between two autoionization states, here $2s2p\ ^1P$ and $2s3d\ ^1D$ of atomic He, is given in Fig. 12.15 [65]. For weak optical fields, the effect of the He $2s2p\ ^1P$ autoionizing state is observed as the usual enhancement of the photoionization cross section in the vicinity of 60.1 eV excitation energy. The resonance profile is described by the typical asymmetric Fano-type lineshape, which is the result of interference between the direct and the resonant ionization process. For higher optical fields, the photoionization cross section is strongly modified by the coupling of the two autoionizing resonances. The opening up of new ionization continua has to be taken into account and results in a very different resonance profile. The sensitivity of the resonance profile to the applied optical field provides thereby a new and very sensitive control on the dynamics of these highly excited autoionization states.

12.6 Online Diagnostics Based on Atomic Physics

Due to the stochastic nature of the SASE process, each FEL pulse is different from all others. In particular, for experiments dealing with nonlinear processes a simple sum-up of the data is not possible. Hence, a single-shot characterization of each single pulse is mandatory. In principle, all possible parameters of each FEL pulse, as the number of photons in the pulse, its spectral distribution, and also its temporal structure should be recorded without disturbing the pulse. In recent years, several methods have been developed, which are based on atomic photoionization, to measure these parameters. Some of these devices have already become a standard at several FEL facilities, for example, the gas monitor detector (GMD) to measure the absolute number of photons in each individual pulse, which is used at FLASH and also at the LCLS in Stanford. In the following sections, the used methods will be described in more detail.

12.6.1 Pulse Energy Analysis

At FLASH, the pulse-resolved radiant power is measured online and nondestructively using four permanently installed calibrated gas monitor detectors [21]. The detector is based on the atomic photoionization of rare gases at a low pressure of around 10^{-5} mbar and the charge detection of photoions and photoelectrons. Figure 12.16 schematically shows the FLASH-GMD. The charged particles (ions and electrons) created upon photoionization are extracted and accelerated in opposite directions by a static homogeneous electric field and simultaneously and separately detected by simple metal electrodes (Faraday cups), which guarantee a linear signal response. The extraction field of about 500 V/cm is high enough to ensure complete separation and collection of the charge particles created within the interaction volume accepted by the particle detectors. The signal electronics provides single-pulse readout for both the electron signal and the ion signal. Moreover, in order to check long term stability of the FEL photon flux, ions may be alternatively detected by a slow averaging ion current with a time constant of up to 25 s which is not affected by any time structure of the radiation.

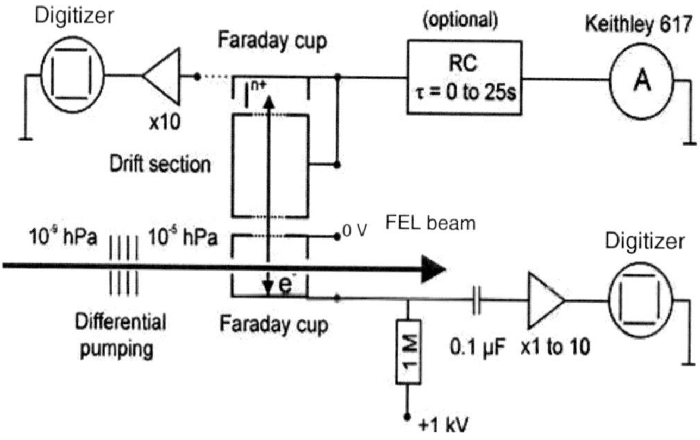

Fig. 12.16 Schematic diagram of the FLASH gas monitor detector and signal detection electronics

The number of photons N_{photon} passing through the detector is determined by the number of measured charged particles (ions or electrons) N_{particle} by:

$$N_{\text{photon}} = \frac{N_{\text{particle}}}{\sigma(\hbar\omega)zn}, \tag{12.7}$$

where $\sigma(\hbar\omega)$ is the known total photoionization cross section of the target gas at the photon energy $\hbar\omega$, z is the effective length of the interaction volume, and n is the density of target gas atoms. The latter is obtained according to $n = p/kT$, where k is the Boltzmann constant, p is the gas pressure determined using a calibrated spinning rotor gauge, and T is the temperature measured by a calibrated Pt100 resistance thermometer. The accurate determination of the FEL beam intensity with an error less than 10% relies on the GMD calibration performed in the VUV spectral range using synchrotron radiation at low intensity in conjunction with a cryogenic radiometer as the primary detector standard as well as on the utilization of the photoionization cross section data to extrapolate the calibration data to the XUV regime [68].

The LCLS-FEL uses an alternative gas detector based on the detection of photoluminescence of nitrogen at a high pressure in the range of 0.015–1 mbar with the help of two photomultiplier tubes [69]. Magnetic coils around a 30-cm-long and 8-cm-wide vacuum chamber confine the photoelectrons produced by the FEL beam; these electrons excite the surrounding nitrogen gas. The deexcitation of the molecules occur by emission of photons in the UV range between 300 and 400 nm. The calibration of the detector was performed by determining the energy loss of the electron beam passing through undulators which enable measuring the average photon pulse energy with an estimated error of about 5%.

12.6.2 Spectral Analysis

Within the GMD, the total electron yield (TEY) is used as a measure for the relative intensities of each signal pulse. In principle, by analyzing the kinetic energy distribution of the emitted electrons also a spectral analysis of each FEL pulse is possible. To record the full spectral information of the SASE pulses on a single-shot basis using photoelectron spectroscopy, some prerequisites have to be fulfilled. For a complete spectrum, a rather large number of electrons have to be recorded; however, the spectra should not be adulterated, for example, by Coulomb interaction between the electrons and the ions. First experiments have been carried out to study the influence of Coulomb repulsion on photoelectron spectra by comparing the photoelectron spectra with the SASE spectra using the special mode of the plane-grating (PG) monochromator beamline [70] in which the PG2 branch is used to record the dispersed SASE spectra in the first order of the grating and taking photoelectron spectra simultanously in the zero order of the grating [54, 71].

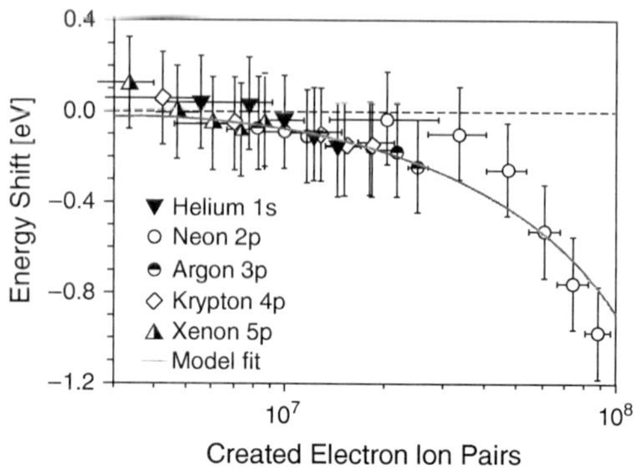

Fig. 12.17 Shift of the kinetic energy of the outgoing *s* or *p* electron of the outer shell for different rare gases as a function of number of created electron–ion pairs. The incident photon energy was 38 eV. The *solid curve* is the result from a simple model as described in the text

The influence of Coulomb repulsion has been studied by measuring the pressure dependence of the kinetic energy distribution of the outer-shell photoelectrons for the rare gases helium to xenon [72]. The Coulomb interaction effects depend on the number of electron–ion pairs created and the effective binding energies. According to (12.7), the number of electron–ion pairs N_{pairs} created depends on the gas-target density $n = p/k_{\text{B}}T$ for an ideal gas, the number of photons N_{photon}, and the cross section σ_{particle} of the rare gas used. In Fig. 12.17, the effective binding energy shift for the photoelectrons of the rare gases is depicted as a function of electron–ion pairs created. The photolines of He, Ar, Kr, and Xe linearly shift to lower energies with increasing number of created electron–ion pairs.

The upper limit for the onset of kinetic energy shift for the given kinetic energy of the photoelectrons of 15–25 eV occurs at 10^7 electron–ion pairs created. This behavior of kinetic energy shift has also been observed in experiments conducted by Pietzsch et al. [73]. The photoline shift can be explained by the outgoing electron cloud being decelerated by the positive charge of the ions. Due to the very short ionization time, the ions are considered as stationary, yielding a stationary electric field wherein the electrons move. Slower electrons are more severely affected by the attractive Coulomb potential of the ions than faster ones. In a simple model of a cylindrical capacitor with the ions on the photon beam axis and the electrons moving radially outbound, the energy loss of the electrons due to Coulomb attraction is on the order of 1 eV for 10^8 electron–ion pairs created and increases linearly with increasing number of electron–ion pairs. This is modeled by the solid curve in Fig. 12.17 [72].

If the number of generated electron–ion pairs is high, additional pulse broadening occurs. In Fig. 12.18, a typical photon energy distribution of the incident photon pulse at 38 eV and the corresponding electron time-of-flight (eTOF) spectrum are depicted. In both spectra, three peaks are clearly visible. The peaks in the photon energy spectrum correspond to three different laser modes present within the single pulse. Due to their statistical origin, these modes vary from shot to shot in their number, their individual intensity, and their individual photon energy within the bandwidth of the FEL (see Sect. 12.2). Due to the repulsive Coulomb force between

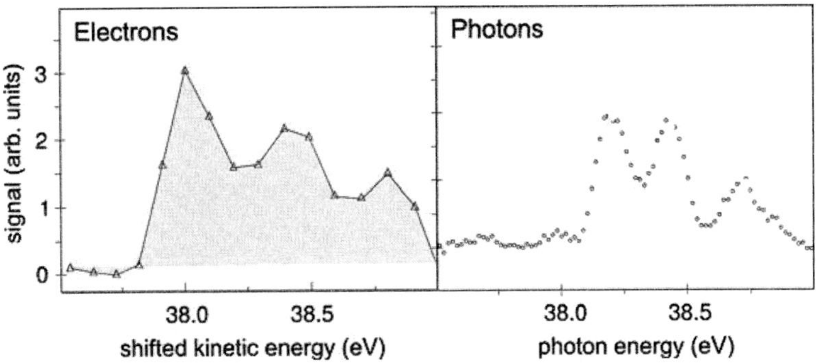

Fig. 12.18 Simultaneously acquired FLASH spectrum of a single shot (*open circles*) and corresponding eTOF spectrum (*open triangles*): for the latter, the binding energy of helium of 24.6 eV has been added to the calculated kinetic energy values to match the photon energy scale

the electrons, the individual electron peaks are broaden and drift apart. Electrons at the head of the cloud with high kinetic energy additionally gain energy as they are repelled from the electrons at the tail of the cloud with lower kinetic energy. As a result, the mean kinetic energy of the broadened electron cloud is shifted to lower kinetic energy. The broadening depends very much on the number of electrons in the pulse and thus the incident number of photons.

These results have proven the feasibility of an online analysis of the spectral distribution of SASE FEL pulses if the effect of Coulomb repulsion is taken into account and corrected. In particular, electron TOF spectrometers are suitable for this purpose because they are able to record spectra also in the multibunch mode of FEL's.

The average photon energy could also be monitored by measuring the ratio of different ionic charge states subsequent to a photoionization process using ion TOF spectroscopy as has been demonstrated by Juranic et al. [74]. This method has the advantage not to be sensitive to Coulomb repulsion, thus combining an ion TOF and an electron TOF would allow to obtain the mean photon energy from ion spectroscopy and the fine structure of SASE pulses from electron spectroscopy.

12.7 Conclusions

Gas-phase experiments on atoms and molecules in the focused beam of FEL radiation have shown a variety of different multiphoton, multielectron photoionization processes, giving evidence that the interaction of photons with matter is strongly affected by inner-atomic nonlinear behavior at irradiance levels beyond $10^{13}\,\mathrm{W\,cm^{-2}}$. The presented results on rare gases in general can be interpreted in terms of one-photon and multiphoton processes. As in one-photon, core level spectroscopy resonances are also of great importance in nonlinear core level spectroscopy; however, their influence is not so obvious, since they can occur in different stages of the multiphoton ionization process. For example, the high charge states of up to 21+ observed by exciting into the Xe $4d \rightarrow 4f$ giant resonance remains still an open question.

Hence, further experimental and theoretical studies, in particular, of the role of inner-shell resonances on the nonlinear photoionization are necessary. The investigation of multi-photon processes on atoms at the new X-ray laser sources has further interesting perspectives. The first aspect concerns the influence of nonlinear processes on photon–matter interaction, and the mechanisms of X-ray induced destruction of materials by electronic excitation, in general. This aspect is significant for all X-ray laser applications in various fields of materials research. The second concerns a fascinating new field of fundamental science at the limits of a quantum mechanical and a semiclassical description of photoionization. The application of atomic photoionization has already been proven to be a valuable tool at the new X-ray laser facilities for online diagnostics of photon beam parameter. It is used for a pulse energy analysis on a shot-to-shot basis and will also be used in the near future for online determination of the spectral distribution by photoelectron spectroscopy and the timing using generated side bands.

Acknowledgements We are thankful for the financial support of the Deutsche Forschungsgemeinschaft MA2541/4-1, RI804/5-1, TI280/3-1, and the BMBF 05KS4GU1/9. Finally, the excellent support of the whole FLASH team is greatly acknowledged. Without their support, the experiments would not have been possible.

References

1. V. Ayvazyan et al., Phys. Rev. Lett. **88**(10), 104802 (2002)
2. W. Ackermann et al., Nat. Photon. **1**(6), 336 (2007)
3. T. Shintake et al., Nat. Photon. **2**(9), 555 (2008)
4. P. Emma et al., Nat. Photon. **4**(9), 641 (2010)
5. T. Sekikawa, A. Kosuge, T. Kanai, S. Watanabe, Nature **432**(7017), 605 (2004)
6. M.C. Chen et al., Phys. Rev. Lett. **105**(17), 173901 (2010)
7. N. Berrah et al., J. Modern Opt. **57**(12), 1015 (2010)
8. V. Ayvazyan et al., Eur. Phys. J. D **37**, 297 (2006)
9. K. Tiedtke et al., New J. Phys. **11**, 023029 (2009)
10. A.J. Nelson et al., Opt. Exp. **17**(20), 18271 (2009)
11. A. Einstein, Ann. Phys. **322**, 132 (1905)
12. M. Protopapas, C.H. Keitel, P.L. Knight, Rep. Prog. Phys. **60**(4), 389 (1997)
13. A.A. Sorokin, M. Wellhöfer, S.V. Bobashev, K. Tiedtke, M. Richter, Phys. Rev. A **75**(5), 051402 (2007)
14. R. Moshammer et al., Phys. Rev. Lett. **98**(20), 203001 (2007)
15. A. Rudenko et al., Phys. Rev. Lett. **101**(7), 073003 (2008)
16. Y.H. Jiang et al., J. Phys. B **42**(13), 134012 (2009)
17. M. Kurka et al., J. Phys. B **42**(14), 141002 (2009)
18. Y.H. Jiang et al., Phys. Rev. Lett. **102**(12), 123002 (2009)
19. K. Motomura et al., J. Phys. B **42**(22), 221003 (2009)
20. A.A. Sorokin et al., Appl. Phys. Lett. **89**(22), 221114 (2006)

21. K. Tiedtke et al., J. Appl. Phys. **103**(9), 094511 (2008)
22. A.A. Sorokin, S. Bobashev, K. Tiedtke, M. Richter, J. Phys. B **39**(14), L299 (2006)
23. P. Lambropoulos, X. Tang, J. Opt. Soc. Am. B **4**(5), 821 (1987)
24. D. Morton, Astrophys. J. Suppl. Ser. **149**, 205 (2003)
25. A.M. Covington et al., Phys. Rev. A **66**(6), 062710 (2002)
26. Y. Nabekawa, H. Hasegawa, E.J. Takahashi, K. Midorikawa, Phys. Rev. Lett. **94**(4), 043001 (2005)
27. H. Hasegawa, E.J. Takahashi, Y. Nabekawa, K.L. Ishikawa, K. Midorikawa, Phys. Rev. A **71**(2), 023407 (2005)
28. M. Richter, S. Bobashev, A. Sorokin, K. Tiedtke, Appl. Phys. A **92**, 473 (2008)
29. S. Düsterer et al., Opt. Lett. **31**(11), 1750 (2006)
30. L.A.A. Nikolopoulos, P. Lambropoulos, J. Phys. B **40**(7), 1347 (2007)
31. E. Foumouo, G.L. Kamta, G. Edah, B. Piraux, Phys. Rev. A **74**(6), 063409 (2006)
32. I.A. Ivanov, A.S. Kheifets, Phys. Rev. A **75**(3), 033411 (2007)
33. L. Feng, H.W. van der Hart, J. Phys. B **36**(1), L1 (2003)
34. S. Laulan, H. Bachau, Phys. Rev. A **68**(1), 013409 (2003)
35. D.A. Horner, F. Morales, T.N. Rescigno, F. Martín, C.W. McCurdy, Phys. Rev. A **76**(3), 030701 (2007)
36. J. Feist, S. Nagele, R. Pazourek, E. Persson, B.I. Schneider, L.A. Collins, J. Burgdörfer, Phys. Rev. A **77**(4), 043420 (2008)
37. I.A. Ivanov, A.S. Kheifets, J. Phys. B **41**(9), 095002 (2008)
38. B. Piraux, E. Foumouo, P. Antoine, H. Bachau, J. Phys. Conf. Ser. **141**(1), 012013 (2008)
39. A. Palacios, D.A. Horner, T.N. Rescigno, C.W. McCurdy, J. Phys. B **43**(19), 194003 (2010)
40. R. Mitzner et al., Opt. Express **16**(24), 19909 (2008)
41. R. Mitzner et al., Phys. Rev. A **80**(2), 025402 (2009)
42. M. Richter, S.V. Bobashev, A.A. Sorokin, K. Tiedtke, J. Phys. B **43**(19), 194005 (2010)
43. M. Nagasono et al., Phys. Rev. A **75**(5), 051406 (2007)
44. V. Richardson et al., Phys. Rev. Lett. **105**(1), 013001 (2010)
45. T. Feigl, S. Yulin, N. Benoit, N. Kaiser, Microelectron. Eng. **83**(4–9), 703 (2006)
46. A.A. Sorokin, S.V. Bobashev, T. Feigl, K. Tiedtke, H. Wabnitz, M. Richter, Phys. Rev. Lett. **99**(21), 213002 (2007)
47. M. Richter et al., Phys. Rev. Lett. **102**(16), 163002 (2009)
48. M.G. Makris, P. Lambropoulos, A. Mihelič, Phys. Rev. Lett. **102**(3), 033002 (2009)
49. U. Becker, D.A. Shirley (eds.), *VUV and Soft X-Ray Photoionization.* Physics of Atoms and Molecules (Plenum Press, New York, 1996)
50. J.P. Connerade, J.M. Esteva, R.C. Karnatak (eds.), *Giant Resonances in Atoms, Molecules and Solids* (Plenum, New York, 1987)
51. M.Y. Amusia, J.P. Connerade, Rep. Prog. Phys. **63**(1), 41 (2000)
52. M. Richter, J. Phys. B **44**(7), 075601 (2011)
53. M. Meyer et al., Phys. Rev. Lett. **104**(21), 213001 (2010)
54. M. Martins, M. Wellhöfer, A.A. Sorokin, M. Richter, K. Tiedtke, W. Wurth, Phys. Rev. A **80**(2), 023411 (2009)
55. C. Bostedt et al., Nucl. Instrum. Meth. A **601**(1–2), 108 (2009)
56. M. Meyer, J.T. Costello, S. Düsterer, W.B. Li, P. Radcliffe, J. Phys. B **43**(19), 194006 (2010)
57. T.E. Glover, R.W. Schoenlein, A.H. Chin, C.V. Shank, Phys. Rev. Lett. **76**(14), 2468 (1996)
58. E.S. Toma et al., Phys. Rev. A **62**(6), 061801 (2000)
59. M. Meyer et al., Phys. Rev. A **74**(1), 011401 (2006)
60. P. Radcliffe et al., Nucl. Instrum. Meth. A **583**(2–3), 516 (2007)
61. P. Radcliffe et al., Appl. Phys. Lett. **90**(13), 131108 (2007)
62. A. Azima et al., Appl. Phys. Lett. **94**(14), 144102 (2009)
63. M. Meyer et al., Phys. Rev. Lett. **101**(19), 193002 (2008)
64. A.I. Magunov, I. Rotter, S.I. Strakhova, J. Phys. B **32**(6), 1489 (1999)
65. S.I. Themelis, P. Lambropoulos, M. Meyer, J. Phys. B **37**(21), 4281 (2004)
66. S.I. Themelis, P. Lambropoulos, F.J. Wuilleumier, J. Phys. B **38**(13), 2119 (2005)

67. N.E. Karapanagioti, D. Charalambidis, C.J.G.J. Uiterwaal, C. Fotakis, H. Bachau, I. Sánchez, E. Cormier, Phys. Rev. A **53**(4), 2587 (1996)
68. M. Richter et al., Appl. Phys. Lett. **83**(14), 2970 (2003)
69. S.P. HauRiege, R.M. Bionta, D.D. Ryutov, J. Krzywinski, J. Appl. Phys. **103**(5), 053306 (2008)
70. M. Martins, M. Wellhöfer, J.T. Hoeft, W. Wurth, J. Feldhaus, R. Follath, Rev. Sci. Instrum. **77**(11), 115108 (2006)
71. M. Wellhöfer, M. Martins, W. Wurth, A.A. Sorokin, M. Richter, J. Opt. A **9**(7), 749 (2007)
72. M. Wellhöfer et al., J. Inst. **3**, P02003 (2008)
73. A. Pietzsch et al., New J. Phys. **10**, 033004 (2008)
74. P.N. Juranić, M. Martins, J. Viefhaus, S. Bonfigt, L. Jahn, M. Ilchen, S. Klumpp, K. Tiedtke, J. Inst. **4**(09), P09011 (2009)

Part IV
Atomic Data Applications and Databases

Chapter 13
Radiation Therapy Using High-Energy Carbon Beams

T. Kanai

Abstract Heavy-ion radiotherapy using high-energy carbon beams has been performed at the National Institute of Radiological Sciences, Japan. The physical frameworks for heavy-ion radiotherapy are established using an understanding of radiation physics. In this chapter, the biophysical and medical physics aspects of heavy-ion radiotherapy are presented. In order to increase the accuracy of heavy-ion radiotherapy, many physical problems should be solved. A calorimeter was developed to measure the absolute dose of the heavy-ion beams. From a comparison of the dosimetry, it was found that the dose indicated by the ionization chamber was underestimated by 3–4%. The clinical results of carbon therapy at heavy-ion medical accelerator in Chiba (HIMAC) are assessed using the linear-quadratic (LQ) model of radiation effect. Development of new scintillation and Rossi counters will allow simultaneous measurement of the radiation dose and quality of heavy-ion beams. Further research is required to provide a comprehensive biophysical model for clinical applications.

13.1 Introduction

It is well known that exposing a malignant tumor to ionizing radiation is an effective treatment method. Just 1 year after the discovery of X-rays, radiation was already being applied to radiation therapy. Since then, various kinds of radiation have been applied to radiation therapy. In the 1960s and 1970s, neutron beams have been expected to be effective at curing radiation-resistant tumors. High linear energy transfer (LET) radiation therapy using neutrons was the focus of many researchers worldwide at that time [1–3]. However, it was found that severe side effects prevented the proper use of the high LET beams.

Wilson [4] proposed using the physical characteristics of high-energy charged particles, the so-called *Bragg peak*, for radiotherapy. It has been recognized that the Bragg peak method has great advantages compared to photon radiotherapy. The Bragg peak therapy, however, requires a high-energy accelerator. Large accelerators

V. Shevelko and H. Tawara (eds.), *Atomic Processes in Basic and Applied Physics*, Springer Series on Atomic, Optical, and Plasma Physics 68,
DOI 10.1007/978-3-642-25569-4_13,

at Lawrence Berkeley National Laboratory (LBNL) at the University of California or at Harvard University, which were originally developed for research physics, have been used for charged particle radiotherapy [5]. At LBNL, ion beams accelerated by BEVALAC have been used for radiation therapy. C.A. Tobias group made great advances in basic researches on the physical and biological characteristics of the high-energy ions [6, 7]. They tested various ions, including Si, Ne, and C, for example, to use for radiation therapy [8]. In some cases, severe side effects from the ion therapy were found.

BEVALAC was shut down in the early 1990s because of the large maintenance cost of keeping a large, old accelerator active. Full-scale radiotherapies using carbon ions at the National Institute of Radiological Sciences using the heavy ion medical accelerator in Chiba (HIMAC) [9] and at the Gesellschaft für Schwerionenforschung mbH [10] have started ion therapy based on the pioneering studies of biophysics and medical physics of heavy ions at LBNL. In both centers for carbon-ion therapy, very good clinical results have been obtained [11, 12].

The two centers have been developed with completely different approaches to carbon-ion therapy. HIMAC has adopted a relatively conservative method for the irradiation system, with a passive scattering method being used to create the irradiation fields. A scatterer and a pair of wobbling magnets are used to broaden the beam, and a ridge filter is used to spread out the Bragg peaks (SOBP). The passive method has an advantage in treating moving targets. Using a technique of synchronized irradiation with respiration [13], various sites of the body, including targets that move during the irradiation, can safely be treated.

In contrast, GSI has adopted a more advanced method [14]. At GSI, the accelerated beam is scanned with a pair of magnets. Irradiation fields with irregular shapes can be created without any collimator in the beam path. Furthermore, the SOBP is performed by actively stacking monoenergetic beams. This is realized by changing the beam energy extracted from the synchrotron, spill by spill. This method has the advantage of achieving excellent dose localization to the target volume. However, this method is only applicable to static targets. In order to apply the scanning beam to moving targets, more research will be necessary [15, 16].

The relative biological effectiveness (RBE) determination is also completely different between the two institutes. At HIMAC, a semiempirical method based on a linear-quadratic model (LQ model) for the survival curve has been adopted [9]. RBE values have been used for all types of tumors and dose levels. Using this fixed RBE system for clinical trials, dose escalation and hypofractionation studies have been performed. In contrast, at GSI, RBE is determined based on the newly developed theory called the local effect model (LEM) [17]. The RBE is different for different tumor types and for different dose levels of treatment. Although their approaches to determining RBE values for each clinical situation differ from each other, a clinical dose, which is defined as the physical dose multiplied by the RBE value, is used in treatment-planning systems in both institutes. Physical dose distributions are designed to have uniform clinical doses in the SOBP region. Using these techniques of heavy-ion radiotherapy, over 5,500 patients at HIMAC and over 300 patients at GSI have been treated with carbon ions. The results of carbon therapy have been excellent, and this therapy has now become a recognized cancer therapy worldwide. Despite the success in clinical applications of the carbon beams, our biophysical understanding is incomplete, and scientific studies should be continued in order to determine the most suitable applications of heavy ions in radiotherapy. In this review chapter, the biophysical aspects of heavy-ion therapy are discussed, and an overview of the unsolved problems in the field of heavy-ion biophysics for clinical trials of heavy-ion radiotherapy will be given.

13.2 Physical Characteristics of Carbon Beam

In order to treat deep-seated tumors, the maximum range of the carbon beam should be larger than 25 cm in water. Energies over 400 MeV/n are required to keep the residual range in patients. As far as electromagnetic interaction is concerned, the physical characteristics of such high-energy carbons are very similar to those of high-energy proton beams. The difference lies in the quantity of the interactions. The stopping power of carbons is much greater than that of protons, and multiple scattering of carbons is much less than that of protons. The stopping power can be calculated quite accurately by the Bethe–Bloch formula. Differences in stopping power greatly affect the biological effects of radiation therapy. We shall next discuss the present status of medical physics in ion therapy in terms of depth-dose distributions and dosimetry of ions.

13.2.1 Dose Distribution in Water Phantom

For the clinical application of carbon beams, one of the most important physical characteristics is depth-dose distributions in water. High-dose localization as well as high biological effectiveness can be achieved simultaneously when using carbon beams for radiotherapy. For these reasons, accurately planned irradiations are especially important.

Figure 13.1 shows the depth-dose distributions in water of carbon beams with energies of 290, 350, and 400 MeV/u. At the deeper region of the Bragg peak, a long tail is observed, the so-called *fragmentation tail*, because the dose includes the contribution of lighter particles fragmented from projectile carbons. The solid curves in the figure show the results of theoretical calculations using a code developed by Sihver et al. [18]. The open circles, triangles, and squares in the figure show the experimental results obtained at HIMAC. With a pair of wobbling magnets and a scatterer, uniform irradiation fields are obtained at the isocenter of the irradiation port. In this situation, the ions are assumed to have flowed out from a point source between the wobbler magnets. The ion fluency is inversely proportional to the square of the distance from the source, source axis distance (SAD). The measured depth-dose distributions were corrected based on

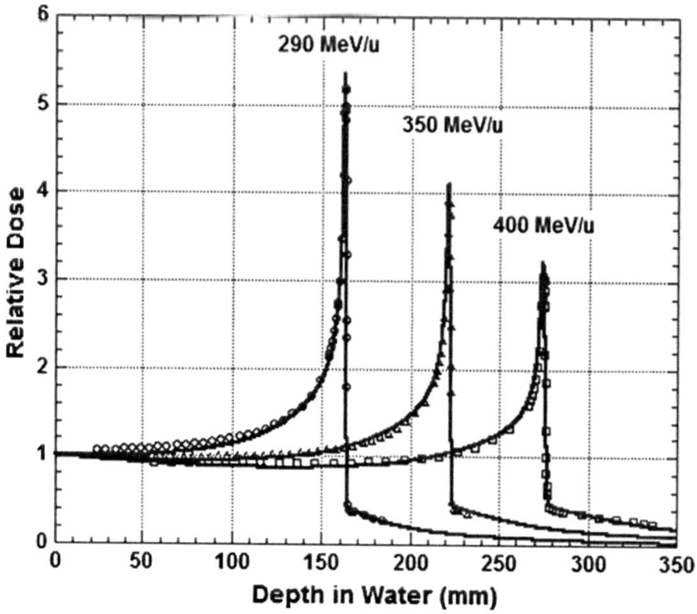

Fig. 13.1 Depth-dose distributions of monoenergetic carbon beams of 290, 350, and 400 MeV/u in water. *Open circles*, *triangles*, and *squares* are experimental values obtained at HIMAC

this beam divergence effect. The depth-dose distributions obtained were shifted so that the measured Bragg-peak positions coincided with the calculated results. The heights of the relative-dose distributions were adjusted so as to coincide with the dose at the surface region of the calculated curves. On the other hand, Gaussian path-length distributions are assumed for the energy straggling in the depth-dose calculation. In order to fit the experimental data, initial residual-range broadenings were introduced in the calculations. The initial residual-range broadening were assumed to be Gaussian shape, and their variance, σ, were 0.4 mm, 0.6 mm, and 0.8 mm for 290 MeV/u, 350 MeV/u, and 400 MeV/u beams, respectively. This may be because of the nonuniform thickness of the scatterer used.

13.2.2 Nuclear Interaction

When a highly energetic particle collides with a nucleus, a nuclear reaction can occur. In this reaction, both the incident ion and the target nucleus break into fragment particles. A cross section of the nuclear reaction is roughly obtained by the geometrical size of the projectile and target nucleus of mass numbers A_{p} and A_{t} [18]:

$$\sigma \sim \pi r_0^2 \left\{ A_{\mathrm{p}}^{1/3} + A_{\mathrm{t}}^{1/3} - (1.581 - 0.786(A_{\mathrm{p}}^{-1/3} + A_{\mathrm{t}}^{-1/3})) \right\}^2 . \tag{13.1}$$

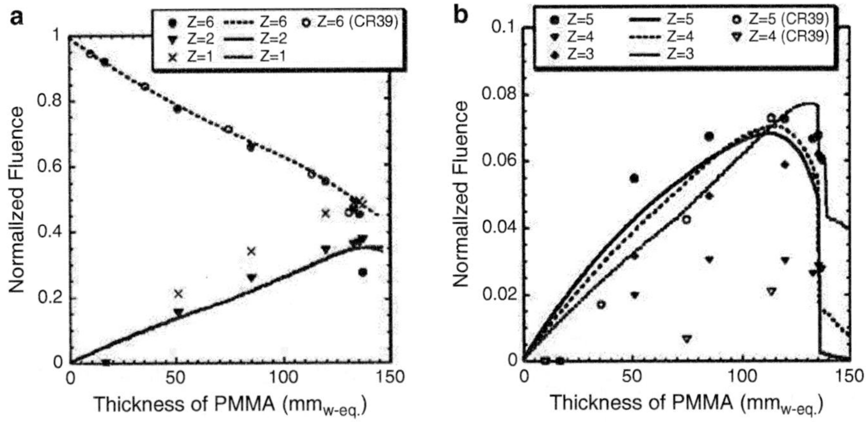

Fig. 13.2 Fluence of primary and fragmented particles when C-290 MeV/u beam was injected in PMMA. From [19]

The fragmentation process can be described by the participant–spectator model. The spectator is emitted from the projectile (projectile fragment) in the forward direction at almost the same speed as that of the projectile. It travels further together with the rest of the primary particles in a therapeutic beam. As the nuclear charge, z, of the projectile fragment is smaller than that of the primary particle, it can travel beyond the Bragg peak. This region where the projectile fragments deposit energy beyond the Bragg peak is called a fragmentation tail. The participant is a highly excited compound nucleus. Nucleons in the compound nucleus are emitted through various nuclear processes. The energy of the nucleons is generally low compared to that of projectile fragments.

Figure 13.2 shows the change in fluence for each particle species included in the C-290 MeV/u beam as a function of the PMMA thickness [19]. About half of the total primary particles could reach the range end without experiencing fragment reactions. The rest were broken into fragment particles. Among them, the fluencies of hydrogen and helium tended to be comparable to that of primary carbon in the vicinity of the Bragg peak.

13.2.3 Spatial Distribution

The spatial distributions of primary carbons and fragmented nuclei are of especially importance in the application of carbon ions in radiation therapy. The lateral intensity distributions of primary carbons can be well described by the theories of [20] or [21]. For the fragmented particles, however, experiments on the spatial fluence distributions are scarce and no analytical expressions of the distribution for use in practical situations exist.

The penumbra is often used to describe the sharpness of the irradiation fields at the isocenter. The width of lateral falloff from 80% of the maximum dose to 20% is expressed as P_{80-20} and is used to describe the penumbra size. The penumbra is composed of scattered primary particles and secondary charged particles. The lateral falloff due to primary particles can be well calculated by a single Gaussian function using the theory of Moliere. Due to the secondary particles, the overall shape of the penumbra can be approximately calculated with an assembly of pencil beams, which can be assumed to be sum of three Gaussian distributions in case of the broad-beam irradiation technique [22, 23]. A complex structure, especially like that for the carbon beam, causes a change in radiation quality in the irradiation field when the field size is small [24]. Even for scanning beams, which pass through only very thin material upstream of the isocenter, lateral distributions can be expressed by the three Gaussian shape [25].

13.2.4 Monte Carlo Simulation Codes

As shown in Fig. 13.1, calculated depth-dose distributions can well reproduce the experimental dose distributions. However, there is presently no fully analytic expression for the behavior of primary carbons and fragmented particles in water or in patients. Monte Carlo is the most powerful tool to obtain detailed information on what kind of particles, energy, and angular distributions are likely to be present in water or a patient body. Monte Carlo simulations in the field of charged particle radiotherapy have undergone remarkable improvements in precision and computing time in recent years. SHIELD-HIT [26], FLUKA [27], Geant4 [28], and PHITS [29] have all been widely applied to solve problems in charged particle radiotherapy. However, some caution should still be taken as to the precision of the outcome. Especially for the angular distributions of fragmented particles, more detailed and comprehensive studies will be necessary.

13.2.5 Dosimetry of Carbon Beams

Accurate dose measurements are essential for ion radiotherapy. Although there is a very large uncertainty in the estimation of the constants in the formula, ionization dosimetry is a very stable and reproducible method for heavy-ion dosimetry.

13.2.6 Dosimetry Using an Ionization Chamber

A formula for clinical ion-beam dosimetry was presented in the IAEA TRS-398 Code of Practice that was published in 2000 [30]. It is based on a calibration factor in terms of the absorbed dose to water in an ionization chamber irradiated by a reference beam of ^{60}Co gamma rays. The TRS 398 applies to heavy-ion beams with atomic numbers between 2(He) and 18(Ar) which have ranges of 2–30 g/cm^2 in water. For a carbon beam, this corresponds to an energy range of 100–450 MeV/u. The protocol for radiotherapy dosimetry encourages comparisons of the delivery of the absorbed dose to patients between the treatment facilities of carbon therapy [31, 32].

According to the IAEA formula for heavy-ion dosimetry, the absorbed dose to water is as follows:

$$D_{\mathrm{w},Q} = M_Q N_{\mathrm{D,w}} k_{Q,Q_0}, \tag{13.2}$$

where Q and Q_0 designate the quality of the radiation. Q is the radiation to be measured, and Q_0 is the reference radiation in which a ^{60}Co photon field is used. Further, M_Q is the reading of the electrometer, $N_{\mathrm{D,w}}$ is the calibration factor of an ionization chamber in the reference radiation field, and k_{Q,Q_0} is the beam quality factor. The beam quality factor k_{Q,Q_0} is given by

$$k_{Q,Q_0} = \frac{(s_{\mathrm{w,air}})_Q (W_{\mathrm{air}})_Q P_Q}{(s_{\mathrm{w,air}})_{Q_0} (W_{\mathrm{air}})_{Q_0} P_{Q_0}}. \tag{13.3}$$

Here, P_Q is a perturbation factor in the measurements of the dose in the Q radiation field. Usually, a heavy-ion radiation field contains various fragmented nuclei, and their energy spectra are not monochromatic. Thus, the stopping-power ratio and the w-value should be averaged over these whole spectra. The stopping-power ratio of water to air, $s_{\mathrm{w,air}}$, is given by

$$s_{\mathrm{w,air}} = \frac{\sum_{\mathrm{i}} \int_0^{\infty} \Phi_{\mathrm{i}}(E) \cdot (S_{\mathrm{i}}(E)/\rho)_{\mathrm{water}}\, \mathrm{d}E}{\sum_{\mathrm{i}} \int_0^{\infty} \Phi_{\mathrm{i}}(E) \cdot (S_{\mathrm{i}}(E)/\rho)_{\mathrm{air}}\, \mathrm{d}E}, \tag{13.4}$$

where i stands for the ith particle in the radiation field and Φ_{i} indicates the energy spectrum of the ith fragmented nuclei. The W-value of air for heavy-ion beams is given by a dose-averaged value of the inverse of the W-value, as follows:

$$\begin{aligned} \left(\frac{\bar{w}}{e}\right)_{\mathrm{HI}} &= \left[\sum_{\mathrm{i}} \int_0^{\infty} \Phi_{\mathrm{i}}(E) \cdot (S_{\mathrm{i}}(E)/\rho)_{\mathrm{air}}\ \mathrm{d}E\right] \\ &\quad \left(\sum_{\mathrm{i}} \int_0^{\infty} \Phi_{\mathrm{i}}(E) \cdot (S_{\mathrm{i}}(E)/\rho)_{\mathrm{air}} (w_{\mathrm{i}}(E)/e)^{-1}\ \mathrm{d}E\right)^{-1} \\ &= D_{\mathrm{air}} \cdot \left(\sum_{\mathrm{i}} \int \frac{e}{w(E)} \cdot d_{\mathrm{i,air}}(E)\ \mathrm{d}E\right)^{-1}. \end{aligned} \tag{13.5}$$

In order to strictly calculate these stopping power ratios and W-values, knowledge of all the constituents in the radiation fields of a heavy-ion beam is required.

Up to now, there has been insufficient experimental and theoretical data concerning these values. IAEA recommends 1.13 for the stopping-power ratio and 34.5 eV for the W-value which is the same as the value for a proton beam.

After IAEA published a comprehensive protocol to measure the absorbed dose to water for various radiations including heavy ions, studies of stopping power and other physical constants have progressed. Many papers were published for estimating the stopping power of heavy ions. The stopping power can be calculated quite accurately by a Bethe–Bloch equation if the mean ionization potential I is accurate. ICRU has provided stopping power tables for protons and alpha particles and for ions heavier than helium using the empirical or theoretical mean ionization potential [33–35]. Codes to determine stopping power for heavy ions have also been developed and improved, such as TRIM2003 and MSTAR [36, 37].

The value of I can be verified experimentally by measuring the range of the ion using the Bethe–Bloch equation. The range of a carbon beam in water will vary about 1 mm when the I-value of water material changes from 67 to 80 eV [38]. Bichsel and Hiraoka [39], Kumazaki et al. [40], Dingfelder [41], or Emfietzoglou [42] have proposed various values for the mean ionization potential of water. Comparing the calculated range of carbon beams in water with experimental values, I-values between 75 and 80 eV agree with experimental results [43] within the experimental error [44]. Up to now, we have not had a finite conclusion about this mean ionization potential for water.

As described before, the stopping-power ratio of water to air for therapeutic heavy-ion beams should be calculated using energy distributions of the projectile and fragmented particles. Monte Carlo codes of SHIELD-HIT, Geant4 or PHITS are used to calculate the energy distributions of the therapeutic beams. Geithner et al. were the first to demonstrate that the stopping-power ratio of water to air greatly depends on the depth in the water when determining the depth-dose distribution by ionization chamber [45]. They calculated depth-dose distributions of monoenergetic carbon beams of 50–450 MeV/u injected in a water phantom using SHIELD-HIT and concluded that the stopping-power ratio near the Bragg peak is 2.3% higher than the recommended value of TRS 398 in the plateau region of the Bragg curve. Another report also discussed the depth dependence of the stopping-power ratio for injecting monoenergetic carbon beams, concluding that the stopping-power ratio is in the range of 1.13–0.02 except very near the Bragg peak [46] and that the stopping-power ratio is 1.119 in the plateau regions and increases by up to about 6% near the Bragg peak [47]. They also calculated the stopping-power ratio for a SOBP beam having homogeneous dose distribution from depths of 130–143.5 mm in water using carbon at an energy of 270 MeV/u [47]. The resulting stopping-power ratio gradually increased from around 1.118 to 1.119 in the plateau region and from 1.121 to 1.128 in the SOBP region.

When the I-value of water is estimated from the range measurements of heavy ions based on the Bethe–Bloch stopping-power equation, more accurate and comprehensive range measurements are required. Also in order to estimate the stopping power ratio of water to air, the I-value of air should be estimated more accurately by means of optical or other novel methods.

13.2.7 Dosimetry Using Calorimeter

The W-value of air for heavy-ion particles is necessary for the dosimetry using ionization chambers. For dosimetry protocols not utilizing heavy ions, parameters such as W-values or stopping power ratios can be easily determined using the plentiful data from calorimeter measurements. For heavy ions, however, measurements using a calorimeter are generally scarce. Absolute dosimetry using a graphite calorimeter or a water calorimeter has been applied for high-energy carbon beams in Germany and Japan [48, 49]. The absorbed doses obtained by the calorimeter were compared with the results obtained by ionization chambers. A comparison of the two dosimetries for carbon beams using a calorimeter and an ionization chamber showed that doses determined by the ionization chamber were underestimated by 3–4%. This underestimation is, we think, because the W-value given in the IAEA protocol, 34.5 eV, is underestimated by approximately 2–3%.

13.3 Therapeutic Application of Carbon Beams

For the applications of high-energy carbon beams to radiotherapy, accelerated beams should be laterally broadened, and the pristine Bragg peak should be broadened in depth, according to the projected size and thickness of the tumor to be treated. In order to modify the accelerated beam, a beam delivery system should be installed at the end of the beam transport system.

13.3.1 Beam Delivery System

For the lateral broadening of the accelerated carbon beam, two methods have been used: scanning a nonscattered beam or wobbling a scattered beam. Figure 13.3 illustrates the two irradiation systems for carbon therapy. A typical scanning system is shown in Fig. 13.3a. The accelerated beam is scanned in the horizontal and vertical directions with a pair of scanning magnets. Any shape of irradiation field can be covered by the accelerated beam. The position and dose of the beam is monitored by the dose and position monitors placed downstream of the irradiation course.

The sharp Bragg peak is slightly broadened in depth by the ripple filter. The depth of the broadened Bragg peak can be controlled using the range shifter just upstream of the patient, or by adjusting the accelerated energy in the synchrotron. The desired two-dimensional dose distribution for a layer can be obtained by a pair of scanning magnets. The dose distributions are stacked in depth, changing the accelerated energy or inserting sheets of range shifters of various thicknesses. Using this irradiation system, any desired three-dimensional dose distribution can be obtained.

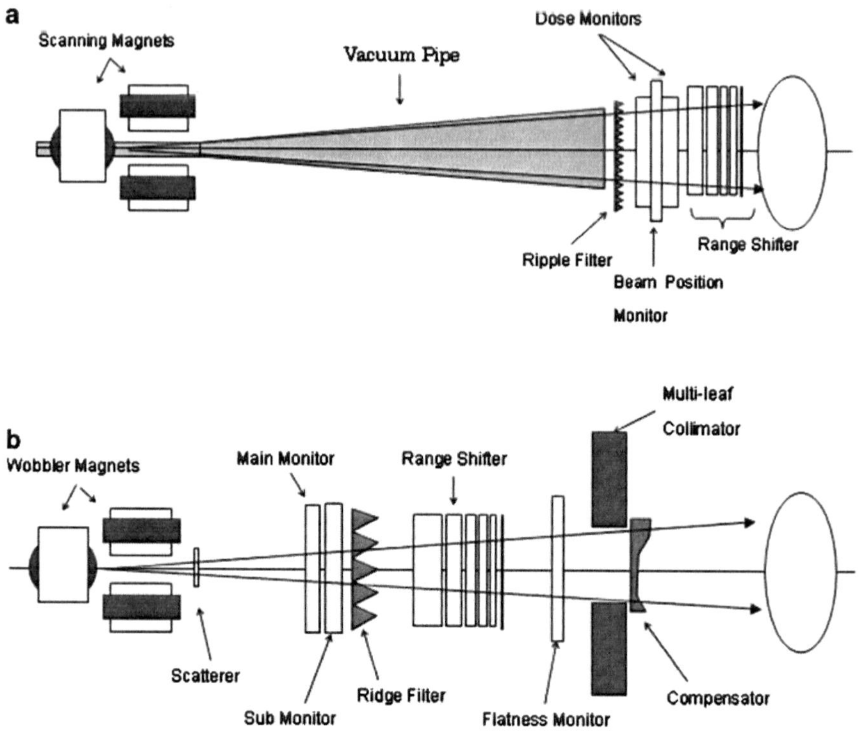

Fig. 13.3 (**a**) Typical illustration of scanning beam delivery system (**b**) Typical illustration of wobbler system

Figure 13.3b shows a typical wobbler system. The irradiation system comprises wobbler magnets, a scatterer, beam monitors, a range shifter, a ridge filter, collimators, patient-positioning devices, and a patient couch. In order to produce uniform irradiation fields, a pair of wobbler magnets and a scatterer are used. The accelerated beam is scanned, and a circular trace is drawn at the isocenter with the wobbler magnets. This wobbling beam is scattered using a scatterer placed just downstream of the wobbler magnets. Large irradiation fields can be obtained by this method, where the difference between the intensities at the central part and those at the peripheral highest part was less than 2% of the average intensity.

The range shifter is used to adjust the residual range of the heavy ions in the patient. A multileaf collimator is used for defining the irradiation fields. A range compensator can be mounted on the housing of the multileaf collimator for shaping the distal part of the irradiation field. The compensator is made using an NC milling machine from a polyethylene block according to the designed shape using a treatment planning system. The ridge filter is used to spread out the Bragg peak in the depth-dose distribution of the heavy ions. Details of the design are explained in Sect. 13.3.3. A secondary emission chamber (SEC) is often used for monitoring the irradiation dose to patients in the beam delivery system. Several aluminum sheets of signal electrodes are sandwiched by high-voltage aluminum electrodes in the SEC. The thickness of the aluminum sheets is a few μm. These electrodes are enclosed by a chamber in which the degree of vacuum is below 10^{-5} Pa. The collected current of the secondary electron is proportional to the fluence rate of the carbons and the energy loss in the aluminum sheets [50].

Only 20–30% of the accelerated particles will be in the uniform area in the case of the wobbling method. For obtaining large fields, a thicker scatterer will be necessary, and the residual range in the patients becomes shorter. The multileaf collimator and range compensator are essential devices for the wobbling method.

A range compensator should be made for each treatment. On the other hand, 100% of the accelerated beam can be used for patient irradiation in the case of the scanning method. Thus, number of neutrons produced in the irradiating system are very low compared to the wobbling method. The hardware of the active irradiating system is very simple. The system is very attractive for radiotherapists. But for moving targets, such as the lung or liver, the scanning system has not been able to be used up to now.

13.3.2 Conversion of Computed Tomography Number to Water-Equivalent Length

In radiation therapy, a medical doctor determines the target regions to be treated using various tomographic images of the patients. The medical doctor draws the target region on each slice of the computed tomography (CT) images by consulting magnetic resonance (MR) images as well as positron emission tomography (PET) images. The ranges in patients should be calculated from the CT images. CT images are images of the absorption coefficient of X-rays in the patients. The absorption coefficient in the patient does not have a linear relationship with the electron density of the patient, which is roughly proportional to the stopping power of the heavy ions. It is necessary to convert the CT values in the CT images to electron density or the water-equivalent length that corresponds to the particular pixel [51, 52]. An empirical relationship between the CT values and the electron densities are obtained by measurements of the CT values. The CT values and electron density for materials having CT values less than water were measured using solutions of C_2H_5OH of various densities. For materials having higher CT values, K_2HPO_4 was used (solid line in Fig. 13.4). The electron density for patients cannot be uniquely obtained from the CT numbers. For real tissues, the values of electron density scatter around the mean values.

Figure 13.4 shows an example of the relationship for real tissues. The error was estimated at about $\pm 1\%$ [52]. Dashed line in Fig. 13.4 shows the relationship based on scattering cross sections of photoelectric, coherent, and incoherent scattering processes (PSI method [51]).

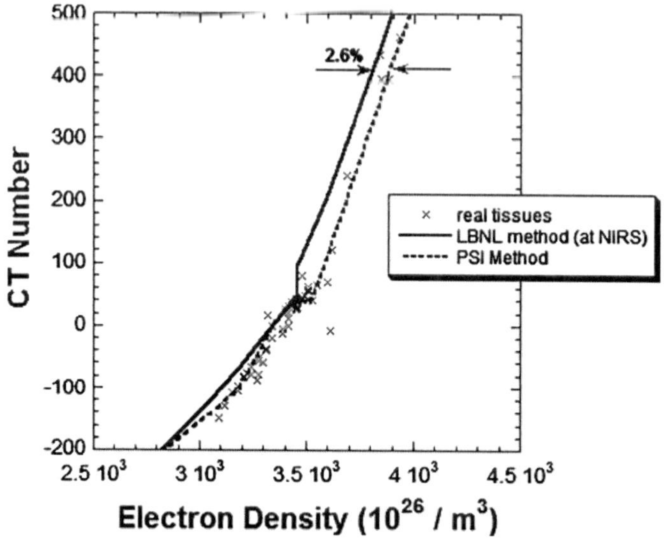

Fig. 13.4 Empirical relationship between the CT number and the electron density. *Solid line* is the experimentally determined relationship using solutions of K_2HPO_4 and C_2H_5OH. *Dashed line* is the relationship obtained by PSI method

In order to reduce these errors in an estimation of the electron density of patients, several ideas have been proposed in the past. For example, the development of a dual-energy X-ray CT, a high-energy X-ray CT using the Compton effect, and a charged-particle CT have been proposed. So far, none of these concepts have been developed to completion.

13.3.3 Design of Spread-Out Bragg Peak

The sharp Bragg peak should be broadened in the depth direction for the carbon therapy (SOBP) in order to kill all of the tumor cells by exposure to the carbon beams. The biological response should be uniform in the SOBP because underdose will cause proliferation of tumor cells after treatment and result in treatment failure. The design of the SOBP is based on the probability of killing the tumor cells. Survival curves of tumor cells subjected to heavy-ion irradiation are expressed by the LQ model as follows:

$$S = \exp(-\alpha D - \beta D^2), \tag{13.6}$$

where S is the surviving fraction of cells for the irradiation of heavy ions and D is the dose of the heavy ions. The coefficients α and β represent the cell killing per unit dose and dose squared, respectively, and depend on the radiation quality of the heavy ions. The simplest expression of the radiation quality is a LET, which is defined by energy absorbed in the surrounding tissues when the ions pass through the unit path length. In Japan, (at NIRS, Hyogo and Gunma), the human salivary gland tumor cell (HSG cell) has been chosen as a representative of various tumor cells because of its moderate radiosensitivity, with various other tumors being either more or less sensitive. In addition, the empirical LET dependence of the coefficients α and β on the ideal tumor cell (HSG cells) were experimentally obtained using 135 MeV/u carbon beams [9].

In order to design the SOBP, the Bragg peaks were shifted toward the shallower depth by inserting an aluminum sheet at the ridge filter position. The shifted Bragg curves were superposed with a determined ratio to make the SOBP. The ridge filter, made up of aluminum, is used for mixing the shifted Bragg curves. The spacing of each bar ridge is 5 mm, and it does not move during irradiation. Due to multiple scattering in the ridge filter and the angular distributions of the wobbled beam, shades of the bar ridge are smeared out at the irradiation site. As discussed in a reference [53], the dose-cell survival for combined high- and low-LET beams could be expressed by a linear-quadratic (LQ) model, in which new coefficients for combined irradiation were obtained by dose-averaging the coefficients α and $\sqrt{\beta}$ for monoenergetic beams over the spectrum of the SOBP beam:

$$S_{\mathrm{mix}}(D) = \exp(-\alpha_{\mathrm{mix}} D - \beta_{\mathrm{mix}} D^2), \tag{13.7}$$

$$\alpha_{\mathrm{mix}} = \sum_{\mathrm{i}} d(L_{\mathrm{i}})\alpha(L_{\mathrm{i}}), \tag{13.8}$$

$$\sqrt{\beta}_{\mathrm{mix}} = \sum_{\mathrm{i}} d(L_{\mathrm{i}})\sqrt{\beta(L_{\mathrm{i}})}, \tag{13.9}$$

where $d(L_{\mathrm{i}})$ is a fraction of dose of the particle having an LET of L_{i} and $\alpha(L)$ and $\beta(L)$ are coefficients of the LQ model for a particle having an LET of L. At HIMAC, experimental values of $\alpha(L)$ and $\beta(L)$ for HSG survival curves are used for the SOBP design. The depth-dose distributions of the therapeutic carbon beams were designed so that the survival level of the HSG cells is 10% in the SOBP region. Up to this stage, biological responses for photon beams have not been necessary as a reference.

In order to guarantee safe treatments, clinical RBE was introduced to the biologically equivalent dose. The biological dose D_{bio} is defined as $D_{\mathrm{bio}} = D_{\mathrm{phys}} \times$ RBE, where D_{phys} is the physical dose and RBE is the ratio of the X-ray physical dose to the carbon physical dose, giving a survival level of 10%.

The clinical RBE of the carbon beam was introduced based on the RBE for the neutron therapy at NIRS. In determining the fractionation schedule and RBE value in the clinical trial of carbon radiotherapy, 20 years of experience of neutron radiotherapy at NIRS was consulted. Around an 80-keV/μm carbon beam at SOBP was found to be equivalent to the NIRS neutron beam in terms of biological

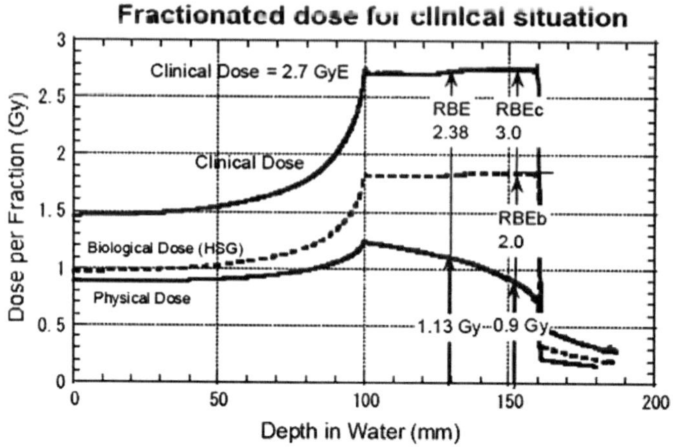

Fig. 13.5 Physical, biological, and clinical dose distributions of therapeutic carbon beams. The width of the SOBP is 6 cm. The design is based on the response of HSG cells. From [9]

responses. The designed dose distributions using HSG survival curves were then normalized so that RBE was equal to 3.0 at a neutron-equivalent position of the carbon beam (LET is 80 keV/μm) [9]. Dose-escalation studies of the clinical trial were performed at HIMAC using this RBE system, which does not vary based on the treatment schedule. Figure 13.5 schematically shows the method for determining the RBE at the center of the SOBP for clinical situations. Usually, in a treatment-planning system, the physical dose at the center of the SOBP is given using an RBE table. In the procedure of dose calibration, the dose monitor is calibrated against the given physical dose at the center of the SOBP. Originally, these RBE values were related to the fractionation schedule and the dose level of the fraction size. In searching for the proper dose level and fraction schedule in clinical trials, however, the RBE value and the neutron-equivalent position are kept constant to avoid confusion, even when the schedule or the fraction size are changed.

The German approach to carbon therapy is very different from the Japanese approach. The LEM was developed for calculation of cell survival curves in association with the German carbon-ion therapy project at GSI [17,54,55]. Instead of macroscopic absorbed dose, it uses local but statistically smoothed doses as a track structure. The target cell is divided into a vast number of tiny voxels, and the LQ model is applied for every voxel to estimate the number of local lesions produced in the voxel. The total number of lesions is derived by summing up the local lesions, and the fate of the cell is determined depending on the number of lesions. Here, α and β parameters used in the LEM are taken from X-ray irradiation, that is, the assumption is that the biological response to various radiations is in principle identical to the response to X-rays, and the microscopic difference in track structure modifies the observed response.

13.3.4 Analysis of Local Control Rate of Carbon Radiotherapy

The tumor control probability (TCP) is calculated according to the following formula:

$$\mathrm{TCP} = \sum_{\mathrm{i}} \frac{1}{\sqrt{2\pi}} \exp\left\{-\frac{(\alpha_{\mathrm{i}} - \alpha)^2}{2\sigma^2}\right\} \times \left[-N\exp\left\{-n\alpha d\left(1 + \frac{d}{\alpha/\beta}\right) + \frac{0.693(T - T_k)}{T_p}\right\}\right], \quad (13.10)$$

where α and β are coefficients of the LQ model of cell-survival curves, and σ is the standard deviation of the coefficient α which reflects patient-to-patient variation or the intertumor heterogeneity of radiosensitivity. N is the number of clonogens in the tumor, and n and d are the total fraction number and fractionated dose, respectively. T, T_k, and T_p are the overall time for treatment, the cell repopulation kickoff time, and the average doubling time of the tumor cells, respectively. In this expression of TCP, the physical dose should be used for the fractionated dose, d.

The local control rates of the clinical trials under the NIRS RBE system were analyzed using the above TCP curves [56]. For each given clinical dose at NIRS clinical trials, treatments using various SOBPs were included. For the analysis of the tumor response, the tumor responses were assumed to be the same as those expected for the clinical dose irrespective of SOBP width. The tumor control rates for treatments using 10-cm SOBP width, for example, are assumed to be the same value as those for 6-cm SOBP width when treated by the same clinical dose of 4 GyE. Then, all treatments could be represented by the treatment using an SOBP of 6 cm. Under this assumption, it was possible to analyze the tumor response using the physical dose at the center of the SOBP of 6 cm. Figure 13.6 shows the local control rate for the treatment of non-small-cell lung cancer (NSCLC) under fractionation schedules of 18 fractions with carbon beams or X-rays.

The symbols shown in Fig. 13.6 are the clinical results at HIMAC for NSCLC. Parameters in the analysis are 0.75 for α, 0.076 for β, 0.15 for σ, 0 for T_k, and 7 days for T_d. Dashed curve in the figure shows photon results. The α value for the NSCLC was very close to the value for the survival curve of the HSG cell.

RBE should depend on the dose level, based on our general understanding of radiation biology. The physical dose in the above TCP analysis is represented by the central dose of the SOBP. The shape of the depth-dose distribution of the SOBP should be different for treatments with different dose levels because RBE is dose-dependent. In our clinical trials, however, we neglected the dose-level dependence of RBE and used the same SOBP shape regardless of the dose level. Nevertheless, based on the results of the analysis, the fixed RBE system is surprisingly valid within the biological variations of the clinical study of heavy-ion therapy, but with deviations at low fraction sizes where the dose exposures per treatment are the highest and the surviving fractions the lowest. The term RBE indicates the biological effectiveness relative to photon exposure.

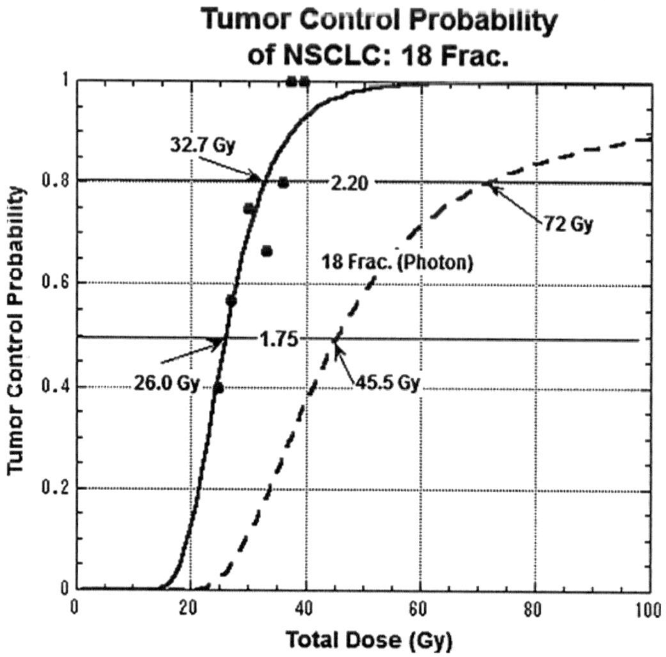

Fig. 13.6 Local control rate of NSCLC for carbon therapy in the case of an 18-fraction schedule. *Solid curves* are the TCP calculated by (13.2). From [56]

In the NIRS analysis described above, a physical carbon dose is used for the TCP calculation. The shape of the SOBP relative to the center of the SOBP is invariant regardless of the dose fractionation schedule. This means that RBE relative to the beam at the center of SOBP may not change much compared to that in the case of photon beams as a reference.

Because of the good dose localization using the particle beam, treatments with hypofractionation have been utilized in radiotherapies using heavy ions including protons. Dose levels over 10 Gy per fraction of photon-equivalent dose can be safely used for therapy. In order to analyze these treatments using the TCP equation, we should know the survival fractions in the region lower than 10^{-4}. However, we do not know the survival curve at these levels, because in vitro experiments cannot be performed at even lower survival fractions; only in vivo experiments can provide data in the necessary clinical range. If we knew the dose-response curve of the survival fraction below the dose level of 10^{-4}, we could adjust the shape of the SOBP and possibly have a much higher TCP than in our results.

Even in case of LEM used at GSI, this situation could not be improved because our approach and the GSI approach have not been proven to be valid at survival levels lower than 10^{-4}. In the region of dose levels up to several Gy in the photon-equivalent dose, we could convert the NIRS GyE to the GSI GyE. The difference between the two units was approximately 15–20%, which is large by any standard of dosimetry quality assurance in radiotherapy: those required in Japan are around 5%.

13.3.5 Differences Between Photon Treatments and Carbon Treatments

As shown in Fig. 13.6, TCP curves are well calculated by the simple equation (13.10). When we compare the TCP curves of carbon therapy and photon therapy, it is clear that the most distinct difference between the clinical results is slope of the curve. The slope of the TCP curve for carbon results is much steeper than that for photon results. Then, RBE at 80% of TCP level is 2.2, although RBE at 50% of TCP level is 1.75. Variance of the sensitivity α in (13.10) for carbon clinical trial was 0.15, which is 20% of the mean sensitivity of 0.75. Variance of the sensitivity from photon treatments was 0.11, which is 33% of the mean sensitivity of 0.33. From the basic studies of radiobiology at LBNL, it is clear that variations of the sensitivity to radiations for heavy-ion beams are much smaller than those for photon beams. Heterogeneity of circumstances of tumor cells affects the proliferation ability much more in case of photon exposure than those in case of heavy-ion exposure. This is the most remarkable characteristics of heavy-ion radiotherapy from the point of view of biophysics.

13.3.6 Measurement of Radiation Quality of Therapeutic Carbon Beams

It is well known that the radiation effects of heavy ions depend not only on LET but also on the kinds of heavy ions. The track structure of the heavy ions influences their biological effect on cells. For low-energy heavy ions, the track width becomes narrow, and the local dose within the track becomes high enough to increase the biological effect. To measure the radiation quality in detail, including the dose and the kinds of ions passing through small voxels in the water material, or nanodosimetry in a water phantom, it is necessary to use a complex detector system [57–59]. Such kinds of measurements to evaluate treatment planning in each case are almost impossible as a practical matter. At present, the radiation quality and clinical dose for heavy-ion radiotherapy is derived through theoretical calculations. The treatment plan is checked by measuring the dose distributions and comparing them with the calculated dose distributions.

Checking the dosimetry by using an ionization chamber for each patient sometimes leads to an inaccurate evaluation of the clinical dose. The lateral dose distribution of an incident carbon beam can be described using multiple scattering theory such as Moliere's theory. Most treatment-planning programs use a Gaussian distribution for beam broadening in the materials through which ions pass. In heavy-ion radiotherapy, however, fragmented particles significantly influence the dose calculation. Fragmented particles have a much greater lateral spread compared with the primary ions. Besides, the fragmented ions usually have a lower LET than the primary carbon ions. The contribution of the fragmented ions to the physical dose and the clinical dose differs due to the dose dependence of their RBE values.

This creates the likelihood that the physical dose measurement does not directly correspond to the clinical dose. In this case, it is important to simultaneously measure the dose and radiation quality

13.3.7 Radiation Quality Measurements Using a Thin Scintillator

As described in the former sections, LET is not a perfect index of the radiation quality. Nevertheless, LET is useful in describing the radiation quality of heavy-ion beams, though with limited accuracy. When LET spectra are measured for therapeutic heavy-ion beams, we can obtain survival curves using the coefficients of the LQ model for the survival curve. From the LET spectrum, the coefficients α and β can be calculated by the following equations:

$$\langle\alpha\rangle = \frac{\sum_{\mathrm{i}} n_{\mathrm{i}} L_{\mathrm{i}} \alpha(L_{\mathrm{i}})}{\sum_{\mathrm{i}} n_{\mathrm{i}} L_{\mathrm{i}}}, \tag{13.11}$$

$$\langle\beta\rangle = \frac{\sum_{\mathrm{i}} n_{\mathrm{i}} L_{\mathrm{i}} \sqrt{\beta(L_{\mathrm{i}})}}{\sum_{\mathrm{i}} n_{\mathrm{i}} L_{\mathrm{i}}}. \tag{13.12}$$

In this approximation, the kind of ion is neglected, and all ions are treated as carbon ions in the case of carbon radiotherapy. An example of this derivation of the radiation quality of the therapeutic carbon beam has been described [60]. A Monte Carlo code, Geant4, was used for the calculation of the LET spectra.

A scintillation counter is the most likely detector for measurements of the LET spectra under the therapeutic conditions of heavy-ion radiotherapy. A proportional counter, which can very accurately measure LET spectra, cannot be used in radiation fields with a high-dose rate. The beam intensity should be reduced in order to accurately measure the spectra. On the other hand, a scintillation counter can be used in circumstances of a high-dose rate when a very small (i.e., 1-mm-diameter) scintillator is used. Inorganic scintillator sheets such as ZnS (Ag) are now being tested for the measurements of LET spectra. When the LET spectrum is integrated, the absorbed dose will be given. The radiation quality and absorbed dose are then obtained in a single measurement of the LET spectrum.

At present, development of this scintillation counter is not complete. For the dosimetry device, the accuracy should be better than 1%. Noise in the LET spectrum should be reduced in order to achieve accuracy. The light output of the inorganic scintillator also has LET dependency.

13.3.8 Radiation Quality Measurements Using a Rossi Counter

The microdosimetric quantity, or z distribution, can also be understood as representing the radiation quality of the heavy-ion beams. Kellerer and Rossi have established the relationship between single-event spectra, or the z_1-distribution, with the LQ model of the survival curves (dual radiation action theory [61]. Hawkins modified the dual radiation theory and established a microdosimetric kinetic model (MKM) [62]. Kase showed that cell-survival curves of mammalian cells can be explained by a modified MKM [63]. In the modified MKM, the survival curve is expressed as follows:

$$S = \exp\left[(-\alpha_0 + \beta \cdot z_{1D}^*)D - \beta D^2\right], \tag{13.13}$$

where z_{1D}^* is the dose mean specific energy, elaborating on the Hawkins saturation effect, α_0 is a parameter of α at the limit of LET $= 0$, and β is a parameter determined by the survival curve for X-ray irradiation. From this equation, survival curves can be obtained by measuring the microdosimetric quantity, z_{1D}^*, with a Rossi counter.

Figure 13.7 shows examples of lineal-energy spectra measured by a Rossi counter with a tissue-equivalent diameter of 1 μm for therapeutic carbon-ion beams with an initial energy of 290 MeV/u and a 6-cm SOBP width [64].

Figure 13.8 shows the clinical dose distribution results of the carbon beam obtained from the single-event spectra measured by the Rossi counter, that is, the microdosimetric z_1-distribution [64]. The designed clinical dose distribution is also shown in Fig. 13.8.

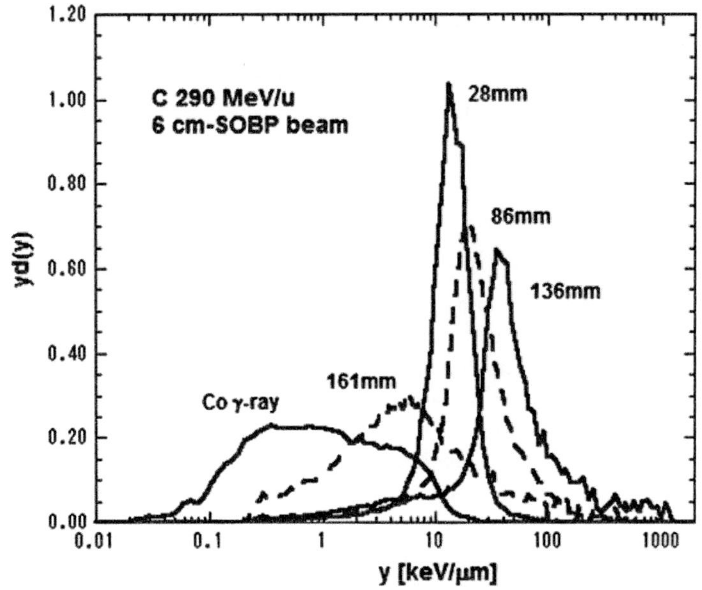

Fig. 13.7 The typical probability of a given dose per logarithmic interval of lineal energy, $yd(y)$, as a function of lineal energy, y, for a therapeutic carbon-ion beam with an initial energy of 290 MeV/u and a 6-cm SOBP width. Measurements were performed with a TEPC with a simulated diameter of 1.0 μm at depths of entrance (3.5 mm H_2O), proximal peak (89 mm H_2O), and distal peak (145 mm H_2O). From [64]

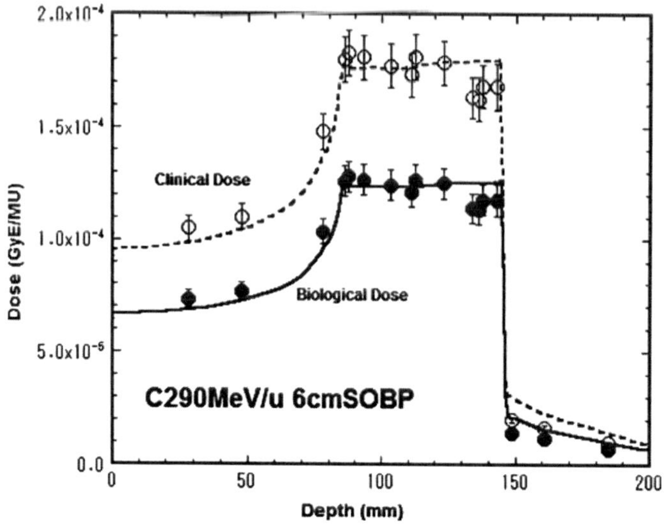

Fig. 13.8 Clinical dose distributions of the therapeutic carbon-ion beam. The plots are obtained from microdosimetric single-event spectra measured with the Rossi counter. The curve is the calculated result used for treatment planning. From [64]

LEM consists of three elementary parts, the target geometry, the photon survival curve, and the track structure. The MKM calculates the lesion number produced in a micrometer-sized region, whereas the LEM calculates the lesion number in a region of nanometer scale. Both models use the LQ model survival curve for low LET radiation to calculate the lesion number. In order to fit the experimental results, minor corrections are made in both models. The modified MKM introduced a saturation number of lesions when a large number of lesions are produced in a micrometer-scale region. The LEM assumes a straight survival curve for the low LET radiation in a high-dose region. It was shown that the survival probabilities of ion-irradiated in vitro mammalian cell lines can be obtained by the modified MKM and also by the LEM using amorphous track structure models [65]. The MKM calculations using the Kiefer–Chatterjee track structure model and also the modified LEM gave results just as well for a wide range of parameters.

It was shown that the results with the two models used in heavy-ion therapy are roughly equivalent in the LET regions. It is very important for quality assurance of carbon therapy that the radiation quality can be obtained by an experimental method. The absorbed dose can be determined by integrating the microdosimetric spectra obtained by the Rossi counter, similar to the case of a scintillation counter. In order to obtain an absolute dose by the Rossi counter, the calibration of the single-event spectra should be traceable to a primary standard dosimetric laboratory.

13.4 Conclusions

Although clinical results have shown that radiotherapy using heavy ions is very effective, especially for advanced cancer, there are still many physical problems to be solved in order to increase the accuracy of heavy-ion radiotherapy. Heavy-ion dosimetry is still an open problem. Comparisons with completely different dosimetries are required. Also, measurements of the radiation quality of heavy-ion beams are very important for the development of a treatment planning system.

Dose escalation studies of carbon ion therapy at NIRS have produced very important data for developing ion therapy. There remains to be discovered a comprehensive biophysical model which is valid throughout the entire dose range used in therapy. A theoretical model for tumor response to the carbon beams and a physical beam model for high-energy heavy-ion particles are required to design the optimal therapeutic beam configurations.

References

1. S. Hornsey, Radiology **97**, 649 (1970)
2. J.F. Fowler, F. Inst, P.J. Denekamp, A.L. Page, A.C. Begg, Br. J. Radiol. **45**, 237 (1972)
3. J.W. Hopewell, W.H. Barnes, M.E.C. Robbins, M. Corp, J.M. Sansom, C.M.A. Young, G. Wiernik, Br. J. Radiol. **63**, 760 (1990)
4. R.R. Wilson, Radiology **47**, 487 (1946)
5. C.A. Tobias, H.O. Anger, J.H. Lawrence, Am. J. Roentgen. Radiat. Ther. Nucl. Med. **67**, 1 (1952)
6. M.C. Pirruccello, C.A. Tobias (eds.), *Biological and Medical Research with Accelerated Heavy Ions at the Bevalac*, 1977–1980 LBL-11220/UC-48 (1980)
7. E.A. Blakely, F.Q.H. NGO, S.B. Curtis and C.A. Tobias. Advances in Radiat. Biol., **11**, 295 (1984)
8. J.R. Castro, M.M. Reimers, Int. J. Radiat. Oncol. Biol. Phys. **14**, 711 (1988)
9. T. Kanai, M. Endo, S. Minohara, et al., Int. J. Radiat. Oncol. Biol. Phys. **44**, 201 (1999)
10. G. Kraft, Nucl. Instrum. Meth. A **454**, 1 (2000)
11. J. Debus, T. Haberer, D. Schultz-Ertner, Strahlenther Onkol. **176**, 211 (2000)
12. H. Tsujii, T. Kamada, M. Baba M, H. Tsuji, H. Kato, S. Kato, S. Yamada, S. Yasuda, T. Yanagi, H. Kato, R. Hara, N. Yamamoto, J. Mizoe, New J. Phys. **10**, 075009 doi:10.1088/1367-2630/10/7/075009 (2008)
13. S. Minohara, T. Kanai, M. Endo: Int. J. Radiat. Oncol. Biol. Phys. **47**, 1097 (2000)
14. T.H. Haberer, W. Becher, D. Schardt, G. Kraft, Nucl. Instrum. Meth. A **330**, 296 (1993)
15. T. Furukawa, T. Inaniwa, S. Sato, T. Tomitani, S. Minohara, K. Noda, T. Kanai, Med. Phys. **34**, 1085 (2007)
16. E. Rietzel, C. Bert, Med. Phys. **37**, 449 (2010)
17. M. Scholz, G. Kraft, Adv. Space Res. **18**, 5 (1996)
18. L. Sihver, C.H. Tsao, R. Silverberg, T. Kanai, A.F. Barghouty, Phys. Rev. C **47**, 1455 (1993)
19. N. Matsufuji, M. Komori, H. Sasaki et al., Phys. Med. Biol. **50**, 3393 (2005)
20. Z. Moliere, Naturforsch **2a**, 133 (1947)
21. V.L. Highland, Nucl. Instrum. Meth. **129**, 497 (1975)
22. Y. Kusano, T. Kanai, Y. Kase et al., Med. Phys. **34**, 193 (2007)
23. Y. Kusano, T. Kanai et al., Med. Phys. **34**, 4016 (2007)

24. H. Nose, Y. Kase, N. Matsufuji, et al., Med. Phys. **36**, 870 (2009)
25. T. Inaniwa, T. Furukawa, A. Nagano, S. Sato, N. Saotome, K. Noda, T. Kanai, Med. Phys. **36**, 2897 (2009)
26. I. Gudowska, N. Sobolevsky, P. Andreo et al., Phys. Med. Biol. **49**, 1933 (2004)
27. A. Fasso, A. Ferrari, J. Ranft et al., Preprint CERN-2005–10 (2005)
28. J. Allison, K. Amako, IEEE Trans. Nucl. Sci. **53**, 270 (2006)
29. K. Niita, T. Sato, H. Iwase et al., Rad. Meas. **41**, 1080 (2006)
30. *International Atomic Energy Agency*. Absorbed Dose Determination in External Beam Radiotherapy. Technical Report Series No. 398 (IAEA, Vienna, 2000)
31. A. Fukumura, T. Hiraoka, K. Omata, M. Takeshita, K. Kawachi, K. Kanai, N. Matsufuji, H. Tomura, Y. Futami, Y. Kaizuka, G.H. Hartmann, Phys. Med. Biol. **43**, 3459–3463 (1998)
32. G.H. Hartmann, O. Jäkel, P. Heeg, C.P. Karger, A. Kriessbach, Phys. Med. Biol. **44**, 1193 (1999)
33. *International Commission on Radiation Units and Measurements*. Stopping Powers and Ranges for Protons and Alpha Particles. ICRU Report, 49 (Oxford University Press, London, 1993)
34. *International Commission on Radiation Units and Measuremets*. Stopping of Ions Heavier than Helium. ICRU Report, 73 (Oxford University Press, London, 2005)
35. P. Sigmund, A. Schinner, H. Paul, J. ICRU vol. 49 no. 1 (Oxford University Press, Oxford, 2009)
36. J.F. Ziegler, SRIM - the stopping and range of ions in matter (2008), Available at http://www.srim.org
37. H. Paul, A. Schinner, MSTAR - stopping power for light ions (2004), Available at http://www.exphys.uni-lins.ac.at/stopping
38. P. Andreo, Phys. Med. Biol. **54**, N205 (2009)
39. H. Bichsel, T. Hiraoka, Nucl. Instrum. Meth. B **66**, 345 (1992)
40. Y. Kumazaki, T. Akagi, T. Yanou, D. Suga, Y. Hishikawa, T. Teshima, Radiat. Meas. **42**, 1683 (2007)
41. M. Dingfelder, D. Hantke, M. Inokuti, H.G. Paretzke, Electron inelastic-scattering cross sections in liquid water. Radiat. Phys. Chem. **53**, 1 (1998)
42. D. Emfietzoglou, R. Garcia-Molina, I. Kyriakou, I. Abril, H. Nikjoo, Phys. Med. Biol. **54**, 3451 (2009)
43. L. Sihver, D. Schart, T. Kanai, Jpn. J. Med. Phys. **18**, 1 (1998)
44. H. Paul, Nucl. Instrum. Meth. Phys. Res. B **225**, 435 (2007)
45. O. Geithner, P. Andreo, N. Sobolevsky, G. Hartmann, O. Jäkel, Phys. Med. Biol. **51**, 2279 (2006)
46. H. Paul, O. Geithner, O. Jäkel, Adv. Quant. Chem. **52**, 289 (2007)
47. K. Henkner, N. Bassler, N. Sobolevsky, O. Jäkel, Med. Phys. **36**, 1230 (2009)
48. H.J. Brede, K.D. Greif, O. Hecker et al., Phys. Med. Biol. **51**, 3667 (2006)
49. M. Sakama, T. Kanai, A. Fukumura, K. Abe, Phys. Med. Biol. **54**, 1111 (2008)
50. M. Sudou, T. Kanai, in *Test of SEC*. Proceedings of the 9th Symposium on Accellerator, Science and Technology (1993), pp. 351–353
51. U. Schneider, E. Pedroni, A. Lomax, Phys. Med. Biol. **41**, 111 (1996)
52. N. Matsufuji, H. Tomura, Y. Futami, H. Yamashita, A. Higashi, S. Minohara, M. Endo, T. Kanai, Phys. Med. Biol. **43**, 3261 (1998)
53. T. Kanai, Y. Furusawa, K. Fukutsu et al., Radiat. Res. **147**, 78 (1997)
54. T. Elsässer, M. Scholz, Radiat. Res. **167**, 319 (2007)
55. T. Elsässer, M. Krämer, M. Scholz, Int. J. Radiat. Oncol. Biol. Phys. **71**, 866 (2008)
56. T. Kanai, N. Matsufuji, T. Miyamoto et al., Int. J. Radiat. Oncol. Biol. Phys. **64**, 650 (2006)
57. J. Llacer, C.A. Tobias, W.R. Holley, T. Kanai, Med. Phys. **11**, 266 (1984)
58. I. Schall, D. Schardt, H. Geissel, H. Irnich, E. Kankeleit, G. Kraft, A. Magel, M.F. Mohar, G. Muenzenberg, F. Nickel, C. Scheidenberger, W. Schwab, Nucl. Instrum. Meth. Phys. Res. B **117**, 221 (1996)
59. N. Matsufuji, A. Fukumura, M. Komori, T. Kanai, T. Kohno, Phys. Med. Biol. **48**, 1605 (2003)

60. Y. Kase, N. Kanematsu, T. Kanai, N. Matsufuji, Phys. Med. Biol. **51**, N467 (2006)
61. A.M. Kellerer, H.H. Rossi, Curr. Top. Radiat. Res. **Q8**, 85 (1972)
62. R.B. Hawkins, Radiat. Res. **140**, 366 (1994)
63. Y. Kase, T. Kanai, Y. Matsumoto et al., Radiat. Res. **166**, 629 (2006)
64. Y. Kase, T. Kanai, M. Sakama, Y. Tameshige, T. Himukai, H. Nose, N. Matsufuji, Microdosimetric approach to NIRS-defined biological dose measurements for carbon-ion treatment beam. J. Radiat. Res. **52**, 59 (2011)
65. Y. Kase, T. Kanai, N. Matsufuji, Y. Furusawa, T. Elsässer, M. Scholz, Phys. Med. Biol. **53**, 37 (2008)

Chapter 14
Atomic and Molecular Data for Industrial Application Plasmas

M.-Y. Song, D.-C. Kwon, W.-S. Jhang, S.-H. Kwang, J.-H. Park, Y.-K. Kang, and J.-S. Yoon

Abstract As interest has increased in the interaction between low-temperature plasmas and materials, the role of modeling and simulation of processing plasmas has become important in understanding the effects of charged particles and radicals in plasma applications. Also, in order to understand the behavior and properties of chemically active plasma, atomic and molecular processes have become a rapidly growing area of scientific endeavor that holds great promise for practical applications for industrial fields. Thus, in this chapter, we briefly introduce the applications of low-temperature plasma, especially plasma processing in semiconductor manufacturing, and what kind of data needed in plasma processing, how to develop the reaction mechanisms, and how it applied to the simulation. 0D global modeling of ICP plasma-etching equipment and development of a two-dimensional fluid simulator for a SiH_4 discharge are given as an example. In addition, we introduce the line-intensity ratio method for plasma diagnostic, it can be a good example how atomic and molecular data can be used plasma diagnostics.

14.1 Introduction

Electron–molecule and electron–atom collisions initiate almost all the relevant chemistry associated with mixed [1–3], the plasma processing of materials for microelectronics [4,5], and modern electric light technology [6]. Especially, electron collision processes play a key role in creating the energetic species that dive the chemistry in extreme environments such as the low-temperature, high-density plasmas used in the etching of semiconductor materials and in plasma-enhanced chemical vapor deposition [7–10].

Since the plasma state is defined as the fourth state of matter as distinct from the solid, liquid, and gas phase. It consists of free positive and negative charges with electrical quasineutrality in addition to feed gas components, and plasma state can be produced electrically, thermally, or optically through the ionization of neutral molecules in the feed gas. In particular, low-temperature plasma, which produced

V. Shevelko and H. Tawara (eds.), *Atomic Processes in Basic and Applied Physics*, Springer Series on Atomic, Optical, and Plasma Physics 68, DOI 10.1007/978-3-642-25569-4_14, © Springer-Verlag Berlin Heidelberg 2012 by the collisional ionization of free electrons under an external electrical power, has a property that the electron energy is much higher than that of the neutral gas. This is one of the primary advantages of low-temperature plasmas for material processing and device fabrication requiring a variety of surface processes and very different reactions among materials adjacent to each other; because of this characteristics of plasma, the key point to achieve efficient studies of such a plasma processing is to perform a multidisciplinary approach of this complex tool. One need to intermix a wide knowledge in physics, chemistry, and material sciences which play major roles in the plasma generation and properties [11, 12]. Thus, plasma chemistry, or atomic and molecular processes in plasma, has become a rapidly growing area of scientific endeavor that holds great promise for practical applications for industrial fields [9, 10].

In early development of plasma, processing had been achieved through trial and error, the increasing demand to shrink the feature dimensions and the demand for diminished defect density produce a corresponding demand for increased sophistication an effectiveness in process control. It is now generally accepted that the traditional method of meeting these requirements by trial and error has reached the point of diminishing returns. Also, to achieve technical goal, complex process equipment and techniques have been developed, and now we have necessitated a better comprehension of the physics standing behind the plasma [13]. Thus, plasma process modeling now becomes a necessary engineering tool in the design of new semiconductor equipment and in process control, and also, the high cost of developing both the plasma equipment and processes has motivated development of less-empirical methods, and modeling/simulation in particular, to speed the time to market and to reduce costs [14, 15].

Following a National Research Council (NRC) report in 1991 which cited the need for science based design of plasma processes, a modeling and simulation infrastructure was developed which can now address a variety of plasma tools and surface processes. In addition, for modeling/simulation to be an effective aid in design, the prototype codes must be accompanied by a reliable data of chemical and physical properties of the gases and surface involved. However, the application of this modeling/simulation infrastructure to industrially relevant problems has been limited by the availability of fundamental atomic and molecular data(e.g., electron impact cross sections, reactive surface sticking coefficients) and validated reaction mechanisms. This situation was described in a second NRC report, "Database Needs for Modeling and Simulation of Plasma Processing." In this second report, target gases were listed which require electron impact cross sections for two etching processes and one deposition process, and also other chemistries and types of processes have come to prominence for which database also required. Thus, for modeling to be an effective aid in design, the prototype codes must be accompanied by a reliable data (or database) of chemical and physical properties of the gases and surface involved. The required data depend on the feed gas, the nature of the plasma processing, and the surface employed. The operating conditions of the reactor also influence the types of cross sections needed in the simulation.

In this chapter, we will briefly introduce the applications of low-temperature plasma, mainly plasma processing in semiconductor manufacturing. Also, here, we will introduce what kind of data needed in plasma processing and how it applied to the simulation and modeling. Finally, as an example of atomic and molecular data, we will discuss about the line-intensity ratio method for plasma diagnostics.

14.2 Application of Low-Temperature Plasma

Since the 4th material state, plasma, is composed of charged particles, it has an advantage on transferring the energy to the material contacts with the plasma by accelerating particles. Therefore, it could change the property of matter and process. The plasma also can be used for various forms of the light source and can easily generate the radical, which has chemically strong reactivity. Therefore, the plasma, which has many applications, is widely applied to the high-value-added industries.

The technique used to plasma is a common technology which is employed in many fields as shown in Fig. 14.1 [7]. Some common examples are plasma

Fig. 14.1 Plasma in everyday life: (1) plasma TV, (2) plasma coated jet turbine blades, (3) plasma manufactured LEDs in panel, (4) diamond like plasma CVD eyeglass coating, (5) plasma ion implanted artificial hip, (6) plasma laser cut cloth, (7) plasma HID headlamps, (8) plasma produced H_2 in fuel cell, (9) plasma aided combustion, (10) plasma muffler, (11) plasma ozone water purification, (12) plasma deposited LCD screen, (13) plasma deposited silicon solar cells, (14) plasma processed microelectronics, (15) plasma sterilization in pharmaceutical production, (16) plasma treated polymers, (17) plasma treated textiles, (18) plasma treated heart stent, (19) plasma deposited diffusion barriers for containers, (20) plasma sputtered window glazing, and (21) compact fluorescent plasma lamp (taken from [7])

sterilization, plasma TV, fluorescent plasma light, or plasma processing for microelectronics. Foremost among these is the electronics industry in which plasma-based processes are indispensable for the manufacture of ultra-large-scale integrated microelectronic circuits.

Low-temperature plasma is that area of plasma addressing partially ionized gases with electron temperatures typically below about 10 eV (100,000 K) [16]. Such plasmas are often known as "collisional plasmas" or "weakly ionized plasmas" because input power first coupled with the charged electrons and ions and then is collisionally transferred to neutral atoms and molecules, creating chemically active species [17, 18]. The technology assisted by low-temperature plasma is generally referred to as plasma processing, and it is classified into plasma enhance chemical vapor deposition, plasma etching, sputtering, ashing, surface modification, and so on, on a size ranging from nanometers to meters.

Plasma etching, deposition, and cleaning are indispensable fabrication techniques in the manufacture of microelectronics components. According to the Semiconductor Industry Association, worldwide sales of microelectronic components were $298.3 billion in 2010, with plasma-etching equipment sales alone generating a market of three to five billion dollars per year and plasma CVD more than five billion dollars per year.

The plasma equipment for these processes typically use partially ionized (fractional ionization less than 1%), low pressure (a few mTorr–1 Torr) plasmas to provide activation energy to dissociate and ionize feedstock gases. The resulting radicals and ions interact with the semiconductor surface, either removing or adding material, to define the desired features or modify the surface.

In this section, we will introduce the general applications of plasma processing in semiconductor manufacturing.

14.2.1 Plasma Processing in Semiconductor Manufacturing

The process of semiconductor manufacturing often requires hundreds of sequential steps. However, here, we only discuss about the plasma processing (see in Fig. 14.2). An excellent introduction to the basic physics involved in plasma processing can be found in the book by Lieberman and Lichtenberg [16].

In general, plasma processing in semiconductor manufacturing can be divided into three categories, depending on whether the goal is to add, remove or modify surface material. Some examples are plasma etching, thin film deposition, protective coating, surface hardening, ion implantation, etc. In order to quantify the ability to accomplish this selective processing, it is useful to have a few figures of merit to describe the process, namely:

Etch rate: Controllable and robust to small deviation in processing conditions. Depending on the application, one may want to higher etch rate for increased throughput or a lower-etch rate for precision.

Uniformity: Both at each feature and across the wafer.

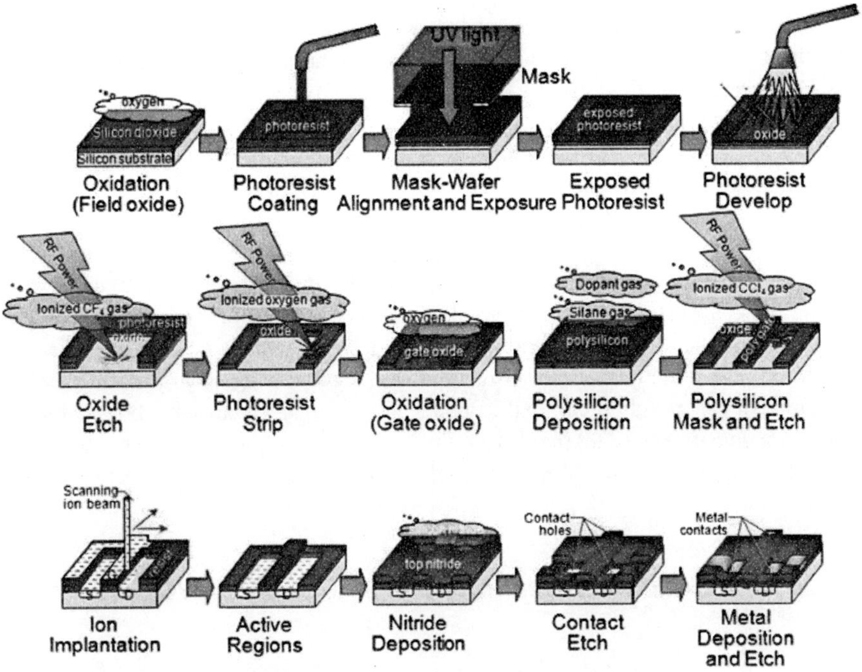

Fig. 14.2 Plasma processing for semiconductor manufacturing

Selectivity: The ability to etch only the desired material, relative to the etching of mask and other substrate materials.

Anisotrophy: The verticality of the etch profile. Also, the nanostructure of this vertical surface is important in many applications.

Damage: Any surface or substrate damage acquired from the processing technique.

Also, every process is a complex interaction between gas phase chemistry, plasma conditions, and surface phase chemistry/conditions. Since, the plasma is a partially ionized gas with a combination of free electrons, ions, radicals, and neutral species. This in order to create and sustain a plasma in the laboratory, energy input is required. Generally, this energy is transferred via coupling of an external electromagnetic field to the plasma constituents. Most general plasma generation method are capacitive coupled plasmas (CCP) and inductively coupled plasmas (ICP).

In a CCP process, energy is supplied as a voltage between an anode and a cathode plate, but time-varying fashion. Most commonly, a radio-frequency (RF) voltage is applied to the plates. The frequency of operation is often at 13.56 MHz, and because of time-varying field, electrons in the plasma oscillate between the anode and the cathode plates. Collisions of rapidly moving electrons with the slowly moving ions cause further ionizations. However, massive ions are less mobile and cannot track the rapidly oscillating electric field charge accumulates on the plate. The resulting potential between the plasma and the negatively charged plate is called the self-bias V_b. The electric field due to V_b drives the positive ions in plasma toward the negatively charged plate. This is the basis for traditional reactive ion etching.

In an ICP process, the excitation is again a time-varying RF source, but is delivered inductively, instead of capacitively, resulting in a changing magnetic field. This changing magnetic field, through the Maxwell–Faraday equation, induces an electric field that tends to circulate the plasma in the plane parallel to the CCP plates. Similarly to a CCP, collisions of the rapidly moving electrons with slowly moving ions cause further ionizations. Loss of electrons from the plasma through the grounded chamber walls tends to create a static voltage, called the plasma voltage V_{plasma}. Inductive coupling is generally realized through a large 4–5 turn coil encircling the plasma chamber. In the typical geometry, this means that one is able to change ion density and other plasma parameters without significantly perturbing the incident energy of the ions. More detail information is given in Makabe et al. [11].

14.2.2 *Physical Vapor Deposition*

Physical vapor deposition (PVD) is a variety of vacuum deposition and is a general term used to describe any of a variety of methods to deposit thin films by the condensation of a vaporized form of the material onto various surfaces. A greater range of deposition layers can be obtained through the use of reactive sputtering. Reactive sputtering is widely used to deposit dielectrics, such as oxides and nitrides, as well as carbides and silicides. Common reactive gases are O_2 and H_2O for O atoms, N_2 and NH_3 for N atoms, CH_4 and C_2H_2 for C atoms, and SiH_4 for Si atoms. Reactive sputtering is often necessary even when the deposited and target materials are the same.

The primary gases for PECVD and CVD applications listed in Table 14.1. Note that this list is not all-inclusive in deposition process and in some cases other gases are used in addition to the major gases listed in Table 14.1.

Table 14.1 Gas data for PECVD and CVD

Film	Gas
a–Si:H	SiH_4/H_2
SiN_x	SiH_4/NH_3, SiH_4/N_2
SiO_2	SiH_4/N_2O, $SiH_4/O_2/Ar$
n+a–Si:H	$SiH_4/H_2/PH_3$
p+a–Si:H	$SiH_4/H_2/B_2H_6$, $SiH_4/H_2/BF_3$
n+a–SiC:H	$SiH_4/CH_4/PH_3$, $SiH_4/C_2H_2/PH_3$, $SiH_4/C_2H_4/PH_3$
p+a–SiC:H	$SiH_4/CH_4/B_2H_6$, $SiH_4/C_2H_2B_2H_6$, $SiH_4/C_2H_4/B_2H_6$
p–SiC	$SiF_4/CF_4/H_2$
3C–SiC	$SiH_4/C_3F_8/Ar/H_2$

14.2.3 Chemical Vapor Deposition

The deposition of elements that can be extracted from gaseous compounds was traditionally achieved by chemical vapor deposition. The desired atoms are freed thermally, so that the major adjustable parameters are temperature, pressure and the flow rates of each component in the mixture gas. Ionized gas provides much more control over the process, producing more radicals as well as ions and avoiding the need for high substrate temperatures, which may lead to damage. This process is usually called plasma-enhanced chemical vapor deposition (PECVD). Figure 14.3 represents general deposition process in plasma-enhanced chemical vapor deposition reactor.

The deposition of hydrogenated amorphous silicon (a-Si:H) play an crucial role in the manufacture of active matrix liquid crystal display(LCD) and in photovoltaic devices, such as solar cells. The common approach is to use SiH_4/H_2 discharge in an RF reactor. In this process, the plasma chemistry is dominated by neutral radicals. The source of these radicals is assumed to be electron impact dissociation of the molecules SiH_4, Si_2H_6, and Si_3H_8. Silane mixtures are also used in the production of dielectric materials by PECVD. For silicon nitride (Si_3N_4), the added gas may be N_2 or NH_3, whereas NO_2 can be added to give the oxynitrides SiN_xO_y. The most commonly used dielectric, SiO_2, can be produced from mixtures of SiH_4 with an oxygen donor such as O_2, NO, N_2O, or H_2O_2, usually dilute in Ar. A typical chemistry model for SiO_2 deposition can be found in the paper by Meeks et al.

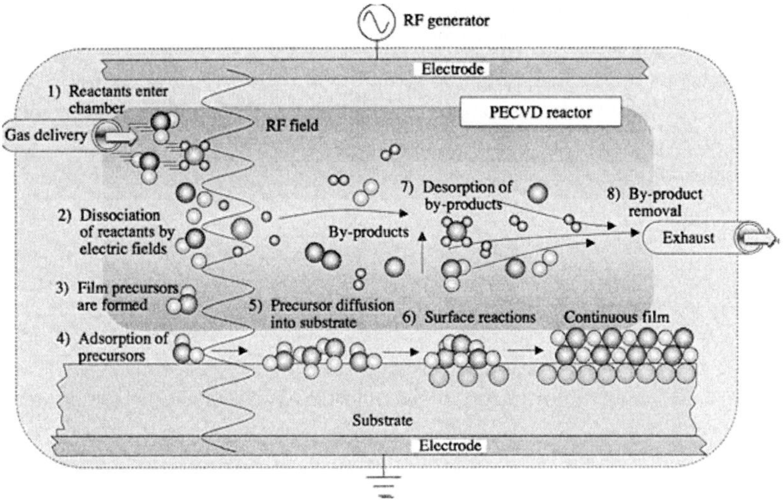

Fig. 14.3 General deposition process in plasma-enhanced chemical vapor deposition (PECVD) reactor. The chemical vapor deposition reaction steps are (1) gas transport to the deposition zone, (2) formation of the film precursors, (3) film precursors at the wafer, (4) precursor adsorption, (5) precursor diffusion, (6) surface reactions, (7) by-product removal from the surface, and (8) by-product removal from the reactor

Since silane is explosive at room temperature, a popular alternative Si source is tetraethooxysilane($Si(O\ CH_2\ CH_3)_4$ or TEOS), which is a relatively inert liquid. A fascinating study of the ions and neutral fragments produced in a TEOS/Ar discharge has been presented by Basner et al. [19].

14.2.4 Dry Etching

Etching is used in microfabrication to chemically remove layers from the surface of a wafer during manufacturing. Etching is a critically important process module, and every wafer undergoes many etching steps before it is complete. The two fundamental types of etchants are liquid-phase ("wet") and plasma-phase ("dry").

The need for dry(plasma) etching, rather than wet(chemical) etching, has increased as the dimensions of semiconductors have shrunk. The creation of narrow trenches or holes with vertical sidewalls demands anisotropic etching, and the need to cut through one layer without damage to very thin underlying layers of different materials requires excellent selectivity in the etching process. Etching rates inside holes or trenches depend on the aspect ratio of the feature as well as on the density of nearby features that are being etched simultaneously. The gas mix is chosen to minimize these effects and to maintain control over feature shape, avoiding notches and protrusions.

Plasma processing is ideal for etching in that the directionality of the incident ions provides a capability to fabricate vertical walls. Plasmas can also be turned off rapidly compared with etching times scale (tens of seconds) so that when one layer is completely etched, the process can be stopped with minimal damage to the underlying layer. In the semiconductor equipment manufacturing companies, etch processes are often subdivided into three categories, silicon, metal, and dielectric.

The primary gases for dry etching applications listed in Table 14.2. Note that this list is not all-inclusive in etching process and in some cases other gases are used in addition to the major gases listed in Table 14.2.

14.2.5 Plasma Cleaning

Control over semiconductor processing requires frequent cleaning, for the reactor itself between wafers as well as for the wafers before or after each processing step. Two problems of special interest are the removal of deposits from reactor walls and the removal of the resists that are used in lithography from the semiconductor wafers. In deposition systems some of the feedstock gas can be deposited on the reactor walls instead of the workpiece, while in etching the wall depositions can either be etch products or the polymer layers that are used for passivation on the substrate.

Table 14.2 Gas data for dry etching

Material	Gas
Si	CF_4/O_2, CF_2CL_2, CF_3Cl, $SF_6/O_2/Cl_2$, $Cl_2/H_2/C_2F_6/CCl_2$, SiF_4/O_2, NF_3, ClF_3, CCl_4, CCl_3F_5, C_2ClF_5/SF_6, C_2F_6/CF_3Cl, Br_2, CF_3Cl/Br_2, CCl_2F_2, Cl_2, Cl_2ClF_5/O_2
SiO_2	CF_4/Ar, C_2F_6, C_3F_8, CHF_3/O_2, C_4F_8/H_2, $C_4F_8/O_2/Ar$, $SF_6/O_2/Ar$, $C_4F_6/O_2/Ar$
Organic solids (resist)	O_2, CF_4/O_2, SF_6/O_2
Silicide	CF_4/O_2, NF_3, SF_6/Cl_2, CF_4/Cl_2. CCl_4/O_2, CCl_2F_2
Al	BCl_3, BCl_3/Cl_2, $CCl_4/Cl_2/BCl_3$, $SiCl_4/Cl_2$
Cr	Cl_2/O_2, CCl_4/Cl_2
Mo, Nb, Ta, Ti, W	CF_4/O_2, SF_6/O_2, NF_3/H_2, Cl_2/O_2
Au	$C_2Cl_2F_4$, Cl_2, $CClF_3$
GaAs	BCl_3/Ar, $Cl_2/O_2/H_2$, $CCl_2F_2/O_2/Ar/He$, CCl_4, $Cl_2/BCl_3/Ar/N_2$, $SiCl_4/SF_6$
GaN	BCl_3/Cl_2, Cl_2/HBr, BCl_3/HBr, HCl/Hbr
InP	CH_4/H_2, C_2H_6/H_2, Cl_2/Ar
ITO, IZO	HBr, HI, Cl_2, BCl_3

Plasma cleaning differs from plasma etching in that the process can be isotropic, but selectivity is still important to avoid damage to the workpiece or reactor. The process is also known as ashing or stripping.

Although the details of cleaning techniques vary with the primary process, the type of reactor, the gas mixture and the wall material, the basic is to deliver low energy etching radicals, such as F or O atoms to the surface. Typical source of these radicals are O_2, NF_3, SF_6, C_xF_y, CF_4/O_2, and CF_2/H_2O, often mixed with rare gases or N_2 to ease plasma formation.

14.3 Plasma Simulation for Low-Temperature Plasma Application

In the sense of physical and chemical reaction between particles and solid surfaces in contact with the plasma. Plasma processes can be subdivided into five subsequent steps [20], which either can take place in the same volume or can be geometrically separated as in remote source processing.

1. Creation of primary plasma, i.e., electrons and ions production. This ionizing plasma can be atomic, as in remote source operation, or molecular, as in in situ processing.
2. Transfer of primary chemistry to secondary chemistry, i.e., dissociation of injected monomers in the recombining plasma from the source. Chemically active radicals and/or ions are formed, which may be characterized by their

reactivity and sticking probability on the growing film: "hard" radicals with a high sticking coefficient (>0.5), "intermediate" radicals with intermediate ($0.05 \sim 0.5$), and "soft" radicals with low sticking coefficient (<0.05).
3. Plasma surface interactions of arriving radicals with a surface lead in first instance to the formation of a chemically bounded surface layer and thus to a (nearly) totally passivated surface. Subsequent-arriving hard radicals will still stick to this surface, whereas intermediate radicals may diffuse across the surface. By finding a vacant site or by reaction they may contribute to deposition and cross-linking. Alternatively, they may abstract, e.g., atomic hydrogen, leading to an open site and desorption of a molecule, or they may reflect as soft radicals. Desorbed molecules can be chemically active, as is clear from the production of ammonia in the presence of atomic hydrogen and nitrogen [21].
4. Nearly always, there is recirculation and new molecules produced in the volume or at the wall will mix with injected monomers [5]. We will distinguish the "forward kinetics" volume with the plasma from the source and the background volume in which recirculation takes place.
5. Clustering of (recirculating) monomers, radicals, and ions and nucleation and formation of dust.

Control of such plasma process can only occur by obtaining a through understanding of the physical and chemical properties of plasmas. Since the radicals and ions produced by the collisions of electron with the feed gas interact with the surface through a series of complex chemical reactions, leading to etching, deposition or cleaning.

14.3.1 General Development of Reaction Mechanisms

Unfortunately, all plasma processes are currently used in the industry with an incomplete understanding of the coupled chemical and physical properties of the plasma involved. Nevertheless, in lack of knowledge for atomic and molecular processes, here we describe general features of the mechanism development process. General approach has been described elsewhere and is shown schematically in Fig. 14.4. This is an iterative process that starts by assembling cross section and kinetic data for gas-phase reactions for the molecules of interest. In the early stages, it is preferable to err on the side of too many rather than too few reactions, so we try to consider all reactions that might be important in the system. But the literature rarely contains all the desired information, so often estimates based on related molecules must be used. The gas-phase data are combined with a set of postulated surface reaction paths and rates. Surface reactions generally have far less information in the literature than gas-phase reactions, which means that most postulated surface species and reactions will have estimated rate parameters. This, in turn, encourages the use of "lumped" and "effective" reactions, which helps to minimize the number of species and reactions, which in turn helps deal with computational resource limitations.

This initial reaction mechanism is compared using a low-order, i.e. low-dimensionality, plasma model, against an experimental data set. The results of

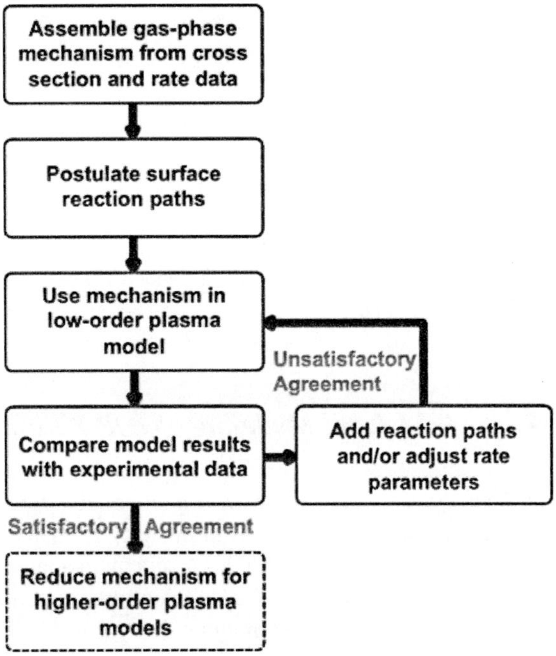

Fig. 14.4 Schematic diagram of reaction mechanism development process

the initial comparisons are invariably highly unsatisfactory, but generally provide some direction as to what reaction paths need to be added and/or have the rate parameters adjusted. The mechanism is then tested against experiment again. These iterations continue until satisfactory agreement is attained, or when new experimental data may not be important under any given set of conditions, but are retained to make the mechanism as broadly applicable and complete as possible. For higher-dimensionality simulations, i.e. 2D simulations of a particular tool, it is best to then reduce the reaction mechanism to a subset of reactions that are most important for the parameter space of interest.

For the gas-phase part of the mechanisms, electron impact reactions provide the driving mechanism behind the plasma processes, since the radicals and ions produced by the collisions of electrons with the feed gas interact with the surface through a series of complex chemical reactions, leading to etching, deposition, or cleaning. Thus electron collision data are just as important as data on heterogeneous surface reactions in understanding and controlling such processes. A number of ion–molecule and ion–ion reactions are also included. However, reactions among neutral species are generally not part of the mechanism, since the very low pressures used in these plasma processes means that the relative importance of gas–wall vs. gas–gas collisions becomes higher than usual in reacting gas systems. Reactions requiring collisional activation of a reactant or collisional stabilization of a production also become unimportant.

The electron impact reactions in the mechanisms generally include vibrational and electronic excitations, dissociative ionizations, dissociations to neutral fragments, and dissociative attachment reactions. These reaction paths are including for electron impact with reactants, major dissociation fragments and by-products of processing.

Developing a surface reaction mechanism is less straightforward than developing a gas-phase reaction mechanism. The rate parameters for surface reactions are often derived by fitting to experimental etch or deposition rate data. Despite this "fitting" procedure, using a set of fundamental reaction path descriptions provides much more accurate and predictive models than would purely phenomenological models. Physically based models are applicable to a wider range of conditions and can predict effects not directly measured in the calibration process.

14.3.2 0D Global Modeling of ICP Plasma Etching Equipment

ICP sources are widely used in semiconductor fabrication processing because a high-density plasma with good uniformity is easily obtained under low pressure without an external magnetic field. A schematic diagram of the ICP reactor is shown in Fig. 14.5. The power absorbed by electrons, including the anomalous skin effect, is calculated by the heating module. The resulting power absorbed by electrons is used to calculate the plasma density and electron temperature in the plasma transport module. The plasma transport module then returns the plasma density and electron temperature to the RF sheath module. Then, the sheath voltage drop is obtained from the RF sheath module which is coupled with the transport module in a self-consistent manner. These processes are repeated until a self-consistent steady state is obtained. The detailed models used in this work are described below subsections.

We numerically solve the spatially averaged continuity equation and electron temperature equation based on the Ar plasma chemistry [22–25] as shown in Table 14.3. The rate coefficients for the electron impact reactions listed in Table 14.3 were calculated by assuming that the electron energy distribution function is Maxwellian. The values of the rate coefficients were fit using the Arrhenius form, we chose the fitting parameters such that the match was best for electron temperatures in the range of 1–7 eV.

The governing equations [16, 26] of the spatially averaged zero-dimensional model are as follows:

$$\frac{\partial n_\mathrm{i}}{\partial t} = \sum_j k_{ij} n_\mathrm{e} n_j + \sum_{j,k} k_{ijk} n_j n_k + Q_\mathrm{inlet} - n_\mathrm{i}\left(\sum_j k_{ji} n_\mathrm{e} + \sum_{i,k} k_{jik} n_k\right) - n_\mathrm{i} \nu_l^i - Q_\mathrm{pump}, \tag{14.1}$$

$$\frac{\partial}{\partial t}\left(\frac{3}{2} n_\mathrm{e} T_\mathrm{e}\right) = \frac{P_\mathrm{abs}}{\Omega} - E_\mathrm{e} - n_\mathrm{e}\varepsilon_{\mathrm{e},l}\nu_l^e, \tag{14.2}$$

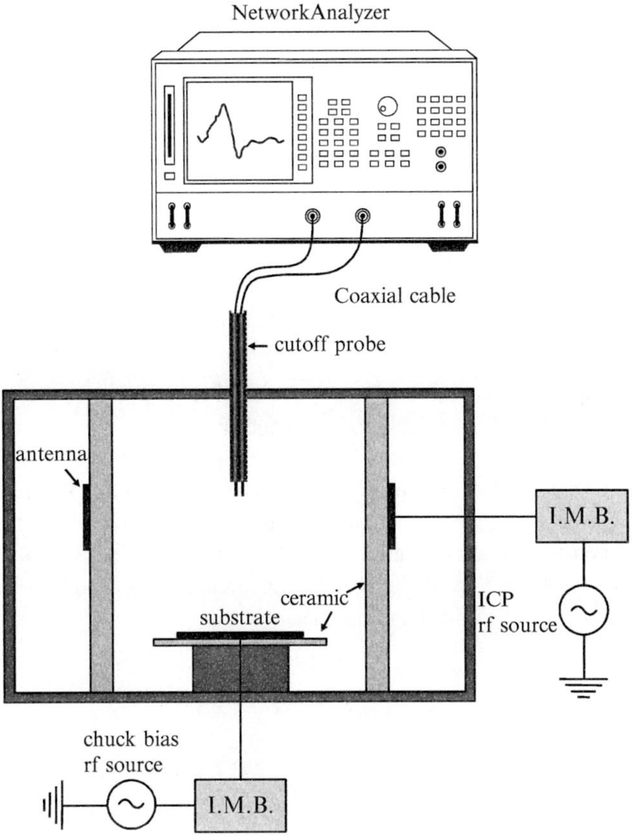

Fig. 14.5 A schematic diagram of the ICP reactor

Table 14.3 Gas-phase reactions included in simulations of Ar plasma and their rate coefficients. The rate coefficients are in the form $k_j = A_j T^{B_j} \exp(-C_j/T_e)$. The units of k_j are in m^3s^{-1} and T_e is in Kelvin

Reactions	*A*	*B*	*C*	ε	Reference
Electron-impact reactions					
e+Ar→Ar* +e	2.783×10^{-16}	0.3287	140, 200.0	11.6	[22]
e+Ar→Ar$^+$ +2e	5.779×10^{-17}	0.6329	186, 400.0	15.76	[23]
e+Ar*→Ar$^+$ +2e	6.225×10^{-14}	0.1072	51, 210.0	4.43	[24]
e+Ar*→Ar+e	2.645×10^{-16}	0.2894	8, 601.0	−11.6	[22]
Metastable reactions					
Ar* +Ar*→Ar$^+$ +Ar+e	1.200×10^{-15}	0.0000	0.0		[25]

where n_i is density of species i particle. k_{ij} is the electron impact rate coefficient of species i generation by electron collision with species j, and k_{ijk} is the rate coefficient of species i generation by reactions between species j and k. Q_{inlet} and Q_{pump} are the source and loss term by gas inlet and pumping out, respectively. n_e is the electron density, T_e is the electron temperature, Ω is the volume of the chamber, and E_e is the collisional energy loss term.

The loss speed of the ions at the plasma-sheath boundary is assumed to be the Bohm velocity, and, thus the loss frequency of the ith ion species becomes $\nu_l^i = S_e\sqrt{T_e/M_i}/\Omega$. Here, S_e is the effective area for particle loss and M_i is the ion mass. Assuming the ambipolar diffusion, the loss frequency of the escaping electron into the wall is given by $\nu_l^e = \sum_i S_e\sqrt{T_e/M_i}/\Omega$ [16].

The effective loss energy into the wall can be given by $\varepsilon_{e,l} = V_{sh} + 2.5T_e$, where V_{sh} is the sheath potential drop [16]. If the bias power is applied to the substrate, V_{sh} can be given by

$$V_{sh} = \begin{cases} V_p & \text{at the chamber wall,} \\ V_p - V_s & \text{at the substrate.} \end{cases}$$

Here, V_p is the plasma potential and V_s is the voltage on the substrate. Assuming that ions and electrons gain and loss the time averaged sheath potential drop, respectively, then the sheath potential drop at the substrate can be given by $V_{sh} \approx V_p - V_{dc}$. V_{dc} is obtained from the RF sheath module which is coupled with the transport module in a self-consistent manner.

14.3.2.1 Current Control Model (Model I)

The one-dimensional spatiotemporal variations of ion density, electron density, ion velocity, and electric potential are described by the ion continuity equation, the Boltzmann relation, the momentum equation, and the Poisson's equation, respectively [27]:

$$\frac{\partial n_i}{\partial t} + \frac{\partial (n_i u_i)}{\partial x} = 0, \tag{14.3}$$

$$\frac{\partial u_i}{\partial t} + u_i \frac{\partial u_i}{\partial x} = -\frac{e}{M_i}\frac{\partial V}{\partial x}, \tag{14.4}$$

$$n_e = n_s \exp\left(\frac{eV}{T_e}\right), \tag{14.5}$$

$$\frac{\partial^2 V}{\partial x^2} = -\frac{e}{\varepsilon_0}(n_i - n_e), \tag{14.6}$$

where u_i is the velocity of ion species, e is the elementary charge, V is the potential, and $n_s = n_0 \exp(eV_1/T_e)$. Here, $V_1 = -T_e/2e$ denotes the potential drop in the presheath and n_0 is the plasma density [28].

Assuming a sinusoidal current source at the electrode, a current balance can be written as $J_i + J_e + J_d = J_0 \sin(2\pi f t)$, where J_i and J_e are the conduction current densities of the ion and electron, respectively [27,29,30]. The displacement current density, J_d, can be derived as

$$J_\mathrm{d} \equiv \frac{C_\mathrm{s}(t)}{A}\frac{\mathrm{d}V_\mathrm{s}(t)}{\mathrm{d}t} - \frac{V_\mathrm{s}(t)}{A}\frac{\mathrm{d}C_\mathrm{s}(t)}{\mathrm{d}t}, \tag{14.7}$$

where $C_\mathrm{s}(t)$ is the sheath capacitance and A is the surface area [27]. The current balance is coupled to the sheath model through the electrode potential and the sheath thickness. By solving the current balance equation with the fourth order Runge–Kutta method, we obtain the potential drop across the sheath. Using the obtained potential $V_\mathrm{s}(t)$, we then solve the fluid equations and the Poission's equation by using a finite difference method with an iterative process.

14.3.2.2 Voltage Control Model (Model II)

As mentioned above, numerically resolving the thin sheath in computer models of the devices requires large computing resources. To reduce the computing resources and compare the our results with CFD-ACE+, we use a simple circuit model based on the previously developed model [28,31].

Based on the previously developed RF sheath model, the following circuit relations can be obtained:

$$C_\mathrm{s}^\mathrm{sh}\frac{\partial}{\partial t}\left(V_\mathrm{p} - V_\mathrm{s}\right) + I_\mathrm{s} + C_\mathrm{b}\frac{\partial}{\partial t}\left(V_\mathrm{RF} - V_\mathrm{s}\right) = 0, \tag{14.8}$$

$$C_\mathrm{w}^\mathrm{sh}\frac{\partial V_\mathrm{p}}{\partial t} + I_\mathrm{w} + C_\mathrm{s}^\mathrm{sh}\frac{\partial}{\partial t}\left(V_\mathrm{p} - V_\mathrm{s}\right) + I_\mathrm{s} = 0, \tag{14.9}$$

$$\frac{\partial \bar{V}_\mathrm{s}}{\partial t} = \frac{\partial V_\mathrm{p}}{\partial t} - \omega_\mathrm{pi}\left(\bar{V}_\mathrm{s} - V_\mathrm{s}\right), \tag{14.10}$$

$$\frac{\partial \bar{V}_\mathrm{w}}{\partial t} = \frac{\partial V_\mathrm{p}}{\partial t} - \omega_\mathrm{pi}\bar{V}_\mathrm{w}, \tag{14.11}$$

where $\bar{V}_\mathrm{s}$ and $\bar{V}_\mathrm{w}$ are damped potential at the substrate and chamber wall, respectively. C_s^sh and C_w^sh are the sheath capacitance on the substrate and chamber wall, respectively, $\omega_\mathrm{pi} = \sqrt{e^2 n_\mathrm{s}/\varepsilon_0 M_\mathrm{i}}$ and C_b is the blocking capacitance. The sheath capacitance C^sh is related with voltage drop across the sheath as $C^\mathrm{sh} = A\varepsilon_0 \partial E/\partial V_\mathrm{sh}$ [29,31], where

$$\frac{\partial E}{\partial V_\mathrm{sh}} = \frac{e n_\mathrm{s} T_\mathrm{e}}{\varepsilon_0 E}\left\{\frac{M^2}{\bar{V}_\mathrm{sh}}\left[\sqrt{1 - \frac{2\bar{V}_\mathrm{sh}}{M^2 T_\mathrm{e}}} - 1\right] + \frac{1}{T_\mathrm{e}}\exp\left(\frac{V_\mathrm{sh}}{T_\mathrm{e}}\right)\right\}. \tag{14.12}$$

Here, $V_{\mathrm{sh}} = V_{\mathrm{s}}\,(\text{or } V_{\mathrm{w}}) - V_{\mathrm{p}}$, $\bar{V}_{\mathrm{sh}} = \bar{V}_{\mathrm{s}}\left(\text{or } \bar{V}_{\mathrm{w}}\right) - V_{\mathrm{p}}$, M is the Mach number, and the electric field is

$$E = -\left(E_{\mathrm{s}}^2 + \frac{2en_{\mathrm{s}}T_{\mathrm{e}}}{\varepsilon_0}\left\{\frac{M^2V_{\mathrm{sh}}}{\bar{V}_{\mathrm{sh}}}\left[\sqrt{1-\frac{2\bar{V}_{\mathrm{sh}}}{M^2T_{\mathrm{e}}}}-1\right]+\left[\exp\left(\frac{V_{\mathrm{sh}}}{T_{\mathrm{e}}}\right)-1\right]\right\}\right)^{1/2}, \tag{14.13}$$

where the electric field at the sheath–presheath boundary is $E_{\mathrm{s}} = -T_{\mathrm{e}}\log(\varepsilon)/\lambda_{\mathrm{i}}$. ε is the ratio of the electron Debye length to the ion mean free path. The total current I to the sheath can be given by $I = A\left(J_{\mathrm{i}} + J_{\mathrm{e}}\right) - \bar{C}^{\mathrm{sh}}\partial\bar{V}_{\mathrm{sh}}/\partial t$, where $\bar{C}^{\mathrm{sh}} = A\varepsilon_0\partial E/\partial\bar{V}_{\mathrm{sh}}$. Here,

$$\frac{\partial E}{\partial\bar{V}_{\mathrm{sh}}} = \frac{en_{\mathrm{s}}T_{\mathrm{e}}}{\varepsilon_0 E}\frac{M^2V_{\mathrm{sh}}}{\bar{V}_{\mathrm{sh}}^2}\left\{1-\sqrt{1-\frac{2\bar{V}_{\mathrm{sh}}}{M^2T_{\mathrm{e}}}}-\frac{\bar{V}_{\mathrm{sh}}/M^2T_{\mathrm{e}}}{\sqrt{1-2\bar{V}_{\mathrm{sh}}/M^2T_{\mathrm{e}}}}\right\}. \tag{14.14}$$

14.3.2.3 Results and Discussions

To validate our model, the experiments were performed at an ICP reactor of Surface Technology Systems as shown in Fig. 14.5. The single-turned antenna coil is wound around the dielectric tube. The Ar gas is fed from the mass flow controller (MFC), and RF source power at 13.56 MHz was supplied to a RF antenna via L-type matching network.

In the present study, we used cutoff probe for the precise measurement of the electron density. It is well known that plasma density can be measured by using the cutoff probe within 2% of uncertainty [32]. The cutoff probe consisted of two tips as shown in Fig. 14.5. The averaged values of the 5 points for the measured electron density were used to compare the simulation results. To acquire the electron temperature and the electron energy probability function (EEPFs), the RF compensated single Langmuir probe (SLP2000) were used.

In this work, we described the self-consistent global model, which consists of the heating model, plasma transport module, and RF sheath module. The simulation results are compared with not only experimental measurements by using the cutoff probe and the SLP but also two-dimensional numerical results by using the CFD-ACE+ [33].

CFD-ACE+ is suitable for modeling a variety of coupled physical phenomena, such as fluid dynamics, electrodynamics, heat, and complex chemical reactions. The governing equations used for the plasma simulation are the Navier–Stokes and continuum equations for gas flow and heat transfer, nonequilibrium plasma transport, and species transport, and the Poisson equation for electrostatic field. Two-dimensional simulations were performed for Ar plasma (see Table 14.3) under the same operating conditions and reactor geometry as used in the experiments.

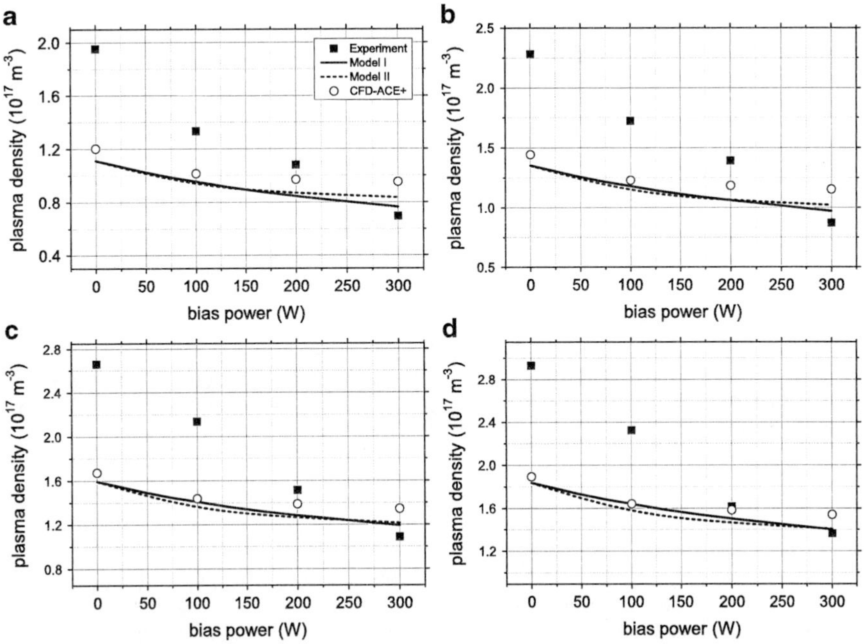

Fig. 14.6 Dependence of plasma densities on the bias power for various ICP source power: (**a**) 500 W, (**b**) 600 W, (**c**) 700 W, and (**d**) 800 W. The *filled square* is experimental data measured by a cutoff probe; the *open circle* is numerical solved data by using a two-dimensional fluid model, and *solid* and *dashed lines* are obtained values from Model I and II, respectively

Figure 14.6 shows the dependence of plasma densities on the bias power for various ICP source powers. Although there is a difference in the absolute values, as the bias power increases, we can see that the numerically solved plasma densities are in a good agreement with the measured results. Here, data numerically solved by using the two-dimensional fluid model were averaged in the same manner as with the experiments.

Dependence of the dc self-bias voltage on he bias power for various ICP source powers is shown in Fig. 14.7. To compare the experimental results, the dc self-bias was defined as one-half of the peak-to-peak voltage. As mentioned in the preceding text, in CFD-ACE+, only the sinusoidal voltage source boundary conditions for the two-dimensional fluid simulation were obtained from Model II. We can see that Model I is in a good agreement with measured results. Although there is a difference in the voltage gradient on the bias power in Model II, it is still useful from the point of view of considering the expansibility to two- or three-dimensional simulation. However, there is a difference between the two-dimensional simulation results and measured data.

In numerical results, the differences of the electron density and the self-bias voltage are due to the fact that Model I is the self-consistent fluid model and Model II is the semianalytic model. We can see that the delicate difference of the

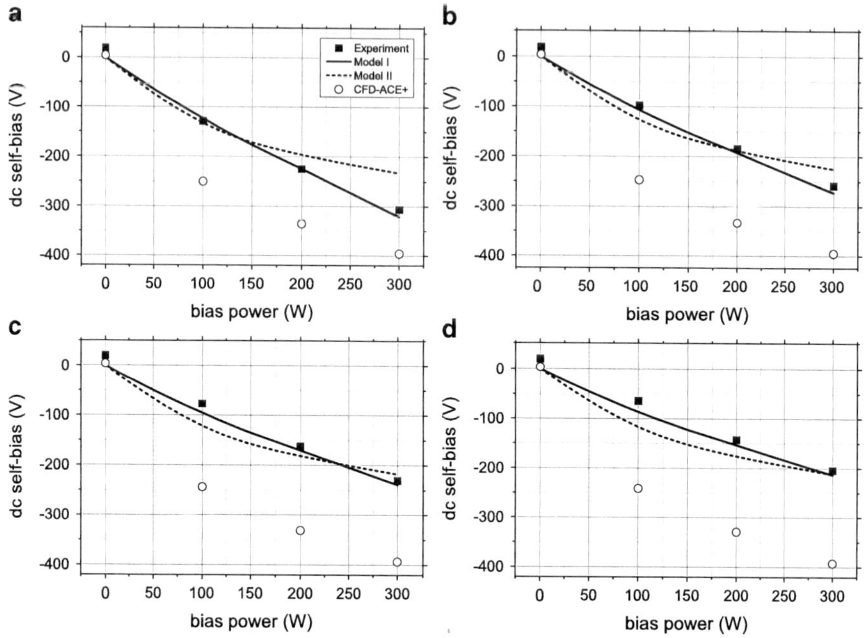

Fig. 14.7 Dependence of dc self-bias on the bias power for various ICP source power: (a) 500 W, (b) 600 W, (c) 700 W, and (d) 800 W. The *filled square* is experimental data measured by a cutoff probe; the circle is numerical solved data by using a two-dimensional fluid model, and *solid* and *dashed lines* are obtained values from Model I and II, respectively

self-bias voltage can affect the plasma density. It may be curious that numerically solved self-bias voltages are shown to agree well with the experimental results while there is a difference in density gradient on the bias power between numerical and experimental data. This is due to the fact that the plasma density were measured at the chamber center region, although the averaged values of 5 points for the measured electron density were used to compare the simulation results, the averaged values can be high than the plasma density obtained by using the spatially averaged global model. However, the plasma density can be similar with the numerical results near the substrate. Therefore, numerically solved self-bias voltages are shown to agree well with the experimental results.

14.3.3 Development of a Two-Dimensional Fluid Simulator for a SiH_4 Discharge

SiH_4 is the most important frequently used constituent in the low-temperature plasmas required in the fabrication process, such as deposition of silicon and amorphous silicon-hybride film, of microelectronic devices, and other semiconducting components [34].

The first detailed fluid study and modeling for a SiH_4 plasma was performed by Kushner for the radiofrequency parallel-plate capacitive discharge [35]. The work summarized the availability of reaction rate coefficients and largely contributed to the fluid modeling of SiH_4 discharges. However, the model includes many crudely estimated reaction coefficients . Meeks et al. compiled the gas-phase and surface reactions for modeling PECVD of SiO_2 from SiH_4, O_2 and Ar gas mixtures in high-density plasma reactors [36]. Although many rate coefficients of chemical reactions were compiled, still several reactions were not considered.

Therefore, in this chapter, we introduce the way how to construct chemical reaction data for SiH_4 discharges and develop a two-dimensional fluid simulator for transformer coupled plasma (TCP) sources. Here, we numerically solve the two-dimensional fluid equations for electrons, positive ions, negative ions, neutrals, and radicals [37]. The absolute power is determined by using an analytic electron heating theory including the anomalous skin effect.

14.3.3.1 Model Descriptions

A schematic diagram of a TCP reactor is shown in Fig. 14.8. The power deposition in the plasma is calculated by using a heating model that includes the anomalous skin effect. The resulting power deposition profile is used to calculate the electron temperature in the plasma transport module. The plasma transport module then returns the electron temperature, density, and collision frequency to the heating module. These processes are repeated until a self consistent steady state is obtained.

The basic equations consist of the continuity, momentum, electron temperature, and Poisson equation are below:

$$\frac{\partial n_{i,j}}{\partial t} + \frac{1}{r}\frac{\partial}{\partial r}\left(r\Gamma_{i,j}^{r}\right) + \frac{\partial \Gamma_{i,j}^{r}}{\partial z} = S_{i,j} - L_{i,j} \tag{14.15}$$

$$\frac{n_{\mathrm{e}}}{\partial t} + \frac{1}{r}\frac{\partial}{\partial r}\left(r\Gamma_{\mathrm{e}}^{r}\right) + \frac{\partial \Gamma_{\mathrm{e}}^{r}}{\partial z} = S_{\mathrm{e}} - L_{\mathrm{e}} \tag{14.16}$$

$$\frac{\partial n_{n,k}}{\partial t} + \frac{1}{r}\frac{\partial}{\partial r}\left(r\Gamma_{n,k}^{r}\right) + \frac{\partial \Gamma_{n,k}^{r}}{\partial z} = S_{n,k} - L_{n,k} \tag{14.17}$$

$$\frac{\partial \Gamma_{i,j}^{r}}{\partial t} + v_{i,j}^{r}\left[\frac{1}{r}\frac{\partial}{\partial r}\left(r\Gamma_{i,j}^{r}\right) + \frac{\partial \Gamma_{i,j}^{r}}{\partial r}\right] + \Gamma_{i,j}^{r}\frac{\partial v_{i,j}^{r}}{\partial r} + \Gamma_{i,j}^{z}\frac{\partial v_{i,j}^{r}}{\partial z}$$
$$= \frac{1}{m_{i,j}}\left(en_{i,j}E^{r} - F_{r,i,j}^{r}\right) + M_{i,j}^{r} \tag{14.18}$$

$$\frac{\partial \Gamma_{i,j}^{z}}{\partial t} + v_{i,j}^{z}\left[\frac{1}{r}\frac{\partial}{\partial r}\left(r\Gamma_{i,j}^{r}\right) + \frac{\partial \Gamma_{i,j}^{r}}{\partial r}\right] + \Gamma_{i,j}^{r}\frac{\partial v_{i,j}^{z}}{\partial r} + \Gamma_{i,j}^{z}\frac{\partial v_{i,j}^{z}}{\partial z}$$
$$= \frac{1}{m_{i,j}}\left(en_{i,j}E^{z} - F_{r,i,j}^{z}\right) + M_{i,j}^{z} \tag{14.19}$$

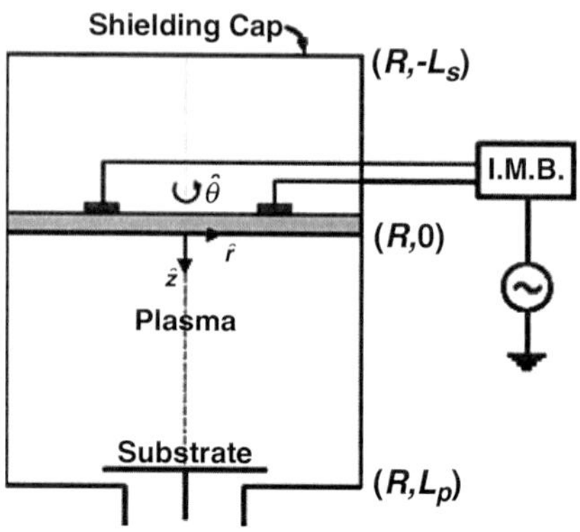

Fig. 14.8 Schematic diagram of TCP source. Here, R is the chamber radius, L_s is the shielding cap height, and L_p is the plasma length

$$\Gamma_e^r = \frac{1}{m_e \nu_{eN}} \left[-e n_e E^r + F_{r,e}^r \right] \quad (14.20)$$

$$\Gamma_e^z = \frac{1}{m_e \nu_{eN}} \left[-e n_e E^z + F_{r,e}^z \right] \quad (14.21)$$

$$\Gamma_{n,k}^r = -D_k \frac{\partial_{n,k}}{\partial r} \quad (14.22)$$

$$\Gamma_{n,k}^z = -D_k \frac{\partial_{n,k}}{\partial z} \quad (14.23)$$

$$\frac{\partial}{\partial t}\left(\frac{3}{2} n_e T_e\right) + \frac{1}{r}\frac{\partial}{\partial t}\left(r \frac{3}{2} n_e T_e v_e^r\right) + \frac{\partial v_e^z}{\partial z} = \frac{1}{r}\frac{\partial}{\partial r}\left(r q_e^r\right) + \frac{\partial q_e^z}{\partial z} - n_e T_e \left[\frac{1}{r}\frac{\partial}{\partial r}\left(r v_e^r\right) + \frac{\partial v_e^z}{\partial z}\right] + P_e, \quad (14.24)$$

$$\frac{1}{r}\frac{\partial}{\partial r}\left(r E^r\right) + \frac{\partial E^z}{\partial z} = 4\pi e \left(\sum_j \alpha_j n_{i,j} - n_e \right). \quad (14.25)$$

Here, $m_{i,j}, n_{i,j}, \Gamma_{i,j}, v_{i,j}, S_{i,j}, L_{i,j}$, and $M_{i,j}$ are the mass, density, flux, velocity, source rate, loss rate, and collisional transfer of momentum from ion species to neutral species of j species ions, respectively. n_e, Γ_e, v_e, S_e, L_e, and T_e are the density, flux, velocity, source rate, loss rate, and temperature of the electrons, respectively. $n_{n,k}, \Gamma_{n,k}, S_{n,k}$, and $L_{n,k}$ are the density, flux, source rate, and loss rate of k species neutrals or radicals, respectively. Superscripts r and z denote the directions of the vectors. eE is the electric force; F_r is the pressure gradient force; ν_{eN} is the elastic collision frequency between an electron and a neutral, and α_j is the charge number of j species ion.

The power absorbed by electrons is determined by using the electron heating model including anomalous skin effect. The power absorbed by electrons is given by

$$P_{abs} = \text{Re}\left[\frac{1}{2}\int_{\Omega_p} J_p^* \cdot E \mathrm{d}r\right], \quad (14.26)$$

where J_p is the plasma current density and Ω_p is the plasma volume.

The diffusion coefficient D_k of a neutral molecule k in a background mixture of SiH_4, H_2 and Si_2H_6 gas is obtained by first determining the separate diffusion coefficients in each of the background gases (SiH_4, Si_2H_6, and H_2) from the classical expression for binary collisions [38]:

$$D_{lk} = 1.858 \times 10^{-3} T^{3/2} \frac{[(M_l + M_k)/M_l M_k]^{1/2}}{p\sigma_{lk}^2 \Omega_\mathrm{D}}, \tag{14.27}$$

where T is the gas temperature, $\sigma_{lk} = (\sigma_l + \sigma_k)/2$ is the binary collision diameter, p is the pressure, and M_l and M_k are the masses of l and k species, respectively. Ω_D is the diffusion collision integral given by

$$\Omega_\mathrm{D} = \frac{A}{T^{*B}} + \frac{C}{\exp[DT^*]} + \frac{E}{\exp[FT^*]} + \frac{G}{\exp[HT^*]}, \tag{14.28}$$

where T^* is kT/ϵ_{lk}, $A = 1.06036$, $B = 0.15610$, $C = 0.19300$, $D = 0.47635$, $E = 1.03587$, $F = 1.52996$, $G = 1.76474$, and $H = 3.89411$. The Lennard-Jones parameters are obtained from [36] and [39], and the parameters are summarized in Table 14.4.

The considered chemical reactions and momentum transfer rate coefficients for electron–neutral collisions are shown in Table 14.5. The momentum transfer rate coefficients are calculated as following equation:

$$k + m(T_\mathrm{e}) = \int f_\mathrm{e}(v) v \sigma_\mathrm{m}(v) \mathrm{d}^3 v, \tag{14.29}$$

where $\sigma_\mathrm{m}(v)$ is the momentum transfer collision cross section and $f_\mathrm{e}(\mathbf{v})$ is the electron velocity distribution, which is assumed to be Maxwellian in the present

Table 14.4 Lennard-Jones parameters used for calculating the diffusion coefficients and the polarizability from the ion mobility

Particle	σ	ε/k_B	α_0	E_ion
SiH_4	4.084	207.6	4.62	11.65
Si_2H_6	4.828	301.3	8.470	9.7
H_2	2.920	38.0	0.808	15.4
H	2.050	145.0		
SiH	3.662	95.8		
SiH_2	3.803	133.1		
SiH_3	3.943	170.3		
Si_2H_2	4.383	323.8		
Si_2H_3	4.494	318.2		
Si_2H_4	4.601	312.6		
Si_2H_5	4.717	306.9		
Si	2.910	3,036.0		

Table 14.5 Chemical reactions and constants for SiH_4 discharges. The rate coefficients are in the form $k(T) = AT^B \exp(-C/T)$, where T is in Kelvin

	Reactions (electron + neutral reactions)	A	B	C	ΔE	Reference
1	$e + SiH_4 \rightarrow SiH_3^+ + H + 2e$	3.06×10^{-02}	−1.3	184,820	12.03	[3]
2	$e + SiH_4 \rightarrow SiH_2^+ + H_2 + 2e$	2.69×10^{-02}	−1.2	179,670	11.57	[3]
3	$e + SiH_4 \rightarrow SiH^+ + H_2 + H + 2e$	1.07×10^{-03}	−1.2	189,440	13.73	[3]
4	$e + SiH_4 \rightarrow H_2^+ + SiH + H + 2e$	1.89×10^{-22}	−0.4	22,610	25.00	[3]
5	$e + SiH_4 \rightarrow H^+ + SiH_2 + H + 2e$	9.49×10^{-23}	−0.4	20,793	24.50	[3]
6	$e + SiH_4 \rightarrow SiH_3 + H + e$	8.96×10^{-03}	−1.0	123,500	8.30	[3]
7	$e + SiH_4 \rightarrow SiH_2 + 2H + e$	1.83×10^{-03}	−1.0	123,500	8.30	[3]
8	$e + SiH_4 \rightarrow SiH_3^- + H$	2.50×10^{-13}	0	0	5.70	[2]
9	$e + Si_2H_6 \rightarrow Si_2H_4 + H_2 + e$	1.23×10^{-10}	0	0	7.00	[2]
10	$e + Si_2H_6 \rightarrow SiH_3^- + SiH_3$	9.45×10^{-14}	0	0	10.80	[2]
11	$e + Si_2H_6 \rightarrow SiH_4 + SiH_2 + e$	2.86×10^{-10}	0	0	5.60	[2]
12	$e + SiH_3 \rightarrow SiH_3^+ + 2e$	2.25×10^{-12}	0.9	94,804	8.74	[3]
13	$e + SiH_4 \rightarrow SiH_3^+ + H + 2e$	1.70×10^{-11}	0.6	133,220	12.09	[3]
14	$e + SiH_2 \rightarrow SiH_2^+ + 2e$	9.16×10^{-12}	0.8	98,701	8.92	[3]
15	$e + SiH_2 \rightarrow SiH^+ + H + 2e$	4.39×10^{-11}	0.6	146,510	11.09	[3]
16	$e + SiH \rightarrow SiH^+ + 2e$	5.25×10^{-11}	0.6	92,419	7.89	[3]
17	$e + H_2 \rightarrow 2H + e$	1.70×10^{-08}	0	119,940	8.09	[3]
18	$e + H_2 \rightarrow H_2^+ + 2e$	1.33×10^{-13}	1.1	197,550	15.40	[3]
19	$e + H \rightarrow H^+ + 2e$	7.33×10^{-12}	0.7	169,360	13.60	[3]

	Reactions (electron + ion reactions)	A	B	C	ΔE	Reference
20	$e + SiH^+ \rightarrow Si + H$	1.69×10^{-07}	0	0	–	[3]
21	$e + SiH_2^+ \rightarrow SiH + H$	1.82×10^{-05}	−0.5	0	–	[6]
22	$e + SiH_3^+ \rightarrow SiH_2 + H$	1.82×10^{-05}	−0.5	0	–	[6]
23	$e + H_2^+ \rightarrow H + H$	6.10×10^{-06}	−0.5	0	–	[6]
24	$e + H^+ \rightarrow H$	1.89×10^{-10}	−0.7	0	–	[7]

	Reactions (neutral + neutral reactions)	A	B	C	ΔE	Reference
25	$H + SiH_4 \rightarrow H_2 + SiH_3$	2.44×10^{-16}	1.9	1,102	–	[3]
26	$SiH_3 + SiH_3 \rightarrow SiH_2 + SiH_4$	2.99×10^{-11}	0	0	–	[3]
27	$SiH_2 + SiH_2 \rightarrow Si_2H_2 + H_2$	1.08×10^{-09}	0	0	–	[3]
28	$SiH_4 + SiH \rightarrow Si_2H_3 + H_2$	2.41×10^{-12}	0	1,006	–	[3]
29	$SiH_2 + H \rightarrow SiH + H_2$	2.31×10^{-11}	0	0	–	[3]
30	$SiH_3 + H \rightarrow SiH_2 + H_2$	2.49×10^{-11}	0	1,258	–	[3]
31	$Si + SiH_4 \rightarrow Si_2H_2 + H_2$	2.49×10^{-10}	0	3,669	–	[3]
32	$H + H + H_2 \rightarrow H_2 + H_2$	2.54×10^{-31}	−0.6	0	–	[3]
33	$SiH_4 + Si_2H_5 \rightarrow Si_2H_6 + SiH_3$	5.69×10^{-13}	0	0	–	[2]
34	$Si_2H_6 + SiH_3 \rightarrow SiH_4 + Si_2H_5$	1.00×10^{-12}	0	0	–	[2]
35	$SiH_3 + SiH_2 \rightarrow Si_2H_5$	3.77×10^{-13}	0	0	–	[2]
36	$SiH_2 + Si \rightarrow Si_2H_2$	4.53×10^{-13}	0	0	–	[2]

(continued)

Table 14.5 continued

	Reactions (neutral + neutral reactions)	A	B	C	ΔE	Reference
37	$SiH_2 + SiH \rightarrow Si_2H_3$	7.22×10^{-13}	0	0	–	[2]
38	$SiH_2 + H \rightarrow SiH + H_2$	7.96×10^{-13}	0	0	–	[2]
39	$SiH_2 + H \rightarrow SiH_3$	1.11×10^{-12}	0	0	–	[2]
40	$SiH + H_2 \rightarrow SiH_3$	1.98×10^{-12}	0	0	–	[2]
41	$Si + H_2 \rightarrow Si_2H_6$	6.59×10^{-12}	0	0	–	[2]
42	$Si_2H_4 + H_2 \rightarrow Si_2H_6$	5.33×10^{-13}	0	0	–	[2]
43	$Si_2H_4 + H_2 \rightarrow SiH_4 + SiH_2$	3.56×10^{-09}	0	0	–	[2]
44	$Si_2H_3 + H_2 \rightarrow Si_2H_5$	1.70×10^{-12}	0	0	–	[2]
45	$Si_2H_2 + H_2 \rightarrow Si_2H_4$	1.40×10^{-11}	0	0	–	[2]
46	$Si_2H_2 + H \rightarrow Si_2H_3$	4.94×10^{-11}	0	0	–	[2]
47	$H + Si_2H_6 \rightarrow Si_2H_5 + H_2$	7.00×10^{-12}	0	0	–	[4]
48	$H + Si_2H_6 \rightarrow SiH_3 + SiH_4$	3.50×10^{-12}	0	0	–	[4]
49	$H_2 + SiH_2 \rightarrow SiH_4$	1.40×10^{-14}	0	0	–	[4]
50	$SiH_4 + SiH_2 \rightarrow Si_2H_6$	1.20×10^{-11}	0	0	–	[4]

	Reactions (ion + neutral and ion+ ion reactions)	A	B	C	ΔE	Reference
51	$H^+ + SiH_4 \rightarrow SiH_3^+ + H_2$	5.00×10^{-09}	0	0	–	[3]
52	$H_2^+ + SiH_4 \rightarrow SiH_3^+ + H_2 + H$	3.60×10^{-09}	0	0	–	[3]
53	$SiH_2^+ + SiH_4 \rightarrow SiH_3^+ + SiH_3$	1.10×10^{-09}	0	0	–	[3]
54	$SiH_2^+ + SiH_4 \rightarrow Si_2H_4^+ + H_2$	2.50×10^{-10}	0	0	–	[4]
55	$SiH_2^+ + H_2 \rightarrow SiH_3^+ + H$	1.00×10^{-10}	0	0	–	[4]
56	$H_2^+ + SiH_4 \rightarrow SiH_2^+ + 2H_2$	6.60×10^{-11}	0	0	–	[4]
57	$SiH_3^- + SiH_3^+ \rightarrow SiH_3 + SiH_3$	5.00×10^{-07}	0	0	–	[4]
58	$SiH_3^- + SiH_2^+ \rightarrow SiH_3 + SiH_2$	1.20×10^{-07}	0	0	–	[5]
59	$SiH_3^- + SiH_4^+ \rightarrow SiH_3 + 2SiH_2$	1.00×10^{-07}	0	0	–	[5]
60	$SiH_3^- + H_2^+ \rightarrow SiH_3 + H_2$	4.80×10^{-07}	0	0	–	[5]

model. The integration is performed using cross sections from the literature [40–42] to generate reaction rates as a function of the electron temperature.

14.3.3.2 Numerical Results

Simulations were performed to observe the dependence of plasma parameters for SiH_4 discharges. The common conditions used in these calculations are $R = 25\,\text{cm}$, $R_{\text{sub}} = 18\,\text{cm}$, $L_{\text{p}} = 10\,\text{cm}$, $r_{\text{c}} = 10\,\text{cm}$, $z_{\text{c}} = -2\,\text{cm}$, $P_{\text{RF}} = 500\,\text{W}$, $\omega/2\pi = 13.56\,\text{MHz}$, 10 mTorr pressure, and $Q = 300$ SCCM.

The profiles of the absorbed power and the electron temperature are shown in Fig. 14.9. We can see that the power absorbed by the electrons is localized near the antenna and that the electron temperature also has a maximum value near the antenna.

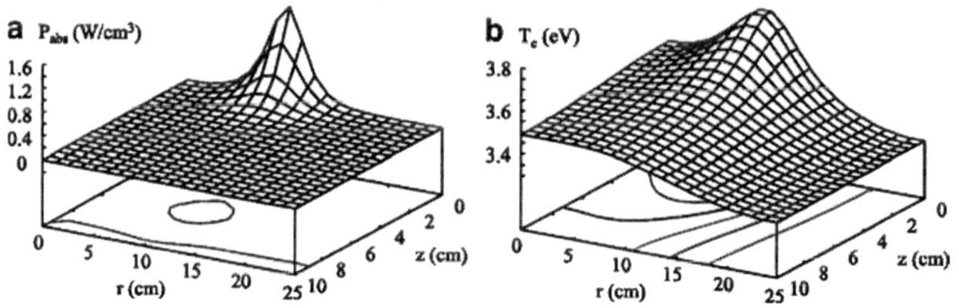

Fig. 14.9 2D profiles of (a) the power absorbed by the electron and of (b) the electron temperature

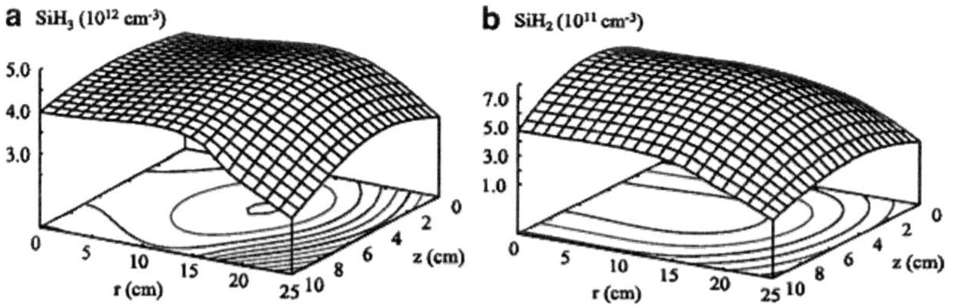

Fig. 14.10 2D profiles of the radical densities of (a) SiH_3 and (b) SiH_2

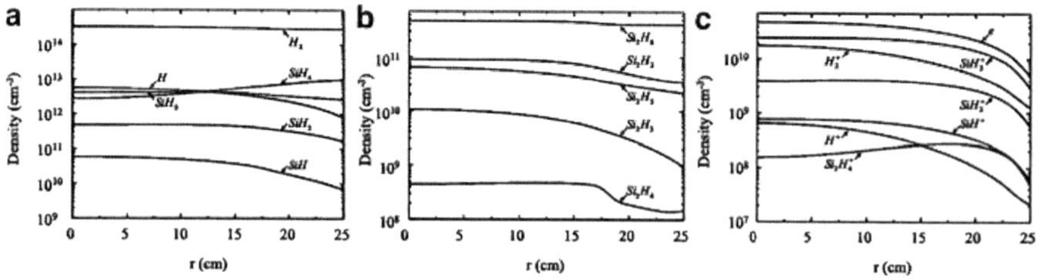

Fig. 14.11 Radial density profiles of radicals and charged species at $z = L_p$

Figure 14.10 shows the spatial profiles of the SiH_3 and the SiH_2 radical densities. We observed that the value of the SiH_3 density is about one order higher than that of the SiH_2 density, and the SiH_3 density has a maximum value near ($r = R/2$, $z = L_p/2$).

In Fig. 14.11, the radical density profiles of radicals and charged species at $z = L_p$ are shown. We can see that compared to the other species densities of Si_2H_6 and Si_2H_4 are considerably uniform at the substrate and that SiH_3^+ and H_2^+ are the dominant charged species. Also, the SiH_4 density is an increasing function of the chamber radius.

In this section, we construct the chemical reaction data for SiH_4 discharges based on previously reported articles and develop a two-dimensional fluid simulator for TCP sources. The power absorbed by electrons is determined by using an analytic electron heating theory including the anomalous skin effect. We observe that H_2

is the dominant radical and that SiH_3^+ and H_2^+ are the dominant charged particles. The results of the present study can be used as a useful tool for designing PECVD sources.

14.4 Line-Intensity Ratio Method for Plasma Diagnostics

Spectroscopy of radiation emitted from a plasma is a powerful technique for studying the nature of the plasma [43, 44]. Since the signals obtained by spectroscopy contain much information about temperature, density and flux of the main species and impurities. A spectroscopic technique based on the relative intensities of helium line is used to measure the electron temperature (T_e) and density (n_e) in the fusion edge plasma [45–48]. This nonintrusive diagnostic relies on the fact that the electron impact excitation rate coefficients for helium singlet and triplet states differ as a function of the electron temperature. In contrast, singlet states differ as a function of the electron density.

Cunningham [49] first applied this line-intensity ratio method to measuring electron temperatures in many plasma conditions. Many subsequent workers used measurement that depend on the same principle, but which are refinements of Cunningham's method [50]. Recently, this techniques has been considerably improved and extended to high-density, fusion edge plasmas [46, 51–53].

At low densities ($n_e < 10^{11}\,cm^{-3}$), the steady-state corona model can be used to determine the electron temperature and density with reasonable accuracy. However, at higher densities ($n_e > 10^{11}\,cm^{-3}$), secondary processes, such as excitation transfer, excitation from metastable are no longer negligible and steady-state corona model can no longer be used. Under these high-density conditions, collisional–radiative equilibrium models must be used to predict the temperature and density. These complex models take into account many of the secondary processes and no longer assume that excitation is exclusively from the ground state.

Thus, in the present work, T_e and n_e measurements using He I emission line-intensity ratios based on the collisional–radiative model are applied to diversified plasma simulator (DiPS) which is for the divertor edge study. A new versatile liner machine, called DiPS has been developed for the divertor edge simulation, space propulsion and electric probe technology and its applications. These measurements have been carried out simultaneously with electric probe system, which give continuous information on the electron temperature and density. Both measurements are complementary.

14.4.1 Principle of Line-Intensity Ratio Method

In general, the intensity of a spectral line is a function of both electron temperature and density. However, it is possible to identify certain spectral lines which are sensitive to electron temperature (density) but do not strongly with electron density (temperature). Such spectral lines can be very useful diagnostics of plasma electron temperature and density. However, the link between measured line-intensity ratio and plasma electron temperature and density is complex and a number of issues must be examined for the diagnostics. In particular, the different processes associated with the formation of excited levels populations responsible for the chosen atomic transitions must be well understood in order to extract information relative to the electron temperature and density of the plasma.

The plasma emissivity ϵ_{qp} (power emitted per unit time per unit volume per unit solid angle) at a specific wavelength λ_{qp} corresponding to an atomic transition from level q to level p can be written as [54]

$$\epsilon_{qp} = \frac{h\nu_{qp}}{4\pi} n_q A_{qp}, \tag{14.30}$$

where $h\nu_{qp}$ is the photon energy associated with the transition, n_q is the population of the emitting level, A_{qp} is the Einstein coefficient for the transition. Assuming a uniform plasma, measured intensity $I_{qp}(\lambda_{qp})$ at specific wavelength λ_{qp} is given by

$$I_{qp}(\lambda_{qp}) = \frac{1}{4\pi} n_q A_{qp} V \Omega T(\lambda_{qp}) \eta(\lambda_{qp}), \tag{14.31}$$

where V is the plasma volume, Ω is the solid angle subtended by the collection optics, $T(\lambda_{qp})$ and $\eta(\lambda_{qp})$ are the transmission factor of the detector system and the quantum efficiency at wavelength λ_{qp}, respectively. Thus, intensity ration for two lines is

$$\frac{I_{qp}(\lambda_{qp})}{I_{ji}(\lambda_{ji})} = \frac{n_q A_{qp} T(\lambda_{qp}) \eta(\lambda_{qp})}{n_j A_{ji} T(\lambda_{ji}) \eta(\lambda_{ji})} = \frac{1}{C} \frac{n_q A_{qp}}{n_j A_{ji}}, \tag{14.32}$$

where C is the relative calibration factor (ratio of the responses of the detection system at the two wavelengths).

At low densities ($n_e < 10^{11}\,\mathrm{cm}^{-3}$), the steady-state corona model can be used to predict the population of excited levels provided the plasma condition satisfies the applicability criteria for the model. The steady-state corona model assumed that line emission is the result of single collisions between electron and atoms in the ground state followed by the direct radiative de-excitation. Thus, in this model, a balance between the rate of collisional excitation from the ground state and the rate of spontaneous radiative decay determines the population densities of the excited levels. Then, the population of the level $q(N_q)$ is given by the expression [50]:

$$n_e n_0 \langle \sigma v \rangle_{0q} = n_q \sum_{p<q} A_{qp}, \tag{14.33}$$

where n_0 is the population of the ground level population, n_e is the electron density, $\sum_{p<q} A_{qp}$ is the total transition probability from level q to all lower states, and

$\langle \sigma v \rangle_{0q}$ is the excitation rate coefficient for the electron impact excitation of the level q from ground state. Thus, the line ratio expressed in terms of excitation rate coefficients in the steady-state corona model becomes

$$\frac{I_{qp}(\lambda_{qp})}{I_{ji}(\lambda_{ji})} = \frac{1}{C} \frac{E_{qp}(T_e)}{E_{ji}(T_e)}, \tag{14.34}$$

where $E_{qp}(T_e)$ is the emission rate coefficient. In this model, the line-intensity ratio depends only on the electron temperature in the plasma. The steady-state corona model is believed to be able to predict the electron temperature with reasonable accuracy for electron densities up to $10^{11}\,\text{cm}^{-3}$. However, at higher densities, this assumption is no longer valid since the occurrence of secondary processes involving collisions with excited or ionized atoms become important. These contributions can be integrated in the line ratio expression by replacing the emission rate coefficients by an apparent or resulting emission rate coefficient $E^*_{qp}(T_e, n_e)$ and $E^*_{ji}(T_e, n_e)$ that include both direct (ground) and indirect (metastables) excitation:

$$\frac{I_{qp}(\lambda_{qp})}{I_{ji}(\lambda_{ji})} = \frac{1}{C} \frac{E^*_{qp}(T_e, n_e)}{E^*_{ji}(T_e, n_e)}. \tag{14.35}$$

Collisional–radiative models are 0-dimensional plasma models, used to calculate atomic state distribution functions for one or more species, as a function of particle temperatures and densities. They normally work under the assumption that the two main types of processes that cause an atom to charge its excited state are radiative processes and electron collisions. These models can be an useful tool in understanding experimental observations, for instance to relate an observed spectrum to an atom state distribution function, and thus to an electron density and temperature. In addition, a collisional–radiative model can be used to calculate effective ionization and recombination rates, and the effective emissivity, as a function of electron temperature and density.

In order to use collisional–radiative models normally assume that the free electrons have a Maxwellian velocity distribution, and the plasma is optically thin, especially to its own resonance radiation. In addition, the ionization process is by electron collision from any bound level and is partially balanced by three-body recombination into any level, excitation transfer between any pair of bound level are induced by electron collisions, and radiation is emitted when a bound electron makes a spontaneous transition to a lower level or when a free electron makes a collisionless transition into a bound level. With this assumptions, we now consider ionization and recombination of a plasma and the population densities of various atomic levels in it under a given plasma condition, which is specified by electron and ion densities, n_e and n_i, respectively, and by the electron temperature, T_e. The time development of the population density of a level p is described by the differential equation:

$$\frac{dn_p}{dt} = -\left\{\sum_{q\neq p} C_{pq}n_e + \sum_{q<p} A_{pq} + S_p n_e\right\} n_p + \sum_{q\neq p} \{C_{qp}n_e + A_{qp}\} n_q + \{\alpha_p n_e + \beta_p\} n_i n_e, \qquad (14.36)$$

where C_{pq} is the rate coefficient for excitation $(p > q)$ or de-excitation $(q < p)$ by electron collisions from a level p to q, A_{pq} is the spontaneous transition probability from p to q, S_p, and α_p are the rate coefficients for ionization by electron collisions from p and its inverse three-body recombination coefficient to this level, respectively, and β_p is the radiative and dielectronic recombination coefficient to this level. The summation $q > p$ means the sum over levels q having higher energy than p. Other processes such as induced emission are neglected. In this paper, we use a collisional–radiative model to evaluate the excited state populations of helium.

14.4.2 Experiment

A new versatile linear machine, called DiPS has been developed for the divertor edge simulation, space propulsion and electric probe technology, and its applications. The schematic drawing of the DiPS device is shown in Fig. 14.12. To verify and develop theories of electric probes with magnetic field, collisionality, various particle source, and wide range of plasma parameters, one needs a plasma source with the following plasma parameters: density = $10^6 \sim 10^{14}\,\text{cm}^{-3}$, electron temperature = $1 \sim 10\,\text{eV}$, magnetic field = $0 \sim 2\,\text{kG}$, and plasma types = RF and DC. For these conditions, the DiPS is developed with two plasma sources, LaB_6 (DC) and Helicon plasma (RF). The LaB_6 cathode used in DiPS consists of disc-type LaB_6 which is indirectly heated by the graphite heater up to 5 kW. The discharge voltage and current are $1 \sim 100$ V and $1 \sim 50$ A, respectively. The helicon plasma source is also developed as a high-density RF plasma source in DiPS, which generates the high-density plasma with the of power of 0–3 kW by using $m = 0$ antenna.

The experimental apparatus is composed of a discharge chamber (test chamber region shown in Fig. 14.12), a spectroscopy system, a fast-scanning probe system with a single probe. The fast-scanning probe system, which is driven by a pneumatic cylinder with stroke of $5 \sim 10$ cm, measures the electron temperature and the plasma density by using a single probe (0.025 mm in diameter, 1.4 mm in length). The spectroscopy system is composed of a monochromator , an optical multichannel analyzer (OMA), and an optical fiber. The monochromator has a 500 mm focal length and a 0.05 nm resolution. The spectral response of the OMA is $300 \sim 900$ nm at $1{,}024 \times 128$ pixels.

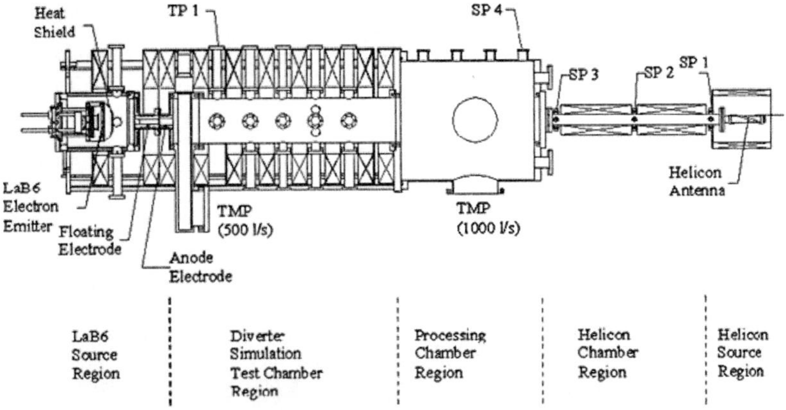

Fig. 14.12 Schematic drawing of the DiPS device

Experiments were performed at test chamber pressure of 2.0×10^{-3} torr, magnetic field of 400 G. The plasma were measured using the fast-scanning probe system and the optical emission spectroscopy system. The fast-scanning probe system with a single probe had an average operating speed of $0.4\sim0.7\,\mathrm{m\,s^{-1}}$ and a maximum speed of 2.2 m s−1, was used to measure the plasma electron temperature and density. The optical emission spectroscopy system with a monochromator received the light emitted from the plasma. The line radiation of the excited He-atoms are measured at three different wavelengths. We selected two transitions in the singlet system ($3d\ ^1D \rightarrow 2p\ ^1P$) at 667.82 nm and ($3s\ ^1S \rightarrow 2p\ ^1P$) at 728.13 nm and one in the triplet system ($3s\ ^3S \rightarrow 2p\ ^3P$) at 706.52 nm. Since the ratio of the line intensities in the singlet system (728.13 nm/667.82 nm) are most sensitive to the electron density whereas the ratio of the line intensities in the triplet and singlet system (728.13 nm/706.52 nm) are strongly depend on the electron temperature. A more detailed line selection procedures are discussed in next section.

14.4.3 Results and Discussions

In order to measure the electron temperature and density from a line ratio, a careful selection of electron of emission lines must be made. First, the line ratio of the selected transitions must insensitive to density variation and therefore applied for the determination of temperature. In contrast, selected transitions must insensitive to temperature variation and used for the determination of density. Also, secondary effects are strongly depend on the density, thus we must select transitions for which secondary effects have limited importance on the resulting intensities for temperature determination.

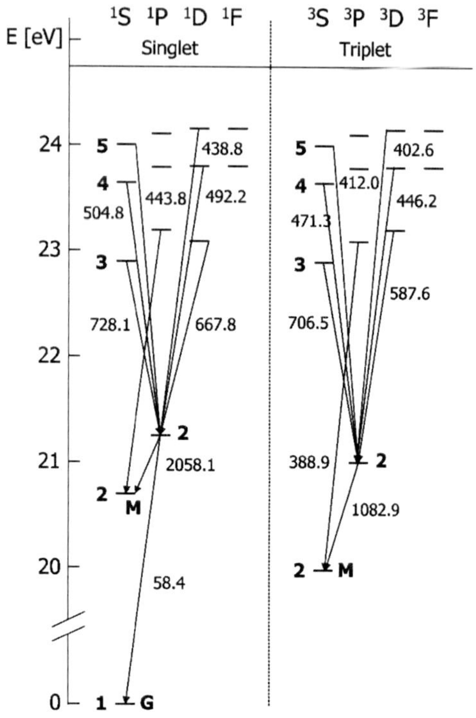

Fig. 14.13 Partial Grotrian diagram of the helium atom with some important transitions (λ in nm). G and M are for ground and metastables states, respectively

Table 14.6 Helium energy levels. Singlet (*left*) and triplet (*right*) states

Level	Energy (eV)	Level	Energy (eV)
2^1S	20.610	2^3S	19.814
2^1P	21.212	2^3P	20.958
3^1S	22.914	3^3S	22.712
3^1P	23.081	3^3P	23.001
3^1D	23.068	3^3D	23.067
4^1S	23.667	4^3S	23.588
4^1P	23.736	4^3P	23.730
4^1D	23.730	4^3D	23.730
4^1F	23.731	4^3F	23.731
5^1S	24.005	5^3S	23.965
5^1P	24.039	5^3P	24.022
5^1D	24.036	5^3D	24.036
5^1F	24.037	5^3F	24.037

The He Grotrian diagram is shown in Fig. 14.13. The left side of the diagram features the singlet states while the triplet states are shown in the right side. The He atom has 2 metastables, namely the $2\ ^1S$ state (singlet) and the $2\ ^3S$ state (triplet). The energy levels of the He atom are given in Table 14.6. As seen in the Grotrian diagram, there are a number of possible transitions that can be used. Transitions originating from a level with principal quantum number $n > 5$ will not be considered since the electron population of these higher levels becomes increasingly smaller with increasing n, resulting in weak transitions. Transitions ending at the ground state $1\ ^1S$ and at metastable level $2\ ^1S$ and $2\ ^3S$ will not be considered since the plasma is not optically thin with respect to these transitions and the resulting intensities are strongly affected by reabsorption.

In order to measure the electron density, $D \rightarrow P$ transition are better suited in two reasons. First, excitation transfer cross sections for allowed transitions are much larger than for non-allowed transitions. Second, the excitation transfer is inversely proportional to the energy difference between levels and strongly dependent on plasma density. For example, the energy difference between the $3\ ^1P$ and $3\ ^1D$ level is only 0.013 eV while the corresponding quantity between the $3\ ^3P$ and $3\ ^3D$ levels 0.066 eV. For the $n = 4$ level, $\Delta E(4\ ^1P/4\ ^1D)$ is only 0.006 eV while $\Delta E(4\ ^3P/4\ ^3D)$ is even smaller. Thus these transitions will be more sensitive to plasma density than electron temperature. In these reasons, we measured HeI line ratios of 728.13 nm ($3s\ ^1S \rightarrow 2p\ ^1P$)/667.82 nm ($3d\ ^1D \rightarrow 2p\ ^1P$) for electron density determination.

Line ratios using $S \rightarrow P$ transitions are better suited to measure electron temperature. The contributions from the metastable $2\ ^1S$ and $2\ ^3S$ states due to excitation transfer are small since these s-transitions are forbidden and the resulting cross section are small. Also, for a given n level, the energy of the S levels is significantly different than the energy of the other P, D, or F levels. Thus excitation transfer cross sections between S and any of these P, D, or F levels are also small compared to cross section involving only P, D, and F levels. In these reasons, 728.13 nm ($3s\ ^1S \rightarrow 2p\ ^1P$)/706.52 nm ($3s\ ^3S \rightarrow 2p\ ^3P$) ratio is attractive for electron temperature determination.

Up to now two transitions in the singlet system ($3d\ ^1D \rightarrow 2p\ ^1P$) at 667.82 nm and ($3s\ ^1S \rightarrow 2p\ ^1P$) at 728.13 nm and one in the triplet system ($3s\ ^3S \rightarrow 2p\ ^3P$) at 706.52 nm are selected. The ratios of the line intensities are compared with those calculated in a collisional–radiative equilibrium model for given plasma parameters. In order to predict the electron temperature and density, here we consider a two-dimensional test function as

$$f(T_e, n_e) = \sum_i \left(\frac{R_i^{exp} - R_i^{cal}(T_e, n_e)}{R_i^{exp}} \right)^2, \tag{14.37}$$

where the summation is over the pairs 728.13 nm/667.82 nm and 728.13 nm/706.52 nm. R_i is the line-intensity ratio and the superscripts "exp" and "cal" indicate the experimental and calculational values, respectively. The electron temperature (T_e) and density (n_e) are simultaneously obtained so as to minimize the test function, (14.37). Figure 14.14 shows the results. It is found that the electron temperature and density profiles which are calculated by line ratio method based on the collisional–radiative equilibrium model have good agreement with the values which are measured by a fast-scanning probe.

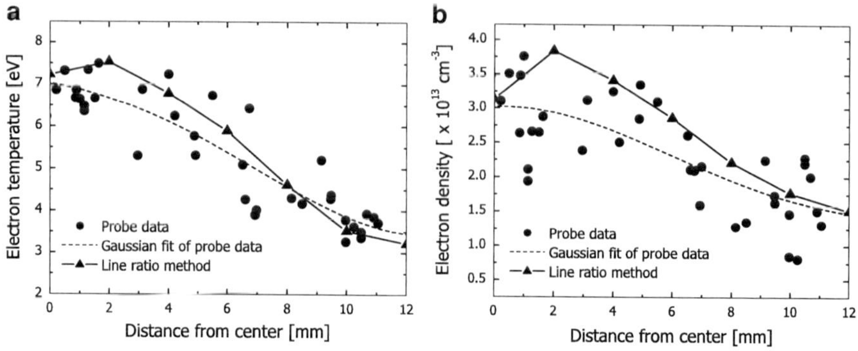

Fig. 14.14 Electron temperature and density profiles obtained with the He I line-intensity ratio method and with fast-scanning probe

14.5 Conclusions

Electron–atom and electron–molecule collisions provide the driving mechanism behind the plasma processes employed in semiconductor manufacturing and related industries. The radicals and ions produced by the collisions of electrons with the feed gas interact with the surface through a series of complex chemical reactions, leading to etching, deposition, or cleaning. Thus, atomic and molecular data are just as important as data on the heterogeneous surface reactions in understanding and controlling such processes.

In early development of plasma processing had been achieved through trial and error, the increasing demand to shrink the feature dimensions and the demand for diminished defect density produce a corresponding demand for increased sophistication an effectiveness in process control. It is now generally accepted that the traditional method of meeting these requirements by trial and error has reached the point of diminishing returns. Also, to achieve technical goal, complex process equipment and techniques have been developed, and now we have necessitated a better comprehension of the physics standing behind the plasma. Thus, plasma process modeling now becomes a necessary engineering tool in the design of new semiconductor equipment and in process control, and also, the high cost of developing both the plasma equipment and processes has motivated development of less-empirical methods, and modeling/simulation in particular, to speed the time to market and to reduce costs.

A self-consistent global model of solenoidal-type ICP discharges including the effects of radio-frequency bias power [55] was introduced in this chapter. Numerically, solve a set of spatially averaged fluid equations for charged particles, neutrals, and radicals. Absorbed power by electrons is determined by using an analytic electron heating model including the anomalous skin effect. To analyze the effects of RF bias power on the plasma properties, simulation model also combines the electron heating and global transport module with an RF sheath module in a self-consistent manner. To validate simulation model, experiments were performed in inductively coupled argon plasmas. The simulation results are also compared using the commercial software package CFD-ACE+. In this work, we observe that the numerical results are in a good agreement with the experimental results, and dependence of the electron density on the bias power cannot be negligible.

As an second simulation example, we introduce how to construct the chemical reaction data for SiH_4 discharges based on previously reported articles and develop a two-dimensional fluid simulator for TCP sources. The power absorbed by electrons is determined by using an analytic electron heating theory including the anomalous skin effect. Here we observed that H_2 is the dominant radical and that SiH_3^+ and H_2^+ are the dominant charged particles. The results of this work can be used as a useful tool for designing PECVD sources.

As a diagnostic tools, line-intensity ration method was introduced in this chapter. The line-intensity ratio method provides a nonintrusive diagnostic for the measurement of electron temperature and density in a plasma. The intensity ratios of He I emission lines are used to determine electron density and temperature in DiPS. The He I line ratios are measured by optical emission spectroscopy and are calculated using the collisional–radiative equilibrium (CRE) model. The measured He I line ratios are 706.52 nm/728.13 nm for electron temperature and 728.13 nm/667.82 nm for electron density determination, respectively. The results obtained from the collisional–radiative equilibrium model are cross-checked with those of electrical probe measurement.

In this chapter, we can not handle every atomic and molecular data applications; however, we hope this should provide easy guide to the atomic and molecular data research and its applications.

Acknowledgements The author acknowledges the collaboration of many colleagues in preparing this chapter. Especially thanks to Prof. Y. Itikawa, Prof. H. Tanaka, and Prof. H. Cho for their helpful discussions and the provision of meaningful information. Also author thanks to Dr. Young-Woo Kim, Dae-Chul Kim, Yong-Hyun. Kim, and Dr. Jong-Sik Kim for their unlimited efforts on A+M data research activities.

This work was supported by the *Basic Plasma Research* (National Fusion Research Institute) and grant funded by the *Ministry of Education, Science and Technology*. Also, partially supported by the *Development of Korea National Standard Reference Data* and program funded by the *Ministry of Knowledge Economy*.

References

1. M. Loidl et al., Mucl. Instr. Meth. A **559**, 769 (2006)
2. D.W. Shoesmith, J. Nucl. Mater. **282**, 1 (2000)
3. G.R. Choppin, M.Kh. Khankhasayev, in *Chemical Separation Technologies and Related Methods of Nuclear Waste Management*. NATO Science Series (Kluwer, Dordrecht, 1999)
4. National Research Council, *Plasma Processing of Materials: Scientific Opportunities and Technological Challenges* (National Academic Press, Washington, DC, 1991)

5. National Research Council: *Database Needs for Modeling and Simulation of Plasma Processing* (National Academic Press, Washington, DC, 1996)
6. J. Meichsner, in *Low Temperature Plasmas in Plasma Physics: Confinement, Transport and Collective Effects*. ed. by A. Dinklage et al. Springer Lecture Notes, vol. 670 (Springer, Berlin, 2005)
7. National Research Council: *Plasma Science: Advancing Knowledge in the National Interest* (National Academy Press, Washington, DC, 2010)
8. Report of the Basic Energy Science Workshop on Electron Scattering for Materials Characterization, *Future Science Needs and Opportunities for Electron Scattering: Next-Generation Instrumentation and Beyond*, Report of a U.S. Department of Energy Office of Basic Energy Science Workshop, March 1–2, 2007, Washington DC
9. N.J. Mason: *Electron Driven Processes: Scientific Challenges and Technological Opportunities* (Springer, Berlin, 2005)
10. D.B. Graves, M.J. Kushner, *Low Temperature Plasma Science: Not only the Fourth State of Matter but All of Them*, Report of the Department of Energy Office of Fusion Energy Sciences Workshop on Low Temperature Plasmas, March 25–27, (2008)
11. T. Makabe, Z. Petrovic, in *Plasma Electronics: Applications in Microelectronic Device Fabrication*. Series in Plasma Physics (Taylor & Francis, London, 2006)
12. A. Fridman *Plasma Chemistry* (Cambridge University Press, New York, 2008)
13. N.J. Mason: J. Phys. D Appl. Phys. **42**, 194003 (2009)
14. H.W. Lee et al., J. Phys. D Appl. Phys. **44**, 053001 (2011)
15. M.J. Kushner, Bull. Am. Phys. Soc. **55**, 107 (2010)
16. M.A. Lieberman, A.J. Lichtenberg, *Principles of Plasma Discharges and Materials Processing*, 2nd edn. (Wiley-Interscience, New York, 2005)
17. R.L. Champion, L.D. Doverspike, in *Electron-Molecule Interactions and Their Applications*, ed. by L.G. Christophorou (Academic, New York, 1984)
18. J.D. Morrison, A.J. Nicholson, J. Chem. Phys. **20**, 1021 (1952)
19. R. Basner, R. Foest, M. Schmidt, F. Hempel, K. Becker, in *Ions and Neutrals in the Ar-TEOS RF Discharge*. Proceedings of the 23rd International Conference on Phenomena in Ionized Gases, vol. IV, Toulouse, July 1997, pp. 196–197
20. D.C. Schram, M.C.M. van de Sanden, R.J. Severens, W.M.M. Kessels, J. Phys. IV **8**, 217–230 (1998)
21. B.F. Gordiets, C.M. Ferreira, M.J. Pinheiro, A. Ricard, Plasma Sources Sci. Technol. **7**, 363–378, 378–388 (1998)
22. M. Hayashi, *Nagoya Institute of Technology Report*, No. IPPJ-AM-19 (1991)
23. D. Rapp, P. Englander-Golden, J. Chem. Phys. **43**, 1464 (1965)
24. S. Tinck, W. Boullart, A. Bogaerts, J. Phys. D **41**, 065207 (2008)
25. A.V. Vasenkov, X. Li, G.S. Oehrlein, M.J. Kushner, J. Vac. Sci. Technol. A **22**, 511 (2004)
26. N.S. Yoon, S.S. Kim, C.S. Chang, D.I. Choi, J. Korean Phys. Soc. **28**, 172 (1995)
27. Z.L. Dai, Y.N. Wang, T.C. Ma, Phys. Rev. E **65**, 036403 (2002)
28. A. Metze, D.W. Ernie, H.J. Oskam, J. Appl. Phys. **60**, 3081 (1986)
29. T. Panagopoulos, D.J. Economou, J. Appl. Phys. **85**, 3435 (1999)
30. E.A. Edelberg, E.S. Aydil, J. Appl. Phys. **86**, 4799 (1999)
31. P.A. Miller, M.E. Riley, J. Appl. Phys. **82**, 3689 (1997)
32. M.A. Sobolewski, J.-H. Kim, J. Appl. Phys. **102**, 113302 (2007)
33. http://www.esi-group.com/products/multiphysics/ace multiphysics-suite/ace-suite/cfd-acc
34. M. Vinodkumar, C. Limbachiya, K. Korot, K.N. Joshipura, Eur. Phys. J. D **48**, 333 (2008)
35. M.J. Kuchner, J. Appl. Phys. **63**, 2532 (1988)
36. E. Meeks, R.S. Larson, P. Ho, S.M. Han, E. Edelberg, E.S. Aydil, J. Vac. Sci. Technol. A **16**, 544 (1998)
37. J.L. Giuliani, V.A. Shamamian, R.E. Thomas, J.P. Apruzese, M. Mulbrandon, R.A. Rudder, R.C. Hendry, A.E. Robson, IEEE Trans. Plasma Sci. **27**, 1317 (1999)
38. O. Leroy, G. Gousset, L.L. Alves, J. Perrin, J. Jolly, Plasma Sources Sci. Technol. **7**, 348 (1998)
39. C.R. Kleijin, Thin Solid Films **365**, 294 (2000)

40. J.S. Yoon, M.Y. Song, J.M. Han, S.H. Hwang, W.S. Chang, B.J. Lee, J. Phys. Chem. Ref. Data **37**, 913 (2008)
41. T. Shimada, Y. Nakamura, Z.L. Petrovic, T. Makabe, J. Phys. D Appl. Phys. **36**, 1936 (2003)
42. N. Sato, Y. Shida, Jpn. J. Appl. Phys. **36**, 4794 (1997)
43. I.H. Hutchinson, *Principles of Plasma Diagnostics*, 2nd edn. (Cambridge University Press, Cambridge, 2002)
44. H.R. Griem, *Plasma Spectroscopy* (McGraw-Hill, New York, 1965), pp. 243–253
45. R. Mewe, Brit. J. Appl. Phys. **18**, 107 (1967)
46. B. Schweer, G. Mank, A. Pospieszczyk, B. Brosda, B. Pohlmeyer, J. Nucl. Mater. **196–198**, 174 (1995)
47. R.F. Biovin, J.L. Kline, E.E. Scime, Phys. Plasmas **8**, 5303 (2001)
48. N.K. Podder et al., Phys. Plasmas **11**, 5436 (2004)
49. S.P. Cunningham, in Conference on Thermonuclear Reactors, Livermore, U.S. Atomic Energy Commission Rep., vol. 279, p. 289 (1955)
50. R.W.P. McWhiter, in *Plasma Diagnostic Techniques*, ed. by R.H. Huddlestone, S.L. Leonard (Academic, New York, 1965), ch. 5
51. S.J. Davies, P.D. Morgan et al., J. Nucl. Mater. **241–243**, 426 (1997)
52. Y. Andrew, S.J. Davies et al., J. Nucl. Mater. **266–269**, 1234 (1999)
53. Y. Andrew, M.G. O'Mullane, Plasma Phys. Contr. Fusion **42**, 301 (2000)
54. W.L. Wiese, M.W. Smith, B.M Glennon, *Atomic Transitions Probabilities*, vol. 1, National Standard Reference Data System NSRDS-NBS-4 (1996)
55. D.C. Kwon, W.S. Chang, M. Park, D.H. You, M.Y. Song, S.J. You, Y.H. Im, J.S. Yoon, J. Appl. Phys. **109**, 073311 (2011)

Chapter 15
Broadening of Hydrogenic Spectral Lines in Magnetized Plasmas: Diagnostic Applications

E. Oks

Abstract In diagnostics based on the broadening of spectral lines in plasmas, magnetic fields are important only if their strength is relatively high—to compete with the Stark and Doppler broadenings. We review the corresponding theories and their diagnostic applications.

15.1 Introduction

In diagnostics based on the broadening of spectral lines in plasmas, magnetic fields are important only if their strength is relatively high—to compete with the Stark and Doppler broadenings. Relevant examples are plasmas produced by ultrashort, high-intensity lasers and edge plasmas of magnetic fusion experiments, as noted in book [1]. The focus of this review is at hydrogenic spectral lines. This is because the generally complicated physics of the Stark–Zeeman broadening can be best understood and used in practice for spectral lines of one-electron systems: hydrogen atoms and hydrogen-like ions. Besides, this subject is also theoretically important for two reasons. First, it deals with a deeply fundamental problem of the simplest, two-particle bound Coulomb system immersed in a multiparticle Coulomb system of free charges (plasma) exhibiting long-range interactions—as noted in review [2]. Second, due to the fact that a bound two-particle Coulomb system possesses a higher algebraic symmetry than its geometrical symmetry, sophisticated analytical advances can be made into the problem of the broadening of spectral lines of such a system in a plasma, thus yielding a profound physical insight. Advanced theories of the Stark broadening of hydrogenic lines are presented in book [3].

Employing hydrogenic spectral lines allows measuring important parameters of magnetized plasmas. Examples are magnetic field strength and pitch angle, effective charge, temperature, electron density, electric field of various kinds of and characteristic frequency of nonlinear processes relevant to one specific electrostatic turbulence—Langmuir turbulence.

V. Shevelko and H. Tawara (eds.), *Atomic Processes in Basic and Applied Physics*, Springer Series on Atomic, Optical, and Plasma Physics 68,
DOI 10.1007/978-3-642-25569-4_15,

As for diagnostic applications of nonhydrogenic spectral lines in magnetized plasmas, some useful theoretical references can be found in Sect. 12.4 of book [1]. On the experimental side, we first note paper [4], where a He-like spectral line CV $1s2s^3S_1$ - $1s2p^3P_2$ at the wavelength 227.09 nm was used for measuring a self-generated magnetic field produced when a pulsed Nd: glass laser was focused onto a carbon foil target. These measurements were made at electron densities 10^{19}–$10^{20}\,\mathrm{cm}^{-3}$ for magnetic fields ~10 T. Second, in as more recent paper [5], the Al III 4p–4s doublet at the wavelengths 569.6 and 572.2 nm was employed for measuring a magnetic field in a plasma plume produced by a laser pulse impinging on the cathode of a coaxial line. These measurements were made at electron densities $\sim 10^{16}\,\mathrm{cm}^{-3}$ for magnetic fields ~1 T based on the fact that different fine-structure components undergo different splitting under the magnetic field.

15.2 Analytical Solutions for Stark–Zeeman Effect Relevant to Plasma Diagnostics

15.2.1 *Atom in Crossed Electric* (F) *and Magnetic* (B) *Fields*

Crossed **F–B** fields appear usually in the following three situations. First, they take place in a magnetized plasma where the electric field **F** is represented by a quasistatic part of the ion microfield and/or by a low-frequency electrostatic turbulence. Second, crossed **F–B** fields also occur when atoms move across a magnetic field: in their reference frame, there is a Lorentz field $\mathbf{F}_L = \mathrm{v} \times \mathbf{B}/c$. Third, in some applications, when an electric field rotates, it is convenient to analyze the problem in the frame rotating with the electric field (see Sects. 15.2.2 and 15.2.3 below). In the rotating frame, there appears an effective magnetic field. An analytical solution for a Coulomb system in crossed **F–B** fields, presented in [6], was made possible by the O(4) symmetry of the hydrogen atom discovered in [7]. This symmetry is intimately connected with an additional integral of motion in a Coulomb field: the Runge–Lenz vector [8,9]:

$$\mathbf{A} = (\mathbf{p} \times \mathbf{L} - \mathbf{L} \times \mathbf{p})/(2m_e) - Z_r e^2 \mathbf{r}/r, \qquad (15.1)$$

where **p** and **L** are linear and angular momenta, respectively; m_e is the electron mass, Z_r is the nuclear charge of the radiating atom/ion. In crossed **F–B** fields, the interaction term in the Hamiltonian can be written in the form:

$$V = \mathbf{J}_+ \cdot \boldsymbol{\omega}_+ + \mathbf{J}_- \cdot \boldsymbol{\omega}_-. \qquad (15.2)$$

Here

$$\mathbf{J}_\pm = [\mathbf{L} \pm \mathbf{A}]/2, \ \boldsymbol{\omega}_\pm = \boldsymbol{\Omega} \pm 3n\hbar\mathbf{F}/(2Zm_e e), \ \boldsymbol{\Omega} = -e\mathbf{B}/(2m_e c), \qquad (15.3)$$

where n is the principal quantum number. The authors of [6] defined wave functions $u_{nn'n''}$, corresponding to states having a definite projection of $\mathbf{J}_+$ onto $\boldsymbol{\omega}_+$ (the projection being represented by the quantum number n'), and a definite projection of $\mathbf{J}_-$ onto $\boldsymbol{\omega}_-$ (the projection being represented by the quantum number n'').

Obviously, these wave functions can diagonalize the interaction V:

$$V = \hbar(n'\omega_+ + n''\omega_-). \tag{15.4}$$

The numbers n' and n'' take half-integer values in the range from $-(n-1)/2$ to $(n-1)/2$.

15.2.2 Exact Solution for the Stark Broadening Based on the Crossed F and B Fields Results: A Binary Model

In paper [10], there was calculated analytically a Stark profile of a hydrogen spectral line emitted during a collision with one charged particle. Here, we call it a binary model, but the following clarification should be made.

The impact broadening theories treat the plasma microfield, created by one sort of charged particles, as a sequence of binary collisions, but the resulting Stark profile of a spectral line is a cumulative effect of a *large number of binary collisions* (see, e.g., [1–3]). In distinction, in the model used in paper [10], the Stark profile was assumed to result from just one binary collision. The elimination of this assumption is presented in Sect. 15.2.3.

In paper [10], the problem of a collision of a hydrogen atom with a classical charged perturber (in the dipole approximation) was reduced to the problem of a hydrogen atom in crossed static electric and magnetic fields [6] as follows. A transformation was made into a reference frame rotating with an angular velocity $\Omega\ (t)$ such that the X-axis follows the direction onto perturber and the z-axis is parallel to Ω and is perpendicular to the collision plane. This resulted in the following Hamiltonian containing an electric interaction and an effective "magnetic" interaction (due to the rotation):

$$H = H_0 + V,\ V = d_x E(t) - \hbar L_z \Omega(t). \tag{15.5}$$

By introducing two new effective angular momenta $\mathbf{J}_\pm$, (that satisfy the usual commutation rules for any angular momentum) and two other vectors having the frequency dimension

$$\boldsymbol{\omega}_\pm(t) = \boldsymbol{\Omega}(t) \pm 3n\hbar\mathbf{E}(t)/(2Z_r m_e e), \tag{15.6}$$

the interaction got represented in the form similar to (15.2), but with the time dependence:

$$V = \mathbf{J}_+ \cdot \boldsymbol{\omega}_+(t) + \mathbf{J}_- \cdot \boldsymbol{\omega}_-(t). \tag{15.7}$$

The wave functions $u_{nn'n''}$ from [6] can diagonalize the "instantaneous" interaction V in (15.7), i.e., the interaction at any "frozen" instant of time. For this to be a useful step toward the exact analytical solution for a hydrogen atom interacting with a *moving* (rather than frozen) charged particle, the direction of the axes $\omega_+(t)$ and $\omega_-(t)$ should remain constant in time. In paper [10], it was shown that this was indeed the case due to the fact that the ratio $\Omega(t)/E(t)$ turns out to be time-independent. As the result, the problem was reduced to the adiabatic formalism, which is equivalent to the classical model of a phase modulation of the atomic oscillator. The wave function in the rotating frame got represented as follows [10]:

$$\chi_{nn'n''}(t) = u_{nn'n''} \exp\left[-\mathrm{i}(n' + n'')B(v,\rho)\int_0^t E(\tau)\mathrm{d}\tau\right], \tag{15.8}$$

where $B(v,\tau)$ is a known function of v and ρ. It should be emphasized that while this solution formally looks like being adiabatic (because the electric field enters it only as the absolute value $E(\tau)$, rather than as a vector $\mathbf{E}(\tau)$), in the reality it incorporates (exactly!) also all nonadiabatic effects.

15.2.3 Exact Solution for the Stark Broadening Based on the Crossed F and B Fields Results: The General Case

In paper [11], the binary assumption used in paper [10] has been removed as follows. In the basis of the wave functions $\chi_{nn'n''}(t)$, the S-matrix (S_u) is diagonal and can be represented in the form:

$$S_u = [\exp(-\mathrm{i}\varphi)]^{n'+n''}, \quad \varphi = \pi[1 + (\rho_W/\rho)^2]^{1/2}, \tag{15.9}$$

where a so-called Weisskopf radius was redefined as $\rho_W = 3n\hbar/(2m_e v)$. For a given collision the S-matrix in the laboratory frame S_L is connected with S-matrix in the rotating frame S_R as

$$S_L = R_0^+[\exp(\mathrm{i}L_z\pi)]S_R R_0. \tag{15.10}$$

Here, R_0 is an operator transforming the wave functions from the laboratory frame to the $(-\mathbf{v}, \boldsymbol{\rho}, \boldsymbol{\rho} \times \mathbf{v})$ reference frame. The XY-plane of the latter was chosen in the collision plane spanned on the vectors $\boldsymbol{\rho}$ and $\mathbf{v}$, the X-axis being aligned with $-\mathbf{v}$. The additional phase factor in (15.10) is due to the fact that the X-axis of the rotating frame changes its direction to the opposite one in the course of a rectilinear path collision.

It should be emphasized that all the angular dependence characterizing a given collision was contained in R_0. So, already at this stage, one could perform the angular averaging as needed for obtaining the impact operator. For the Lyman lines the impact operator contains only the S-matrix for the upper level—so in [11] it was found that the averaged S-matrix in the basis of the spherical wave functions had a diagonal form with the following matrix elements:

$$< lm|S|lm >= [1/(2l+1)] \sum_{m'} (-1)^{m'} < lm'|S_R|lm' > . \qquad (15.11)$$

In order to bridge the gap between (15.9) and (15.11), one should perform the following unitary transformations connecting the basis of the spherical wave functions $|lm>$ with the basis of the wave functions $|n'n''>$ from [6].

1. Transformation from the basis $|n'n''>$ to the O(4)-basis $|i'i''>$:

$$< i'i''|n'n'' >= d_{n'i'}(\beta_1)\, d_{n''i''}(\beta_2), \qquad (15.12)$$

   where $d_{ni}(\beta)$ are the Wigner functions, $\beta_1 + \beta_2 = \pi$, $\tan\beta_2 = \rho/\rho_W$.
2. Transformation from the O(4)-basis to the spherical basis $|nlm>X$ with the quantization along the X-axis via the Clebsh–Gordon coefficients:

$$< i'i''|nlm >= C(j, j, j; i', i'', m), \ j = (n-1)/2. \qquad (15.13)$$

3. Rotation about the Y-axis by $\pi/2$ to express the result in terms of the spherical wave functions with the quantization along the Z-axis:

$$< nlm'|nlm >_X = d_{mm'}(\pi/2). \qquad (15.14)$$

Starting from the diagonal S_u matrix in (15.9) and performing three unitary transformations (15.12)–(15.14), the authors of [11] used (15.11) for calculating the angular average of the. Next they substituted this result into the S-matrix impact operator and integrated over impact parameters and velocities. In this way, they obtained the exact impact operator and then arrived at the analytical expression for the lineshape.

For $n = 2$, by introducing a scaled impact parameter $z \equiv \rho/\rho_W$, so that $\varphi(z) = \pi(1 + 1/z^2)^{1/2}$ (see (15.9)), they obtained explicitly the following angular averages of the S-matrices. For $l = 0$

$$1 - \{< 200|S|200 >\}_{\text{ang.av.}} = \left[2/\left(1 + z^2\right)\right] \sin^2(\varphi/2). \qquad (15.15)$$

For $l = 1$; $m = 1, 0, +1$:

$$1 - \{< 2lm|S|2lm >\}_{\text{ang.av.}} = \left[2 + 3z^2 + (2 + z^2)\cos\varphi\right] / \left[3\left(1 + z^2\right)\right]. \qquad (15.16)$$

It is instructive to compare the above exact analytical results with the simplest semiclassical theory of the dynamical Stark broadening (see, e.g., book [12])—sometimes called the standard theory (ST). The ST limits itself with the second order of the Dyson perturbation expansion of the evolution operator. For the above case, the perturbative approach from the ST would yield:

$$1 - \{< nlm|S|nlm >\}_{\text{ang.av.}} = 2(n^2 - l^2 - l - 1)/(3z^2). \tag{15.17}$$

After obtaining (15.15) and (15.16), the authors of [11] integrated them over impact parameters and introduced functions $g_l(Z) = \int_0^Z (1 - S_l)zdz$. It was possible to perform the integration analytically and to express the results in terms of tabulated special functions of the argument $Z = \rho_D/\rho_W$:

$$g_0(Z) = -\text{ci}(\varphi) - [\text{ci}(\varphi - \pi) + \text{ci}(\varphi + \pi)]/2 + \ln(Z\varphi/\pi), \tag{15.18}$$

$$g_1(Z) = -[\text{ci}(\varphi) + \{\text{ci}(\varphi - \pi) + \text{ci}(\varphi + \pi)\}/2$$
$$+\pi\{\text{si}(\varphi - \pi) - \text{si}(\varphi + \pi)\}/2 - Z^2(1 + \cos\varphi) - \ln(Z\varphi/\pi)]/3. \tag{15.19}$$

Here, si(ϵ) and ci(ϵ) are the integral sine and cosine, respectively.

They also presented asymptotic expressions for the case where $Z = \rho_D/\rho_W \ll 1$:

$$g_0(Z) \approx \left\{\frac{C}{2} + \text{ci}\pi + \frac{\text{ci}(2\pi)}{2} + \ln\pi\right\} + 2\ln Z \approx 0.578 + 2\ln Z, \tag{15.20}$$

$$g_1(Z) \approx \frac{1}{6}\left\{C + 2\text{ci}\pi + \text{ci}(2\pi) + \pi\,\text{si}(2\pi) + \ln\frac{\pi}{2}\right\} + \frac{2}{3}\ln Z \approx 0.550 + \frac{2}{3}\ln Z, \tag{15.21}$$

where C is the Euler constant. The corresponding conventional perturbative expression would be:

$$g_{\text{ST}} \approx (2/3)(n^2 - l^2 - l - 1)[\alpha Z_{\min}^2 + \ln(Z/Z_{\min}], \tag{15.22}$$

where $Z_{\min}$ is an arbitrarily defined lower cutoff and $\alpha = (0–2)$ is an arbitrarily defined strong collision constant. In distinction to this, the authors of [11] were able to rigorously deduce the exact value of the strong collision term comparing (15.21) with (15.22).

Figure 15.1 shows a comparison of the exact solution for the S-matrix (solid line) with the perturbative solution by the ST (dashed line) for the $l = 0$ case. It is seen that the ST breaks down at $z \approx 1$ ($\rho \approx \rho_W$) and has a $1/z^2$-singularity at small z. In distinction, the exact solution is bounded between 0 and 2 in accordance with the unitarity of the S-matrix. Both solutions coincide at large z where the parameter of perturbation expansion, $1/z$, becomes much smaller than unity. The oscillatory character of the exact solution in the region of small impact parameters is due to

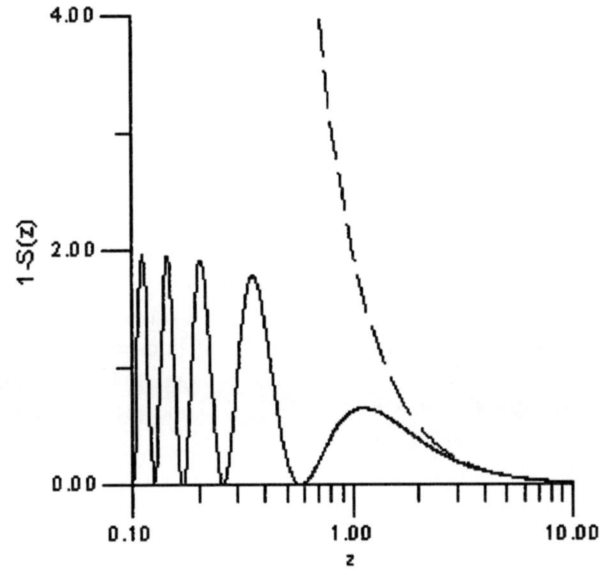

Fig. 15.1 Exact solution for the Lyα line from [11]: $1- < 21m|S|21m >_{\rm ang}$ vs. the scaled impact parameter $Z = \rho/\rho_{\rm W}^{\rm exact}$ (*solid line*). Here, the quantum numbers are those of the spherical quantization: $|n, l, m >$. The quantity $\rho_{\rm W}^{\rm exact}$ for any principal quantum number n is defined as $\rho_{\rm W}^{\rm exact} = 3n\hbar/(2m_{\rm e}v)$. This quantity arises naturally in the exact solutions from both paper [10] and paper [11]: it can be interpreted as the exact Weisskopf radius—in distinction to the Weisskopf radius of the ST defined only by the order of magnitude. In particular, for the Lyα line ($n = 2$): $\rho_{\rm W}^{\rm exact} = 3\hbar/(m_{\rm ev})$. The corresponding approximate result for $\{1- < 21m|S|21m >\}_{\rm ang}$ obtained by the perturbation theory in the ST is shown for comparison by the *dashed line*

the fact that the closer/stronger the collision, the more transitions between atomic sublevels it causes.

Finally, the integration over perturber velocities was performed in [11], by noting that the function $W_{\rm M}(v)/v$ (where $W_{\rm M}(v)$ is the Maxwell distribution) has its maximum at $v = (T_{\rm e}/m_{\rm e})^{1/2}$ and that

$$\int_0^\infty W_{\rm M}(v)v^{-1}{\rm d}v = \sqrt{\frac{2m_{\rm e}}{\pi T_{\rm e}}}. \tag{15.23}$$

Using this, the authors of [11] arrived at the following final expression for the impact operator:

$$\Phi_l = -[9(2\pi)^{1/2}n^2\hbar^2/(m_{\rm e}^{3/2}T_{\rm e}^{1/2})]g_l[2\rho_{\rm D}(m_{\rm e}T_{\rm e})^{1/2}/(3n\hbar)]. \tag{15.24}$$

At zero quasistatic electric field, the shape of the L_α line is the following Lorentzian:

$$I(\delta\omega) = (1/\pi)\gamma_l/[\gamma_1^2 + \delta\omega^2],\ \gamma_l = -{\rm Re}\Phi_l. \tag{15.25}$$

Thus, they obtained the exact analytical solution for a multiparticle, purely dynamic Stark profile.

At relatively small quasistatic fields F, such that $3nhF/(2m_e e) < 2\gamma_l$ (see, e.g., [13]), there appears also a forbidden component. At $F \gg 4m_e e\gamma_l/(3n\hbar)$, its Lorentzian profile is centered at $\Delta\omega = 0$ and has the HWHM $\gamma_0 = -\text{Re}\Phi_0$. It should be noted that in paper [13], it was considered that the ratio of widths of these two components was always about 3. In reality, the ratio $\gamma_0/\gamma_1 \approx 3$ corresponds only to the limit $Z = \rho_D/\rho_W \ll 1$. However, from (15.18), it can be shown that these two widths may equalize as Z becomes smaller than 10.

It is also worth mentioning that in papers [14, 15] it was shown that by averaging the binary result from [10] over the ensemble of collisions one can obtain an exact multiparticle profile of any hydrogenic line. The authors of [14, 15] performed such averaging explicitly for the case of the He II L_α line [15]. However, they did analytically only the angular part of the averaging—the rest of the averaging was done numerically.

15.3 Exact Analytical Solutions for Stark–Zeeman Effect in the Fields of an Arbitrary Strength and their Diagnostic Applications

First, we briefly discuss an analytical classical description of circular Rydberg states (CRS) in collinear electric (**F**) and magnetic (**B**) fields—following paper [16]. CRS of hydrogen-like systems correspond to $|m| = n - 1 \gg 1$, where m and n are magnetic and principal quantum numbers, respectively. CRS have been extensively studied both theoretically and experimentally for several reasons (see, e.g., [17–20] and references therein). First, CRS have long radiative lifetimes and highly anisotropic collision cross sections, thus enabling experimental works on inhibited spontaneous emission, cold Rydberg gases, etc. [21–23]. Second, classical CRS correspond to quantal coherent states that are objects of fundamental importance. Third, a classical description of CRS serves as the primary term in the quantal method based on the 1/n-expansion (see, e.g., [24] and references therein).

Paper [16] presented a study of CRS of an electron in the field of two stationary Coulomb centers (TCC) of charges Z and Z' separated by a distance R. The study was performed in terms of the following scaled quantities:

$$b \equiv Z'/Z,\ f \equiv FM^4/Z^3,\ \omega \equiv \Omega M^3/Z^2,\ \epsilon \equiv EM^2/Z^2. \tag{15.26}$$

Here, M = const is the projection of the angular momentum on the magnetic field **B**, $\Omega = B/(2c)$ is the Larmor frequency, a practical formula for which reads: $\Omega\,(\text{s}^{-1}) \approx 8.794 \cdot 10^6\, B(G)$.

One of the most important outcomes of paper [16] is an analytical dependence of the classical ionization threshold on the electric and magnetic fields. The electric

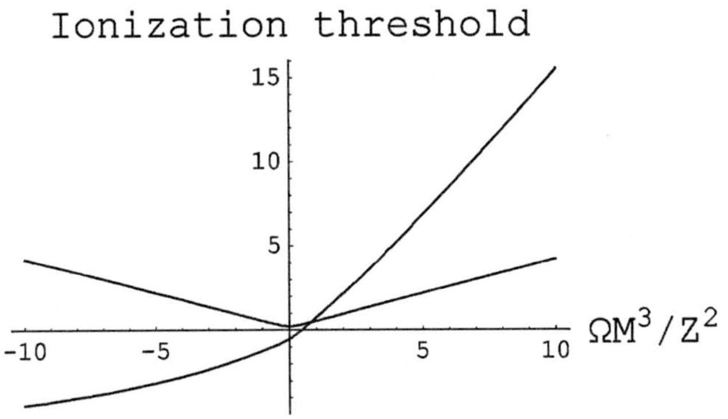

Fig. 15.2 Dependence of the critical value of the scaled electric field $f_c \equiv F_c M^4/Z^3$ at the classical ionization threshold (symmetric curve) and of the critical value of the scaled energy $\epsilon_c \equiv E_c M^2/Z^2$ at the classical ionization threshold (asymmetric curve) on the scaled magnetic field $\omega \equiv \Omega M^3/Z^2$

field increases the radius of the orbit and shifts the plane of the orbit. Thus, it works as a destabilizing factor: as the electric field reaches some critical value f_c, the motion becomes unstable, and the system gets ionized. The magnetic field decreases the radius of the orbit. Thus, it works as a stabilizing factor: the greater the magnetic field, the greater the electric field is required for reaching the classical ionization threshold.

Figure 15.2 shows the dependence the critical value of the scaled electric field $f_c \equiv F_c M^4/Z^3$ at the classical ionization threshold and of the critical value of the scaled energy $\epsilon_c \equiv E_c M^2/Z^2$ at the classical ionization threshold on the scaled magnetic field $\omega \equiv \Omega M^3/Z^2$.

The above results are practically important especially for such rapidly developing area of experimental research as cold Rydberg plasmas [25–27]. Indeed, in plasmas, including cold Rydberg plasmas, the intrinsic electric microfield causes a phenomenon of "continuum lowering" (discussed below), which significantly affects radiative properties of these media. Our results could be used as a theoretical basis for setting up experiments on a *magnetic control of the continuum lowering in cold Rydberg plasmas*.

Now, we briefly discuss paper [28] presenting an analytical classical description of CRS of a system consisting of a hydrogenic atom/ion of the nuclear charge Z under the field of the nearest plasma ion of the charge Z' and a magnetic field **B**. It is treated as the two-Coulomb-center system in a magnetic field **B**, the latter is considered to be parallel to the internuclear axis for simplicity. Thus the unperturbed system consists of two nuclei of charges Z and Z', separated by a distance R, and one electron and is denoted as ZeZ'. Analytical results for the electronic terms $E(R)$ of the ZeZ'-system for the field-free case were obtained in [40,41] from first principles within a purely classical approach. The classical approach reproduces [29,30] several electronic terms and two of these terms undergo a V-shape crossing at some separation $R*$, so that CRS cannot exist for $R < R*$.

In paper [28], an exact analytical classical solution was obtained for the electronic terms $E(R, B)$ for CRS of the ZeZ'-system in the presence of a magnetic field **B**. The solution is exact and is valid for any strength of the magnetic field. It allowed studying how the classical electronic terms are influenced by the magnetic field, including the case of a strong field. This is a fundamental problem in its own right. These results have also an important application to the phenomenon of the continuum lowering (CL) in plasmas.

The CL in plasmas was studied for almost half a century—see, e.g., books and reviews [1, 29, 30] and references therein. The CL plays a key role in calculations of the equation of state, partition function, bound-free opacities, and other collisional atomic transitions in plasmas (see, e.g., [31, 32] and references therein). The latest development was based on the employment of dicenter models of the plasma state [29, 33–38]. These models are more advanced than previously used ion sphere models (referred to in [1, 29, 30]).

Specific calculations of the CL on the basis of a dicenter model were presented in [29, 35] and called a "percolation theory of the CL." The starting point of this theory is the observation that a bound electron is localized within the volume of a given ion as long as its binding energy is well below the lowest potential barrier that separates its ion from the neighboring ones. When the energy approaches this lowest potential barrier, the electron may tunnel out from the ionic potential well. When this happens, its wave function overlaps two or more ions. Electronic states of increasingly higher energy overlap greater and greater clusters of ions. Above some critical value of the electron energy, which is noticeably higher than the top of the potential barrier U_{top} separating the radiating ion from its nearest neighbor, one of the clusters percolates, that is, the wave function of an electron having this or higher energy overlaps a macroscopic portion of the plasma ions. When this occurs, the electron can be regarded as a negative-energy continuum electron.

Apart from a questionable boundary condition used in this model (which was criticized in [37]), there is a serious conceptual flaw in this model. Indeed, at some value of energy slightly higher than U_{top}, the quasimolecule, consisting of the two ion centers plus an electron, can get ionized in a true sense of this word before the electron would be shared by more than two ions. In other words, the electron belonging to the dicenter can go directly into a true (positive-energy) continuum bypassing the multistage process of being shared by more and more ions.

Paper [39] presented a study of the ionization of a molecule consisting of two Coulomb Centers (TCC) plus an electron. In that paper there was derived analytically the value of the CL in this ionization channel for an arbitrary ratio of charges of the TCC at the absence of the magnetic field. (We note that specific values of the CL in the percolation theory [29, 35] were obtained only for dicenters consisting of two identical ions.) The results in paper [39] were obtained in a purely classical approach and derived from *first principles*. In particular, it was shown that, whether the electron is bound primarily by the smaller or by the larger out of two positive charges, makes a dramatic qualitative and quantitative difference for this ionization channel. The treatment in paper [39] was similar to previous papers [40, 41], where it was demonstrated that a paradigm, in which level crossings and

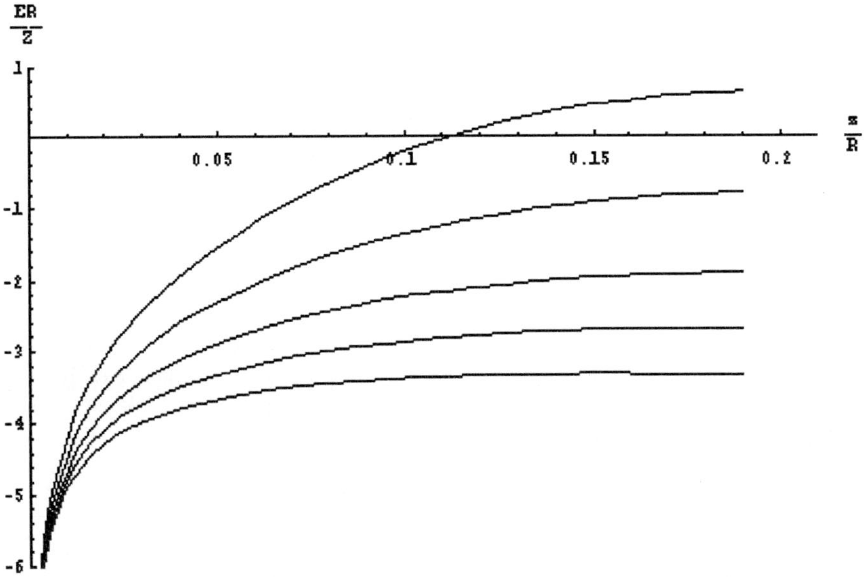

Fig. 15.3 The rescaled electronic energy ER/Z vs. the scaled equilibrium position z/R of the orbital plane for $b = 2$ at several values of the scaled magnetic field ω [28]. For each value of ω, only one electronic term is shown. The five curves, in order of the increasing energy, correspond to the following five values of the scaled magnetic field ω: 0, 2, 3, 4, and 5, respectively

charge exchange in plasmas were considered as inherently quantum phenomena, is generally incorrect and that these phenomena actually have classical roots.

Figure 15.3 shows the rescaled electronic energy ER/Z vs. the scaled equilibrium position $w_0 = z/R$ of the orbital plane for $b = 2$ at several value of the scaled magnetic field ω. For each value of ω, only one electronic term is shown in Fig. 15.3. The bottom curve in Fig. 15.3 corresponds to the absence of the magnetic field. At the point w_c, where the maximum occurs, the scaled energy $(\epsilon_0(w_c)$ differs from zero—it is negative. Thus, for the ionization of the hydrogen-like ion of the nuclear charge Z_{min} perturbed by the charge Z_{max}, it is sufficient to reach the scaled energy $\epsilon_0(w_c) < 0$—at that point, the electron switches to the unstable motion and the radius of its orbit increases without a limit. This constitutes a CL by the amount of $E = Z < 1/R > |\epsilon_0(w_c)|$, where $< 1/R >$ is the value of the inverse distance of the nearest neighbor ion from the radiating ion averaged over the ensemble of perturbing ions.

The next four curves in Fig. 15.3, in order of the increasing energy, correspond to the following four values of the scaled magnetic field ω: 2, 3, 4, and 5, respectively [28]. It is seen that as ω increases from 0 to 4, the value of $|\epsilon_0(w_c)|$ progressively decreases. This means that as the magnetic field increases, the value of the CL in the ionization channel decreases.

From Fig. 15.3, it is also seen that for $\omega = 5$, the value of $(0(wc))$ becomes positive. This means that there is no more CL in the ionization channel. Instead of getting lower, the continuum in this channel becomes higher than for the isolated hydrogen-like ion of the nuclear charge Z_{min}.

Thus, as the magnetic field increases from zero, it decreases the competitiveness of the CL in the ionization channel compared to the percolation channel. Eventually, as the magnetic field reaches some threshold, it takes the ionization channel totally out of the competition with the percolation channel.

We point out that plasmas having ions of different nuclear charges are quite common for a variety of applied projects. Some important examples are (1) laser fusion, where a thermonuclear fuel is a D–T mixture and/or where a low-Z thermonuclear fuel is doped by a high-Z material for diagnostic purposes; (2) plasmas resulting from a laser interaction with composite solid targets; (3) powerful Z-pinches used as plasma radiation sources for various applications. Finally, we note that this classical description of the problem can be generalized—by allowing for the Debye screening as well as by going beyond the circular states of the TCC problem.

15.4 Stark Broadening in Magnetic Fusion Plasmas Without Turbulence and Its Diagnostic Applications

All basic theoretical works on the Stark broadening can be divided in two groups: *analytical theories* and *simulation models*. As a preface to Sects. 15.4.1–15.4.3, we provide here just a brief description of these two groups. Further details and the corresponding references can be found, e.g., in books [1, 3, 12].

The most "user-friendly" are *semiclassical theories*. In the semiclassical theories, the radiator is described quantally, while the perturbing charges are described classically. Within the semiclassical theories, the simplest (and the oldest) is a so-called *Standard Theory* (ST)—already referred to in Sect. 15.2.3 above. In the ST, it is assumed that from the viewpoint of the radiator, the ion microfield (i.e., the electric field due to ionic perturbers) is quasistatic, while the electron microfield is treated dynamically in the so-called *impact approximation*. Roughly speaking, the impact approximation considers a sequence of binary collisions of the perturbing electrons with the radiator, the collisions being completed. The ST combines the impact approximation with an approximate solution of the time-dependent Schrödinger equation (for the radiating bound electron) obtained in the second (the first nonvanishing) order of the Dyson perturbation expansion. This Stark broadening theory can be extended to the limited-quantal formulation.

A more accurate version of the collisional theory of the Stark broadening is called the *unified theory*. Physically, the primary distinction of the unified theory from the ST is the allowance for incomplete collisions. The second distinction is that the unified formalism allows in principle a transition to the nearest-neighbor quasistatic result in the wings of the spectral line—this is a less important distinction because numerically the unified theory does not always yield such transition correctly.

It should be mentioned that mechanisms of spectral line broadening are also divided in two types: *homogeneous* and *inhomogeneous*. A particular broadening mechanism is *homogeneous* when it is the same for all radiators. A typical example is the Stark broadening by the electron microfield (within the ST). In distinction, the Stark broadening by the quasistatic part F_{qs} of the ion microfield is *inhomogeneous* because different radiators are subjected to generally different values of F_{qs}.

The most advanced, but still user-friendly semiclassical theory is called the *Generalized Theory* (GT). The GT has two versions. The more general version allows for the incomplete collisions and thus can be called the *generalized unified theory*. The simpler version considers only completed collisions and thus can be called the *generalized impact theory*. More details on the GT and the corresponding references are given in Sect. 15.4.2.

As for the simulation models, perhaps the oldest one is the *model microfield method*. In this method, the true dynamics of the electric microfield is modeled by choosing a simple approximate form of the field correlation function in such a way as to correctly fit the long and short time limits. Another simulation model was developed to meet the need of calculating hundreds of spectral line profiles at one time (as required for studies of opacities, detailed radiative transfer, etc.), in which case more rigorous calculations might be too time-consuming. It is called the *frequency fluctuation model*. It is based on the idea that fluctuation of the ion microfield during the radiative transition can be modeled by the process of the emission switching from one Stark component to another, resulting into a mixing of these components of the spectral line.

Lastly, among the simulation models we mention the approaches called the *molecular dynamics* and *independent quasiparticle* technique. They engage full-blown numerical simulations of the motion of plasma electrons and, separately, of plasma ions, surrounding the radiator. Through fully numerical solutions of the corresponding Schrödinger equations, they produce the simulated electron and ion microfields and calculate numerically their effects on the radiator.

These simulation techniques are useful for testing the results of other approaches for those ranges of plasma and radiator parameters where there are no results from benchmark experiments. However, most of the above simulation techniques do not take into account one of the major couplings between the electron and ion microfields. Second, as any fully numerical method, they lack the physical insight.

Besides, all simulation models have problems in calculating Stark shifts, especially in dense plasmas of relatively low temperatures (such plasmas are sometimes called "warm dense matter"). Moreover, the frequency fluctuation model cannot calculate Stark shifts at all.

The bottom line is that *semiclassical theories* are the most user-friendly and physically transparent. Their applications to magnetized plasmas are discussed in detail in subsequent sections.

Finally, we note that most of the semiclassical theories were developed in a so-called *no-coupling approximation*. This approximation amounts to the assumption that the electron microfield and the ion microfield are statistically independent from each other and that so are their effects on the radiator. In other words, in this approximation, first the motion of perturbing electrons is considered independent of the motion of perturbing ions. In reality, such a *direct coupling* between the electron and ion microfields does exist (this effect exists also in the multiparticle description of the ion microfield). It is represented by the acceleration of the perturbing electrons by the ion nearest to the radiator. The direct coupling results in a narrowing of the lines (as well as in a reduction of the shift), which becomes quite dramatic as the temperature T decreases and/or the density N_e increases.

Second, in the no-coupling approximation, it is assumed that the Stark profile due to electrons and the Stark profile due to the ions can be calculated independent of each other and that the combined Stark profile of a spectral line would be simply a convolution of these two. In reality, there is such an *indirect coupling* between the electron and ion broadenings, the coupling being carried out via the radiating atom/ion acting as an intermediary. An attempt to study this phenomenon analytically within the ST resulted only in a very weak (logarithmic) indirect coupling between the electron and ion microfields. Only with the development of the GT, it was demonstrated for the first time that the indirect coupling of the electron and ion microfields can be strong. The higher the electron density N_e and/or the principal quantum number n of the upper level of the radiator, the more important the indirect coupling becomes. The GT allows also for the direct coupling of the electron and ion microfields.

15.4.1 Stark Broadening of Intense (Low-n) Lines Under Weak (or Zero) Magnetic Fields

In plasmas of relatively low electron densities Ne and of relatively high temperatures T, the homogeneous Stark width of Hydrogen/Deuterium Spectral Lines (HDSL), originating from the levels of a relatively low principal quantum number n, is controlled by the dynamic part of the ion microfield. In the ST terminology, this requires that the number of perturbing particles within a sphere of a so-called Weisskopf radius should be much smaller than one. This number scales as $n^6 N_e / T^{3/2}$. Typically, at plasma temperatures of few electron volt, this situation occurs for α, β and sometimes γ-lines of the Lyman and Balmer series at electron densities below $N_e \sim 10^{15}\,\mathrm{cm}^{-3}$. As long as magnetic fields are much smaller than 1 T, the Stark broadening is a predominant factor.

For the situation, where the broadening is due to just one sort of charged particles and can be treated in the impact approximation, a significant analytical advance (in the no-coupling approximation) was presented in papers [42–44]. For a long time, it was generally accepted that within both the ST and the unified theory, the Stark profile of a hydrogenic spectral line, caused by either the impact electrons at the absence of the ion microfield or by ions if they can be treated in the impact approximation, is much more complicated than a superposition of Lorentzians, corresponding to Stark components of the line. Contrary to this belief, the authors of [42–44] found that the Stark profile in this case actually *reduces to a single Lorentzian* and derived a simple analytical result for the Stark width of an arbitrary hydrogenic line in this case. This important result was totally counter intuitive. The corresponding expression for the half-width at the half-maximum (HWHM), caused by perturbers of a sort p, has the form

$$(\mathrm{HWHM})_\mathrm{p} = \left(2N_\mathrm{p}Z_\mathrm{p}^2 a_0^2 I_\mathrm{H}/m_\mathrm{e}\right)\left(2\pi\mu_\mathrm{p}/T_\mathrm{p}\right)^{1/2} k_{nn'}(b_{\max}, b_\mathrm{c}), \tag{15.27}$$

where a_0 is the Bohr radius, I_H is the hydrogen ionization potential, μ_p is the reduced mass of the perturber–radiator pair, b_c is the strong collision cutoff (for the impact parameters) defined as

$$b_\mathrm{c} = a_0 Z_\mathrm{p}\left[\mu_\mathrm{p} I_\mathrm{H} k_{nn'}/(m_\mathrm{e}T_\mathrm{p})\right], \tag{15.28}$$

and the function f has the form (for rectilinear trajectories, i.e., for H-lines):

$$f(b_{\max}, b_\mathrm{c}) = 1 - C + 2\ln(b_{\max}/b_\mathrm{c}). \tag{15.29}$$

Here, $b_{\max}$ is the Debye radius for perturbers of the sort p and C is the Euler constant. (For hyperbolic trajectories, i.e., for HL-lines, a more complicated expression for the function f is given by (23) of paper [60].) The primary achievement by the authors of [42–44] was an analytical result for the quantity $k_{nn'}$ appearing in (15.27) and (15.28):

$$k_{nn'} = \left[3/(2Z_r^2)\right]\left[n^4 + n'^4 - 2n^2n'^2 - n^2 - n'^2\right], \tag{15.30}$$

where n and n' are the principal quantum numbers of the upper and lower levels, respectively.

15.4.2 Stark Broadening of Intense (Low-n) Lines Under Strong Magnetic Fields

In the conditions typical for the edge plasmas of magnetic fusion devices (e.g., in the divertor region of tokamaks), the homogeneous Stark width of HDSL is controlled by the dynamic part of the ion microfield—as already noted in Sect. 15.4.1. A strong magnetic field **B** (up to ~10 T) in such plasmas significantly affects shapes of HDSL and thus provides opportunities for spectroscopic diagnostics.

Paper [45] presented a study of this situation based on the most advanced semiclassical theory of the Stark broadening: the generalized theory (GT). The GT is based primarily on a *generalization* of the formalism of dressed atomic states (DAS) in plasmas. DAS is the formalism initially designed to describe the interaction of a *monochromatic* (or quasimonochromatic) field, e.g., laser or maser radiation—with gases. Later, it was applied for the interaction of a laser or maser radiation with plasmas [46]. The employment of DAS led to the enhancement of the accuracy of analytical calculations and to more robust codes.

The *generalization* of DAS is based on using atomic states dressed by a *broadband* field of plasma electrons or ions [3, 47–49]. Therefore, generalized DAS is a more complicated concept than the usual DAS, where the dressing was due to a monochromatic field.

The employment of the generalized DAS in the GT allowed describing analytically various *couplings*, such as, e.g., the coupling between the dynamic electron microfield and the quasistatic part of the ion microfield or the coupling between the dynamic part of the ion microfield and the external static magnetic field. In both cases the coupling is facilitated by the radiating atom/ion—therefore, it is called *indirect coupling*. The indirect coupling between the dynamic part of the ion microfield and the external static magnetic field *increases with the principal quantum number n* and *with the magnetic field* **B**. Besides, it *increases as the temperature T decreases*.

The GT allows analytically in the exact way (in all orders) for the component of the dynamic plasma microfield parallel to the additional static (electric or magnetic) field. Thus the GT made a significant advance compared to the simplest semiclassical theory of the dynamical Stark broadening—the ST—since the ST allowed for the same component of the dynamic plasma microfield only in the second order of the Dyson perturbation expansion. In distinction to the ST, the GT is not divergent at small impact parameters. In paper [50], it was shown analytically that for the overwhelming majority of hydrogenic lines the GT does not violate the unitarity of the S-matrix at any impact parameter and therefore does not have to separate collisions into "weak" and "strong"—in distinction to the ST. The latter has to separate collisions into "weak" and "strong" for all hydrogenic lines (to avoid the divergence) and defines the boundary between the "weak" and "strong" collisions only by the order of magnitude. Only for few hydrogenic lines (e.g., for L_α, and to a lesser extent for L_β and H_α) the GT might violate the unitarity at small impact parameters (as discussed in paper [48]) and could use the separation into "weak" and "strong" collisions for enhancing the accuracy. More details on this are given below.

In semiclassical theories of the dynamical Stark broadening of *nonhydrogenic* spectral lines by electron or ion microfields, the broadening is controlled by *virtual transitions* between different sublevels characterized by the same principal quantum number *n* (see, e.g., book [12]). This is a *nonadiabatic* contribution—since it involves different sublevels. For hydrogenic spectral lines (including HDSL), under the usual assumption that the fine structure can be neglected, in addition to the nonadiabatic contribution there is also a significant *adiabatic* contribution—see, e.g., review [2] and book [3]. Physically, the adiabatic contribution is due to the fact that for hydrogenic spectral lines, the overwhelming majority of sublevels have permanent electric dipole moments (which is especially clear in the parabolic quantization). Even an isolated sublevel having a permanent electric dipole moment would produce some adiabatic contribution—in other words, the latter does not require virtual transitions between different sublevels. Classically, the adiabatic contribution originates from the phase modulation of the atomic oscillator by the dynamic microfield of electrons or ions (as already noted in Sect. 15.2.2).

In the GT, contributions to the dynamical Stark width due to ions are explicitly separated into adiabatic and nonadibatic. The main result of the GT for magnetic fusion plasmas is the following. At values of the magnetic field **B** typical for magnetic fusion plasmas, practically the entire dynamical Stark width due to ions is due only to the *adiabatic* contribution. This is because, as the magnetic field B increases, causing the increase of the separation ω_B between the Zeeman sublevels of hydrogenic atoms, the nonadiabatic contribution to the dynamical Stark broadening by ions decreases—specifically, it decreases dramatically at magnetic fields typical for magnetic fusion plasmas.

In accordance to [3, 47–49], the adiabatic contribution to the dynamical Stark width due to ions γ_{ad} has the following form, which is the *exact, nonperturbative* analytical result:

$$\gamma_{ad} = 18(\hbar/m_e)^2(X_{\alpha\beta}/Z_r)^2 Z_i^2 N_i (2\pi M/T_i)^{1/2} I(R_i), \quad X_{\alpha\beta} = |n_\alpha q_\alpha - n_\beta q_\beta|. \tag{15.31}$$

Here, Z_i, N_i, and T_i are the charge, the density, and the temperature of the plasma ions, respectively; M is the reduced mass of the pair "radiator–perturbing ion"; the function $I(R_i)$ is defined as follows:

$$I(R_i) = \left\{R_i^2[3 - \cos(1/R_i)] + (R_i - 2R_i^3)\sin(1/R_i) - \mathrm{ci}(1/R_i)\right\}/6. \tag{15.32}$$

In (15.32), ci(1/R$_i$) is the cosine integral function, the quantity R_i being

$$R_i = r_D/r_{Wa}, \tag{15.33}$$

where

$$r_D = \left[T_e/\left(4\pi e^2 N_e\right)\right]^{1/2} = 743.40\left[T_e(\mathrm{eV})/N_e\left(\mathrm{cm}^{-3}\right)\right]^{1/2}, \tag{15.34}$$

is the Debye radius and

$$r_{Wa} = 3X_{\alpha\beta}\hbar/(Z_r m_e v_i) = 3.5486\times10^{-6}(X_{\alpha\beta}/Z_r)(M/M_p)^{1/2}/[T_i(\mathrm{eV})]^{1/2},\ \mathrm{cm} \tag{15.35}$$

is the adiabatic Weisskopf radius (M_p is the proton mass). The quantity r_{Wa} naturally arises in the GT with the exact coefficient given in (15.33)—in distinction to the Weisskopf radius of the ST defined only by the order of magnitude. The practical part of formula (15.33) was obtained using the fact that the averaging over ion velocities is performed with the effective statistical weight factor $W_M(v_i)/v_i$, where $W_M(v_i)$ is the Maxwell distribution, and that $W_M(v_i)/v_i$ has the maximum at $v_i = (T_i/M)^{1/2}$.

In 2009, a group of authors published a paper [51], where they revisited the subject studied in paper [45] in 1994: the dynamical Stark broadening of hydrogen/deuterium lines by ions in magnetized plasmas. Paper [51] presented some analytical results and some simulations. The authors of paper [51] based their analytical results on the ST (despite they knew that a more advanced study—based on the GT—has been already published 15 years earlier). Below we show that the analytical results from [51] yield a very dramatic inaccuracy—up to two orders of magnitude.

For comparing the adiabatic Stark widths of the GT and of the ST, it is convenient to introduce the adiabatic broadening cross-section $\sigma_a(v_i)$ related to the adiabatic width γ_a as follows:

$$\gamma_a = N_i \int_0^{\infty} dv_i W(v_i) v_i \sigma_a(v_i), \tag{15.36}$$

where $W(v_i)$ is the velocity distribution. In the GT, the adiabatic broadening cross-section $\sigma_{aGT}(v_i)$ is

$$\sigma_{aGT}(v_i) = 2\pi [r_{Wa}(v_i)]^2 I[R_i(v_i)], \tag{15.37}$$

where $I[R_i(v_i)]$ is given by (15.32) and $r_{Wa}(v_i)$ is given by the first equality in (15.35). This is the exact analytical result equivalent to the summation of *all orders* of the Dyson perturbation expansion.

In paper [51], based on the *second order* of the Dyson perturbation expansion of the ST, the adiabatic broadening cross section $\sigma_{aRos}(v_i)$ is (the subscript Ros is the abbreviated name of the first author, Rosato, of paper [51]):

$$\begin{aligned}\sigma_{aRos}(v_i) &= \pi \left[r_{str}(v_i)\right]^2 + 2\pi \left[r_{W,Ros(v_i)}\right]^2 ln[r_D/r_{str}(v_i)] \\ &\approx 2\pi \left[r_{W,Ros}(v_i)\right]^2 \{1/2 + ln[r_D/r_{W,Ros}(v_i)\}, \end{aligned} \tag{15.38}$$

where $r_{str}(v_i)$ is a so-called "strong collision radius" (i.e., the boundary between weak and strong collisions); $r_{str}(v_i) \approx r_{W,Ros}(v_i)$ for the adiabatic contribution. The Weisskopf radius in the ST is defined only by the order of magnitude (which is one of the major sources of the inaccuracy of the ST): it is $\sim n_\alpha^2 \hbar/(Z_r m_e v_i)$. The authors of paper [51] arbitrarily chose the following numerical coefficient in the Weisskopf radius of the ST:

$$r_{W,Ros}(v_i) = (2/3)^{1/2} n_\alpha^2/(m_e v_i). \tag{15.39}$$

We note that they set $Z_r = 1$ because their work was limited to HDSL. Therefore, in the comparison below, we also set $Z_r = 1$.

We denote the ratio of the adiabatic broadening cross sections as follows:

$$\kappa = \sigma_{aRos}/\sigma_{aGT}. \tag{15.40}$$

Below, we provide examples of the values of the ratio κ for several hydrogen lines in a hydrogen plasma(so that $M = M_p/2$) at the conditions typical for tokamak divertors. The components of a particular line are identified by the parabolic quantum numbers: $(n_1 n_2 m)_\alpha - (n_1 n_2 m)_\beta$; we also indicate the polarization of the component (π or σ). The ratio κ is calculated at $T = 4\,\text{eV}$ and $N_e = (1-3) \times 10^{13}\,\text{cm}^{-3}$.

For the Paschen-alpha line, for the component (102)–(101), which is one of the two most intense σ-components: $\kappa = 80$.

For the Balmer-gamma line, for the intense σ-component (220)–(010): $\kappa = 50$.

For the Balmer-alpha line, for the component (101)–(100), which is one of the two most intense σ-components: $\kappa = 30$.

The above shows that the *authors of paper [51] overestimated the primary, adiabatic contribution to the dynamical Stark broadening by ions in magnetic fusion plasmas by up to two orders of magnitude*.

For completeness, let us now discuss also the secondary, nonadiabatic contribution. The nonadiabatic contribution to the dynamical Stark width due to ions $\gamma_{\rm na}$, calculated for magnetic fusion plasmas using the GT, has a more complicated form than (15.31), as can be seen from (5) to (8) of paper [45] or (4.4.5)–(4.4.6) of book [3]. It is controlled by the integral of a so-called width function $A_{-}(\chi, Y, Z)$ over scaled (dimensionless) impact parameters Z:

$$a_{-}(\chi, Y, Z_{\rm D}) = \int_0^{Z_{\rm D}} {\rm d}Z A_{-}(\chi, Y, Z)/Z. \tag{15.41}$$

The scaled impact parameter Z is defined as

$$Z(\rho) = \rho/\rho_{\rm B} = 2m_{\rm e}cv\rho/(eB). \tag{15.42}$$

The upper limit of the integration in (15.41) is $Z_{\rm D} = Z(r_{\rm D})$, where $r_{\rm D}$ is the Debye radius. Typically, $r_{\rm D} > r_{\rm Wa}$, which is assumed in (15.41).

Compared to the ST, there are two new parameters that enter the width function. The first one χ stands for

$$\chi = (n_\alpha q_\alpha - n_\beta q_\beta)/n_\alpha. \tag{15.43}$$

The second new parameter Y is physically the most important: it is a coupling parameter defined as

$$Y = 3n_\nu Z_{\rm i}\hbar eB/(2m_{\rm e}^2 cv_{\rm i}^2) = 0.31885 n_\nu Z_{\rm i}(M/M_{\rm p})B(T)/T_{\rm i}({\rm eV}), \quad \nu = \alpha \text{ or } \beta, \tag{15.44}$$

where $Z_{\rm i}$ is the charge of the plasma ions.

For example, for the D_α or L_β line ($n_\nu = 3$) from a deuterium plasma ($M/M_{\rm p} = 1, Z_{\rm i} = 1$), (15.44) yields: $Y = 0.95655B(T)/T_{\rm i}({\rm eV})$. For the typical parameters of tokamak divertors, the ratio $B(T)/T_{\rm i}({\rm eV})$ is greater or of the order of unity, so that Y is also greater or of the order of unity. At these values of the coupling parameter, first, the ST becomes quite inaccurate, and second (but most importantly), there occurs a dramatic decrease of the nonadiabatic contribution to the dynamical Stark width due to ions. Thus, the total contribution to the dynamical Stark width due to ions can be well represented by the adiabatic contribution given by (15.31).

The finding, that the nonadiabatic contribution significantly decreases with the increase of the magnetic field, was quite clear already in 1994: from the results of paper [45] (where (15.41)–(15.44) were first presented) complemented by the results of paper [47] (where it was shown that the function $a_{-}(\chi, Y, Z_{\mathrm{D}})$, controlling the nonadiabatic contribution, significantly decreases with the increase of the coupling parameter Y). Therefore, the claim by the authors of paper [51] that they were the first to "discover" this effect in their paper [51] published in 2009 is without merit.

Finally, let us discuss the relation between the unitarity of the S-matrix and the nonadibatic contribution calculated by the GT or by the ST. In both theories, the nonadibatic contribution is calculated via $\{1 - S_{\mathrm{na}}\}_{\mathrm{ang}}$, which is the angular average of the nonadiabatic part of the S-matrix. It is calculated up to the second order of the Dyson perturbation expansion, but using the different basis: the basis of the dressed atomic states in the GT as opposed to the usual atomic basis in the ST. At small impact parameters, the ST would violate the unitarity of the S-matrix. To avoid the violation, the ST has to separate collisions into "weak" and "strong," the boundary between them being defined from the condition:

$$|\{1 - S_{\mathrm{na}}\}_{\mathrm{ang}}| = C, \ 0 < C < 2. \tag{15.45}$$

The uncertainty in the choice of the constant C in (15.45) is yet another major source of inaccuracy of the ST.

In paper [50], it was shown analytically that for the overwhelming majority of hydrogenic spectral lines, the nonadiabatic contribution calculated by the GT does not violate the unitarity of the S-matrix—in distinction to the ST (as already noted above). Therefore, for the overwhelming majority of hydrogenic spectral lines the lower limit of the integration in (15.41) can remain to be zero.

As an illustration of this important distinction between the GT and ST, we present Fig. 15.4. For the most intense π-component (400)–(100) of the Balmer-gamma line, Fig. 15.4 shows the dependence of the integrand A_{-}/Z in (15.41) vs. the scaled impact parameter Z: by the GT (solid curve) and by the ST (dashed curve). The solid curve is calculated by the GT for the coupling parameter $Y = 0.85$, which corresponds, e.g., to $B = 4\,\mathrm{T}$ and $T = 1.5\,\mathrm{eV}$, or $B = 6\,\mathrm{T}$ and $T = 2.25\,\mathrm{ev}$, or $B = 8\,\mathrm{T}$ and $T = 3\,\mathrm{eV}$. Two possible unitarity restrictions are presented by straight lines. The solid straight line corresponds to the choice $C = 1$ in (15.45), the dashed straight line corresponds to the choice $C = 2$ in (15.45).

Figure 15.4 clearly demonstrates the following:

1. The ST violates the unitarity of the S-matrix and has to separate collisions into weak and strong at the value of Z somewhere between 0.8 and 1.
2. The GT does not need to engage the unitarity cutoff: the integrand A_{-}/Z strongly oscillates at small Z and thus practically "kills" the contribution from the small impact parameters to the integral.
3. Even after engaging the unitarity cutoff, the ST significantly overestimates the nonadiabatic contribution—by several times (in addition to dramatically overestimating the adiabatic contribution by up to two orders of magnitude).

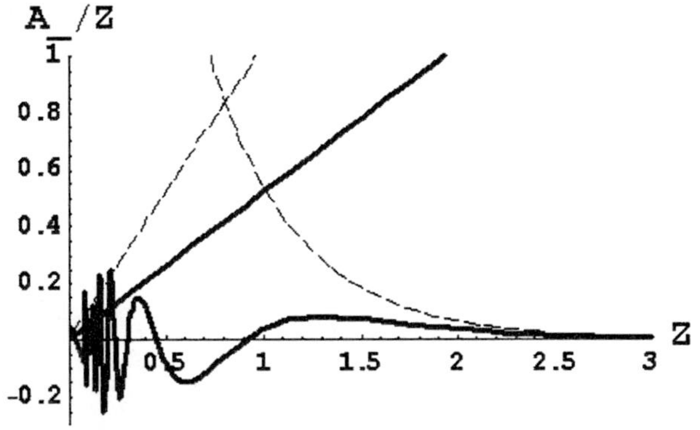

Fig. 15.4 Dependence of the integrand A_/Z in (15.41) vs. the scaled impact parameter Z: by the GT (*solid curve*) and by the ST (*dashed curve*). The *solid curve* is calculated by the GT for the coupling parameter $Y = 0.85$, which corresponds, e.g., to $B = 4\,\mathrm{T}$ and $T = 1.5\,\mathrm{eV}$, or $B = 6\,\mathrm{T}$ and $T = 2.25\,\mathrm{ev}$, or $B = 8\,\mathrm{T}$ and $T = 3\,\mathrm{eV}$. Two possible unitarity restrictions are presented by *straight lines*. The *solid straight line* corresponds to the choice $C = 1$ in (15.45), the *dashed straight line* corresponds to the choice $C = 2$ in (15.45). The entire illustration is for the most intense π-component (400)–(100) of the Balmer-gamma line

The fact that at the presence of an additional static field, the width function of the GT exhibits oscillations at small impact parameters, just like in the exact solution at the absence of the additional static field, can be explained as follows. At small impact parameters, at the closest approach of the perturber to the radiator, the interaction with the electric field of the perturber is much stronger than the interaction with an additional static field, which makes the situation similar to the case where the additional static field is absent. The oscillatory character of the exact solution in the region of small impact parameters in both situations is due to the fact that the closer/stronger the collision, the more transitions between atomic sublevels it causes.

We note that for the Lyman-alpha line, *as an exception*, the GT might need to engage the unitarity restriction and therefore separate collisions into weak and strong. Figure 15.5 presents the plot for the same conditions as in Fig. 15.4, but for the σ-components of the Lyman-alpha line: (001)–(000), (00-1)–(000). Figure 15.5 shows that for the chosen plasma conditions, both the ST and the GT need engaging the unitarity cutoff. It shows also that, after engaging the unitarity cutoff for both theories, the ST still overestimates the nonadiabatic contribution to the broadening, though only slightly.

It should be emphasized that the Lyman-alpha line has zero or little practical importance for diagnostics of magnetic fusion plasmas because additional broadening mechanisms (opacity broadening and/or Doppler broadening) would usually dominate over the Stark broadening of this line. Therefore, the fact that for the Lyman-alpha line, as an exception, the GT might need engaging the unitarity cutoff (just like the ST) for describing the secondary, nonadiabatic contribution to the broadening has zero or little practical importance.

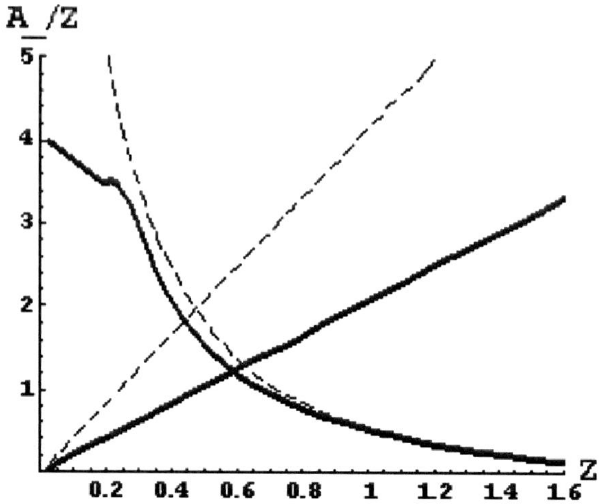

Fig. 15.5 The same as in Fig. 15.4, but for the σ-components of the Lyman-alpha line: (001)–(000), (00-1)–(000)

15.4.3 Stark Broadening of Highly Excited Lines ($n \gg 1$)

Experimental profiles of highly excited HDSL of Balmer and Paschen series are used for diagnostics of tokamak plasmas [52–54]. From the theoretical point of view, the ion dynamics becomes less and less important as the principal quantum number increases. The interpretation of the early experiments [52, 60], where high-n HDSL were employed, was done within the ST—thus assuming the Stark broadening by ions to be quasistatic. However, for typical conditions of tokamak divertors, this assumption is valid only for the HDSL originating from levels of $n > 8$, but needs to be corrected for the HDSL originating from slightly lower levels. The latter has been recognized in later simulations [55] and analytical approaches [56]. As the principal quantum number approaches approximately 13, a part of the electron microfield becomes quasistatic, and for the dynamical part, one has to account for incomplete collisions in spirit of the unified theory. The corresponding simulations were presented in [57]. More recently, some approximate analytical methods for faster calculations of the profiles of high-n HDSL were presented in paper [58].

The Stark broadening gets significantly enhanced with the increase of the principal quantum number n: its approximate scaling is n^4 in the impact regime or n^2 in the quasistatic regime. Therefore, magnetic fields usually have only a very minor effect on the experimental HDSL in the conditions of tokamak divertors.

Figure 15.6 shows the experimental FWHM of the deuterium Balmer lines, originating from the levels of $n = 6, 7, 8, 9$, and 10, observed from the divertor region of the tokamak DIII -D [54]. Fits by Stark broadening theories are shown for the generalized theory, GT (called "advanced theory" in the legend) with $N_e = 5.0 \times 10^{20}\,\mathrm{m}^{-3}$ and $T = 1.5\,\mathrm{eV}$ (solid line), and the standard theory, ST (called "conventional theory" in the legend) with $N_e = 3.9 \times 10^{20}\,\mathrm{m}^{-3}$ (dashed line) and $N_e = 5.1 \times 10^{20}\,\mathrm{m}^{-3}$ (dotted line).

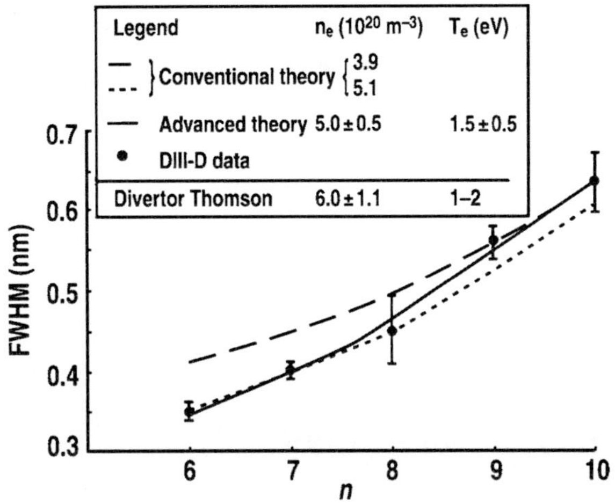

Fig. 15.6 Experimental full width at half-maximum (FWHM) of the deuterium Balmer lines, originating from the levels of $n = 6, 7, 8, 9$, and 10, observed from the divertor region of the tokamak DIII -D [54]. Fits by Stark broadening theories are shown for: the generalized theory (called "advanced theory" in the legend) with $N_e = 5.0 \times 10^{20}\,\mathrm{m}^{-3}$ and $T = 1.5\,\mathrm{eV}$ (*solid line*), and the standard theory (called "conventional theory" in the legend) with $N_e = 3.9 \times 10^{20}\,\mathrm{m}^{-3}$ (*dashed line*) and $N_e = 5.1 \times 10^{20}\,\mathrm{m}^{-3}$ (*dotted line*)

From Fig. 15.6, it is seen that the GT more accurately reproduces the shape of the experimental FWHM dependence than the ST. The shape of the FWHM dependency in the GT is more complicated than that in the ST because of the competition of two additional effects included in the GT, but neglected in the ST, namely, the indirect coupling and the ion-dynamical broadening. Broadening due to the indirect coupling increases with n, while that due to ion-dynamical broadening decreases with n.

The theoretical results presented in Fig. 15.6 were obtained from the best fit to the shapes of the Balmer lines observed in the experiment [54]. The best fit by the ST yields $N_e = (4.5 \pm 0.6) \times 10^{20}\,\mathrm{m}^{-3}$, but no information about the temperature. The most probable values of the densities, deduced from individual line profiles fitted by the ST, vary by over 15%. The best fit by the GT yields $N_e = (5.0 \pm 0.5) \times 10^{20}\,\mathrm{m}^{-3}$ and the temperature $T = (1.5 \pm 0.5)\,\mathrm{eV}$. The most probable values of the densities, deduced from individual line profiles fitted by the GT, vary by less than 5%.

The plasma parameters deduced from the Stark broadening of the experimental high-n Balmer lines have been compared in [54] with the corresponding parameters determined from the experimental Thomson scattering data: $N_e = (6.0 \pm 1.1) \times 10^{20}\,\mathrm{m}^{-3}$ and $T = (1 - 2)\,\mathrm{eV}$. It is seen that the latter agrees with the plasma parameters deduced by the GT better than with the plasma parameters deduced by the ST. Besides, the analysis employing the GT allows obtaining both and while the ST can determine only Ne and with a much higher error than the Ne deduced via the GT.

15.5 Stark Broadening in Turbulent Magnetic Fusion Plasmas and Its Diagnostic Applications

Electrostatic turbulence frequently occurs in various kinds of laboratory and astrophysical plasmas [59, 60]. It affects transport phenomena in magnetized plasmas [61]. It is represented by oscillatory electric fields (OEFs) sometimes called also collective electric fields: they correspond to collective degrees of freedom in plasmas—in distinction to the electron and ion microfield that correspond to individual degrees of freedom of charged particles.

15.5.1 *Low-Frequency Electrostatic Turbulence*

At the absence of a magnetic field, there is only one type of a low-frequency electrostatic turbulence: ion acoustic waves—frequently called ionic sound. The corresponding OEF is a broadband field, whose frequency spectrum is below or of the order of the ion plasma frequency:

$$\omega_{\mathrm{pi}} = (4\pi e^2 N_{\mathrm{i}} Z^2/m_{\mathrm{i}})^{1/2} = 1.32 \times 10^3 Z(N_{\mathrm{i}} m_{\mathrm{p}}/m_{\mathrm{i}})^{1/2}, \tag{15.46}$$

where N_{i} is the ion density, Z is the charge state; m_{p} and m_{i} are the proton and ion masses, respectively. In the "practical" parts of (15.46)–(15.49), CGS units are used.

In magnetized plasmas in addition to the ionic sound, propagating along the magnetic field **B**, two other types of low-frequency electrostatic turbulence are possible. One is *electrostatic ion cyclotron wave*, whose wave vector is nearly perpendicular to **B**. Its frequency is close to the ion cyclotron frequency:

$$\omega_{\mathrm{ci}} = ZeB/(m_{\mathrm{i}}c) = 9.58 \times 10^3 ZB(m_{\mathrm{p}}/m_{\mathrm{i}})^{1/2}. \tag{15.47}$$

Another type is lower hybrid oscillations having the wave vector perpendicular to **B**. Its frequency is

$$\omega = 1/\left[(\omega_{\mathrm{ci}}\omega_{\mathrm{ce}})^{-1} + (\omega_{\mathrm{pi}})^{-2}\right]^{1/2}, \tag{15.48}$$

where ω_{ce} is the electron cyclotron frequency

$$\omega_{\mathrm{ci}} = eB/(m_{\mathrm{e}}c) = 1.76 \times 10^7 B. \tag{15.49}$$

From (15.48), it is seen that $\omega < \omega_{\mathrm{pi}}$ always. This means that frequencies of both ionic sound and lower hybrid oscillations are below or of the order of the ion plasma frequency ω_{pi}.

It is usually assumed that hydrogenic radiators perceive OEFs, associated with a low-frequency plasma turbulence as *quasistatic*. Let us discuss this assumption in more detail. The discussion is based on papers [62, 63], the results of which were recently summarized in paper [64].

The physics of the spectral line broadening in plasmas containing OEFs is very rich and complex due to the interplay of a large number of characteristic times and frequencies. There are seven characteristic frequencies, which can be considered as "elementary" parameters. For our specific discussion here, the following four frequencies are important: (1) $\Delta\omega$—detuning from the unperturbed position of a given spectral line of the radiator (it affects the characteristic value of the argument τ of the correlation function); (2) ω—QEF frequency; (3) γ—homogeneous width of the power spectrum of OEF, which is also the inverse of the QEF *coherence time* τ_F; (4) $\delta_s(E_0)$—instantaneous Stark shift at the amplitude value E_0 of OEF—for example, $\delta_s(E_0) = a_1 E_0$ in the case of the linear Stark effect or $\delta_s(E_0) = a_2 E_0^2$ in the case of the quadratic Stark effect ($a_1(k), a_2(k)$ are Stark constants that depend on the set of quantum numbers of the particular states of the radiator). *Here and below, the set of quantum numbers is denoted by k.*

On the basis of the above "elementary" frequencies, there occur two composite parameters that are characteristic times as follows.

1. $\tau_{QS}(k, E_0, \omega)$—characteristic time of the formation of quasienergy states (QS):

$$\tau_{QS}(k, E_0, \omega) \sim \min(1/(\omega^2 \delta_s)^{1/3}, 1/\omega). \tag{15.50}$$

   Being subjected to OEF, the states of the radiator can oscillate with the OEF frequency ω. This effect is described as the emergence of QS, which were introduced in 1967 in papers [65] and [66] (independently of each other). The above formula for $\tau_{QS}(k, E_0, \omega)$ was derived in paper [62]. So, for relatively weak OEF, the QS are formed at the timescale of the order of the period of the OEF $1/\omega$. However, for relatively strong OEF, the QS are formed at a much shorter time scale proportional to $1/E_0^{1/3}$ or to $1/E_0^{2/3}$ in the cases of the linear or quadratic Stark effect, respectively.

2. $\tau_{\text{life}}(k, N_e, T_e, N_i, T_i, \gamma, \omega, E_0, \Delta\omega)$—the lifetime of the exited state of the radiator:

$$\tau_{\text{life}}(k, N_e, T_e, N_i, T_i, \gamma, \omega, E_0, \Delta\omega) \sim 1/\Gamma, \tag{15.51}$$

$$\Gamma = \gamma_e(k, N_e, T_e, \Delta\omega) + \gamma_i(k, N_i, T_i, \Delta\omega) + \gamma_F(k, \gamma, \omega, E_0). \tag{15.52}$$

   Here, Γ is the sum of the *homogeneous* Stark widths due to electrons, dynamic part of ions, and OEF. The contribution $\gamma_F(k, \gamma, \omega, E_0)$ from OEF was calculated in paper [67].

   A criterion for OEF to be considered as quasistatic is the following:

$$\tau_{QS}(k, E_0, \omega) \gg \min[1/\delta_s(E_0), 1/\Delta\omega, \tau_{\text{life}}(k, N_e, T_e, N_i, T_i, \gamma, \omega, E_0, \Delta\omega)]. \tag{15.53}$$

For Balmer and Paschen lines emitted from magnetic fusion plasmas, the condition (15.53) is usually fulfilled for OEFs of low-frequency electrostatic turbulence. Therefore, from the spectroscopic point of view, one deals here with a hydrogenic atom/ion in crossed static electric (**F**) and magnetic (**B**) fields—the problem allowing an exact solution (presented above in Sect. 15.2.1) for each value of **F** and **B**. If the emission is observed from a relatively small volume, within which the magnetic field can be considered homogeneous, then the spectral line profiles can be obtained by averaging the solution from Sect. 15.2.1 over the ensemble distribution $W(\mathbf{F})$ of the quasistatic field **F**. In other words, the key part of the problem becomes the calculation of $W(\mathbf{F})$.

In magnetic fusion plasmas, typically Zeeman splitting in the field **B** is much greater than Stark splitting in the Lorentz field $\mathbf{v} \times \mathbf{B}/c$ because the following condition is fulfilled:

$$v_{T_a} \ll c/(205.5n), \tag{15.54}$$

where v_{T_a} is the thermal velocity of the radiating atoms; n is the principal quantum number of the upper level, from which the spectral line originates. Equation (15.54) can be rewritten in terms of the temperature T_a of the radiating atoms as follows:

$$T_a \ll (104/n^2)\,\mathrm{eV}. \tag{15.55}$$

Therefore, the distribution $W(\mathbf{F})$ of the quasistatic fields is either the distribution $W_t(\mathbf{E}_t)$ of the turbulent fields or a convolution of $W_t(\mathbf{E}_t)$ with the distribution $W_t(\mathbf{F}_t)$ of the ion microfield, if the latter is also quasistatic.

The situation, where the turbulent field is quasistatic, but the ion microfield is not quasistatic, is possible even if their frequencies are of the same order of magnitude—as long as the average turbulent field E_0 is much greater than the characteristic ion microfield $<F_i>$. Indeed, if the average amplitude E_0 of the OEF is sufficiently large, then the criterion (15.53) reduces to $1/[\omega^2\delta_s(E_0)]^{1/3} \gg 1/\delta_s(E_0)$. For example, for the linear Stark effect (hydrogenic atoms/ions) this inequality is equivalent to

$$\omega \ll \delta_s(E_0) = a_1(k)E_0, \tag{15.56}$$

which can be fulfilled even if the corresponding criterion for the ion microfield $\omega_i \ll a_1(k) < F_i >$ would not be met. In magnetic fusion plasmas, this kind of situation can occur for low-n HDSL at densities $\ll 10^{15}\,\mathrm{cm}^{-3}$ if the average field of a low-frequency electrostatic turbulence would be of the order of or greater than $10\,\mathrm{kV\,cm}^{-1}$.

In magnetized plasmas the distribution of the turbulent field should be axially symmetric, the axis of symmetry being along the magnetic field. It is Sholin-Oks' distribution [68] of the following form:

$$\begin{aligned} W(E,\cos\theta)\mathrm{d}E\,\mathrm{d}(\cos\theta) &= \left[4/\left(\pi E_{\mathrm{par}}^2\right)\right]^{1/2}\left(E^2/E_{\mathrm{perp}}^2\right) \\ &\times \exp\left[-E^2/E_{\mathrm{perp}}^2 - \cos^2\theta\left(E^2/E_{\mathrm{par}}^2 - E^2/E_{\mathrm{perp}}^2\right)\right]\mathrm{d}E\,\mathrm{d}(\cos\theta), \end{aligned} \tag{15.57}$$

where E_{par} and E_{perp} are the root-mean-square values of the turbulent field components parallel and perpendicular to **B**, respectively; θ is the angle between **E** and **B**.

In the situation, where both the turbulent field and the ion microfield are quasistatic, the distribution of the *total* quasistatic field $\mathbf{E}_t + \mathbf{F}_i$ has been calculated analytically in paper [69] for the case where the turbulent fields are isotropic. The distribution of the ion microfield was the Holtsmark distribution [70]:

$$W(\beta) = \frac{2}{\pi}\int_0^\infty dx \sin(\beta x)\exp(-x^{3/2}), \quad \beta = F/F_0, \quad F_0 \approx 2.603eN^{2/3}. \tag{15.58}$$

The Holtsmark distribution describes a transition from the Gaussian distribution for weak fields $\beta \ll 1$ to the binary distribution (nearest-neighbor distribution) of strong fields $\beta \gg 1$, as noted in review [2]. Indeed, weak-field part of the Holtsmark distribution $W_H \approx 4\beta^2/(3\pi)$ is due to a cumulative effect of large number of perturbers—therefore, like any sum of a large number of random quantities, it follows the Gaussian distribution (i.e., its starting part $\sim \beta^2$). In the opposite limit of $\beta \gg 1$, where $W_H \approx 1.496\beta^{-5/2}$, only the nearest neighbor controls the distribution.

An approximate method for calculating the distribution of the *total* quasistatic field $\mathbf{E}_t + \mathbf{F}_i$ in the case, where the turbulent field is anisotropic (and described by Sholin-Oks' distribution (15.57)) was presented in paper [71]. It was based on the fact that the weak-field parts of the Holtsmark and Sholin-Oks' distributions are alike, being both proportional to the square of the field. In accordance to paper [71], the correction to the Rayleigh distribution was controlled by the behavior of the ion microfield distribution $W_i(\beta)$ at *small fields* ($\beta \ll 1$).

However, this was a misconception, as demonstrated in paper [87]. In this paper, it was shown that in the situation where the average turbulent field is much greater than the characteristic ion microfield, the correction to the Rayleigh distribution is controlled by the second moment of the ion microfield distribution $W_i(\beta)$. The integral defining the second moment of $W_i(\beta)$ accumulates most of its value at $\beta \gg 1$: it converges only because at very large values of β, the ion–ion correlations (the repulsion between the radiating and perturbing ions) "kill" the integral. Thus, in reality the correction to the Rayleigh distribution is controlled by the behavior of the ion microfield distribution $W_i(\beta)$ *at large fields* rather than at small fields.

While in paper [87], this was shown by the example of the isotropic distribution of the turbulent fields, the result is valid qualitatively also for the anisotropic case. It should be also noted that the authors of paper [87], after presenting the general result, used then the well-known asymptotic of $W_i(\beta)$ at $\beta \gg 1$ at a charged point and obtained a universal analytical result for the second moment of $W_i(\beta)$ and thus for the correction to the Rayleigh distribution.

15.5.2 Langmuir Turbulence

1. The highest-frequency electrostatic turbulence is Langmuir turbulence. Its connection to anomalous transport phenomena is known for many decades (see, e.g., paper [72]). The magnetic fusion research community is interested to find out whether Langmuir turbulence develops in magnetic fusion plasmas and, if it does, to determine its parameters. It is desirable to have spectroscopic diagnostics for this purpose, because they are "nonintrusive": they do not perturb parameters to be measured.

   A number of spectroscopic methods for diagnosing Langmuir turbulence/oscillations in different kinds of plasmas have been developed and practically implemented by the author of this review and his collaborators, as presented in book [46]. All these methods related to situations where the radiator (e.g., a hydrogen or deuterium atom) is subjected to a quasistatic electric field—in addition to the oscillatory electric field of Langmuir turbulence and to the broadband dynamic microfield due to plasma electrons. The quasistatic electric field was usually represented by the ion microfield (in the case where the latter was mostly quasistatic) and/or by a low-frequency electrostatic turbulence (e.g., by ionic sound).

   In this situation, there occur the following two major effects of Langmuir turbulence on profiles of hydrogenic spectral lines. The first effect is an appearance of dips/depressions at distances from the unperturbed line position (in the frequency scale) that are proportional to the plasma electron frequency ω_{p}, the proportionality coefficients being rational numbers (expressed via the corresponding quantum numbers). Langmuir-wave-caused dips (hereafter, *L-dips*) in profiles of hydrogenic spectral lines were discovered experimentally in 1977 [73] and explained theoretically in papers [73–77]. This effect was observed and used for diagnostics in a large number of experiments conducted by various experimental groups at different plasma sources (see, e.g., book [46]). The latest experimental results (obtained in a laser-produced plasma) can be found in paper [78].

   The second effect is an additional dynamical Stark broadening [84, 96] (presented also in book [62]). In distinction to the first one, it was not widely used for diagnostics.

   In all of the above experiments, magnetic fields did not play any substantial role. Therefore, for magnetic fusion plasmas, characterized by a strong magnetic field of several (up to 10) tesla, possible effects of Langmuir turbulence on hydrogenic lines have been analyzed afresh in recent papers [80, 81]. Below, we follow these papers to present such analysis and methods for the spectroscopic diagnostics of Langmuir turbulence in magnetic fusion plasmas. The primary focus will be the additional dynamical broadening—for reasons explained below.
2. In 1975, in paper [67], there were derived analytically additional contributions to the width and shift of hydrogenic spectral lines due to Langmuir turbulence for the case where the separation ω_{F} between sublevels of the principal quantum number n is caused by a quasistatic electric field F (hereafter, the "electric"

case, for brevity). In this case, the separation between the Stark sublevels in the frequency scale is

$$\omega_F = 3n\hbar F/(2Z_r m_e e), \tag{15.59}$$

where Z_r is the nuclear charge of the radiator. The stochastic electric field of Langmuir turbulence was represented in [84] in the form:

$$\mathbf{E}_p(t) = \sum_{j=1}^{J} \mathbf{E}_j(t)\cos[\omega_j t + \varphi_j(t)], \tag{15.60}$$

where the phase $\varphi_j(t)$ and the amplitude $\mathbf{E}_j(t)$ change their values with the every change of the state of a Poisson process characterized by the average change frequency γ_p. Between the changes, the quantities $\varphi_j(t)$ and the components $\mathbf{E}_j^\sigma(t)$ are constant taking random values characterized by a certain distribution. In particular, the phase $\varphi_j(t)$ has a uniform distribution in the interval (0, 2π) with the density 1/(2π). The stochastic function $\mathbf{E}_j(t)$ in (15.60) is the realization of a kangaroo-type uniform Markovian stationary stochastic process. A convenient characteristic is the root-mean-square average $E_0 = (<|\mathbf{E}_j(t)|^2>)^{1/2}$, which is called for brevity the average amplitude.

The main frequencies ω_j are all approximately equal to the plasma electron frequency:

$$\omega_p = \left(4\pi e^2 N_e/m_e\right)^{1/2} = 5.641 \times 10^4 \left[N_e(\mathrm{cm}^{-3})\right]^{1/2}. \tag{15.61}$$

The frequency $\gamma_p < \omega_p$ is the largest of the characteristic frequencies of various nonlinear processes in the plasma—the processes such as, e.g., the generation of the Langmuir waves, the induced scattering of the Langmuir waves on the charges particles, the nonlinear decay into ionic sound, and so on. The frequency γ_p is assumed to control the width of the power spectrum of Langmuir turbulence.

The additional contributions to the width and shift of hydrogenic spectral lines due to Langmuir turbulence, derived analytically in paper [67], depend on the separation between the Stark sublevels ω_F caused by a quasistatic electric field. However, for the conditions typical for magnetic fusion plasmas—in particular, in the tokamak divertor region—the ion microfield is not quasistatic for the most intense (low-n) HDSL. Therefore, at the absence of a strong low-frequency electrostatic turbulence, the separation between sublevels of the principal quantum number n is caused by a relatively strong magnetic field $\mathbf{B}$ (so that in this case, these are Zeeman sublevels rather than the Stark sublevels):

$$\omega_B = eB/(2m_e c). \tag{15.62}$$

Thus, in this "magnetic" case, the Langmuir-wave-caused contributions to the diagonal elements $\Gamma_{\alpha\beta} = -\mathrm{Re}\Phi_{\alpha\beta}$ and $D_{\alpha\beta} = -\mathrm{Im}\Phi_{\alpha\beta}$ of the impact broadening operator Φ can be obtained from the corresponding results from [67] by substituting ω_F by ω_B. Here, α and β label sublevels of the upper (a) and lower (b) levels involved in the radiative transition. For brevity, we call $\Gamma_{\alpha\beta}$ and $D_{\alpha\beta}$ the width and the shift, respectively. In this way, we obtain the following expressions for $\Gamma_{\alpha\beta}$ and $D_{\alpha\beta}$ (the corresponding expressions for nondiagonal elements of the operator Φ will be published elsewhere):

$$\Gamma_{\alpha\beta} = \Gamma_\alpha + \Gamma_\beta - d_{\alpha\alpha} d_{\beta\beta} E_0^2 \gamma_p / \left[3\hbar^2 \left(\gamma_p^2 + \omega_p^2\right)\right], D_{\alpha\beta} = D_\alpha + D_\beta, \quad (15.63)$$

where

$$\Gamma_\alpha = \left[E_0^2 \gamma_p/(12\hbar^2)\right] \left(2\mathbf{d}_{\alpha\alpha}^2 / \left(\gamma_p^2 + \omega_p^2\right) + \left(|\mathbf{d}_{\alpha,\alpha-1}|^2 + |\mathbf{d}_{\alpha,\alpha+1}|^2\right) \times \left[1/\left(\gamma_p^2 + (\omega_B - \omega_p)^2\right) + 1/\left(\gamma_p^2 + (\omega_B + \omega_p)^2\right)\right]\right), \quad (15.64)$$

$$D_\alpha = \left[E_0^2 \gamma_p/(12\hbar^2)\right] \left(|\mathbf{d}_{\alpha,\alpha-1}|^2 - |\mathbf{d}_{\alpha,\alpha+1}|^2\right) \times \left[(\omega_B - \omega_p)/\left(\gamma_p^2 + (\omega_B - \omega_p)^2\right) + (\omega_B + \omega_p)/\left(\gamma_p^2 + (\omega_B + \omega_p)^2\right)\right]. \quad (15.65)$$

Here the matrix elements of the dipole moment operator are

$$\mathbf{d}_{\alpha,\alpha}^2 = \left[3ea_0 n_\alpha q_\alpha/(2Z_r)\right]^2, \; |\mathbf{d}_{\alpha,\alpha-1}|^2 - |\mathbf{d}_{\alpha,\alpha+1}|^2 = \mathbf{d}_{\alpha\alpha}^2/q_\alpha^2, \quad |\mathbf{d}_{\alpha,\alpha-1}|^2 + |\mathbf{d}_{\alpha,\alpha+1}|^2 = \mathbf{d}_{\alpha\alpha}^2 \left(n^2 - q^2 - m^2 - 1\right)_\alpha / \left(2q_\alpha^2\right), \quad (15.66)$$

where a_0 is the Bohr radius, Z_r is the nuclear charge of the radiator; $q = n_1 - n_2$; n_1, n_2 and m are the parabolic quantum numbers. In (15.64) and (15.65) in the subscripts, we used the notation $\alpha+1$ and $\alpha - 1$ for the Zeeman sublevels of the energies $+\hbar\omega_B$ and $-\hbar\omega_B$, respectively (compared to the energy of the sublevel α). Formulas for Γ_β and D_β entering (15.63) can be obtained from (15.64) and (15.65) by substituting the superscript α by β.

Let us analyze the above results for the width—because it is often practically more important than the shift. The expressions for the width demonstrate the following two characteristic features. For relatively large magnetic fields, such that

$$\omega_B \gg \omega_p, \quad (15.67)$$

the term containing the diagonal matrix elements $\mathbf{d}_{\alpha\alpha^2} + \mathbf{d}_{\beta\beta^2}$ predominates, so that the other term can be neglected. The dominating term is the *adiabatic* contribution: it does not couple (by virtual transitions) different Zeeman sublevels—in distinction to the neglected term. Under the same condition (15.67), the nondiagonal matrix elements of the impact broadening operator become much smaller than the diagonal elements, so that the quantity $\Gamma_{\alpha\beta}$ from (15.63) becomes a "true width."

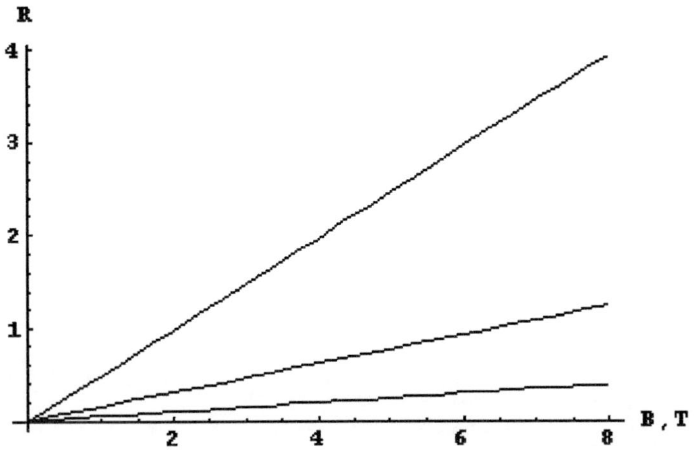

Fig. 15.7 The ratio of the frequencies $R = \omega_B/\omega_p$ vs. the magnetic field **B** in tesla for three different electron densities N_e : $10^{13}\,cm^{-3}$ (the *upper line*), $10^{14}\,cm^{-3}$ (the *middle line*), and $10^{15}\,cm^{-3}$ (the *lower line*)

Figure 15.7 shows the ratio $R = \omega_B/\omega_p$ vs. the magnetic field B for three different electron densities N_e. It is seen that even for $N_e = 10^{13}\,cm^{-3}$, which is usually considered as the lowest electron density relevant to magnetic fusion plasmas, the fulfillment of the condition (15.67) requires magnetic fields greater than 10 T.[1]

The most interesting is another scenario, where

$$\omega_B = \omega_p. \tag{15.68}$$

This resonance can occur exactly or approximately for a number of pairs (B, N_e) typical for the conditions of tokamak divertors. Indeed, from

[1]In principle, there might exist also another adiabatic effect of the stochastic electric field of Langmuir turbulence if $\gamma_p \ll \omega_p$: the formation of satellites separated by $\pm\kappa\omega_p$ (in the frequency scale) from each component of the Zeeman triplet ($k = 1, 2, 3, \ldots$). For the case, where Langmuir turbulence develops anisotropically in such a way, that its electric field is linearly polarized, the satellite intensities were calculated analytically in paper [82] (see also book [46]). However, the satellite intensities are relatively small. Even for the most intense satellite ($k = 1$), the ratio of its intensity I_s to the intensity of the corresponding component of the Zeeman triplet I_0 is

$$I_s/I_0 \sim \left(n^2 T_e/U_{Hi}\right)\left[E_0^2/(8\pi N_e T_e)\right].$$

Here, $U_{Hi} = 13.6\,eV$ is the ionization potential hydrogen/deuterium atoms, Te is the electron temperature; the quantity $E_0^2/(8\pi N_e T_e)$, which is called the degree of turbulence, is the ratio of the energy density of the Langmuir turbulence to the thermal energy density of the plasma. The latter ratio is always much smaller than unity: usually it is in the range 10^{-2}–10^{-4}. Given that for spectroscopic experiments related to tokamak divertors, where the most intense hydrogenic lines are used ($L_\alpha, L_\beta, H_\alpha, H_\beta$) one has $n^2T_e/U_{Hi} \sim 1$, it is seen that indeed $I_s/I_0 \sim (10^{-2} - 10^{-4}) \ll 1$. Thus, these satellites do not seem to be useful for diagnostics of magnetic fusion plasmas unless highly excited hydrogenic lines ($n \gg 1$) are employed.

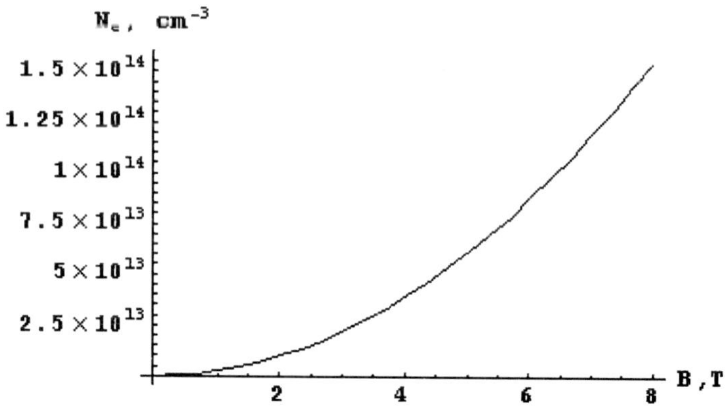

Fig. 15.8 The line (the geometric set of points) in the plane (B, Ne) corresponding to the resonance: $\omega_B = \omega_p$. Here **B** is the magnetic field in tesla, N_e is the electron density in cm^{-3}

Fig. 15.8, which shows the line (the geometric set of points) in the plane (B, N_e) corresponding to the resonance (15.68), it follows that the resonance takes place, e.g., for $B = 2\,\mathrm{T}$ and $N_e = 10^{13}\,\mathrm{cm}^{-3}$, for $B = 5\,\mathrm{T}$ and $N_e = 6 \times 10^{13}\,\mathrm{cm}^{-3}$, or for $B = 8\,\mathrm{T}$ and $N_e = 1.6 \times 10^{14}\,\mathrm{cm}^{-3}$. In the conditions close to the resonance (15.68), the Langmuir-wave-caused Stark width dramatically increases. Neglecting the nonresonance terms in (15.63) and (15.64), it can be represented in the form:

$$\Gamma_{\alpha\beta} = \left(|\mathbf{d}_{\alpha,\alpha-1}|^2 + \mathbf{d}_{\alpha,\alpha+1}|^2 + |\mathbf{d}_{\beta,\beta-1}|^2| + |\mathbf{d}_{\beta,\beta+1}|^2\right) E_0^2 / \left(12\hbar^2\gamma_p\right). \quad (15.69)$$

We note that all terms in (15.69) correspond to the *nonadiabatic* contribution: they couple by virtual transitions different Zeeman sublevels.

Now let us compare $\Gamma_{\alpha\beta}$ from (15.69) to the width due to the competing Stark broadening mechanism. For the typical parameters of tokamak divertors, the magnetic field causes a significant decrease of the nonadiabatic contribution to the dynamical Stark width due to ions—as discussed in detail above in Sect. 15.4.2. Thus, the total contribution to the dynamical Stark width due to ions can be well represented by the adiabatic contribution given by (15.41) of the generalized theory.

The ratio of $\Gamma_{\alpha\beta}$ from (15.69) to the corresponding contribution due to the dynamical broadening by ions from (15.41) can be represented as the product of five dimensionless factors as follows:

$$\begin{aligned}\Gamma_{\alpha\beta}/\gamma_{\alpha\beta} &= \left\{\pi^{1/2}\left(|\mathbf{d}_{\alpha,\alpha-1}|^2 + |\mathbf{d}_{\alpha,\alpha+1}|^2 + |\mathbf{d}_{\beta,\beta-1}|^2 + |\mathbf{d}_{\beta,\beta+1}|^2\right) / \left[27 \cdot 2^{1/2}\left(ea_0 X_{\alpha\beta}\right)^2 I(R_i)\right]\right\} \\ &\times (m_e/M)^{1/2}\left[T_e\rho_D/e^2\right](\omega_p/\gamma_p)\left[E_0^2/(8\pi N_e T_e)\right]. \quad (15.70)\end{aligned}$$

The quantity $E_0^2/(8\pi N_e T_e)$ is the degree of turbulence: the ratio of the energy density of Langmuir turbulence to the thermal energy density of the plasma. We note that the right side of (15.70) can be simplified to a more explicit scaling: $\Gamma_{\alpha\beta}/\gamma_{\alpha\beta}$ is proportional to $E_0^2 T_e^{1/2}/(N_e \gamma_p M^{1/2})$, if $T_i = T_e$. However, the representation of $\Gamma_{\alpha\beta}/\gamma_{\alpha\beta}$ as the product of the five dimensionless factors in (15.70) provides a better physical understanding and is more convenient for estimates. A practical formula for the product of the second and third factors in the right side of (15.70) is

$$(m_e/M)^{1/2}\left[T_e\rho_D/e^2\right] = 1.204\times 10^8[T_e(\mathrm{eV})]^{3/2}\left[N_e(\mathrm{cm}^{-3})\right]^{-1/2}(M_p/M)^{1/2}. \tag{15.71}$$

Let us estimate the ratio $\Gamma_{\alpha\beta}/\gamma_{\alpha\beta}$ for a hydrogen plasma (so that $M = M_p/2$) of the electron density $N_e = 6\times 10^{13}\,\mathrm{cm}^{-3}$ and of the temperature $T_e = 5\,\mathrm{eV}$. From (15.71), we get: $(m_e/M)^{1/2}[T_e\rho_D/e^2] = 246 \gg 1$, so that $\Gamma_{\alpha\beta}/\gamma_{\alpha\beta} \sim 2\times 10^2(\omega_p/\gamma_p)[E_0^2/(8\pi N_e T_e)]$. The ratio ω_p/γ_p is a large quantity—typically in the range of (10^2–10^4), while the degree of turbulence $E_0^2/(8\pi N_e T_e)$ is a small quantity—typically in the range of (10^{-4}–10^{-2}). Since the first factor in (15.69) is typically $\sim$0.5 for magnetic fusion plasmas, we obtain the following range: $\Gamma_{\alpha\beta}/\gamma_{\alpha\beta} \sim$ (1–10^4).

This example shows that for magnetic fusion plasmas, the contribution to the dynamical Stark width due to Langmuir turbulence can dominate over the competing dynamical Stark broadening by ions, so that the half-width at half-maximum of a hydrogenic line will be $\delta\lambda_{1/2} = [\lambda_0^2/(2\pi c)]\Gamma_{\alpha\beta}/\gamma_{\alpha\beta}$, where λ_0 is the unperturbed wavelength and λ_0 is given by (15.69). Thus, it can be used for diagnostics of Langmuir turbulence. Specifically, from the experimentally measured Stark width of hydrogenic spectral lines in the conditions close to the resonance, it is possible to determine the quantity E_0^2/γ_p—as it is seen from (15.69).

3. The above results are obtained for the case where Langmuir turbulence developed *isotropically*. Now, we address the situation where it develops anisotropically. In this case, the spectroscopic diagnostic can be significantly enhanced by the polarization analysis, allowing to obtain an additional information—such as the *degree of anisotropy* of the distribution of Langmuir turbulence.

In 1973, in paper [68], there was presented a theory of optical polarization measurements of low-frequency electrostatic turbulence in plasmas. It was successfully implemented in a later experiment [83]. In 1977, in paper [79], there was developed a theory of optical polarization measurements of Langmuir turbulence for the "electric" case, i.e., for the situation where the separation ω_F between sublevels of the principal quantum number n is caused by a quasistatic electric field F. Based on the results from [68, 79], we present here a theory of optical polarization measurements of Langmuir turbulence for the "magnetic" case, i.e., for the situation where the separation ω_B between sublevels of the principal quantum number n is caused by a relatively strong magnetic field **B**.

Following paper [68], we consider a typical situation where the directionality diagram of Langmuir turbulence has the axial symmetry. Profiles of a chosen spectral line are observed at two directions of a linear polarizer: (1) parallel to the axis of symmetry and (2) perpendicular to the axis of symmetry. Hydrogenic spectral lines consist of π- and σ-components, having different polarization properties. Therefore, in accordance to paper [68], there are the following four functions controlling the angular dependence of the characteristics of the π- and σ-components of the spectral line observed at the polarizer directions (1) and (2):

$$f_{1\pi} = \cos^2\theta, \; f_{2\pi} = f_{1\pi} = 3(1-\cos^2\theta)/2, \; f_{2\pi} = 3(1+\cos^2\theta)/4, \quad (15.72)$$

where θ is the polar angle with respect to the axis of symmetry, which is the direction of the magnetic field **B**. The axially symmetric Sholin-Oks' distribution $W(E, \cos\theta)$ of the amplitude of the electric field of Langmuir turbulence from (15.70) can be re-written in the form:

$$W(E, \cos\theta) = \left[2E^2/\left(\pi^{1/2}\eta E_{\text{perp}}^3\right)\right] \times \exp\left\{-\left(E/E_{\text{perp}}\right)^2\left[1-\cos^2\theta\left(1-1/\eta^2\right)\right]\right\}. \quad (15.73)$$

Here, η is the degree of anisotropy defined as

$$\eta = 2^{1/2} E_{\text{par}}/E_{\text{perp}}, \quad (15.74)$$

where E_{par} and E_{perp} are the root-mean-squared values of the amplitude in the directions parallel and perpendicular to the axis of symmetry, respectively.

Using (15.71)–(15.74), we obtain the following results for the relative difference of the Langmuir-wave-caused contributions to the dynamical Stark width of π- and σ-components of the spectral line in two perpendicular polarizations:

$$P_\pi(\eta) = (\Gamma_{1\pi} - \Gamma_{2\pi})/(\Gamma_{1\pi} + \Gamma_{2\pi}) = (\eta^2 - 1)/(\eta^2 + 1), \quad (15.75)$$

$$P_\sigma(\eta) = (\Gamma_{1\sigma} - \Gamma_{2\sigma})/(\Gamma_{1\sigma} + \Gamma_{2\sigma}) = (1 - \eta^2)/(\eta^2 + 3). \quad (15.76)$$

Thus, from the experimental polarization difference of the Langmuir-wave-caused dynamical Stark width, it is possible to deduce the degree of anisotropy η—in addition to the quantity E_0^2/γ_{p}.

4. Let us now briefly discuss L-dips. They were discovered experimentally and explained theoretically for dense plasmas, where one of the electric fields experienced by hydrogenic radiators is quasistatic— due to the ion microfield and/or a low-frequency electrostatic turbulence (see [73–78]). In this situation, the central point of the L-dip phenomenon was a resonant coupling between a quasistatic electric field **F** and an oscillatory electric field of the Langmuir wave. In the profile of the Stark component of the Lyman line originating from the sublevel q, the resonance could manifest, generally, as two dips (L^+-dip

and L^--dip) located at the following distances $\Delta\lambda_{\pm}^{\text{dip}}$ from the unperturbed wavelength λ_0 of this Lyman line:

$$\Delta\lambda_{\pm}^{\text{dip}} = -\left[\lambda_0^2/(2\pi c)\right] q\omega_{\text{p}} + \left[2\omega_{\text{p}}^3/\left(27n^3 Z_{\text{r}} Z_{\text{i}}\omega_{at}\right)\right]^{1/2} \times \left[n^2\left(n^2 - 6q^2 - 1\right) + 12n^2q^2 \pm 6n^2 q\right]. \qquad (15.77)$$

Here, $\omega_{\text{at}} = m_{\text{e}}e^4/\pi^3 \approx 4.14 \times 10^{16}\,\text{s}^{-1}$ is the atomic unit of frequency, n is the principal quantum number. The first, primary term in braces reflects the dipole interaction with the ion microfield. The second term in braces takes into account—via the quadrupole interaction—a spatial nonuniformity of the ion microfield. This second term is, generally speaking, a correction to the first term—except for the case of the central Stark component ($q = 0$), for which the first term vanishes. We note that in the profile of the central Stark component there could be only one L-dip (hereafter, "central L-dip") since the term $\pm 6n^2q$ vanishes. A formula for the L-dip positions in profiles of hydrogenic lines from other spectral series (Balmer, etc.) can be found in [46, 77].

It is important to emphasize the following. For a given electron density Ne, the value of the plasma electron frequency ω_{p} is fixed—in accordance to (15.57). The resonance occurs when the separation between the Stark sublevels of the principal quantum number n caused by the field F:

$$\omega_{\text{F}} = 3n\hbar F/(2Z_{\text{r}}m_{\text{e}}e) \qquad (15.78)$$

is equal to ω_{p}:

$$\omega_{\text{F}} = \omega_{\text{p}}. \qquad (15.79)$$

The quasistatic electric field in plasmas has a broad distribution over the ensemble of radiators - regardless of whether this field represents the ion microfield or the low-frequency electrostatic turbulence. Therefore, if the ion microfield is mostly quasistatic or a low-frequency electrostatic turbulence has been developed in the plasma, then there would always be a fraction of radiators, for which the resonance condition (15.79) is satisfied.

However, for the conditions typical for magnetic fusion plasmas—in particular, in the tokamak divertor region—the ion microfield is not quasistatic. Therefore, at the absence of a low-frequency electrostatic turbulence, the separation $\omega_{\text{B}} = eB/(2m_{\text{e}}c)$ between sublevels of the principal quantum number n is caused by a relatively strong magnetic field $\mathbf{B}$ (so that in this case these are Zeeman sublevels rather than the Stark sublevels). Then, the resonance condition is given by (15.78) instead of (15.79).

In this situation, the following two conditions are necessary for observing L-dips. First, the magnetic field should have a noticeable nonuniformity ΔB across the region, from which a particular hydrogenic line is emitted:

$$\Delta B/B > [\lambda_0/(2\pi c)]n^2\hbar E_0/(m_{\text{e}}eZ_{\text{r}}). \qquad (15.80)$$

Second, the Langmuir electric field should not be too strong:

$$n^2\hbar E_0/(m_e e Z_r) < \gamma_p. \tag{15.81}$$

Under conditions (15.80) and (15.81), it could be possible to observe an L-dip in the profile of each component of the Zeeman triplet. The halfwidth of the L-dip $\delta_{1/2}$ would be controlled only by one parameter of Langmuir turbulence—by the averaged amplitude E_0

$$\delta_{1/2} \approx (3/2)^{1/2}\lambda_0^2 n^2\hbar E_0/(8\pi m_e e c Z_r), \tag{15.82}$$

so that the other parameter, namely γ_p, would not enter formula (15.82).

Therefore, the following diagnostic method can be proposed. If L-dips are observed in the profiles of the components of the Zeeman triplet, one can first deduce the averaged amplitude E_0 of the Langmuir electric field from the experimental halfwidth of the L-dip using (15.82). Then from the experimental halfwidth of the components of the Zeeman triplet, one can infer the quantity E_0^2/γ_p via (15.69) and thus (since E_0 would be already determined) the characteristic frequency γ_p of the nonlinear process controlling the width of the power spectrum of Langmuir turbulence. Finally, the polarization analysis can allow deducing the degree of anisotropy of Langmuir turbulence $\eta = 2^{1/2}E_{par}/E_{perp}$.

Thus, the combination of these diagnostics can allow the experimental determination of the following parameters of Langmuir turbulence: the average amplitude E_0, the characteristic frequency γ_p of the nonlinear processes controlling the width of the power spectrum of Langmuir turbulence, and the degree of anisotropy $\eta = 2^{1/2}E_{par}/E_{perp}$.

We emphasize that one of the experimental manifestations of the dynamical Stark broadening by Langmuir turbulence would be unusually long Lorentzian-shape wings of HDSL. Such wings of the D_α line were observed recently at the Tore Supra [84].

15.6 Conclusions

We reviewed the latest advances in the Stark broadening and the Stark–Zeeman broadening of hydrogenic spectral lines in magnetized plasmas (advances with respect, e.g., to the results presented in [85]). We also discussed diagnostic applications, being especially focused at magnetic fusion plasmas—with and without turbulence. From the experimental point of view, for implementing these diagnostics passive spectroscopic methods could be sufficient.

As for active spectroscopic methods, in magnetic fusion plasmas they are implemented through the injection of neutral beams. In paper [86], by using the generalized theory of the Stark broadening, it was proposed a dual purpose diagnostic. Namely, it was shown how a polarization analysis of spectral lines emitted from a hydrogen or deuterium neutral beam allows to measure both the magnetic pitch angle γ_{pitch} and the effective charge Z_{eff}. The physics behind this dual purpose diagnostic was based on the fact that the ion dynamical broadening of magnetically splitted levels of beam atoms is highly anisotropic, resulting in a pronounced angular dependence of widths of π- and σ-components of spectral lines. The highly anisotropic character of the broadening was revealed by applying the generalized theory (the standard theory would not be able to do so).

It is worth mentioning also another active method for magnetic fusion plasmas based on the beam spectroscopy: the motional Stark effect diagnostic. Recently, it has been enhanced by using laser-induced fluorescence (see, e.g., paper [87] and references therein) and by employing B-Stark technique [88].

Acknowledgements The author would like to thank Chris Klepper for sharing one recent experimental result at the Tore Supra.

References

1. H.R. Griem, *Principles of Plasma Spectroscopy* (Cambridge University Press, Cambridge, 1997)
2. V.S. Lisitsa, Sov. Phys. Uspekhi **122**, 449 (1977)
3. E. Oks, *Stark Broadening of Hydrogen and Hydrogenlike Spectral Lines in Plasmas: The Physical Insight* (Alpha Science International, Oxford, 2006)
4. E.A. McLean, J.A. Stamper, C.K. Manka, H.R. Griem, D.W. Droemer, B.H. Ripin, Phys. Fluids **27**, 1327 (1984)
5. E. Stambulchik, K. Tsigutkin, Y. Maron, Phys. Rev. Lett. **98**, 225001 (2007)
6. Yu. Demkov, B. Monozon, V. Ostrovsky, Sov. Phys. JETP **30**, 775 (1970)
7. V. Fock, Z. Phys. **98**, 145 (1935)
8. L.D. Landau, E.M. Lifshitz, *Mechanics* (Pergamon, Oxford, 1960)
9. L.D. Landau, E.M. Lifshitz, *Quantum Mechanics* (Pergamon, Oxford, 1965)
10. V.S. Lisitsa, G.V. Sholin, Sov. Phys. JETP **34**, 484 (1972)
11. A. Derevianko, E. Oks, in *Physics of Strongly Coupled Plasmas*, ed. by W.D. Kraeft, M. Schlanges (World Scientific, Singapore, 1996), p. 286, 291
12. H.R. Griem, *Spectral Line Broadening by Plasmas* (Academic, New York, 1974)
13. M.L. Strekalov, A.I. Burshtein, Sov. Phys. JETP **34**, 53 (1972)
14. R.L. Greene, J. Cooper, E.W. Smith, J. Quant. Spectr. Rad. Tran. **15**, 1025 (1975)
15. R.L. Greene, J. Cooper, J. Quant. Spectr. Rad. Tran. **15**, 1037, 1045 (1975)
16. E. Oks, Europ. Phys. J. D **28**, 171 (2004)
17. E. Lee, D. Farrelly, T. Uzer, Optic. Express **1**, 221 (1997)
18. T.C. Germann, D.R. Herschbach, M. Dunn, D.K. Watson, Phys. Rev. Lett. **74**, 658 (1995)
19. C.H. Cheng, C.Y. Lee, T.F. Gallagher, Phys. Rev. Lett. **73**, 3078 (1994)
20. L. Chen, M. Cheret, F. Roussel, G. Spiess, J. Phys. B **26**, L437 (1993)
21. S.K. Dutta, D. Feldbaum, A. Walz-Flannigan, J.R. Guest, G. Raithel, Phys. Rev. Lett. **86**, 3993 (2001)
22. R.G. Hulet, E.S. Hilfer, D. Kleppner, Phys. Rev. Lett. **55**, 2137 (1985)
23. K.B. MacAdam, E. Horsdal-Petersen, J. Phys. B **36**, R167 (2003)
24. V.M. Vainberg, V.S. Popov, A.V. Sergeev, Sov. Phys. JETP **71**, 470 (1990)
25. T.C. Killian, M.J. Lim, S. Kulin, R. Dumke, S.D. Bergeson, S.L. Rolston, Phys. Rev. Lett. **86**, 3759 (2001)

26. T.C. Killian, S. Kulin, S.D. Bergeson, L.A. Orozco, C. Orzel, S.L. Rolston, Phys. Rev. Lett. **83**, 4776 (1999)
27. M.P. Robinson, B.L. Tolra, M.W. Noel, T.F. Galagher, P. Pillet, Phys. Rev. Lett. **85**, 4466 (2000)
28. M.R. Flannery, E. Oks, Phys. Rev. A **73**, 013405 (2006)
29. D. Salzmann, *Atomic Physics in Hot Plasmas* (Oxford University Press, Oxford, 1998), Chaps. 2, 3
30. M.S. Murillo, J.C. Weisheit, Phys. Rep. **302**, 1 (1998)
31. C. Stehle, J.M. Caillol, A. Escarguel, D. Gilles, J. Phys. IV France **10**, Pr0–901 (2000)
32. Ďjachkov, J. Quant. Spectr. Rad. Tran. **59**, 65 (1998)
33. J. Stein, I.B. Goldberg, D. Shalitin, D. Salzmann, Phys. Rev. A **39**, 2078 (1989)
34. D. Salzmann, J. Stein, I.B. Goldberg, R.H. Pratt, Phys. Rev. A **44**, 1270 (1991)
35. J. Stein, D. Salzmann, Phys. Rev. A **45**, 3943 (1992)
36. P. Malnoult, B. ĎEtat, H. Nguen, Phys. Rev. A **40**, 1983 (1989)
37. Y. Furutani, K. Ohashi, M. Shimizu, A. Fukuyama, J. Phys. Soc. Jpn. **62**, 3413 (1993)
38. P. Sauvan, E. Leboucher-Dalimier, P. Angelo, H. Derfoul, T. Ceccotti, A. Poquerusse, A. Calisti, B. Talin, J. Quant. Spectr. Rad. Tran. **65**, 511 (2000)
39. E. Oks, Phys. Rev. E **63**, 057401 (2001)
40. E. Oks, Phys. Rev. Lett. **85**, 2084 (2000)
41. E. Oks, J. Phys. B Atom. Mol. Opt. Phys. **33**, 3319 (2000)
42. C. Stehle, A. Mazure, G. Nollez, N. Feautrier, Astron. Astrophys. **127**, 263 (1983)
43. C. Stehle, N. Feautrier, J. Phys. B **17**, 1477 (1984)
44. C. Stehle, Astron. Astrophys. **305**, 677 (1996)
45. A. Derevianko, E. Oks, Phys. Rev. Lett. **73**, 2079 (1994)
46. E. Oks, *Plasma Spectroscopy: The Influence of Microwave and Laser Fields*. Springer Series on Atoms and Plasmas, vol. 9 (Springer, New York, 1995)
47. Ya. Ispolatov, E. Oks, J. Quant. Spectr. Rad. Tran. **51**, 129 (1994)
48. E. Oks, A. Derevianko, Ya. Ispolatov, J. Quant. Spectr. Rad. Tran. **54**, 307 (1995)
49. E. Oks, in *Spectral Line Shapes*, vol. 18. AIP Conf. Proc. 874, Auburn, Alabama, USA, 2006, p. 19
50. J. Touma, E. Oks, S. Alexiou, A. Derevianko, J. Quant. Spectr. Rad. Tran. **65**, 543 (2000)
51. J. Rosato, Y. Marandet, H. Capes, S. Ferri, C. Mosse, L. Godbert-Mouret, M. Koubiti, R. Stamm, Phys. Rev. E **79**, 046408 (2009)
52. B.L. Welch, H.R. Griem, J. Terry, C. Kurz, B. LaBombard, B. Lipschultz, E. Marmar, G. McCracken, Phys. Plasmas **2**, 4246 (1995)
53. B.L. Welch, H.R. Griem, J.L Weaver, J.L. Terry, R.L. Boivin, B. Lipschultz, D. Lumma, E.S. Marmar, G. McCracken, J.C. Rost, in 10th Conf. "Atomic Processes in Plasmas", AIP Conf. Proceedings 381, New York, 1996, p. 159
54. N.H. Brooks, S. Lisgo, E. Oks, D. Volodko, M. Groth, A.W. Leonard, the DIII-D Team, Plasma Phys. Rep. **35**, 112 (2009)
55. B.L. Welch, H.R. Griem, J.L Weaver, J.U. Brill, J.L. Terry, B. Lipschultz, D. Lumma, G. McCracken, S. Ferri, A. Calisti, R. Stamm, B. Talin, R.W. Lee, in 13th Int. Conf. "Spectral Line Shapes", AIP Conf. Proceedings 386, New York, 1997, p. 113
56. E. Oks, Phys. Rev. E, Rapid Comm. **60**, R2480 (1999)
57. S. Ferri, A. Calisti, R. Stamm, B. Talin, R.W. Lee, in 14th Int. Conf. "Spectral Line Shapes", AIP Conf. Proceedings 467, New York, 1999, p. 115
58. E. Stambulchik, Y. Maron, J. Phys. B Atom. Mol. Opt. Phys. **41**, 095703 (2008)
59. V.N. Tsytovich, *Theory of Turbulent Plasmas* (Consultants Bureau, New York, 1977)
60. B.B. Kadomtsev, *Plasma Turbulence* (Academic, New York, 1965)
61. Turbulence and Anomalous Transport in Magnetized Plasmas, in *Proc. Int. Workshop at "Institut d'Etudes Scientifiques de Cargese"*, Corse de Sud, France, ed. by D. Gresillon, M.A. Dubois (Editions de Physique, Les Ulis, 1987)
62. E. Oks, Sov. Phys. Doklady **29**, 224 (1984)
63. E. Oks, Meas. Tech. **29**, 805 (1986)
64. P. Sauvan, E. Dalimier, E. Oks, O. Renner, S. Weber, C. Riconda, J. Phys. B **42**, 195501 (2009)

65. Ja.B. Zeĺdovich, Sov. Phys. JETP **24**, 1006 (1967)
66. V.I. Ritus, Sov. Phys. JETP **24**, 1041 (1967)
67. E. Oks, G.V. Sholin, Sov. Phys. JETP **41**, 482 (1975)
68. G.V. Sholin, E. Oks, Sov. Phys. Doklady **18**, 254 (1973)
69. E. Oks, G.V. Sholin, Sov. Phys. Tech. Phys. **21**, 144 (1976)
70. J. Holtsmark, Ann. Phys. **58**, 577 (1919)
71. E. Sarid, Y. Maron, L. Troyansky, Phys. Rev. E **48**, 1364 (1993)
72. V.N. Tsytovich, L. Stenflo, Phys. Scripta **12**, 323 (1975)
73. A.I. Zhuzhunashvili, E. Oks, Sov. Phys. JETP **46**, 2142 (1977)
74. E. Oks, V.A. Rantsev-Kartinov, Sov. Phys. JETP **52**, 50 (1980)
75. V.P. Gavrilenko, E. Oks, Sov. Phys. JETP **53**, 1122 (1981)
76. V.P. Gavrilenko, E. Oks, Sov. Phys. Plasma Phys. **13**, 22 (1987)
77. E. Oks, St. Böddeker, H.-J. Kunze, Phys. Rev. A **44**, 8338 (1991)
78. O. Renner, E. Dalimier, E. Oks, F. Krasniqi, E. Dufour, R. Schott, E. Förster, J. Quant. Spectr. Rad. Tran. **99**, 439 (2006)
79. E. Oks, G.V. Sholin, Opt. Spectrosc. **42**, 434 (1977)
80. E. Oks, J. Phys. B Atom. Mol. Opt. Phys. Fast Track Comm. **44**, 101004 (2011)
81. E. Oks, Intern. Rev. Atom. Mol. Phys. **1**, 169 (2010)
82. E.M. Lifshitz, Sov. Phys. JETP **26**, 570 (1968)
83. M.V. Babykin, A.I. Zhuzhunashvili, E. Oks, V.V. Shapkin, G.V. Sholin, Sov. Phys. JETP **38**, 86 (1974)
84. C.C. Klepper, private communication (2011)
85. M.L. Adams, R.W. Lee, H.A. Scott, H.K. Chung, and L. Klein, Phys. Rev. E **66**. 066413 (2002).
86. A. Derevianko, E. Oks, Rev. Sci. Instr. **68**, 998 (1997)
87. E.L. Foley, F.M. Levinton, Rev. Sci. Instr. **75**, 3462 (2004)
88. N.A. Pablant, K.H. Burrell, R.J, Groebner, C.T. Holcomb, D.H. Kaplan, Rev. Sci. Instrum. **81**, 10D729 (2010)

Chapter 16
Approach to Ultralow Ion-Beam Temperatures by Beam Cooling

A. Noda, M. Grieser, and T. Shirai

Abstract "Stochastic cooling" invented based on the requirement from particle physics to create vector bosons by proton–antiproton collision, has attained the reduction of beam emittance in 6 dimensional phase space. "Electron cooling" also enables detailed research of fine structures in electron–molecular ion reactions and realizes one-dimensional ordering of ion beams. Based on these attainments, three-dimensional laser cooling with much stronger cooling force to realize a "Crystalline beam" has been studied. In this chapter, an approach to attain an ultra-low-temperature beam by heat reduction with "Beam cooling" is explained as the ultimate goal.

16.1 Introduction

The beam of an accelerator is, in general, in a gaseous state with a rather high temperature, where the constituent particles collides between each other due to the random velocity distribution of the beam. The temperature of the beam depending on its energy is, in general, rather high from several hundreds to several millions Kelvin. "Beam cooling" takes away the heat possessed by the beam and reduces its temperature.

"Beam cooling" has been started in 1960s with the big motive force to accumulate antiprotons in a storage ring to realize the creation of vector bosons by pair annihilation at proton and antiproton collisions. Antiproton as anti-matter does not exist in the natural world. Antiprotons, for particle physics experiments, are produced in collisions of protons, with energies higher than the threshold energy, with a tungsten target. The kinetic energy of the produced antiprotons decreases with the increase of the production angle, measured from the initial proton beam direction, which can be easily given by the kinematics of two particles collision. In order to collect a large number of antiprotons, it is required to enlarge the solid angle in the antiprotons production, which inevitably increases the energy spread of the collected antiprotons.

V. Shevelko and H. Tawara (eds.), *Atomic Processes in Basic and Applied Physics*, Springer Series on Atomic, Optical, and Plasma Physics 68,
DOI 10.1007/978-3-642-25569-4_16,

This was the bottleneck in the realization of enough luminosity in $p-\bar{p}$ collision experiments. With "Beam cooling" a reduction of *the volume occupied by the beam in the 6-dimensional phase space* (called as "Emittance") can be realized. Beam cooling is a process to reduce the temperature of the beam, composed of the same kind of charged particles. For the ideal gas with Maxwell velocity distribution, its temperature is related to the average kinetic energy:

$$\frac{3}{2}k_{\mathrm{B}}T = \frac{1}{2}m\left(v_x^2 + v_y^2 + v_z^2\right), \tag{16.1}$$

where k_{B} is the Boltzmann constant and v_x, v_y, and v_z are the velocity components in x, y, and z directions, respectively. Here, a right-handed coordinate system, x, y, and z, is utilized, where z denotes the beam direction.

The longitudinal temperature of the beam, $T_{\parallel}$, and the transverse temperature in the direction perpendicular to the beam, $T_{\perp}$, can be given as:

$$k_{\mathrm{B}}T_{\parallel} = m_0c^2\beta^2\left(\frac{\delta p}{p}\right)^2, \tag{16.2}$$

$$k_{\mathrm{B}}T_{\perp} = \frac{1}{2}m_0c^2\beta^2\gamma^2\left(y'^2 + z'^2\right), \tag{16.3}$$

where $\delta p/p$ satisfying, $\delta p/p = \frac{1}{\gamma^2}\cdot\delta v/v$, utilizing Lorentz factor, γ, y', and z' are fractional momentum spread, declination angles of the beam orbit to the equilibrium closed orbit in the horizontal and vertical directions, respectively.

16.2 Various Cooling Methods

16.2.1 Electron Beam Cooling

A scheme to take out the heat from an ion beam by overlapping the ion beam with a very cold electron beam in a certain part of an accelerator ring has been invented by Dr. G. Budker at Budker Institute for Nuclear Physics in Novosivirsk, Russia in 1966 and is called "electron beam cooling" which was also experimentally confirmed [1, 2]. The concept of the "electron beam cooling", is illustrated in Fig. 16.1.

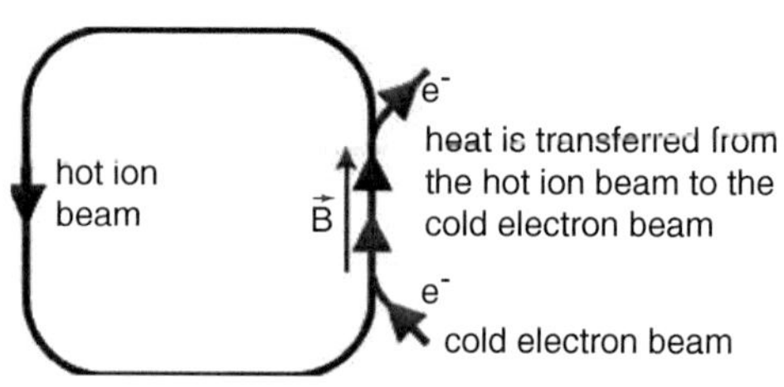

Fig. 16.1 Principle of "electron beam cooling"

Trials to cool a antiproton beam produced in an accelerator has been eagerly performed using the ICE ring at CERN in the time from 1970s to 1980s. This scheme, however, is not suited for the handling of a rather hot beam with an energy spread of a few percent ~10 and more percent [3] as is anticipated for the collected antiprotons produced in a large solid angle. "Stochastic cooling" plays the major role in the creation of a antiproton beam. However, it should be noted that a scheme to sweep the relative velocity between the ion beam and electron beam to cool down a hot ion beam with electron cooling has been proposed [4] and experimentally demonstrated using the TSR ring at MPI-K in Heidelberg [5].

16.2.2 Laser Cooling

The energy transfer between the irradiated laser photons and the ion beam can occur through excitation of the atom to the upper state under the condition that the energy level distance coincides with the photon energy taking Doppler shift into account as shown in Fig. 16.2.

After excitation, the ion has gained a momentum in the direction of the irradiated laser beam. The recoil momentum of the ion, due to emission of a photon, will average to zero because of the isotropy of the angular distribution of the emitted photons. In case of head on collision of the laser with the ion beam, the ions are decelerated by the photon momentum transfer, while the ion is accelerated by the

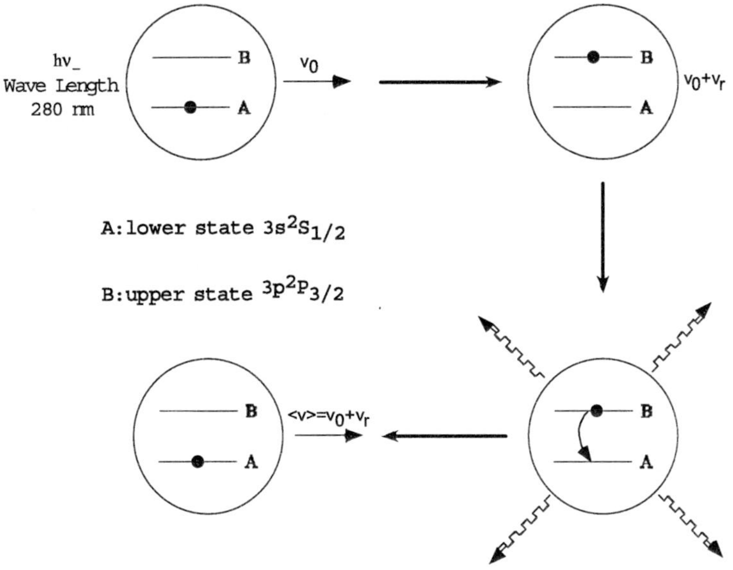

Fig. 16.2 Illustration of laser-cooling scheme

momentum transfer in case of the parallel passage of the ion beam with the laser. By sweeping the laser frequency through the energy spread of the ion beam, the whole ion beam can be compressed in a very limited velocity spread in the mK region.

16.2.3 Stochastic Cooling

"Stochastic cooling" is well known because of the contribution by Carlo Rubia and S. van der Meer, who were awarded with Nobel Physics Prize in 1984, by realization of $p - \bar{p}$ collision experiments and discovery of vector bosons, W and Z. Without the presence of "Maxwell's Demon," it is impossible to detect a single particle with a unit charge with enough S/N ratio. In order to conquer such a situation, this scheme utilizes signals produced by a subensemble of particles passing a pickup for monitoring at the same time. This signal and the subensemble of ions simultaneously arrive at the kicker which corrects the betatron and synchrotron oscillations of the subensemble ions. It should be noted that the subensemble passing through the pickup at the same time changes at each passage due to the velocity difference among the constituent particles (called as "Mixing";) this mixing plays an essential role in the stochastic cooling scheme. It is assumed the sub-ensemble does not change during its passage from the pickup to the kicker (without mixing) due to the small distance between these two equipments compared with the whole circumference.

Assuming the ideal case where the feedback electronics system has a constant gain over its bandwidth W and zero gain outside of this bandwidth, the cooling rate $1/\tau$ is approximately described as:

$$\frac{1}{\tau} = \frac{W}{2N}, \tag{16.4}$$

where N is the number of the beam particles. With the mixing factor M ($M = 1$ for perfect mixing), the cooling rate is given as:

$$\frac{1}{\tau} = \frac{W}{2N}\left[2g - g^2\left(M + U\right)\right], \tag{16.5}$$

where g is the gain of the feedback electronics system and U is a thermal noise factor given by the noise/signal power ratio. This can be optimized by adjusting the gain to be $1/(M + U)$ resulting the value of $1/\tau$ as $W/[2N\,(M + U)]$ [6]. It can be understood that negative feedback is an extreme case where noise factor, U, is zero and with perfect mixing, $M = 1$. This scheme has played an essential role for collecting hot antiproton beams, but it is not, in general, suited for realizing ultra low temperature beam as D. Möhl has pointed out [3].

16.3 One-Dimensional Ordering by Electron Beam Cooling

16.3.1 Survey of the World Attainments of One-Dimensional Ordering

The group from NAP-M at BINP has reported their experimental results of electron cooling for a 65 MeV proton beam at the workshop ECOOL84 held in Karlsruhe [7], as shown in Fig. 16.3. Although a momentum spread of about 10^{-4} was expected by simulation for a 65 MeV proton beam, momentum spreads in the order of 10^{-6} was measured, with electron cooling, at intensities below 10 μA. This observation was explained by ordering effects of the beam particles, where the spacing between the beam particles is kept constant and the momentum transfer from the transverse direction to the longitudinal one is forbidden. This result caused a heavy discussion at the workshop because of its contradiction against the intrabeam scattering (IBS) theory. Due to today's reconsiderations, the data shown in Fig. 16.3 presents not "one-dimensional ordering" because the data does not show any sharp jump as shown in the later measurements. This momentum spread report, however, triggered various approaches at ESR and SIS at GSI, Darmstadt, CRYRING at Manne Siegbahn Laboratory (MSL), Stockholm and S-LSR at ICR, Kyoto University to realize an ordered ion beam.

M. Steck et al., at GSI, Darmstadt, Germany, had found a sudden jump in the momentum spread of an electron cooled multicharge heavy-ion beam (Ar^{18+}, Ni^{28+}, Kr^{36+}, Xe^{54+}, Au^{79+}, U^{92+}) when the ion number was reduced to around several thousands ions, as shown in Fig. 16.4 [8]. Similar phenomena are also observed with SIS at GSI and CRYRING at MSL for heavy multicharge ions Xe^{36+} [9]. Now, this phenomenon is called "Ordering of constituent particles of the beam." This ordering has also been realized for rather lighter ions like C^{6+} at ESR [10], but a proton beam with a single charge has not been "ordered" before the approach at S-LSR because of its weaker electron cooling force and weaker Schottky noise to detect the ion beam.

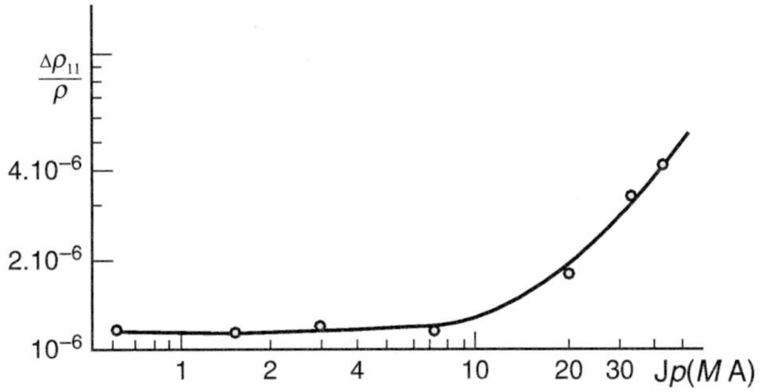

Fig. 16.3 Dependence of the momentum spread of an electron cooled 65 MeV proton beam on the beam current Jp (from [7]). Measurements done at NAP-M storage ring at BINP in Novosibirsk

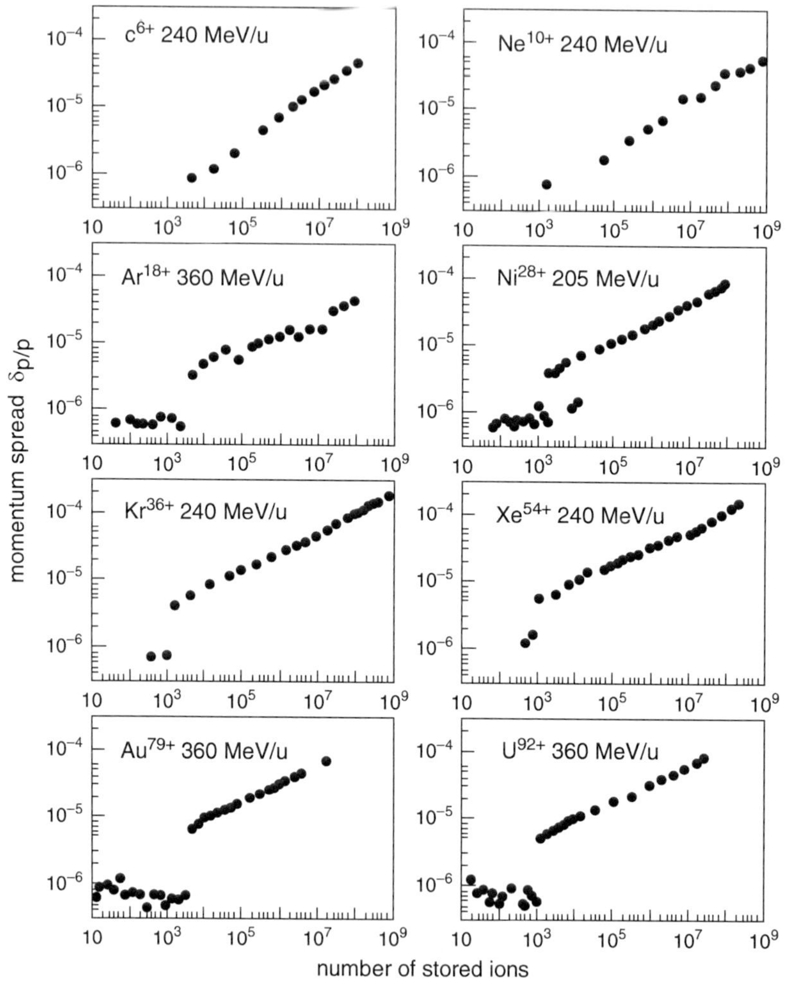

Fig. 16.4 Data of one-dimensional ordering of heavy-ion beams obtained at ESR of GSI [8]

16.3.2 Recent Attainment at S-LSR

With S-LSR at ICR, Kyoto University, the ordering of protons as the lightest ion, has been investigated to check the universality of "ion beam ordering." A proton beam of 7 MeV has been electron cooled at different particle numbers. The momentum spread and the transverse beam size was determined at different beam intensity by measuring the Schottky noise of the beam and the scraper position where the beam loss starts. In Fig. 16.5, the layout of S-LSR and its beam cooling systems are shown [11].

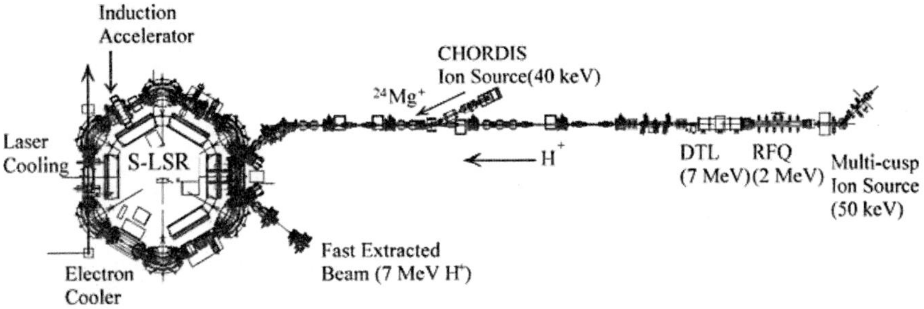

Fig. 16.5 Layout of S-LSR and its beam cooling and injection systems

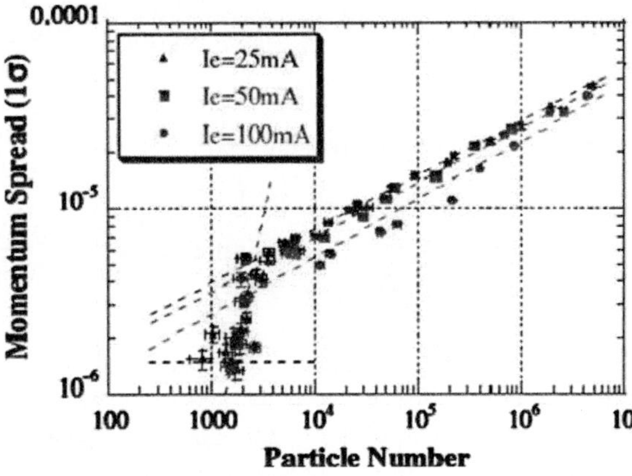

Fig. 16.6 Correlation between momentum spread of 7 MeV proton beam after electron cooling and particle number of the beam

The dependence of the momentum spread and the transverse beam size of the 7 MeV proton beam as a function of the particle number has been studied by observation of the Schottky noise of the ion beam and the scraper position where the beam loss starts. A typical example of the results are given in Fig. 16.6. The momentum spread is suddenly reduced at the particle number of ∼2,000, demonstrating the one-dimensional ordering phenomenon for 7 MeV protons, which was already realized for heavier ions. The S-LSR lattice with a superperiodicity of 6 fulfills, in contrast to other storage rings, the so-called formation and maintenance condition to create a crystalline ion beam. From the momentum spread at the transition point, a reduction of the longitudinal beam temperature from 0.17 meV (2 K) to 26 μeV (0.3 K) can be estimated which is close to the estimated temperature (20 μeV) of the electron beam. Due to the suppression of IBS the ion beam temperature is reduced close to the theoretical electron temperature limit [12].

For these phenomena, R. W. Hasse proposed a model, where the ions cannot penetrate the Coulomb barrier of each other when the ion temperature is sufficiently low. Each ion is reflected by the adjacent ions, thus they cannot take over other ions in longitudinal direction [13]. By the reflections among adjacent ions, the diffusion rate due to IBS is reduced and the momentum spread of the ion beam is

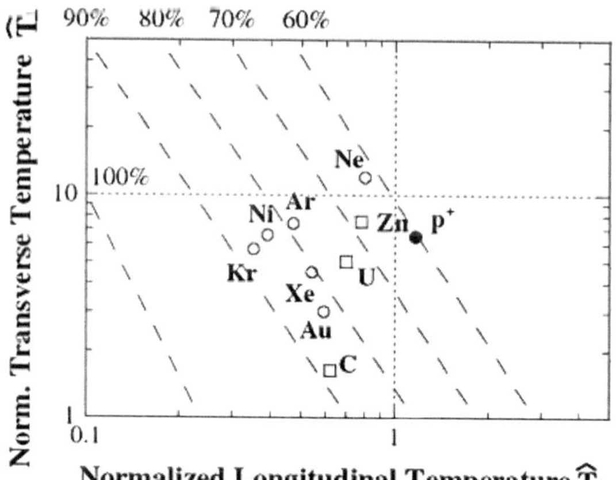

Fig. 16.7 Reflection probabilities of heavy ions at GSI (*open circle* and *square* are taken from [15] and [10], respectively, and *filled circle* is taken at S-LSR [12])

suddenly reduced at the transition point. H. Okamoto et al. defined the normalized temperatures as

$$\hat{T}_i = \frac{2k_B}{mc^2}\left(2r_p\beta\gamma\frac{\nu_0}{R}\right)^{-2/3} T_i \quad (i = \parallel or \perp), \tag{16.6}$$

where $\hat{T}_\parallel$ and $\hat{T}_\perp$ are normalized beam temperatures in the longitudinal and transverse directions, respectively, and r_p, ν_0 and R are the classical radius of the proton, betatron tune of the beam and the average radius of the ring, respectively. They proposed a scheme where the reflection probability for a single ion on adjacent ions is obtained from the phase-transition temperature [14]. In Fig. 16.7, the calculated reflection probabilities are plotted for various normalized temperatures.

Open circles and open squares denote the measurements on heavy ions at GSI [10, 15], and the data for 7 MeV protons at S-LSR is marked with a filled circle [12]. These data show that both, the heavy-ion beam with high-charge state and the single charge protons make phase transition when the reflection probability exceeds 60%, indicating that the one-dimensional ordering is a global phenomenon independent on ion species and charge state.

16.4 Crystalline Ion Beams by Laser Cooling

Laser cooling is required for all experimental observation of ion crystals in traps, which has been also extended to higher velocity ion beams in storage rings. The basic ideas of laser cooling, especially with respect to stored ion beams at high velocities is presented in this section. Because laser cooling based on repeated photon absorbtion and remission, a requirement for this application is a sufficiently strong optical transition in a wavelength range accessible to available laser technology. In ion traps, Be^+, Mg^+, Sr^+, Ba^+, In^+, Yb^+, and Hg^+ ions have been laser cooled. Other possible ions are metastable $^{6,7}Li^+$ ions. In storage rings, metastable $^{6,7}Li^+$ ions and Be^+, Mg^+, and $^{12}C^{3+}$ ions have been laser cooled up to now.

16.4.1 Attainments at Ion Traps

Ions stored and cooled in traps or storage rings can arrange themselves in a crystal-like ordered structure if the kinetic energy in their thermal motion is sufficiently low [16]. Early molecular dynamic calculations showed that a crystal-like ion beam can be reached at a plasma parameter $\Gamma \approx 170$ [17]:

$$\Gamma = \frac{q^2/\left(4\pi\epsilon_0 a\right)}{k_B T}, \tag{16.7}$$

where $a = \left(\frac{3}{4\pi\rho}\right)^{1/3}$, ρ is the ion density, and T is the temperature of the ions. Basic properties of ion crystals have been studied during the past decades in Penning and Paul traps. After sufficient cooling, the ions arrange themselves in ordered structures with typical distances of $a = 10\,\mu m$, corresponding to densities 10^{15} times smaller than usual crystals. Ion traps store ions with an arrangement of electric and/or magnetic fields. The most often used traps is the Penning trap and the Paul trap (Fig. 16.8). In a Penning trap, a static electric quadrupole potential and a homogenous magnetic field are used. This trap was developed by Dehmelt [18]. The electrodes consisting of a ring and two endcaps are hyperboloid in shape. The endcaps repelled the ions resulting in axial ion oscillations. The strong homogeneous axial magnetic field confines the particle in radial direction. In Paul traps, confinement is achieved by oscillating electrical fields without a magnetic field. The shape of the electrodes is the same as the one for the Penning trap. The oscillating

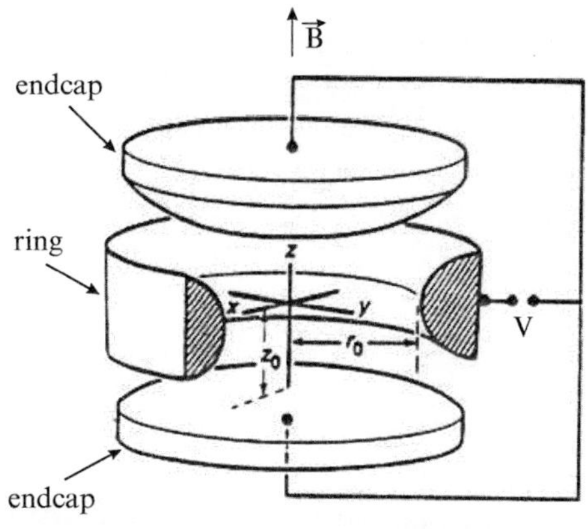

Fig. 16.8 Electrodes of a Penning or a Paul trap. The uniform magnetic field (**B**) is used in the Penning trap

electrical field focuses the particles in a half period in a certain direction and defocuses in the other part, resulting in an overall focusing force toward the trap center that is comparable to alternating gradient focusing in a storage ring.

16.4.2 Laser Cooling in a Storage Ring

In the laser cooling process taking place in a storage ring, the ion beam is irradiated with parallel or antiparallel laser light. During the directed absorption, the momentum of the photon is transferred to the ion. In the following spontaneous emission, which occurs isotropically, on average no momentum is transferred statistically. Therefore, in one absorption–emission cycle, the photon momentum $\hbar\mathbf{k}$ is transferred, where $\hbar$ is the reduced Planck constant. (Planck constant divided by 2π.) The rate at which these cycles can take place is given by the product of the excitation probability, p_e, and the spontaneous decay rate, Γ_{sp}, inverse lifetime, of the upper state. Therefore, the average force [19] resulting from the repeated scattering of photons is given as

$$\mathbf{F}_{sp} = \hbar\mathbf{k} \cdot \Gamma_{sp} \cdot p_e. \tag{16.8}$$

This force is called the spontaneous force, as shown in Fig. 16.9. The excitation probability, p_e, can be described by the Lorentzian line shape:

$$p_e = \frac{S/2}{\left(2\Delta/\Gamma_{sp}\right)^2 + 1 + S}, \tag{16.9}$$

where $\Delta = \left(1 - v_{\parallel}/c\right)\gamma\omega - \omega_0$ is the frequency detuning seen by the ion, $v_{\parallel}$is the longitudinal ion velocity, S is the optical saturation parameter defined by the ratio between laser intensity I and saturation intensity I_{sat} of the transition, and $\gamma = 1/\sqrt{1 - v^2/c^2}$ is the relativistic factor. The Doppler effect in (16.9) makes the laser force extremely sensitive to the ion velocity. The spontaneous force of a single laser beam can accelerate or decelerate the ions in a narrow velocity interval, but cannot cool the ion beam. For cooling, a stable point with $\mathbf{F}(v_s) = 0$ and $\partial\mathbf{F}/\partial v|_{v=v_s} < 0$ is needed. Under the influence of such a force, ions which are moving at velocities $v_{\parallel} < v_s$ receive a positive force and therefore are accelerated, while faster ions with $v_{\parallel} > v_s$ are decelerated. The ion distribution will finally gather

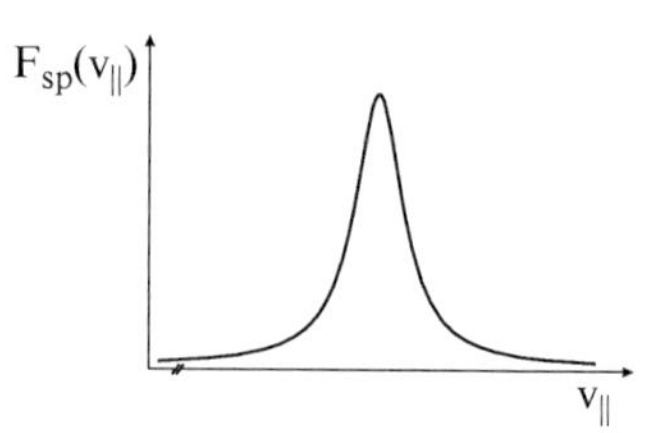

Fig. 16.9 The spontaneous laser force F_{sp} as a function of the longitudinal ion velocity

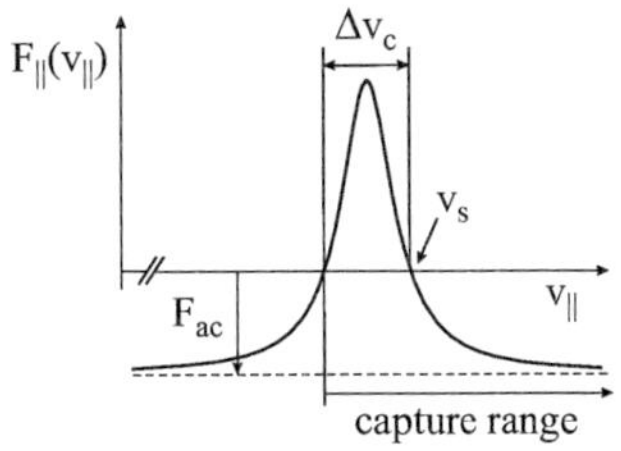

Fig. 16.10 The stable point of the laser-cooling force $F_{\|}$ in the presence of velocity independent counteracting auxiliary force F_{ac}

at the stable point v_s, where the total force vanishes. In this case, the ion motion is damped toward the stable point v_s. For the case of a constant auxiliary force, it can be seen from Fig. 16.10 that an ion which undergoes a (negative) velocity change $|\Delta v_{\|}| > \Delta v_c$ can be transferred to an unstable region of the total cooling force and will be lost. Experiments at storage rings show that the critical capture range plays a crucial role in laser cooling of high-velocity ions, whereas it is of minor importance for thermal atomic beams. The laser cooling rate, defined by: $\Lambda = -\frac{1}{T_{\|}}\frac{\partial T_{\|}}{\partial t}$, where $T_{\|}$ is the longitudinal ion temperature, related to the friction coefficient $-\partial F_{\|}/\partial v|_{v=v_s}$:

$$\Lambda = -\frac{2}{m}\frac{\partial F_{\|}}{\partial v}|_{v=v_s}, \qquad (16.10)$$

where m is the ion mass. The heating are caused by the random nature of spontaneous emission and photon absorption and can be approximated by the corresponding rate [19, 20]:

$$D_l \approx \frac{7}{10}\frac{\hbar^2 k^2 \Gamma}{k_B m} \cdot p_e, \qquad (16.11)$$

where p_e is the excitation probability defined in (16.9) and k_B is the Bolzmann's constant. In storage-ring experiments, the laser-cooling rate as well as the laser-induced heating rate is reduced by the fraction, η_L of the ring circumference because the laser beam and ion beam are merged only in one straight section of the storage ring; this leads to the ring average quantities which are: $\bar{F} = \eta_L \cdot F_{\|}$, $\bar{\Lambda} = \eta_L \cdot \Lambda$, and $\overline{D_L} = \eta_L \cdot D_L$. Stored ion beams are heated by IBS processes, described by D_{IBS} which are often much stronger, depending on the ion density, than the laser heating. With the D_{IBS}-heating rate, the equilibrium temperature, $T_{\|}$ of the ion beam can be expressed by

$$T_{\|} = \frac{D_{IBS} + \overline{D_L}}{\bar{\Lambda}}. \qquad (16.12)$$

With an induction accelerator [21] a constant axially force can be created. The induction accelerator acts like a transformer, where a ramping current in the primary windings induces a voltage in the secondary windings. Here, the secondary winding is the ion beam itself, which can, thus, be energy shifted by

$$\Delta E = qef\Delta\phi, \qquad (16.13)$$

where qe is the ion charge, f the revolution frequency of the ion, and $\Delta\phi$ the flux change caused by ramping current. An electronic feedback system guarantees a flux change with constant rate, and therefore a constant, ring averaged, auxiliary force, $\overline{F_{ac}}$ is given by

$$\overline{F_{ac}} = \frac{qe}{C_0}\frac{\Delta\phi}{\Delta t}, \tag{16.14}$$

where C_0 is the circumference of the storage ring.

16.4.3 Laser-Cooling Experiments at the TSR

As an example of laser cooling using an induction accelerator, experiments performed at the TSR on 7.3 MeV Be^+ ions are discussed. The ions are accelerated by a tandem accelerator and typically $3 \cdot 10^7$ ions are stored with multi-turn injection. With electron-cooling the initial transverse temperature of $T_\perp \approx 10^6$ K after injection is reduced to $T_\perp \approx 5,000$ K. The typically electron-cooling time is about 7 s corresponding to about half of the beam lifetime limited by stripping processes in the residual gas. The laser cooling experiments are carried out with copropagating laser beams in one straight section with a length of approximate 5 m ($\eta_L \approx 0.09$). Due to the hyperfine splitting of the ground state of 1.3 GHz two beams of Argon ion lasers are combined to provide a two frequency laser field with a frequency separation of 1.3 GHz [19]. The lasers used in the $^9Be^+$ experiments are fixed frequency Argon ion laser (wavelength of 300.3 nm) with a typical power of 100 mW. For diagnostics, the fluorescence light emitted by the ions perpendicular to the beam is detected by a photo multiplier tube [19]. The highest cooling rate, which could be achieved at TSR, using a induction accelerator, was $\Lambda = 3 \cdot 10^4\,s^{-1}$ ($S = 0.8$, $\overline{F_{ac}} = -6.3$ meV/m). The $^9Be^+$ ion number in this experiment was reduced to $\approx 10^6$ ions, in order to limit IBS effects. From fluorescence light diagnostics [19], a longitudinal equilibrium ion temperature $T_\parallel < 5$ mK could be measured. The longitudinal heating rate D_{IBS} resulting from IBS could be determined indecently by switching off the laser cooling and measuring the temperate increase by laser diagnostics. A IBS rate of $D_{IBS} = 4\,K\,s^{-1}$ was measured. This IBS rate and the cooling rate of $\bar{\Lambda} = 3 \cdot 10^4\,s^{-1}$together with the equilibrium condition, (16.9), predicts a temperature of 3 mK, which agrees well with the upper limit of $T_\perp$<5 mK measured with fluorescence light diagnostics [19].

16.4.4 Laser Cooling by Two Counterpropagating Laser Beams

A drawback of this laser cooling scheme, using an auxiliary force, however, is the existence of an unstable region [22], where the auxiliary force decelerates the ions away from the narrow region, which resonates with the laser force. These ions are lost for cooling. This disadvantage could be avoided by using co-propagate and

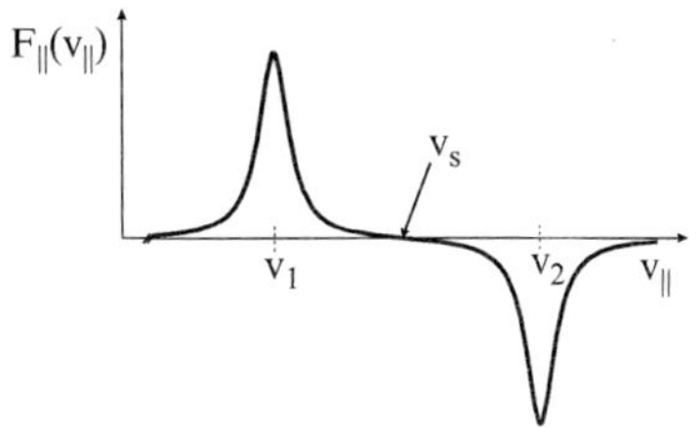

Fig. 16.11 Laser-cooling force and stable point created by two counterpropagating laser beams

a counterpropagate laser beams simultaneously. This scheme, shown in Fig. 16.11 was applied in experiments reported in [24]. The first laser beam copropagates with the ion beam and accelerates ions of velocities $v < v_s$, and the second one is counterpropagating and decelerates ions of velocity $v > v_s$ to the stable point v_s. Usually, the initial ion-velocity distribution is much larger than this narrow laser force. Therefore, the cooling in this scheme is usually started with a large $v_2 - v_1$, including all ions between v_1 and v_2. Then $v_2 - v_1$ is reduced by scanning one or both laser frequencies in order to compress the ion ensemble into a narrow velocity interval.

16.4.5 Bunched Beam Laser Cooling

The third cooling method having a large capture range is bunched beam laser cooling explored both in the ASTRID [25] and TSR [26]. Bunched beam laser cooling exploits the standard technique of radio-frequency (RF) beam bunching. The resulting synchrotron oscillations in the longitudinal RF potential are damped by resonant interaction with the laser light. This scheme readily leads to longitudinally space-charge-dominated beams. In experiments [16] at the TSR, a longitudinal cooling rate of $\Lambda \approx 25,000\,\mathrm{s}^{-1}$ was reached in the bunched scheme. Although the ion density in the RF buckets was very high when total of $\approx 2 \cdot 10^7$ ions were spatially compressed in the RF buckets by a factor of ≈ 5 with respect to a coasting beam, the cooling produced a very longitudinally cold ion beam ($T_{\|} < 100\,mK$).

16.4.6 Transverse Laser Cooling

Laser cooling reaches extremely high-longitudinal cooling rate adequate for the formation of an ion crystal, but it does not directly affect the transverse degrees of freedom. At high phase-space densities relaxation due to intra-beam Coulomb collisions leads to efficient indirect cooling of the transverse motion [27]. Measurements shown in Fig. 16.12, performed at the TSR storage ring with $^9\mathrm{Be}^+$ ions of the intensity $\approx 2 \cdot 10^7$ ($E = 7.3\,\mathrm{MeV}$), demonstrates the three-dimensional beam cooling. In the experiment, electron cooling is applied for the first 5 s after injection and leads to temperatures $T_\mathrm{H} \approx 4,000\,\mathrm{K}$, $T_\mathrm{V} \approx 400\,\mathrm{K}$ and $T_{\|} \approx 500\,\mathrm{K}$. After

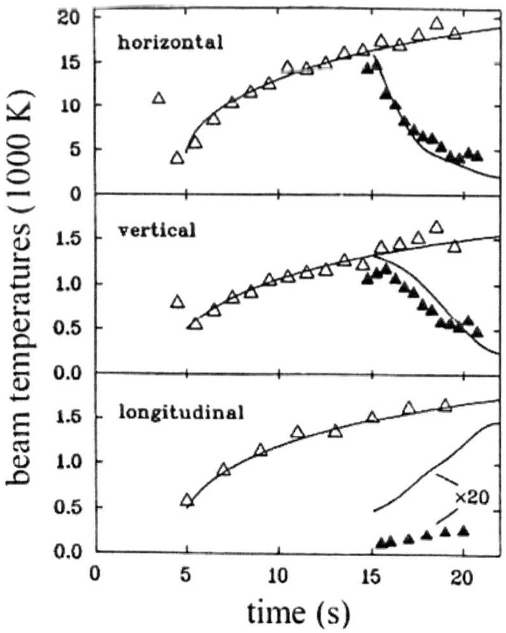

Fig. 16.12 Measurements on beam blowup (*open triangles*) and indirect transverse laser cooling (*filled triangles*) on 7.3 MeV $^9Be^+$ ions stored at the TSR. Beam temperatures shown as a function of time after injection $\approx 2 \cdot 10^7$ ions into the storage ring (initial beam current $\approx 0.7\,\mu A$). The 1/e beam life time is 25 s. The *solid lines* are fit to the experimental data using standard intrabeam scattering theory

switching off the electron cooler, a drastic heating of the ion beam is observed in all three degrees of freedom (see open triangles in Fig. 16.12). Within 10 s, this blowup already leads to temperatures $T_H \approx 16{,}000\,K$, $T_V \approx 1{,}300$ Kelvin and $T_\parallel \approx 1{,}500\,K$. When laser cooling is then applied to the beam (filled triangles in Fig. 16.12), the longitudinal temperature decreases very rapidly, on a time scale of the inverse cooling rate$1/\Lambda_\parallel \approx 10\,ms$, down to $T_\parallel \approx 6\,K$, and also reduction of the transverse temperatures are sets in immediately. Within 5 s, the transverse cooling reaches final temperatures close to the equilibrium of electron cooling. A slow increase of the longitudinal temperature to $T_\parallel \approx 15\,K$ is observed. The solid line in Fig. 16.12 is a calculation based on standard IBS theory. The temporal evolution of the laser cooling process is calculated with a cooling rate $\Lambda_\parallel = 100\,s^{-1}$. When laser cooling is applied, the transverse temperatures reach equilibrium within 5 s. From this measurement, one derives for both transverse degrees of freedom an indirect transverse cooling rate of $\Delta_\perp \approx 1\,s^{-1}$, but these rates are still far below typical longitudinal cooling rates. To stabilize crystalline order of an ion beam against shear and other destructive effects induced by the ring lattice, the longitudinal cooling force should be a function of the transverse displacement [28]. In [29], transverse cooling, solely based on single-particle interaction of ions with a laser beam, was demonstrated. The cooling scheme exploits longitudinal–horizontal coupling of the particle motion arising from the storage-ring dispersion, that is, the dependence of the ion's horizontal position on the longitudinal momentum, in combination with a transverse gradient of the light force ("dispersive cooling"). Thus, damping of the longitudinal momentum fluctuations is transferred to the horizontal degree of freedom. Transverse cooling rates in the order of $1\,s^{-1}$ have been reached at the TSR, with dispersive cooling, comparable to IBS cooling. Under the past experimental conditions, it was unrealistic to attain a sufficient longitudinal–transverse coupling for 3D ion crystals in the TSR.

16.4.7 Results at PALLAS

PALLAS, shown in Fig. 16.13, at the Ludwig-Maximilian University in Munchen is a circular RFQ. At PALLAS, a circulating $^{24}Mg^{+}$ ion beam with a velocity of $2{,}800\,m\,s^{-1}$ corresponding to a beam energy of 1 eV is stored. With Doppler-laser cooling utilizing $3s^2S_{1/2} - 3p^2P_{3/2}$ transition of the moving $^{24}Mg^{+}$ ions, it was possible to achieve phase transition, resulting in a continuous ring-shaped crystalline ion beam (Fig. 16.14 [29]).

This is the first realization of a crystalline ion beam in a "storage ring," where the stored ion beam has a finite velocity. This crystalline structure, however, was destroyed immediately when the stored ion beam is accelerated from this energy by an RF voltage applied to the drift tubes due to shear heating [31].

16.4.8 Recent Challenge at S-LSR

As it was shown in a previous section, phase transition was observed by electron cooling at very low ion numbers. In a similar manner as static materials transit from gaseous state to solid one through liquid state, a moving ion beam with high velocity

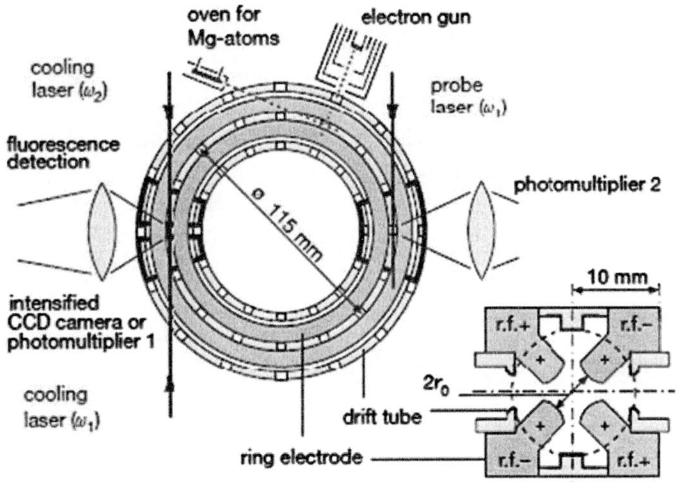

Fig. 16.13 Structure of circular RFQ, PALLAS [Schramm, private communication]

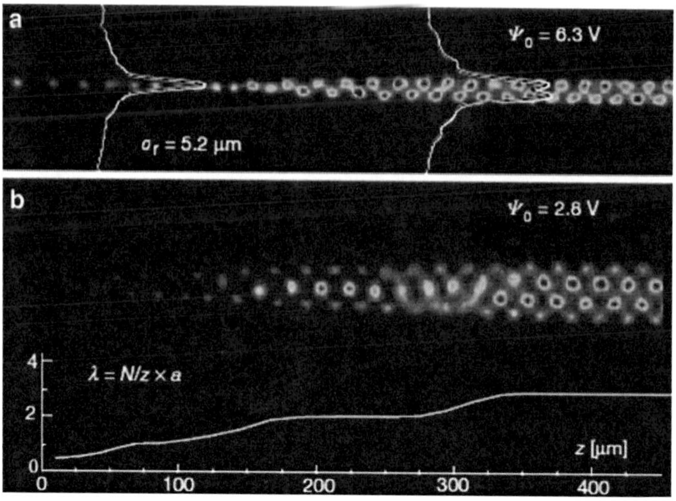

Fig. 16.14 Images of ion crystals at rest in PALLAS [29]

is also expected to change its existing state by cooling. With electron beam cooling, temperatures, corresponding to the temperature of the electron beam itself, which is in the order of Kelvin, should be possible for the ion beam. To overcome this limitation, laser cooling, which has a much stronger cooling force, can be applied.

16.4.9 Longitudinal Laser Cooling

At S-LSR, one-dimensional laser cooling in the longitudinal direction has been applied for a $^{24}Mg^+$ beam with a kinetic energy of 40 keV. The acceleration by the laser photon is balanced with the deceleration force of an induction accelerator as shown in Fig. 16.5. A typical example of laser cooling at S-LSR with about 10^6 ions is shown in Fig. 16.15. The correlation between the beam intensity and the equilibrium cooled temperature in the longitudinal direction is presented in Fig. 16.16. The lowest-equilibrium-cooled beam temperature was 3.6 K at the beam intensity of about 10^4 ions in the equilibrium between IBS and longitudinal laser cooling [32].

16.4.10 Transverse Laser Cooling with Synchro-Betatron Coupling

Following the results of transverse laser cooling at TSR using intrabeam scattering and "dispersive cooling" described above, "Synchro-betatron coupling" proposed

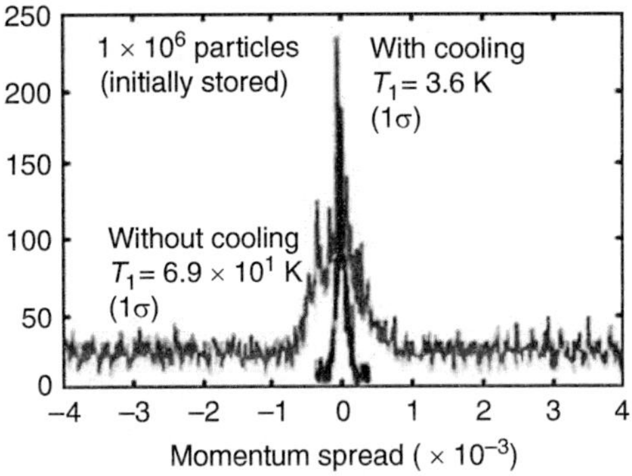

Fig. 16.15 Example of longitudinal laser cooling for a coasting beam at a beam intensity of $1 \cdot 10^6$ [32]

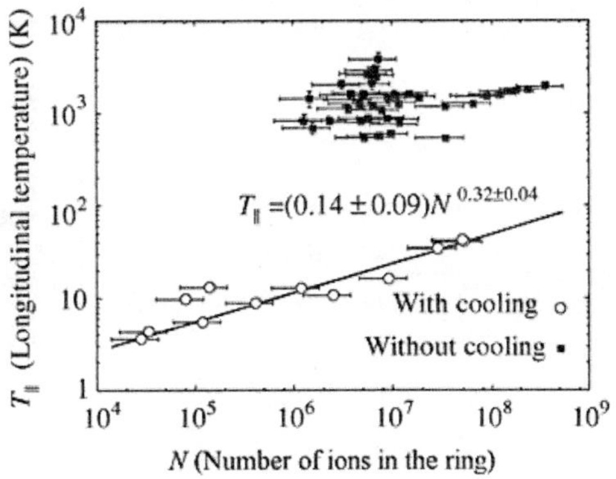

Fig. 16.16 Intensity dependence of the longitudinal-cooled temperature by the laser cooling for a coasting beam [32]

by Okamoto et al. [33], has been applied at S-LSR. In order to couple the longitudinal motion with the horizontal one, RF bunching at a finite dispersion ($D = 1.1$ m) was used at S-LSR. The main parameters for this laser cooling scheme at S-LSR are listed up in Table 16.1 [34]. For synchro-betatron coupling, the condition:

$$\nu_s - \nu_H = m \quad (m : integer), \tag{16.15}$$

has to be fulfilled, where ν_s and ν_H are synchrotron tune and horizontal betatron tune, respectively, In this cooling scheme, kinetic energy is transferred to the longitudinal direction resulting in reduction of the horizontal beam size.

In Fig. 16.17, the dependence of the betatron and synchrotron tunes (a) and the observed horizontal beam size by a CCD camera, and the measured momentum spread by a Schottky pickup (b) as a function of the applied RF voltage are given [34, 35].

It is shown that the horizontal beam size has its local minimum at the RF voltage of approximately 30 V, where the "synchro-betatron resonance" condition given by (16.15) is satisfied. From the values of the momentum spread and the horizontal beam size shown in Fig. 16.17 together with the values, β_H and α_H of 0.89 m and 0, respectively, at the observation point of the horizontal beam size, a temperature for

Table 16.1 Parameters of laser cooling at S-LSR

Circumference	22.557 m
Average radius	3.59 m
Length of straight section	1.86 m
Radius of curvature	1.05 m
Superperiodicity	6
Ion species	$^{24}Mg^{+}$:40 keV
Initial momentum spread	$1 \cdot 10^{-3}$
Initial particle number	$3 \cdot 10^{7}$
Betatron tunes	(2.068, 1.105)
Synchrotron tune	$0.0376 \approx 0.1299$
Laser frequency	$1,074,110.3 \pm 0.05$ GHz
Detuning	-0.2 ± 0.005 GHz
Laser power at exit window	$11 \sim 20$ mW

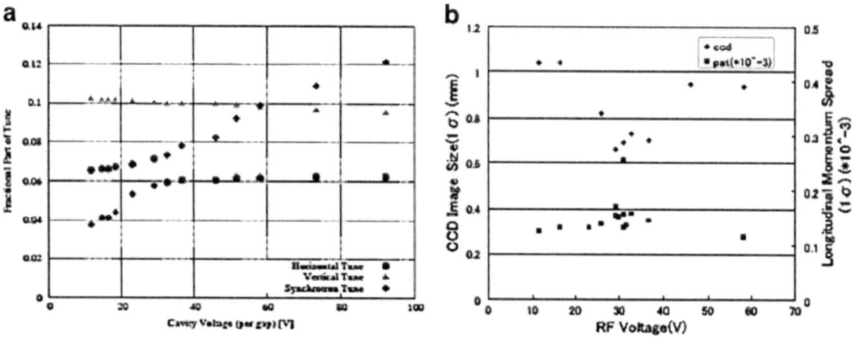

Fig. 16.17 Dependence of various tunes (**a**) and horizontal beam size and momentum spread (**b**) on the applied RF voltage

the 40 keV $^{24}Mg^{+}$ ion beam at the resonant condition is estimated to be around 60 and 500 K for the longitudinal and transverse directions, respectively. From the data presented in Fig. 16.17b, the equilibrium longitudinal temperature by laser cooling is estimated to be ~15 K, which is consistent with the former result of 18 K, obtained by a bunched longitudinal laser cooling at S-LSR [36].

16.4.11 Further Approach Toward Ultralow Temperature at S-LSR

From the experimental results up to now, the $^{24}Mg^{+}$ with a kinetic energy of 40 keV injected into S-LSR has a temperature of about 660 K in the longitudinal direction [36] and ≈2,000 K in the transverse direction (estimated from the data in Fig. 16.17b). Due to the rather high initial transverse temperature after injection, the

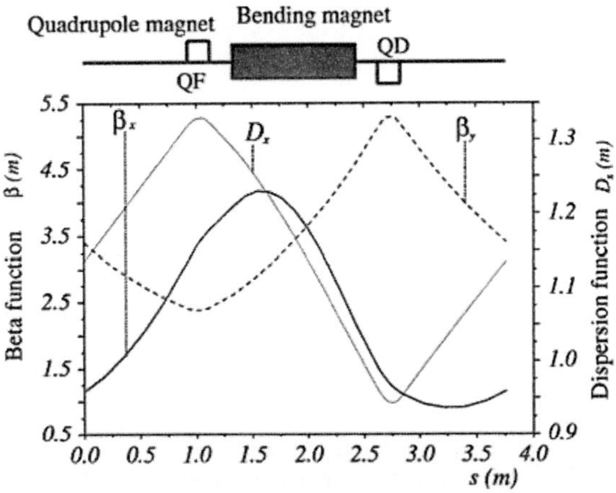

Fig. 16.18 Beta and dispersion functions of S-LSR with the operation point (2.07, 1.07, 0.07) satisfying the maintenance condition and synchro-betatron resonance

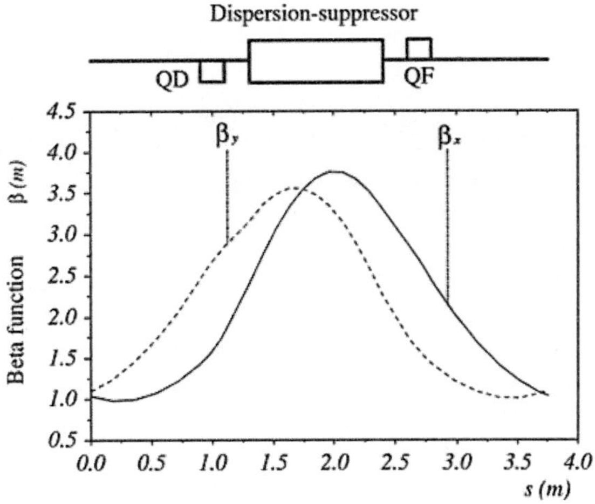

Fig. 16.19 Beta and dispersion functions of dispersion-free mode of S-LSR with the operation point of (2.07, 2.07) [37]

equilibrium longitudinal temperature remains at 60 K, which should be improved by collimation of the injection beam and/or application of preelectron cooling (Fig. 16.18).

Lattice of S-LSR has a superperiodicity, $N_{sp} = 6$, satisfying the so-called "maintenance condition" given by the following relations:

$$\gamma \leq \gamma_t, \tag{16.16}$$

$$N_{sp} \geq 2\sqrt{\nu_H{}^2 + \nu_V{}^2}, \tag{16.17}$$

where γ_t, ν_H and ν_V represent transition γ, horizontal and vertical tunes, respectively [30, 36]. S-LSR also can be operated with dispersion-free mode as shown in Fig. 16.19, where the dipole magnetic field to bend the ion beam is overlapped with an electric field (called as "dispersion suppressor") as shown in Fig. 16.20 [37, 38].

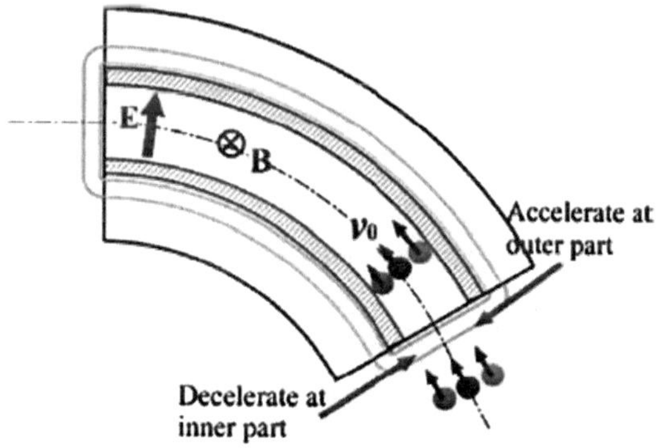

Fig. 16.20 Deflection element adopted at S-LSR which has such a characteristic as dispersion suppressor

The dispersion suppressor consists of crossed electric and magnetic fields, satisfying the relation

$$\left(1 + 1/\gamma^2\right) E = -v_0 \times B, \tag{16.18}$$

where B, E, and v_0 are the magnetic field, the electric field, and the velocity of the reference particle in the beam, respectively. The idea of dispersion suppressor was originally published much earlier [39, 40] and its capability for avoiding shear was also proposed by Pollock [41]; our S-LSR, however, is the first realization of such a storage ring utilizing dispersion supressors. In the dispersion suppressor, the beam is bent without linear dispersion. Such a mode is effective for realization of a three-dimensional crystalline beam by suppression of "shear heating." The realization of three-dimensional crystalline beams as well as the realization of strong threedimensional cooling rates [42] to overcome IBS are the major research subjects at S-LSR from now on.

16.5 Conclusion

Stochastic cooling was used to realize an antiproton beam used to create weak bosons, contributing very much to elementary particle physics. By the application of electron beam cooling, one dimensional ordered state has been realized and beam temperature has reached to the order of 0.1 K in the longitudinal direction although the transverse temperature remains to be higher ($\sim$10 K). With laser cooling, the lowest longitudinal temperature such as about 3 mK was realized, but the transverse temperature remains rather hot ($\sim$500 K). The creation of an ultracold ion beam with a temperature in the similarly cold region in all degrees of freedom is our future research goal in the beam crystal research.

References

1. G.I. Budker et al., Part. Accel. **7**, 197 (1976)
2. M. Bell et al., Nucl. Instr. Meth. **190**, 237 (1981)
3. D. Mohl, *A Comparison between Electron Cooling and Stochastic Cooling*, Proc. of ECOOL84, p. 293 (1984)
4. Y.A.S. Derebenev, A.N. Skrinsky, Sov. Phys. Rev. **1**, 165 (1979)
5. H. Fadil et al., Nucl. Instr. Meth. **A517**, 1 (2004)
6. S. Van der Meer, Rev. Mod. Phys. **57**, 689 (1985)
7. V.V. Parkhomchuk, *Physics of Fast Electron Cooling*, Proc. of ECOOL84, p. 71 (1984)
8. M. Steck et al., Phys. Rev. Lett. **77**, 3803 (1996)
9. H. Danared et al., J. Phys. B At. Mol. Opt. Phys. **36**, 1003–1010 (2003)
10. M. Steck et al., Nucl. Instrum. Meth. **A532**, 357 (2004)
11. A. Noda, Nucl. Instrum. Meth. **A532**, 150 (2004)
12. T. Shirai et al., Phys. Rev. Lett. **98**, 204801 (2007)
13. R.W. Hasse, Phys. Rev. Lett. **83**, 3430 (1999)
14. H. Okamoto et al., Phys. Rev. **E69**, 066504 (2004)
15. R.W. Hasse, Nucl. Instrum. Meth. **A532**, 382 (2004)
16. D. Habs, R. Grimm, Annu. Rev. Nucl. Part. Sci. **45**, 391 (1995)
17. J. Schiffer, Proceedings of the workshop on crystalline ion beams, (*Wertheim*, 1988), p. 2
18. H. Dehmelt, Angew. Chem. **102**, 774 (1990)
19. W. Pertrich et al., Phys. Rev. **A48**, 2127 (1993)
20. R. Grimm et al., Workshop on Beam Cooling and Related Topics(Montreux, Switzerland, 1993), CERN94-03, p39
21. C. Ellert et al., Nucl. Instrum. Meth. **A314**, 399 (1992)
22. H.-J. Miesner et al., Workshop on Beam Cooling and Related Topics(Montreux, Switzerland, 1993), CERN94-03, p349
23. S. Schroder et al., Phys. Rev. Lett. **64**, 2901 (1990)
24. J. Hangst et al., *Workshop on Beam Cooling and Related Topics* (Montreux, Switzerland, 1993), CERN94-03, p. 343
25. H.-J. Miesner et al., Nucl. Instrm. Meth. **A393**, 634 (1996)
26. H.-J. Miesner et al., Phys. Rev. Lett. **77**, 623 (1996)
27. J. Wei et al., Phys. Rev. Lett. **80**, 2606 (1998)
28. I. Lauer et al., Phys. Rev. Lett. **81**, 2052 (1998)
29. T. Schatz, U. Schramm, D. Habs, Crystalline ion beams. Nature **412**, 717 (2001)
30. J. Wei, X.-P. Li, A.M. Sessler, Phys. Rev. Lett. **73**, 3089 (1994)
31. M. Tanabe et al., Appl. Phys. Express **1**, 028001 (2008)
32. H. Okamoto, A.M. Sessler, D. Mohl, Phys. Rev. Lett. **72**, 397 (1994)
33. M. Nakao et al., Proc. of IPAC'10, Kyoto, Japan, pp. 858–860 (2010)
34. H. Souda et al., Proc. of IPAC'10, Kyoto, Japan pp. 861–863 (2010)
35. M. Nakao, Master Thesis at Kyoto University (2008)
36. X.-P. Li, H. Enokizono, H. Okamoto, Y. Yuri, A.M. Sessler, J. Wei, Phys. Rev. ST Accel. Beams **9**, 034201 (2006)
37. A. Noda, M. Ikegami, T. Shirai, New J. Phys. **8**, 288 (2006)
38. M. Ikegami, A. Noda, M. Tanabe, M. Grieser, H. Okamoto, Phys. Rev. ST Accel. Beams **7**, 120101 (2004)
39. W. Henneberg, Ann. Phys. Lpz. **19**, 335 (1934)
40. W.E. Millet, Phys. Rev. **74**, 1058 (1948)
41. R.E. Pollock, Z. Phys. A, Hadrons and Nuclei **341**, 95 (1991)
42. J. Wei, H. Okamoto, A.M. Sessler, Phys. Rev. Lett. **80**, 2606 (1998)

Chapter 18
Atomic and Molecular Data on Internet

D. Humbert and R.E.H. Clark

Abstract Getting reliable atomic and molecular data for applications is a major challenge. In some instances, data can be obtained using analytical functions as proposed earlier. There are also some atomic and molecular codes available through the Internet, which can generate data online. However, the major trend currently is to download the information directly through the Internet.

Web-based atomic and molecular databases have the advantage, when available, to readily provide reliable original data and possibly evaluated and/or recommended data. In this chapter, the current state of atomic and molecular and online calculation tools in early 2011 is described, with special focus on databases with a convenient user web-interface, as well as information distributed through data centers.

18.1 Data Quality and Data Assessment

The Internet is the major source of information worldwide. But this information is expanding so rapidly, it is difficult to have a high level of confidence in information or to perform ones own evaluation of such information. Atomic and molecular data follow the same trend and a researcher must be critical regarding the data retrieved from the web. Any user has to do an assessment of the data to be able to use them with confidence. Some data centers provide data with only a little quality control while some databases, like atomic spectra database (ASD) [1] at NIST or Basecol [2] at Observatoire de Paris, contain exclusively evaluated data. We present here some general criteria to quickly evaluate numerical data retrieved from the web.

Any information should be traceable. Without traceability, it is very difficult to assess any data. The source of the reference gives important information on the data quality:

- Type of paper: Refereed paper, lab report...
- Year of publication
- Nature of the institution originally reporting the data
- Authors

V. Shevelko and H. Tawara (eds.), *Atomic Processes in Basic and Applied Physics*, Springer Series on Atomic, Optical, and Plasma Physics 68,
DOI 10.1007/978-3-642-25569-4_18,

Information on the data provider is also of importance:

- Nature of the data provider: Data center, research laboratory…
- A system for error reporting on the part of users

Data quality also depends on some additional information provided by the data providers:

- Link to the original bibliographic data
- Evaluated and/or recommended data
- Accuracy and/or error bars
- Type of numerical data: Raw data, fitting function, interpolation…
- Information on the method used
  - Experimental, theoretical, compilation…
  - Additional information on the experiment and diagnostics, method used for calculation, what code…
- Year of data production

Among these criteria, none is dominant, only their combination permits a quick evaluation. For example, it would be a mistake to think systematically that a data center, wellknown for the quality of their work, always provide the best data. To be more explicit, the database, ASD at NIST, includes only evaluated data and is recognized as the most accurate atomic spectroscopic database. Even so, in ASD, some data may be outdated.

Transitions for Ar III (4s → 4p) between 3,000 and 3,100 A, for example, are retrieved from ASD, Fig. 18.1. For some lines, wavelengths and transition probabilities, referenced as T1619 and L7493, are given with estimated accuracies D and E (≤50% and >50%, respectively). Thanks to the traceability of the data provided by NIST, source references T1619 and L7493 are known to have been published in 1969 and 1937, respectively, and the method used CA and LS refer to:

- CA: Calculated using the Coulomb approximation.
- LS: the transition probability for this line was calculated from the mul-tiplet value assuming a pure LS-coupling.

3 002.6407	3 002.6399	16			211 063.760 - 244 358.081	$3s^23p^3(^2P^\circ)4s - 3s^23p^3(^2P^\circ)4p$	$^1P^\circ - {}^1D$	1 - 2	3 - 5			L11520
	3 005.22		4.3e-02	D	0.000 - 33 265.7	$3s^23p^4 - 3s^23p^4$	$^3P - {}^1S$	2 - 0	5 - 1	E2	T761	
3 010.02	3 009.985	10			210 212.72 - 243 425.7974	$3s^23p^3(^2P^\circ)3d - 3s^23p^3(^2P^\circ)4p$	$^3D^\circ - {}^3P$	3 - 2	7 - 5			L7493
3 023.9801	3 023.9806	5	2.6e+08	D	207 233.003 - 240 292.3714	$3s^23p^3(^2P^\circ)4s - 3s^23p^3(^2P^\circ)4p$	$^3P^\circ - {}^3D$	2 - 3	5 - 7		T1619CA,LS	L11520
3 027.16	3 027.0864	5	6.4e+07	D	207 233.003 - 240 258.453	$3s^23p^3(^2P^\circ)4s - 3s^23p^3(^2P^\circ)4p$	$^3P^\circ - {}^3D$	2 - 2	5 - 5		T1619CA,LS	L7493
3 036.96	3 036.8767	3	7.0e+06	E	207 233.003 - 240 151.990	$3s^23p^3(^2P^\circ)4s - 3s^23p^3(^2P^\circ)4p$	$^3P^\circ - {}^3D$	2 - 1	5 - 3		T1619CA,LS	L7493
3 054.7736	3 054.775	3	1.9e+08	D	207 532.33 - 240 258.453	$3s^23p^3(^2P^\circ)4s - 3s^23p^3(^2P^\circ)4p$	$^3P^\circ - {}^3D$	1 - 2	3 - 5		T1619CA,LS	L11520
3 064.77	3 064.745	10	1.0e+08	D	207 532.33 - 240 151.990	$3s^23p^3(^2P^\circ)4s - 3s^23p^3(^2P^\circ)4p$	$^3P^\circ - {}^3D$	1 - 1	3 - 3		T1619CA,LS	L7493
3 078.15	3 078.1323	10	1.4e+08	D	207 674.192 - 240 151.990	$3s^23p^3(^2P^\circ)4s - 3s^23p^3(^2P^\circ)4p$	$^3P^\circ - {}^3D$	0 - 1	1 - 3		T1619CA,LS	L7493
3 083.64	3 083.598	3			211 005.56 - 243 425.7974	$3s^23p^3(^2P^\circ)3d - 3s^23p^3(^2P^\circ)4p$	$^3D^\circ - {}^3P$	2 - 2	5 - 5			L7493

Fig. 18.1 NIST ASD output, Ar III (4s → 4p)

Information given on the year of publication, 1937 and 1969, as well as the method used, drives to the conclusion that more accurate data may be found on the web or may be computed by using some online calculation tools, like the LANL Atomic Physics Codes [3].

To conclude on data quality regarding A+M data on the web, except maybe for atomic spectroscopic data, one should be happy if his quest is successful. This also implies to stay critical about the information obtained.

18.2 Atomic and Molecular Numerical Databases

Most of the dissemination of the atomic and molecular data is done though data centers. Some research labs or personal pages may have their own web database, but this brings the problem of maintenance and often infrequent or no data updating or corrections. Some data centers also provide online calculation tools.

From the following two Tables 18.1 and 18.2, one may notice that the atomic field is better covered than the molecular one. Also, it is easier to find spectroscopic data than collisional ones. Additional information regarding hot plasma applications may be found in [4] for some of these databases and calculation tools.

Curtin University http://atom.curtin.edu.au/CCC-WWW/

The CCC database provides excitation including magnetic sublevel and ionization cross sections by electron impact for a few chemical elements. Data are the results of convergent-close-coupling calculations (CCC). The CCC method yields accurate excitation and ionization cross sections for atomic and ionic targets which are well modeled by one or two valence electrons above a Hartree–Fock core.

Harvard-CfA http://www.cfa.harvard.edu/amp/tools.html

Several databases are maintained at the Harvard Smithsonian Center for Astrophysics. Most relevant for atomic and molecular applications are the two spectroscopic databases: Kelly Atomic Linelist, the interface allows selective searches on atomic spectral lines, from X-ray to visible, compiled by R. L. Kelly. Data provided include energy levels, vacuum wavelengths, intensity estimates, and source references Kurucz Atomic Linelist, spectral lines from 250 nm to 0.4 mm are taken from Bob Kurucz' CD-ROM 23 of spectroscopic line calculations [5]. Among data provided are wavelengths, transitions probabilities, oscillator strengths, and source references.

IAEA http://www-amdis.iaea.org

The Atomic and Molecular Data Unit of the IAEA supports many international projects for production, collection, evaluation, and dissemination of atomic and molecular data for use in fusion energy research and other plasma applications. The ALADDIN database provides recommended collisional data, cross sections, and rate coefficients, on various atomic and molecular processes such

Table 18.1 Numerical databases

Institution: databases	A+M data	Spectroscopic data	Collisional data
Curtin University, Australia			
CCC database [6]	A		XS
Harvard-Smithsonian CfA, USA			
Kelly [7]	A	EL, TP	
Kurucz [5]	A	EL, TP	
IAEA, Austria			
ALADDIN [8]	A+M		RC, XS
GENIE [9]	A	EL, WL, TP	RC, XS
IAPCM, China			
CAMDB [10]	A+M	EL, WL, TP	RC, XS
KAERI, Korea			
AMODS [11]	A+M	EL, WL, TP	XS
NASA, USA			
uaDB [12]	A+M	EL, WL, TP	RC, XS
NIFS, Japan			
AMDIS [13]	A+M	RC, XS	
NIST, USA			
ASD [1]	A	EL, WL, TP	
MCHF/MCDHF [14]	A	EL, WL, TP	
Molecular collisions [15]	A+M		XS
Observatoire de Paris, France			
BASECOL [2]	M		RC, XS
ADAS, United Kingdom			
OPEN-ADAS [16]	A	EL, WL, TP	RC
ORNL, USA			
MIRF results [17]	A		XS
Opacity and iron project			
TIPTOPbase [18]	A	EL, WL, TP	RC, XS
V NIITF-VNIIFTRI, Russia			
Spectr-W3 [19]	A	EL, WL, TP	RC, XS

Keys: *A*, atomic data, *M*, molecular data, *EL*, energy levels, *TP*, transition probabilities, *WL*, wavelength, *RC*, rate coefficients, *XS*, cross sections

Table 18.2 Calculation tools

Institution: databases	A+M data	Spectroscopic data	Collisional data
IAEA, Austria			
LANL physics codes [3]	A	EL, WL, TP	RC, XS
Heavy particle [20]	A		XS
AAEXCITE [21]	A		XS
KAERI, Korea			
AMODS [11]	A	EL, WL, TP	
NIFS, Japan			
AMDIS [13]	A		RC, XS

Keys are similar to Table 18.1

as electron-impact, photon and heavy-particle collisions. An interface to particle–surface interactions is also available.

IAEA also maintains a General Internet Search Engine (GENIE) which allows simultaneous data retrievals from several atomic databases. This is a very useful tool to quickly retrieve spectroscopic data, such as wavelengths and energy levels, or collisional data such as rate coefficients and cross sections.

IAEA and LANL http://aphysics2.lanl.gov/tempweb/lanl/

The gateway to the Los Alamos Atomic Physics Codes is a joint-project between the IAEA and the Atomic and Optical Theory group at Los Alamos National Laboratory. This interface is a powerful online calculation tool for atomic spectroscopic data and collisional data by electron impact.

The atomic structure is calculated using the Cowan package of atomic codes based on the Hartree–Fock method [22]. Electron-impact cross sections can be determined by different methods, in particular, the excitation cross sections by the distorted-wave method or the first-order many-body theory, while the ionization cross sections may use the scaled hydrogenic, binary encounter, or distorted-wave methods. Calculation of photoionization cross sections or autoionization probabilities can also be performed.

IAPCM http://www.camdb.ac.cn/e/

The China Research Association of Atomic and Molecular Data (CRAAMD) of the IAPCM, brings together scientists from 10 Chinese institutions. The aim of CRAAMD is to collect, assess, and compile atomic and molecular data for various elementary processes, with a focus on data needed in plasma simulation and diagnosis. More than 850,000 records are currently available within two databases, the Atomic Database and Molecular Database. These comprise atomic and molecular structure and spectra, as well as numerous atomic and molecular processes. A special effort is underway to assess the data and to improve the consistency of the data presentation.

KAERI http://amods.kaeri.re.kr

The information system AMODS at the Korea Atomic Energy Research Institute (KAERI) proposes spectroscopic and collisional data. The collisional database includes electron-impact ionization cross sections for several neutral and singly ionized atoms and molecules. Online calculations can be performed using a multi-configuration Dirac–Fock code developed by Desclaux [23].

NASA http://heasarc.gsfc.nasa.gov/uadb/

The Universal Atomic Database (uaDB) provides atomic data for astrophysics and other plasma applications. Initially devoted to the X-ray astrophysics, the aim of uaDB is now to be as comprehensive as possible to reach a wider range of atomic data users. It includes spectroscopic data as well as some collisional ones. Source reference and information are given with the data.

NIFS http://dbshino.nifs.ac.jp

The NIFS in Japan, supports a program for the production, collection, evaluation, and dissemination of atomic and molecular data relevant to the fusion energy research. The Atomic and Molecular Information System (AMDIS) of NIFS is a useful interface to a wide range of atomic and molecular collision data. AMDIS also provides data on sputtering yields. Finally, some calculation tools are available online to calculate ionization cross sections and rate coefficients by electron impact using the Lotz formula, charge-exchange cross sections, sputtering yields, and back-scattering coefficients.

NIST http://www.nist.gov/pml/data/

The National Institute of Standards and Technology (NIST) provides both spectroscopic and collisional data for use in atomic applications. The ASD is probably the most extensive and comprehensive tool for atomic spectroscopic data. It includes recommended and evaluated data on atomic structures and spectral lines. Physical quantities are given with the full method description and linked the NIST bibliographic databases.

MCHF/MCDHF, the Multiconfiguration Hartree–Fock and Multiconfiguration Dirac–Hartree–Fock database is hosted at NIST. This database contains theoretical atomic data computed either in the LSJ Breit-Pauli or the multi-configuration Dirac–Hartree–Fock approximation.

Regarding collisional data, NIST provides also some other useful atomic and molecular data [24] as the electron-impact cross sections for ionization and excitation. Major data are ionization cross sections of molecules by electron impact, but the database also includes cross sections for some atoms and energy distributions of ejected electrons for H, He, and H2. The cross sections are calculated using the binary-encounter-bethe (BEB) model, which combines the Mott cross section with the high-incident energy behavior of the Bethe cross section. Selected experimental data are included. Electron-impact excitation cross sections are also included for some selected atoms.

Observatoire de Paris http://basecol.obspm.fr/

This database BASECOL is devoted to collisional rovibrational excitation of molecules by colliders such as atom, ion, molecule, or electron. Focus is at present on collisional systems of interest for various astrophysical media. Data are evaluated with a full quality control, including full description on methods and bibliographic references. Numerical outputs are linked to the relevant reference information. Rate coefficients for the systems are currently available via a restricted access.

Opacity Project, Iron Project http://cdsweb.u-strasbg.fr/OP.htx

The Opacity Project (OP) and Iron Project (IP) are two international collaborations to calculate the extensive atomic data required to estimate stellar envelope opacities and to compute Rosseland mean opacities and other related quantities. The TOPbase and TIPbase are the produced databases from these respective projects. Both databases provide energy levels, transition probabilities and oscillator strengths calculated in pure LS-coupling for the TOPbase and intermediate coupling for the TIPbase data. Additionally, the TIPbase includes (effective) electron-impact excitation collision strengths, while the TOPbase contains photoionization cross sections.

OPEN-ADAS http://open.adas.ac.uk/

OPEN-ADAS is the free web access to ADAS data. The Atomic Data and Analysis Structure (ADAS) is an interconnected set of computer codes and data collections for modeling the radiating properties of ions and atoms in plasmas. Data available in ADAS to assist in the analysis and interpretation of spectral emission and to support detailed plasma models are freely accessible through OPEN-ADAS. Provided data sets can address plasmas ranging from the interstellar medium through the solar atmosphere and laboratory thermonuclear fusion devices to technological plasmas.

ORNL URL http://www-cfadc.phy.ornl.gov

One objective of the Controlled Fusion Atomic Data Center (CFADC) at the Oak Ridge National Laboratory (ORNL) is to produce, collect, evaluate, and compile numerical atomic and molecular data for use for controlled fusion energy research. The web site provides various collisional data, covering various types of collisions. MIRF experimental results include electron-impact excitation and ionization cross sections.

VNIITF-VNIIFTRI http://spectr-w3.snz.ru

Atomic spectroscopic and collisional databases are maintained at VNIITF. The SPECTR-W3 database contains about 450,000 data describing energy levels, spectral lines, transition probabilities, and ionization potentials. The collisional section presently includes cross sections on electron-impact excitation, electron-impact ionization, and dielectronic recombination. For both databases, all data are linked to the relevant bibliographical source.

18.3 Bibliographical Data

While most of the numerical databases provide a direct link to the source references, it may be of interest, when starting a new project or making some technology watch, to look for purely bibliographic data. This can be done using bibliographic databases or specialized scientific search engines.

The advantage of bibliographic databases over search engines is that with the data stored in a structures manner, it is possible to search the data on different sets of criteria. Unfortunately this implies an extensive and expensive commitment of expertise to identify and analyze the data. Therefore, only a few remaining databases are still active in this field, in particular, for atomic spectroscopic data the NIST bibliographical databases [25] and for atomic and molecular collisions the IAEA AMBDAS [26] and ISAN BIBL [27].

NIST bibliographic atomic databases http://www.nist.gov/pml/data/asbib/index.cfm

The NIST bibliographic atomic databases is the most relevant bibliographic databases for atomic spectroscopy:

- Atomic Transition Probabilities Bibliographic Database
- Atomic Spectral Line Broadening Bibliographic Database
- Atomic Energy Levels and Spectra Bibliographic Database

AMBDAS http://www-amdis.iaea.org/AMBDAS/
The bibliographic database AMBDAS covers both collisional and spectroscopic data, as well as particle surface interactions, with the literature references dating from 1918. The database is regularly updated, thanks to the data mostly provided by NIST for the spectroscopic data and by the CFADC at ORNL for the collisional data. The data structure permits easy search on various criteria.

ISAN http://das101.isan.troitsk.ru/bibl.htm
The Institute of Spectroscopy (ISAN), Russia, maintains BIBL, a bibliographic atomic database. The database is regularly updated and presents a comprehensive coverage of the literature for collisional processes and spectroscopic data.

18.3.1 Google Scholar and CrossRef

The indexing of the bibliographical data requires an important effort in expertise, and some gaps or delay in updating are not uncommon, especially for the collisional data. It is then often useful to complete the search, using available scientific search engines. The difference between Google Scholar [28] and CrossRef [29], both supported by Google, resides in their searching scope. Google Scholar searches data in any information considered as scientific (scientific paper, laboratory reports, slideshows, etc.), while CrossRef only checks scientific publications with a Digital Object Identifier (DOI), being a unique ID for a electronic information (in this case, a scientific publication through a publisher).

Google Scholar also provides an advanced search tool, which is definitely missing for CrossRef. Well-known publishers usually provide a portal to a CrossRef search.

To complete this section, it must be mentioned that a list of numerical and bibliographic web sites on atomic and plasma physics is maintained at the Weizmann Institute of Science [30].

18.4 New Trends

With the emergence of the world wide web, masses of Data in atomic and molecular data are available. Today, as shown above, a multitude of data sources which can satisfy most of users information needs are available. Being able to manage these large sets of data or information sources potentially available on line and in a distributed environment has become a major necessity.

Since this information is generally independently built and managed, data sources can be highly heterogeneous. This heterogeneity is obvious while using the IAEA web search engine GENIE [9]. In a single search, thousands of lines may be retrieved from nine different databases. Unfortunately, the output format differs from one site to the others, making the utilization of these data cumbersome.

Efforts, mainly through data centers, are underway to deal with the numerous, distributed, highly heterogeneous, replicated, and changing information. Some of these efforts are:

- The Data centre Network (DCN) [31], coordinated by the International Atomic Energy Agency (IAEA). The goal of the DCN is to establish priorities in compilation, evaluation, and generation of atomic and molecular data for controlled fusion energy research.
- The XSAMS project [32, 33], also coordinated by the IAEA. XSAMS (Xml schema for atoms, molecules and solids) is an XML standard for the dissemination and exchange of atomic and molecular data.
- The Virtual Atomic and Mocular Center (VAMDC) [34] is a European project to develop a universal portal to atomic and molecular data.
- The Non-LTE Workshop [35].

References

1. Atomic Spectra Database (ASD), National Institute of Science and Technology, NIST, http://www.nist.gov/pml/data/asd.cfm
2. BASECOL, Observatoire de Paris, France, http://basecol.obspm.fr/
3. LANL Atomic Codes, Los Alamos National Laboratory, USA, http://aphysics2.lanl.gov/tempweb/lanl/
4. Y. Ralchenko, A guide to internet atomic databases for hot plasmas. J. Quant. Spectros. Radiative Tran. **99**, 499–510 (2006). doi:10.1016/j.jqsrt.2005.05.040
5. Atomic spectral line database from CD-ROM 23 of R. L. Kurucz, http://www.cfa.harvard.edu/amp/ampdata/kurucz23/sekur.html
6. CCC, Convergent Close Coupling database, Curtin University, Australia, http://atom.curtin.edu.au/CCC-WWW/
7. Kelly, http://www.cfa.harvard.edu/ampcgi/kelly.pl
8. IAEA ALADDIN database, http://www-amdis.iaea.org/ALADDIN
9. GENIE, General Search Engine for Atomic Data, http://www-amdis.iaea.org/GENIE
10. CAMDB, China Atomic and Molecular Database, IAPCM, China, http://www.camdb.ac.cn/e
11. AMODS, Korean Atomic Energy Research Institute (KAERI), Korea, http://amods.kaeri.re.kr
12. Universal Atomic Database (uaDB), NASA, USA, http://heasarc.gsfc.nasa.gov/uadb/
13. AMDIS, National Institute for Fusion Science (NIFS), Japan, http://dbshino.nifs.ac.jp
14. MCHF/MCDHF, http://nlte.nist.gov/MCHF/
15. Molecular Collisions at NIST, http://www.nist.gov/pml/data/ionization/index.cfm
16. OPEN-ADAS, ADAS project, http://open.adas.ac.uk
17. MIRF Database, Oak Ridge Natioanal Laboratory, USA, URL http://www-cfadc.phy.ornl.gov
18. TIPTOPbase, Opacity Project, http://cdsweb.u-strasbg.fr/OP.htx

19. Spectr-W3, VNIITF-VNIIFTRI, Russia, http://spectr-w3.snz.ru
20. IAEA Heavy Particles Collisions, http://www-amdis.iaea.org/HEAVY/
21. AAEXCITE, Electron impact excitation cross sections for atomic ions using the average approximation of Peek and Mann, http://www-amdis.iaea.org/AAEXCITE/
22. R.D. Cowan, *The Theory of Atomic Structure and Spectra* (University of California Press, Berkeley, 1981)
23. J.P. Desclaux, A multiconfiguration relativistic Dirac-Fock program. Comp. Phys. Comm. **9**, 31–45 (1975)
24. Physical Reference Data, NIST, USA, http://www.nist.gov/pml/data/
25. NIST bibliographic atomic databases, NIST, USA, http://www.nist.gov/pml/data/asbib/index.cfm
26. AMBDAS, International Atomic Energy Agency (IAEA), http://www-amdis.iaea.org/AMBDAS/
27. BIBL, ISAN, Russia, http://das101.isan.troitsk.ru/bibl.htm
28. Google Scholar, http://scholar.google.com/
29. CrossRef, http://www.crossref.org/
30. Databases for Atomic and Plasma Physics, Weissmann Institute, Israel, http://plasma-gate.weizmann.ac.il/directories/databases/
31. Data Centre Network (DCN), International Atomic Energy Agency (IAEA), http://www-amdis.iaea.org/DCN/
32. XSAMS, International Atomic Energy Agency, IAEA, Austria, http://www-amdis.iaea.org/XSAMS/
33. Y. Ralchenko, D. Humbert, R.E.H. Clark, D.R. Schultz, T. Kato, Y.J. Rhee, Development of the atomic and molecular data markup language for internet data exchange. J. Plasma Fusion Res. **7**, 338 (2006)
34. Virtual Atomic and Mocular Center (VAMDC), http://www.vamdc.eu/
35. NLTE Workshop, http://nlte.nist.gov/NLTE7/

Index

V. Shevelko and H. Tawara (eds.), *Atomic Processes in Basic and Applied Physics*,
Springer Series on Atomic, Optical, and Plasma Physics 68,
DOI 10.1007/978-3-642-25569-4, © Springer-Verlag Berlin Heidelberg 2012